T0345200

Bayesian Methods
for Data Analysis

Third Edition

CHAPMAN & HALL/CRC
Texts in Statistical Science Series

Series Editors

Bradley P. Carlin, *University of Minnesota, USA*
Julian J. Faraway, *University of Bath, UK*
Martin Tanner, *Northwestern University, USA*
Jim Zidek, *University of British Columbia, Canada*

Bayesian Methods
for Data Analysis
Third Edition

Bradley P. Carlin

Univesity of Minnesota

Minneapolis, MN, U.S.A.

Thomas A. Louis

Johns Hopkins Bloomberg School of Public Health

Baltimore, MD, U.S.A.

CRC Press
Taylor & Francis Group
Boca Raton London New York

CRC Press is an imprint of the
Taylor & Francis Group, an **informa** business

A CHAPMAN & HALL BOOK

Chapman & Hall/CRC
Taylor & Francis Group
6000 Broken Sound Parkway NW, Suite 300
Boca Raton, FL 33487-2742

ISBN 13: 978-1-58488-697-6 (hbk)

Library of Congress Cataloging-in-Publication Data

Carlin, Bradley P.
    Bayesian methods for data analysis / authors, Bradley P. Carlin and Thomas A.
    Louis. -- 3rd ed.
        p. cm. -- (Chapman & Hall/CRC texts in statistical science
    series ; 78)
    Originally published: Bayes and Empirical Bayes methods for data analysis. 1st ed.
    Includes bibliographical references and index.
    ISBN 978-1-58488-697-6 (alk. paper)
    1. Bayesian statistical decision theory. I. Louis, Thomas A., 1944- II. Carlin, Bradley
P. Bayes and Empirical Bayes methods for data analysis. III. Title. IV. Series.

QA279.5.C36 2008
519.5'42--dc22
                                                               2008019143

Visit the Taylor & Francis Web site at
http://www.taylorandfrancis.com

and the CRC Press Web site at
http://www.crcpress.com

TO

CAROLINE, SAMUEL, JOSHUA, AND NATHAN

AND

GERMAINE, MARGIT, AND ERICA

Contents

Preface to the Third Edition

As has been well-discussed, the explosion of interest in Bayesian methods over the last ten to twenty years has been the result of the convergence of modern computing power and efficient Markov chain Monte Carlo (MCMC) algorithms for sampling from posterior distributions. Practitioners trained in traditional, frequentist statistical methods appear to have been drawn to Bayesian approaches for two reasons. One is that Bayesian approaches implemented with the majority of their informative content coming from the current data, and not any external prior information, typically have good frequentist properties (e.g., low mean squared error in repeated use). But perhaps more importantly, these methods as now readily implemented in WinBUGS and other MCMC-driven packages now offer the *simplest* approach to hierarchical (random effects) modeling, as routinely needed in longitudinal, frailty, spatial, time series, and a wide variety of other settings featuring interdependent data.

This book represents the third edition of a book originally titled *Bayes and Empirical Bayes Methods for Data Analysis*, first published in 1996. This original version was primarily aimed at advanced students willing to write their own Fortran or C++ code to implement empirical Bayes or fully Bayes–MCMC analyses. When we undertook our first revision in 2000, we sought to improve the usefulness of the book for the growing legion of applied statisticians who wanted to make Bayesian thinking a routine part of their data analytic toolkits. As such, we added a number of new techniques needed to handle advanced computational and model selection problems, as well as a variety of new application areas. However, the book's writing style remained somewhat terse and mathematically formal, and thus potentially intimidating to those with only minimal exposure to the traditional approach. Now, with the WinBUGS language freely available to any who wish to try their hands at hierarchical modeling, we seek to further broaden the reach of our book to practitioners for whom statistical analysis is an important component of their work, but perhaps not their primary interest.

As such, we have made several changes to the book's structure, the most significant of which is the introduction of MCMC thinking and related data

analytic techniques right away in Chapter 2, the basic Bayes chapter. While
the theory supporting the use of MCMC is only cursorily explained at this
point, the aim is get the reader up to speed on the way that a great deal of
applied Bayesian work is now routinely done in practice. While a probabilist
might disagree, the real beauty of MCMC for us lies not in the algorithms
themselves, but in the way their power enables us to focus on *statistical
modeling* and *data analysis* in a way impossible before. As such, Chapter 2
is now generously endowed with data examples and corresponding R and
WinBUGS code, as well as several new homework exercises along these same
lines. The core computing and model criticism and selection material, formerly in Chapters 5 and 6, has been moved up to Chapters 3 and 4, in
keeping with our desire to get the key modeling tools as close to the front
of the book as possible. On a related note, new Sections 2.4 and 4.1 contain
explicit descriptions and illustrations of hierarchical modeling, now com-
monplace in Bayesian data analysis. The philosophically related material
on empirical Bayes and Bayesian performance formerly in Chapters 3 and
4 has been thinned and combined into new Chapter 5. Compensating for
this, the design of experiments material formerly (and rather oddly) tacked
onto Chapter 4 has been expanded into its own chapter (Chapter 6) that includes more explicit advice for clinical trialists and others requiring a basic
education in Bayesian sample size determination, as well as the frequentist
checks still often required of such designs (e.g., by regulatory agencies) before they are put into practice. Finally, the remaining chapters have been
updated as needed, including a completely revised and expanded Subsection 7.1 on ranking and histogram estimation, and a new Subsection 8.3
case study on infectious disease modeling and the 1918 flu epidemic.

As with the previous two editions, this revision presupposes no previous exposure to Bayes or empirical Bayes (EB) methods, but readers with
a master's-level understanding of traditional statistics – say, at the level
of Hogg and Craig (1978), Mood, Graybill, and Boes (1974), or Casella
and Berger (1990) – may well find the going easier. Thanks to the rear-
rangements mentioned above, a course on the basics of modern applied
Bayesian methods might cover only Chapters 1 to 4, since they provide
all that is needed to do some pretty serious hierarchical modeling in standard computer packages. In the Division of Biostatistics at the University
of Minnesota, we do essentially this in a three-credit-hour, single-semester
(15 week) course aimed at master's and advanced undergraduate students
in statistics and biostatistics, and also at master's and doctoral students
in other departments who need to know enough hierarchical modeling to
analyze their data. For those interested in fitting advanced models beyond the scope of standard packages, or in doing methodological research
of their own, the material in the latter chapters may well be crucial. At
Minnesota, we also have a one-semester course of this type, aimed at
doctoral and advanced master's students in statistics and biostatistics.

See `http://www.biostat.umn.edu/~brad/` on the web for many of our datasets and other teaching-related information.

We owe a debt of gratitude to those who helped in our revision process. Haijun Ma and Laura Hatfield did an enormous amount of work on the new examples and homework problems in Chapters 2 and 4. Speaking of new homework problems, many were authored by Sudipto Banerjee, as part of his teaching an early master's-level version of this material at the University of Minnesota. The Bayesian clinical trial material in Sections 6.2 and 6.3 owes a great deal to the relentless Brian Hobbs. Much of Subsection 7.1.1 is due to Rongheng Lin, and virtually all of Section 8.3 is due to Anny-Yue Yin, both recent graduates from Johns Hopkins Biostatistics. Gareth Roberts patiently explained the merits of Langevin-Hastings sampling in terms so plain and convincing that we had no choice but to include it with the other Metropolis-Hastings material in Subsection 3.4.2. The entire University of Minnesota 2008 spring semester "Introduction to Bayesian Analysis" class, co-taught with Prof. Banerjee and ably assisted by Ms. Hatfield (who is also currently co-developing an instructor's manual for the book) served as aggressive copy-editors, finding flawed homework problems, missing references, and an embarrassing number of other goof-ups. Rob Calver, David Grubbs, and the legendary Bob Stern at Chapman and Hall/CRC/Taylor and Francis Group (or whatever the company's called now) were pillars of strength and patience, as usual. Finally, we thank our families, whose ongoing love and support made all of this possible.

BRADLEY P. CARLIN   Minneapolis, Minnesota
THOMAS A. LOUIS   Baltimore, Maryland

April 2008

CHAPTER 1

Approaches for statistical inference

1.1 Introduction

The practicing statistician faces a variety of challenges: designing complex studies, summarizing complex data sets, fitting probability models, drawing conclusions about the present, and making predictions for the future. Statistical studies play an important role in scientific discovery, in policy formulation, and in business decisions. Applications of statistics are ubiquitous, and include clinical decision making, conducting an environmental risk assessment, setting insurance rates, deciding whether (and how) to market a new product, and allocating federal funds. Currently, most statistical analyses are performed with the help of commercial software packages, most of which use methods based on a classical, or *frequentist*, statistical philosophy. In this framework, maximum likelihood estimates (MLEs) and hypothesis tests based on *p*-values figure prominently.

Against this background, the *Bayesian* approach to statistical design and analysis is emerging as an increasingly effective and practical alternative to the frequentist one. Indeed, due to computing advances that enable relevant Bayesian designs and analyses, the philosophical battles between frequentists and Bayesians that were once common at professional statistical meetings are being replaced by a single, more eclectic approach. The title of our book makes clear which philosophy and approach we prefer; that you are reading it suggests at least some favorable disposition (or at the very least, curiosity) on your part as well. Rather than launch headlong into another spirited promotional campaign for Bayesian methods, we begin with motivating vignettes, then provide a basic introduction to the Bayesian formalism followed by a brief historical account of Bayes and frequentist approaches, with attention to some controversies. These lead directly to our methodological and applied investigations.

1.2 Motivating vignettes

1.2.1 Personal probability

Suppose you have submitted your first manuscript to a journal and have assessed the chances of its being accepted for publication. This assessment uses information on the journal's acceptance rate for manuscripts like yours (let's say around 30%), and your evaluation of the manuscript's quality. Subsequently, you are informed that the manuscript has been accepted (congratulations!). What is your updated assessment of the probability that your *next* submission (on a similar topic) will be accepted?

The direct estimate is of course 100% (thus far, you have had one success in one attempt), but this estimate seems naive given what we know about the journal's overall acceptance rate (our external, or *prior*, information in this setting). You might thus pick a number smaller than 100%; if so, you are behaving as a Bayesian would because you are adjusting the (unbiased, but weak) direct estimate in the light of your prior information. This ability to formally incorporate prior information into an analysis is a hallmark of Bayesian methods, and one that frees the analyst from ad hoc adjustments of results that "don't look right."

1.2.2 Missing data

Consider Table 1.1, reporting an array of stable event prevalence or incidence estimates scaled per 10,000 population, with one value (indicated by "⋆") missing at random. The reader may think of them as geographically aligned disease prevalences, or perhaps as death rates cross-tabulated by clinic and age group.

79	87	83	80	78
90	89	92	99	95
96	100	⋆	110	115
101	109	105	108	112
96	104	92	101	96

Table 1.1 *An array of well-estimated rates per 10,000 with one estimate missing.*

With no direct information for ⋆, what would you use for an estimate? Does 200 seem reasonable? Probably not, since the unknown rate is surrounded by estimates near 100. To produce an estimate for the missing cell you might fit an additive model (rows and columns) and then use the model to impute a value for ⋆, or merely average the values in surrounding cells. These are two examples of *borrowing information*. Whatever your approach, some number around 100 seems reasonable.

Now assume that we obtain data for the $\star$ cell and the estimate is, in fact, 200, based on 2 events in a population of 100 ($200 = 10000 \times 2/100$). Would you now estimate $\star$ by 200 (a very unstable estimate based on very little information), when with no information a moment ago you used 100? While 200 is a perfectly valid estimate (though its uncertainty should be reported), some sort of *weighted average* of this direct estimate (200) and the indirect estimate you used when there was no direct information (100) seems intuitively more appealing. The Bayesian formalism allows just this sort of natural compromise estimate to emerge.

Finally, repeat this mental exercise assuming that the direct estimate is still 200 per 10,000, but now based on 20 events in a population of 1000, and then on 2000 events in a population of 100,000. What estimate would you use in each case? Bayes and empirical Bayes methods structure this type of statistical decision problem, automatically giving increasing weight to the direct estimate as it becomes more reliable.

1.2.3 Bioassay

Consider a carcinogen bioassay where you are comparing a control group (C) and an exposed group (E) with 50 rodents in each (see Table 1.2). In the control group, 0 tumors are found; in the exposed group, there are 3, producing a non-significant, one-sided Fisher exact test p-value of approximately 0.125. However, your colleague, who is a veterinary pathologist, states, "I don't know about statistical significance, but three tumors in 50 rodents is certainly *biologically* significant!"

	C	E	Total
Tumor	0	3	3
No Tumor	50	47	97
	50	50	100

Table 1.2 *Hypothetical bioassay results; one-sided $p = 0.125$.*

This belief may be based on information from other experiments in the same lab in the previous year in which the tumor has never shown up in control rodents. For example, if there were 400 historical controls in addition to the 50 concurrent controls, none with a tumor, the one-sided p-value becomes 0.001 (see Table 1.3). Statistical and biological significance are now compatible. In general, it can be inappropriate simply to pool historical and concurrent information. However, Bayes and empirical Bayes methods may be used to structure a valid synthesis; see, for example, Tarone (1982), Dempster et al. (1983), Tamura and Young (1986), Louis and Bailey (1990), and Chen et al. (1999).

	C	E	Total
Tumor	0	3	3
No Tumor	450	47	497
	450	50	500

Table 1.3 *Hypothetical bioassay results augmented by 400 historical controls, none with a tumor; one-sided $p = 0.001$.*

1.2.4 Attenuation adjustment

In a standard errors-in-variables simple linear regression model, the least squares estimate of the regression slope (β) is biased toward 0, an example of *attenuation*. More formally, suppose the true regression is $Y = \beta x + \epsilon$, $\epsilon \sim N(0, \sigma_\epsilon^2)$, but Y is regressed not on x but on $X \equiv x + \delta$, where $\delta \sim N(0, \sigma_\delta^2)$. Then the least squares estimate $\hat{\beta}$ has expectation $E[\hat{\beta}] \approx \rho\beta$, with $\rho = \frac{\sigma_\epsilon^2}{\sigma_\epsilon^2 + \sigma_\delta^2} \leq 1$. If ρ is known or well-estimated, one can correct for attenuation and produce an unbiased estimate by using $\hat{\beta}/\rho$ to estimate β.

Though unbiasedness is an attractive property, especially when the standard error associated with the estimate is small, in general it is less important than having the estimate "close" to the true value. The expected squared deviation between the true value and the estimate (*mean squared error*, or MSE) provides an effective measure of proximity. Fortunately for our intuition, MSE can be written as the sum of an estimator's sampling variance and its squared bias,

$$\text{MSE} = \text{variance} + (\text{bias})^2 . \tag{1.1}$$

The unbiased estimate sets the second term to 0, but it can have a very large MSE relative to other estimators; in this case, because dividing $\hat{\beta}$ by ρ inflates the variance as the price of eliminating bias. Bayesian estimators typically strike an effective tradeoff between variance and bias.

1.3 Defining the approaches

Three principal approaches to inference guide modern data analysis: frequentist, Bayesian, and likelihood. We now briefly describe each in turn.

The *frequentist* evaluates procedures based on imagining repeated sampling from a particular model (the *likelihood*), which defines the probability distribution of the observed data conditional on unknown parameters. Properties of the procedure are evaluated in this repeated sampling framework for fixed values of unknown parameters; good procedures perform well over a broad range of parameter values.

The *Bayesian* requires a sampling model and, in addition, a *prior* distribution on all unknown quantities in the model (parameters and missing data). The prior and likelihood are used to compute the conditional distribution of the unknowns given the observed data (the *posterior* distribution), from which all statistical inferences arise. Allowing the observed data to play some role in determining the prior distribution produces the *empirical Bayes* (EB) approach. The Bayesian evaluates procedures over repeated sampling of unknowns from the posterior distribution for a given data set. The empirical Bayesian may also evaluate procedures under repeated sampling of *both* the data and the unknowns from their joint distribution.

Finally, the *likelihoodist* (or *Fisherian*) develops a sampling model but not a prior, as does the frequentist. However, inferences are restricted to procedures that use the data only as reported by the likelihood, as a Bayesian would. Procedure evaluations can be from a frequentist, Bayesian, or EB point of view.

As presented in Appendix B, the frequentist and Bayesian approaches can be connected in a *decision-theoretic* framework, wherein one considers a model where the unknown parameters and observed data have a joint distribution. The frequentist conditions on parameters and replicates (integrates) over the data; the Bayesian conditions on the data and replicates (integrates) over the parameters. EB (or *preposterior*) evaluations integrate over both parameters and data, and can be represented as frequentist performance averaged over the prior distribution, or as Bayesian posterior performance averaged over the marginal sampling distribution of the observed data (the sampling distribution conditional on parameters, averaged over the prior). Preposterior properties are relevant to both Bayesians and frequentists (see Rubin, 1984).

Historically, frequentists have criticized Bayesian procedures for their inability to deal with all but the most basic examples, for overreliance on computationally convenient priors, and for being too fragile in their dependence on a specific prior (i.e., for a lack of *robustness* in settings where the data and prior conflict). Bayesians have criticized frequentists for failure to incorporate relevant prior information, inefficiency, inflexibility, and *incoherence* (i.e., a failure to process available information systematically, as a Bayesian approach would). Another common Bayesian criticism is that, while frequentist methods do avoid dependence on any single set of prior beliefs, the resulting claims of "objectivity" are often illusory since such methods still require myriad assumptions about the underlying data generating mechanism, such as a simple (often normal) model free from confounding, selection bias, measurement error, etc. Bayesians often remark that the choice of prior distribution is only one assumption that should be explicitly declared and checked in a statistical analysis. Greenland (2006) points out that statistically significant frequentist findings in observational epidemiology rarely come with the requisite "truth-in-packaging" caveats that prior selection forces Bayesians to provide routinely.

Recent computing advances have all but eliminated constraints on priors and models, but leave open the more fundamental difficulties of prior selection and possible non-robustness. In this book, therefore, we shall often seek the middle ground: Bayes and EB procedures that offer many of the Bayes advantages, but do so without giving up too much frequentist robustness. Procedures occupying this middle ground can be thought of as having good performance over a broad range of prior distributions, but not so broad as to include all priors (which would in turn require good performance over all possible parameter values).

1.4 The Bayes-frequentist controversy

While probability has been the subject of study for hundreds of years (most notably by mathematicians retained by rich noblemen to advise them on how to maximize their winnings in games of chance), statistics is a relatively young field. Linear regression first appeared in the work of Francis Galton in the late 1800s, with Karl Pearson adding correlation and goodness-of-fit measures around the turn of the last century. The field did not really blossom until the 1920s and 1930s, when R.A. Fisher developed the notion of likelihood for general estimation, and Jerzy Neyman and Egon Pearson developed the basis for classical hypothesis testing. A flurry of research activity was energized by the World War II, which generated a wide variety of difficult applied problems and the first substantive government funding for their solution in the United States and Great Britain.

By contrast, Bayesian methods are much older, dating to the original 1763 paper by the Rev. Thomas Bayes, a minister and amateur mathematician. The area generated some interest by Laplace, Gauss, and others in the 19th century, but the Bayesian approach was ignored (or actively opposed) by the statisticians of the early 20th century. Fortunately, during this period several prominent non-statisticians, most notably Harold Jeffreys (a physicist) and Arthur Bowley (an econometrician), continued to lobby on behalf of Bayesian ideas (which they referred to as "inverse probability"). Then, beginning around 1950, statisticians such as L.J. Savage, Bruno de Finetti, Dennis Lindley, and many others began advocating Bayesian methods as remedies for certain deficiencies in the classical approach. The following example discusses the case of interval estimation.

Example 1.1 Suppose $X_i \stackrel{iid}{\sim} N(\theta, \sigma^2)$, $i = 1, \ldots, n$, where N denotes the normal (Gaussian) distribution and iid stands for "independent and identically distributed." We desire a 95% interval estimate for the population mean θ. Provided n is sufficiently large (say, bigger than 30), a classical approach would use the confidence interval

$$\delta(\mathbf{x}) = \bar{x} \pm 1.96s/\sqrt{n} \, ,$$

where $\mathbf{x} = (x_1, \ldots, x_n)$, $\bar{x}$ is the sample mean, and s is the sample standard deviation. This interval has the property that, on average over repeated applications, $\delta(\mathbf{x})$ will fail to capture the true mean θ only 5% of the time. An alternative interpretation is that, *before* any data are collected, the probability that the interval contains the true value is 0.95. This property is attractive in the sense that it holds for *all* true values of θ and σ^2.

On the other hand, its use in any single data-analytic setting is somewhat difficult to explain and understand. After collecting the data and computing $\delta(\mathbf{x})$, the interval either contains the true θ or it does not; its coverage probability is not 0.95, but either 0 or 1. After observing $\mathbf{x}$, a statement like, "the true θ has a 95% chance of falling in $\delta(\mathbf{x})$," is not valid, though most people (including most statisticians irrespective of their philosophical approach) interpret a confidence interval in this way. Thus, for the frequentist, "95%" is not a conditional coverage probability, but rather a tag associated with the interval to indicate either how it is likely to perform before we evaluate it, or how it would perform over the long haul. A 99% frequentist interval would be wider, a 90% interval narrower, but, conditional on $\mathbf{x}$, all would have coverage probability 0 or 1. ■

By contrast, Bayesian confidence intervals (known as "credible sets," and discussed further in Subsection 2.3.2) are free of this awkward frequentist interpretation. For example, conditional on the observed data, the probability is 0.95 that θ is in the 95% credible interval. Of course, this natural interpretation comes at the price of needing to specify a (possibly quite vague) prior distribution for θ.

The Neyman-Pearson testing structure can also lead to some very odd results, which have been even more heavily criticized by Bayesians.

Example 1.2 Consider the following simple experiment, originally suggested by Lindley and Phillips (1976), and reprinted many times. Suppose in 12 independent tosses of a coin, I observe 9 heads and 3 tails. I wish to test the null hypothesis $H_0 : \theta = 1/2$ versus the alternative hypothesis $H_a : \theta > 1/2$, where θ is the true probability of heads. Given only this much information, two choices for the sampling distribution emerge:

1. *Binomial:* The number $n = 12$ tosses was fixed beforehand, and the random quantity X was the number of heads observed in the n tosses. Then $X \sim Bin(12, \theta)$, and the likelihood function is given by

$$L_1(\theta) = \binom{n}{x} \theta^x (1 - \theta)^{n-x} = \binom{12}{9} \theta^9 (1 - \theta)^3 . \qquad (1.2)$$

2. *Negative binomial:* Data collection involved flipping the coin until the third tail appeared. Here, the random quantity X is the number of heads required to complete the experiment, so that $X \sim NegBin(r = 3, \theta)$,

with likelihood function given by

$$L_2(\theta) = \binom{r + x - 1}{x} \theta^x (1 - \theta)^r = \binom{11}{9} \theta^9 (1 - \theta)^3 . \qquad (1.3)$$

Under either of these two alternatives, we can compute the p-value corresponding to the rejection region, "Reject H_0 if $X \geq c$." Doing so using the binomial likelihood (1.2), we obtain

$$\alpha_1 = P_{\theta = \frac{1}{2}}(X \geq 9) = \sum_{j=9}^{12} \binom{12}{j} \theta^j (1 - \theta)^{12-j} = .075 ,$$

while for the negative binomial likelihood (1.3),

$$\alpha_2 = P_{\theta = \frac{1}{2}}(X \geq 9) = \sum_{j=9}^{\infty} \binom{2 + j}{j} \theta^j (1 - \theta)^3 = .0325 .$$

Thus, using the "usual" Type I error level $\alpha = .05$, we see that the two model assumptions lead to two different decisions: we would reject H_0 if X were assumed negative binomial, but not if it were assumed binomial. But there is no information given in the problem setting to help us make this determination, so it is not clear which analysis the frequentist should regard as "correct." In any case, assuming we trust the statistical model, it does not seem reasonable that how the experiment was monitored should have any bearing on our decision; surely only its *results* are relevant! Indeed, the likelihood functions tell a consistent story, since (1.2) and (1.3) differ only by a multiplicative constant that does not depend on θ. ∎

A Bayesian explanation of what went wrong in the previous example would be that the Neyman-Pearson approach allows *unobserved* outcomes to affect the rejection decision. That is, the probability of X values "more extreme" than 9 (the value actually observed) was used as evidence against H_0 in each case, even though these values did not occur. More formally, this is a violation of a statistical axiom known as the *Likelihood Principle*, a notion present in the work of Fisher and Barnard, but not precisely defined until the landmark paper by Birnbaum (1962). In a nutshell, the Likelihood Principle states that once the data value x has been observed, the likelihood function $L(\theta|x)$ contains all relevant experimental information delivered by x about the unknown parameter θ. In the previous example, L_1 and L_2 are proportional to each other as functions of θ, hence are equivalent in terms of experimental information (recall that multiplying a likelihood function by an arbitrary function $h(x)$ does not change the MLE $\hat{\theta}$). Yet in the Neyman-Pearson formulation, these equivalent likelihoods lead to two different inferences regarding θ. Put another way, frequentist test results actually depend not only on what x was observed, but on how the experiment was stopped.

Some statisticians attempt to defend the results of Example 1.2 by arguing that all aspects of the design of an experiment *are* relevant pieces of information even after the data have been collected, or perhaps that the Likelihood Principle is itself flawed (or at least should not be considered sacrosanct). We do not delve deeper into this foundational discussion, but refer the interested reader to the excellent monograph by Berger and Wolpert (1984) for a presentation of the consequences, criticisms, and defenses of the Likelihood Principle. Violation of the Likelihood Principle is but one of the possibly anomalous properties of classical testing methods; more are outlined later in Subsection 2.3.3, where it is also shown that Bayesian hypothesis testing methods overcome these difficulties.

If Bayesian methods offer a solution to these and other drawbacks of the frequentist approach that were publicized several decades ago, it is perhaps surprising that Bayesian methodology did not make bigger inroads into actual statistical practice until only recently. There are several reasons for this. First, the initial staunch advocates of Bayesian methods were primarily *subjectivists*, in that they argued forcefully that all statistical calculations should be done only after one's own personal prior beliefs on the subject had been carefully evaluated and quantified. But this raised concerns on the part of classical statisticians (and some Bayesians) that the results obtained would not be objectively valid, and could be manipulated in any way the statististician saw fit. (The reply of some subjectivists that frequentist methods were invalid anyway and should be discarded did little to assuage these concerns.) Second, and perhaps more important from an applied standpoint, the Bayesian alternative, while theoretically simple, required evaluation of complex integrals in even fairly rudimentary problems. Without inexpensive, high-speed computing, this practical impediment combined with the theoretical concerns to limit the growth of realistic Bayesian data analysis. However, this growth finally did occur in the 1980s, thanks to a more objective group of Bayesians with access to inexpensive, fast computing. The objectivity issue is discussed further in Chapter 2, while computing is the subject of Chapter 3.

These two main concerns regarding routine use of Bayesian analysis (i.e., that its use is not easy or automatic, and that it is not always clear how objective results may be obtained) were raised in the widely-read paper by Efron (1986). While the former issue has been largely resolved in the intervening years thanks to computing advances, the latter remains a challenge (see Subsection 2.2.3 below). Still, we contend that the advantages in using Bayes and EB methods justify the increased effort in computation and prior determination. Many of these advantages are presented in detail in the popular textbook by Berger (1985, Section 4.1), to wit:

1. Bayesian methods provide the user with the ability to formally incorporate prior information.

2. Inferences are conditional on the *actual* data.

3. The reason for stopping the experimentation does not affect Bayesian inference (a concern in Example 1.2).

4. Bayesian answers are more easily interpretable by nonspecialists (a concern in Example 1.1).

5. All Bayesian analyses follow *directly* from the posterior; no separate theories of estimation, testing, multiple comparisons, etc. are needed.

6. Any question can be *directly* answered through Bayesian analysis.

7. Bayes and EB procedures possess numerous optimality properties.

As an example of the sixth point, to investigate the bioequivalence of two drugs, we would need to test $H_0 : \theta_1 \neq \theta_2$ versus $H_a : \theta_1 = \theta_2$. That is, we must reverse the traditional null and alternative roles, because the hypothesis we hope to reject is that the drugs are *different*, not that they are the same. This reversal turns out to make things quite awkward for traditional testing methods (see e.g., Berger and Hsu, 1996), but not for Bayesian testing methods, since they treat the null and alternative hypotheses equivalently; they really can "accept" the null, rather than merely "fail to reject" it. Finally, regarding the seventh point, Bayes procedures are typically consistent, can automatically impose parsimony in model choice, and can even define the class of optimal *frequentist* procedures, thus "beating the frequentist at his own game." We return to this final issue (frequentist motivations for using Bayes and EB procedures) in Chapter 5.

1.5 Some basic Bayesian models

The most basic Bayesian model has two stages, with a likelihood specification $Y|\theta \sim f(y|\theta)$ and a prior specification $\theta \sim \pi(\theta)$, where either Y or θ can be vectors. In the simplest Bayesian analysis, π is assumed known, so that by probability calculus, the posterior distribution of θ is given by

$$p(\theta|y) = \frac{f(y|\theta)\pi(\theta)}{m(y)} \ , \tag{1.4}$$

where

$$m(y) = \int f(y \mid \theta)\pi(\theta)d\theta \ , \tag{1.5}$$

the *marginal* density of the data y. Equation (1.4) is a special case of *Bayes' Theorem*, the general form of which we present in Section 2.1. For all but rather special choices of f and π (see Subsection 2.2.2), evaluating integrals such as (1.5) used to be difficult or impossible, forcing Bayesians into unappealing approximations. However, recent developments in Monte Carlo computing methods (see Chapter 3) allow accurate estimation of such integrals, and thus have enabled advanced Bayesian data analysis.

1.5.1 A Gaussian/Gaussian (normal/normal) model

We now consider the case where both the prior and the likelihood are Gaussian (normal) distributions, namely,

$$
\begin{aligned}
\theta &\sim N(\mu, \tau^2) \\
Y \mid \theta &\sim N(\theta, \sigma^2).
\end{aligned}
\tag{1.6}
$$

The marginal distribution of Y given in (1.5) turns out to be $N(\mu, \sigma^2 + \tau^2)$, and the posterior distribution (1.4) is also Gaussian with mean and variance

$$
\begin{aligned}
E(\theta \mid Y) &= B\mu + (1 - B)Y \tag{1.7} \\
Var(\theta \mid Y) &= (1 - B)\sigma^2, \tag{1.8}
\end{aligned}
$$

where $B = \sigma^2/(\sigma^2 + \tau^2)$. Since $0 \leq B \leq 1$, the posterior mean is a weighted average of the prior mean μ and the direct estimate Y; the Bayes estimate is pulled back (or *shrunk*) toward the prior mean. Moreover, the weight on the prior mean B depends on the relative variability of the prior distribution and the likelihood. If σ^2 is large relative to τ^2 (i.e., our prior knowledge is more precise than the data information), then B is close to 1, producing substantial shrinkage. If σ^2 is small (i.e., our prior knowledge is imprecise relative to the data information), B is close to 0 and the estimate is moved very little toward the prior mean. As we show in Chapter 5, this shrinkage provides an effective tradeoff between variance and bias, with beneficial effects on the resulting mean squared error; see equation (1.1).

If one is willing to assume that the structure (1.6) holds, but that the prior mean μ and variance τ^2 are unknown, then hierarchical or empirical Bayes methods can be used; see Chapters 2 and 5, respectively.

1.5.2 A beta/binomial model

Next, consider applying the Bayesian approach given in (1.4)–(1.5) to estimating a binomial success probability. With Y the number of events in n independent trials and θ the event probability, the sampling distribution is

$$
P(Y = y \mid \theta) = f(y \mid \theta) = \binom{n}{y} \theta^y (1 - \theta)^{n-y}.
$$

In order to obtain a closed form for the marginal distribution, we use the *Beta*(a, b) prior distribution for θ (see Section A.2 in Appendix A). For convenience we reparametrize from (a, b) to (μ, M) where $\mu = a/(a+b)$, the prior mean, and $M = a + b$, a measure of prior precision. More specifically, the prior variance is then $\mu(1 - \mu)/(M + 1)$, a decreasing function of M. The marginal distribution of Y is then referred to as *beta-binomial*, and can be shown to have mean and variance satisfying

$$
E\left(\frac{Y}{n}\right) = \mu \;\; \text{and} \;\; Var\left(\frac{Y}{n}\right) = \frac{\mu(1 - \mu)}{n}\left[1 + \frac{n - 1}{M + 1}\right].
$$

The term in square brackets is known variously as the "variance inflation factor," "design effect," or "component of extra-binomial variation."

The posterior distribution of θ given Y in (1.4) is again *Beta* with mean

$$
\begin{aligned}
\hat{\theta} &= \frac{M}{M+n}\mu + \frac{n}{M+n}\left(\frac{Y}{n}\right), \\
&= \mu + \frac{n}{M+n}\left[\left(\frac{Y}{n}\right) - \mu\right]
\end{aligned}
\tag{1.9}
$$

and variance $Var(\theta \mid Y) = [\hat{\theta}(1 - \hat{\theta})]/(M + n)$. Note that, similar to the posterior mean (1.7) in the Gaussian/Gaussian example, $\hat{\theta}$ is a weighted average of the prior mean μ and the maximum likelihood estimate Y/n, with weight depending on the relative size of M (the information in the prior) and n (the information in the data).

Discussion

Statistical decision rules can be generated by any philosophy under any collection of assumptions. They can then be evaluated by any criteria, even those arising from an utterly different philosophy. We contend (and the subsequent chapters will show) that the Bayesian approach is an excellent "procedure generator," even if one's evaluation criteria are frequentist provided that the prior distributions introduce only a small amount of information. This agnostic view considers features of the prior (possibly the entire prior) as "tuning parameters" that can be used to produce a decision rule with broad validity. The Bayesian formalism will be even more effective if one desires to structure an analysis using either personal opinion or objective information external to the current data set. The Bayesian approach also encourages documenting assumptions and quantifying uncertainty. Of course, no approach automatically produces broadly valid inferences, even in the context of the Bayesian models. A procedure generated by a high-information prior with most of its mass far from the truth will perform poorly under both Bayesian and frequentist evaluations.

Statisticians need design and analysis methods that strike an effective tradeoff between efficiency and robustness, irrespective of the underlying philosophy. For example, in estimation, central focus should be on reduction of MSE and related performance measures through a tradeoff between variance and bias. This concept is appropriate for both frequentists and Bayesians. In this context, our strategy will be to use the Bayesian formalism to reduce MSE even when evaluations are frequentist.

Importantly, the Bayesian formalism properly propagates uncertainty through the analysis enabling a more realistic (typically inflated) assessment of the variability in estimated quantities of interest. Also, the formalism structures the analysis of complicated models where intuition may produce faulty or inefficient approaches. This structuring becomes especially important in multiparameter models in which the Bayesian approach requires that the joint posterior distribution of all parameters structure all analyses. Appropriate marginal distributions focus on specific parameters, and this integration ensures that all uncertainties influence the spread and shape of the marginal posterior.

1.6 Exercises

1. Let θ be the true proportion of men in your community over the age of 40 with hypertension. Consider the following "thought experiment":

   (a) Though you may have little or no expertise in this area, give an initial point estimate of θ.

   (b) Now suppose a survey to estimate θ is established in your community, and of the first 5 randomly selected men, 4 are hypertensive. How does this information affect your initial estimate of θ?

   (c) Finally, suppose that at the survey's completion, 400 of 1000 men have emerged as hypertensive. Now what is your estimate of θ?

   What guidelines for statistical inference do your answers suggest?

2. Repeat the journal publication thought problem from Subsection 1.2.1 for the situation where

   (a) you have won a lottery on your first try.

   (b) you have correctly predicted the winner of the first game of the World Series (professional baseball).

3. Assume you have developed predictive distributions of the length of time it takes to drive to work, one distribution for Route A and one for Route B. What summaries of these distributions would you use to select a route

   (a) to maximize the probability that the drive takes less than 30 minutes?

   (b) to minimize your average commuting time?

4. For predictive distributions of survival time associated with two medical treatments, propose treatment selection criteria that are meaningful to you (or if you prefer, to society).

5. Here is an example in a vein similar to that of Example 1.2, and originally presented by Berger and Berry (1988). Consider a clinical trial established to study the effectiveness of vitamin C in treating the common cold. After grouping subjects into pairs based on baseline variables such as gender, age, and health status, we randomly assign one member of each pair to receive vitamin C, with the other receiving a placebo. We then count how many pairs had vitamin C giving superior relief after 48 hours. We wish to test $H_0 : P(\text{vitamin C better}) = \frac{1}{2}$ versus $H_a : P(\text{vitamin C better}) \neq \frac{1}{2}$.

(a) Consider the expermental design wherein we sample $n = 17$ pairs, and observe $x = 13$ preferences for vitamin C. What is the p-value for the above two-sided test?

(b) Now consider a *two-stage* design, wherein we first sample $n_1 = 17$ pairs, and observe x_1 preferences for vitamin C. In this design, if $x_1 \geq 13$ or $x_1 \leq 4$, we stop and reject H_0. Otherwise, we sample an *additional* $n_2 = 27$ pairs, and subsequently reject H_0 if $X_1 + X_2 \geq 29$ or $X_1 + X_2 \leq 15$. (This second stage rejection region was chosen because, under H_0, $P(X_1 + X_2 \geq 29$ or $X_1 + X_2 \leq 15) = P(X_1 \geq 13$ or $X_1 \leq 4)$, the p-value in part (a) above.)

If we once again observe $x_1 = 13$, what is the p-value under this new design? Is your answer consistent with that in part (a)?

(c) What would be the impact on the p-value if we kept adding stages to the design, but kept observing $x_1 = 13$?

(d) How would you analyze these data in the presence of a necessary but unforeseen change in the design – say, because the first five patients developed an allergic reaction to the treatment, and the trial was stopped by its clinicians.

(e) What does all of this suggest about the claim that p-values constitute "objective evidence" against H_0?

6. In the Normal/Normal example of Subsection 1.5.1, let $\sigma^2 = 2$, $\mu = 0$, and $\tau^2 = 2$.

(a) Suppose we observe $y = 4$. What are the mean and variance of the resulting posterior distribution? Sketch the prior, likelihood, and posterior on a single set of coordinate axes.

(b) Repeat part (a) assuming $\tau^2 = 18$. Explain any resulting differences. Which of these two priors would likely have more appeal for a frequentist statistician?

7. In the basic diagnostic test setting, a disease is either present $(D = 1)$ or absent $(D = 0)$, and the test indicates either disease $(T = 1)$ or no disease $(T = 0)$. Represent $P(D = d|T = t)$ in terms of test sensitivity, $P(T = 1|D = 1)$, specificity, $P(T = 0|D = 0)$, and disease prevalence, $P(D = 1)$, and relate to Bayes' theorem (1.4).

8. In analyzing data from a $Bin(n, \theta)$ likelihood, the MLE is $\hat{\theta}_{MLE} = Y/n$, which has MSE $= E_{y|\theta}(\hat{\theta}_{MLE} - \theta)^2 = Var_{y|\theta}(\hat{\theta}_{MLE}) = \theta(1 - \theta)/n$. Find the MSE of the estimator $\hat{\theta}_{Bayes} = (Y + 1)/(n + 2)$ and discuss in what contexts you would prefer it over $\hat{\theta}_{MLE}$. ($\hat{\theta}_{Bayes}$ is the estimator from equation (1.9) with $\mu = 1/2$ and $M = 2$.)

## CHAPTER 2

The Bayes approach

2.1 Introduction

We begin by reviewing the fundamentals introduced in Chapter 1. The Bayesian approach begins exactly as a traditional frequentist analysis does, with a *sampling model* for the observed data $\mathbf{y} = (y_1, \ldots, y_n)$ given a vector of unknown parameters $\boldsymbol{\theta}$. This sampling model is typically given in the form of a probability distribution $f(\mathbf{y}|\boldsymbol{\theta})$. When viewed as a function of $\boldsymbol{\theta}$ instead of $\mathbf{y}$, this distribution is usually called the *likelihood*, and sometimes written as $L(\boldsymbol{\theta}; \mathbf{y})$ to emphasize our mental reversal of the roles of $\boldsymbol{\theta}$ and $\mathbf{y}$. Note that L need *not* be a probability distribution for $\boldsymbol{\theta}$ given $\mathbf{y}$; that is, $\int L(\boldsymbol{\theta}; \mathbf{y})d\boldsymbol{\theta}$ need not be 1; it may not even be finite. Still, given particular data values $\mathbf{y}$, it is very often possible to find the value $\widehat{\boldsymbol{\theta}}$ that maximizes the likelihood function, i.e.,

$$\widehat{\boldsymbol{\theta}} = argmax_{\boldsymbol{\theta}} \ L(\boldsymbol{\theta}; \mathbf{y}) \ .$$

This value is called the *maximum likelihood estimate* (MLE) for $\boldsymbol{\theta}$. This idea dates to Fisher (1922; see also Stigler, 2005) and continues to form the basis for many of the most commonly used statistical analysis methods today.

In the Bayesian approach, instead of supposing that $\boldsymbol{\theta}$ is a fixed (though unknown) parameter, we think of it as a *random* quantity as well. This is operationalized by adopting a probability distribution for $\boldsymbol{\theta}$ that summarizes any information we have about it not related to that provided by the data $\mathbf{y}$, called the *prior* distribution (or simply the *prior*). Just as the likelihood had parameters $\boldsymbol{\theta}$, the prior may have parameters $\boldsymbol{\eta}$; these are often referred to as *hyperparameters*, in order to distinguish them from the likelihood parameters $\boldsymbol{\theta}$. For the moment we assume that the hyperparameters $\boldsymbol{\eta}$ are known, and thus write the prior as $\pi(\boldsymbol{\theta}) \equiv \pi(\boldsymbol{\theta}|\boldsymbol{\eta})$. Inference concerning $\boldsymbol{\theta}$ is then based on its *posterior* distribution, given by

$$\begin{aligned} p(\boldsymbol{\theta}|\mathbf{y}) &= \frac{p(\mathbf{y}, \boldsymbol{\theta})}{p(\mathbf{y})} = \frac{p(\mathbf{y}, \boldsymbol{\theta})}{\int p(\mathbf{y}, \boldsymbol{\theta}) \, d\boldsymbol{\theta}} \\ &= \frac{f(\mathbf{y}|\boldsymbol{\theta})\pi(\boldsymbol{\theta})}{\int f(\mathbf{y}|\boldsymbol{\theta})\pi(\boldsymbol{\theta}) \, d\boldsymbol{\theta}} \ . \end{aligned} \quad (2.1)$$

This formula is known as *Bayes' Theorem*, and first appeared (in a somewhat simplified form) in Bayes (1763). Notice the contribution of both the experimental data (in the form of the likelihood f) and prior opinion (in the form of the prior π) to the posterior in the last expression of equation (2.1). The posterior is simply the product of the likelihood and the prior, renormalized so that it integrates to 1 (and is thus itself a valid probability distribution).

Readers less comfortable with the probability calculus needed to handle the continuous variables in (2.1) may still be familiar with a discrete, set theoretic version of Bayes' Theorem from a previous probability or statistics course. In this simpler formulation, we are given an event of interest A and a collection of events B_j, $j = 1, \ldots, J$ that are mutually exclusive and exhaustive (that is, exactly one of them must occur). Given the probabilities of each of these events $P(B_j)$, as well as the conditional probabilities $P(A|B_j)$, from fundamental rules of probability, we have

$$
\begin{aligned}
P(B_j|A) &= \frac{P(A, B_j)}{P(A)} = \frac{P(A, B_j)}{\sum_{j=1}^{J} P(A, B_j)} \\
&= \frac{P(A|B_j)P(B_j)}{\sum_{j=1}^{J} P(A|B_j)P(B_j)} ,
\end{aligned}
\tag{2.2}
$$

where $P(A, B_j)$ indicates the *joint* event where both A and B_j occur; many textbooks write $P(A \cap B_j)$ for $P(A, B_j)$. The reader will appreciate that all four expressions in (2.2) are just a discrete finite versions of the corresponding expressions in (2.1), with the B_j playing the role of the parameters $\boldsymbol{\theta}$ and A playing the role of the data $\mathbf{y}$.

This simplified version Bayes' Theorem (referred to by many textbook authors as *Bayes' Rule*) may appear too simple to be of much practical value, but interesting applications do arise:

Example 2.1 Ultrasound tests done near the end of the first trimester of a pregnancy are often used to predict the sex of the baby. However, the errors made by radiologists in reading ultrasound results are not symmetric, in the following sense: girls are virtually always correctly identified as girls, while boys are sometimes misidentified as girls (in cases where the penis is not clearly visible, perhaps due to the child's position in the womb). More specifically, a leading radiologist states that

$$P(test + |G) = 1 \ \text{ and } \ P(test + |B) = .25 \ ,$$

where "test $+$" denotes that the ultrasound test predicts the child is a girl. Thus, we have a 25% false positive rate for girl, but no false negatives.

Suppose a particular woman's test comes back positive for girl, and we wish to know the probability she is actually carrying a girl. Assuming 48% of babies are girls, we can use (2.2) where "boy" and "girl" provide the $J = 2$ mutually exclusive and exhaustive cases B_j. Thus, with A being the event of a positive test, we have

$$P(G \mid test+) = \frac{P(test + |G)P(G)}{P(test + |G)P(G) + P(test + |B)P(B)}$$

$$= \frac{(1)(.48)}{(1)(.48) + (.25)(.52)} = .787 \;,$$

or only a 78.7% chance the baby is, in fact, a girl. ∎

Now let us return to the general case, where expressions on either side of the conditioning bar can be continuous random variables. Apparently the greatest challenge in evaluating the posterior lies in performing the integral in the denominator of (2.1). Notice we are writing this as a single integral, but in fact it is a *multiple* integral, having dimension equal to the number of parameters in the $\boldsymbol{\theta}$ vector.

The denominator integral is sometimes written as $m(\mathbf{y}) \equiv m(\mathbf{y}|\boldsymbol{\eta})$, the m denoting that this is the *marginal* distribution of the data $\mathbf{y}$ given the value of the hyperparameter $\boldsymbol{\eta}$. This distribution plays a crucial role in model checking and model choice, since it can be evaluated at the observed $\mathbf{y}$; we will say much more about this in Subsections 2.3.3 and 2.5 below.

In some cases this integral can be evaluated in closed form, leading to a closed form posterior:

Example 2.2 Consider the normal (Gaussian) likelihood, where $f(y|\theta) = \frac{1}{\sigma\sqrt{2\pi}} \exp(-\frac{(y-\theta)^2}{2\sigma^2})$, $y \in \Re$, $\theta \in \Re$, and σ is a known positive constant. Henceforth we employ the shorthand notation $f(y|\theta) = N(y|\theta, \sigma^2)$ to denote a normal density with mean θ and variance σ^2. Suppose we take $\pi(\theta|\boldsymbol{\eta}) = N(\theta|\mu, \tau^2)$, where $\mu \in \Re$ and $\tau > 0$ are known hyperparameters, so that $\boldsymbol{\eta} = (\mu, \tau)$. By plugging these expressions into equation (2.1), it is fairly easy to show that the posterior distribution for θ is given by

$$p(\theta|y) = N\left(\theta \,\Big|\, \frac{\sigma^2\mu + \tau^2 y}{\sigma^2 + \tau^2} \,,\; \frac{\sigma^2\tau^2}{\sigma^2 + \tau^2}\right) \;; \qquad (2.3)$$

see the end of this chapter, Exercise 2. Writing $B = \sigma^2/(\sigma^2 + \tau^2)$, this posterior has mean

$$B\mu + (1 - B)y \;.$$

Since $0 < B < 1$, the posterior mean is a weighted average of the prior mean and the observed data value, with weights that are inversely proportional to the corresponding variances. For this reason, B is sometimes called a *shrinkage factor*, because it gives the proportion of the distance that the posterior mean is "shrunk back" from the ordinary frequentist estimate y toward the prior mean μ. Note that when τ^2 is large relative to σ^2 (i.e., vague prior information), B is small and the posterior mean is close to the data value y. On the other hand, when τ^2 is small relative to σ^2 (i.e., a highly informative prior), B is large and the posterior mean is close to the prior mean μ.

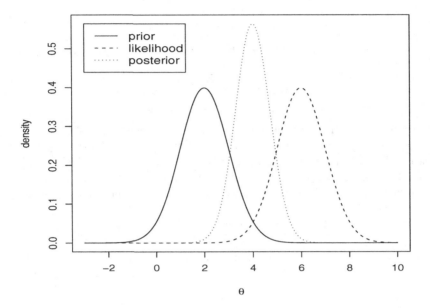

Figure 2.1 *Prior, likelihood, and posterior distributions, elementary normal/normal model with a single observation y = 6.*

Turning to the posterior variance, which from (2.3) is

$$B\tau^2 \equiv (1 - B)\sigma^2 ,$$

note that it is smaller than the variance of either the prior or the likelihood. It is easy to show that the precision in the posterior is the sum of the precisions in the likelihood and the prior, where *precision* is defined to be the reciprocal of the variance; again see Exercise 2. Thus the posterior distribution offers a sensible compromise between our prior opinion and the observed data, and the combined strength of the two sources of information leads to increased precision in our understanding of θ.

As a concrete example, suppose that $\mu = 2, \tau = 1, y = 6$, and $\sigma = 1$. Figure 2.1 plots the prior (centered at $\theta = 2$), the likelihood (centered at $\theta = 6$), and the posterior arising from (2.3). Note that due to the equal weighting of the prior and the likelihood in this problem ($\tau = \sigma$), the posterior mean is 4, the unweighted average of the prior and the likelihood means. The posterior is not however "twice as tall" as either the prior or the likelihood since it is *precisions* (not density heights) that are additive here: the posterior has precision $1 + 1 = 2$, hence variance $1/2$, hence standard deviation $\sqrt{1/2} \approx .707$. Thus, the posterior covers a range of roughly $4 \pm 3(.707) \approx (1.88, 6.12)$, as seen in Figure 2.1. ∎

Given a sample of n independent observations, we can obtain $f(\mathbf{y}|\boldsymbol{\theta})$ as $\prod_{i=1}^{n} f(y_i|\boldsymbol{\theta})$, and proceed with equation (2.1). But evaluating this expression may be simpler if we can find a statistic $S(\mathbf{y})$ that is *sufficient* for $\boldsymbol{\theta}$ (that is, for which $f(\mathbf{y}|\boldsymbol{\theta}) = h(\mathbf{y})g(S(\mathbf{y})|\boldsymbol{\theta})$). To see this, note that for $S(\mathbf{y}) = s$,

$$
\begin{aligned}
p(\boldsymbol{\theta}|\mathbf{y}) &= \frac{f(\mathbf{y}|\boldsymbol{\theta})\pi(\boldsymbol{\theta})}{\int f(\mathbf{y}|\mathbf{u})\pi(\mathbf{u})\,d\mathbf{u}} \\
&= \frac{h(\mathbf{y})g(S(\mathbf{y})|\boldsymbol{\theta})\pi(\boldsymbol{\theta})}{\int h(\mathbf{y})g(S(\mathbf{y})|\mathbf{u})\pi(\mathbf{u})\,d\mathbf{u}} \\
&= \frac{g(s|\boldsymbol{\theta})\pi(\boldsymbol{\theta})}{m(s)} \;=\; p(\boldsymbol{\theta}|s) \;,
\end{aligned}
$$

provided $m(s) > 0$, since $h(\mathbf{y})$ cancels in the numerator and denominator of the middle expression. So if we can find a sufficient statistic, we may work with it instead of the entire dataset $\mathbf{y}$, thus reducing the dimensionality of the problem.

Example 2.3 Consider again the normal/normal setting of Example 2.2, but where we now have an independent sample of size n from $f(\mathbf{y}|\theta)$. Since $S(\mathbf{y}) = \bar{y}$ is sufficient for θ, we have that $p(\theta|\mathbf{y}) = p(\theta|\bar{y})$. But, because we know that $f(\bar{y}|\theta) = N(\theta, \sigma^2/n)$, equation (2.3) implies that

$$
\begin{aligned}
p(\theta|\bar{y}) &= N\left(\theta \left| \frac{(\sigma^2/n)\mu + \tau^2\bar{y}}{(\sigma^2/n) + \tau^2} \;,\; \frac{(\sigma^2/n)\tau^2}{(\sigma^2/n) + \tau^2} \right.\right) \\
&= N\left(\theta \left| \frac{\sigma^2\mu + n\tau^2\bar{y}}{\sigma^2 + n\tau^2} \;,\; \frac{\sigma^2\tau^2}{\sigma^2 + n\tau^2} \right.\right) \;.
\end{aligned}
\tag{2.4}
$$

Returning to the specific setting of Example 2.2, suppose we keep $\mu = 2$ and $\tau = \sigma = 1$, but now let $\bar{y} = 6$. Figure 2.2 plots the prior distribution, along with the posterior distributions arising from two different sample sizes, 1 and 10. As already seen in Figure 2.1, when $n = 1$, the prior and likelihood receive equal weight, and hence the posterior mean is $4 = \frac{2+6}{2}$. When $n = 10$, the data dominate the prior, resulting in a posterior mean much closer to $\bar{y}$. Notice that the posterior variance also shrinks as n gets larger; the posterior collapses to a point mass on $\bar{y}$ as n tends to infinity.

Plots of posterior distributions like Figures 2.1 and 2.2 are easily drawn in the R package; see www.r-project.org. This software, the freeware heir to the S and S-plus languages, has become widely popular with statisticians for its easy blending of data manipulation and display, graphics, programmability, and both built-in and user-contributed packages for fitting and testing a wide array of common statistical models. The reader is referred to Venables and Ripley (2002) for an extensive and popular R tutorial; here as an initial illustration we provide the code to draw Figure 2.2. First, we set up a function called `postplot` to calculate the posterior mean and variance in our normal/normal model, as indicated by equation (2.4):

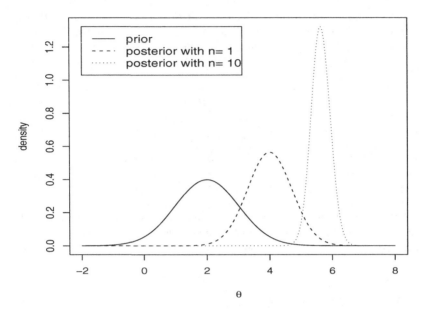

Figure 2.2 *Prior and posterior distributions based on samples of size 1 and 10, normal/normal model.*

R code
```
postplot <- function(mu, tau, ybar, sigma, n){
    denom <- sigma^2/n + tau^2
    mu <- (sigma^2/n*mu + tau^2*ybar)/denom
    stdev <- sqrt(sigma^2/n*tau^2/denom)
    return(c(mu,stdev))
    }
```

Next we set up the horizontal plotting axis, and plot the prior using the dnorm function, which gives the density of a normal with the given mean and variance over the grid:

R code
```
x <- seq(-2,8,length.out=100)
plot(x, dnorm(x, mean=2, sd=1), ylim=c(0, 1.3),
    xlab=expression(theta), ylab="density", type= "l")
```

Here the ylim option sets the vertical scale, while the type="l" option indicates we want a connected line graph (not merely points plotted at each grid value). Next, we may add the posterior densities for $n = 1$ and $n = 10$ using the lines command:

R code
```
param1 <- postplot(2,1,6,1,1)
lines(x, dnorm(x, mean=param1[1], sd=param1[2]), lty=2)
param10 <- postplot(2,1,6,1,10)
lines(x, dnorm(x, mean=param10[1], sd=param10[2]), lty=3)
```

where we request dashed (1ty=2) or dotted (1ty=3) line types for the two posteriors, since we've already used a solid line (1ty=1) for the prior. Finally, we can add a legend to the figure by typing

R code

```
legend(-2, 1.3, legend=c("prior", "posterior with n=1",
    "posterior with n=10"), lty=1:3, ncol=1)
```

This completes Figure 2.2. ∎

The R language also allows the user to draw random samples from a wide variety of probability distributions. This is sometimes called *Monte Carlo sampling*, after the city known for its famous casinos (and presumably rarely visited by probabilists who named the technique). In the previous example, we may draw 2000 independent random samples directly from the posterior (say, in the $n = 1$ case) using the rnorm command:

R code

```
y1 <- rnorm(2000, mean=param1[1], sd=param1[2])
## param1: [1] 4.0000000 0.7071068
```

A histogram of these samples can be added to our previous plot using hist and lines as follows:

R code

```
r1 <- hist(y1, freq=F, breaks=20, plot=F)
lines(r1, lty=2, freq=F, col="gray90")
```

producing Figure 2.3. We remark that hist can be used without first storing its result in r1, but this will start the plot over again, erasing the three existing curves. The empirical mean and standard deviation of our Monte Carlo samples may be obtained simply by typing mean(y1) and sd(y1) in R; for our sample we obtained 4.03 and 0.693, very close to the true values of 4.0 and 0.707, respectively. These estimates could be made more accurate by increasing the Monte Carlo sample size (say, from 2000 to 10,000 or 100,000); we defer details about the error inherent in Monte Carlo estimation to Section 3.3.

Of course, Monte Carlo methods are not necessary in this simple normal/normal example, since the integral in the denominator of Bayes' Theorem can be evaluated in closed form. In this case, we would likely prefer the resulting smooth curve and corresponding exact answers for the posterior mean and variance to the bumpy histogram and the estimated mean and variance produced by Monte Carlo. However, if the likelihood f and the prior π do *not* permit evaluation of the denominator integral, Monte Carlo methods are generally the preferred method for estimating the posterior. The reason for this is the approach's great generality: samples can typically be drawn from any posterior regardless of how high-dimensional the parameter vector $\boldsymbol{\theta}$ is. Thus, while we no longer get a smooth, exact functional form for the posterior $p(\theta|\mathbf{y})$, we gain the ability to work problems of essentially unlimited complexity.

Chapter 3 contains an extensive discussion of the various computational methods useful in Bayesian data analysis; for the time being, we give only

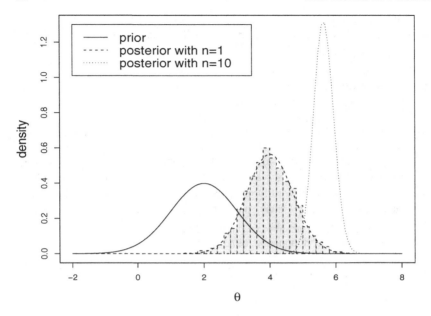

Figure 2.3 *Prior and posterior distributions for the normal/normal model, with a histogram of Monte Carlo draws superimposed for the $n = 1$ case.*

the briefest indication of how these methods can be implemented in the WinBUGS language. WinBUGS is a freely available program developed by statisticians and probabilistic expert systems researchers at the Medical Research Council Biostatistics Unit at the University of Cambridge, England. In a nutshell, it allows us to draw samples from any posterior distribution, freeing us from having to worry overmuch about the integral in (2.1). This allows us instead to focus on the statistical modeling, which is after all our primary interest. WinBUGS uses syntax very similar to that of R, and in fact can now be called from R using the BRugs package, a subject to which we return below.

WinBUGS implements a particular form of Monte Carlo algorithm known as *Markov chain Monte Carlo*, or MCMC for short. Again, while we return to this subject in excruciating detail in Section 3.4, for now we simply note that the approach essentially amounts to sampling sequentially from each parameter's *full conditional* distribution, i.e., the distribution of each model parameter given every other model parameter *and* the data. So for example, in a 10-parameter model, the full conditional for θ_1 is $p(\theta_1|\theta_2, \theta_3, \ldots, \theta_{10}, \mathbf{y})$. It turns out these full conditional distributions are often easy to sample from even when the full *joint* posterior distribution (in this case, $p(\theta_1, \ldots, \theta_{10}|\mathbf{y})$) is not. Because of the sequential nature of the

Monte Carlo updating, an MCMC algorithm produces *correlated* (not independent) draws from the true joint posterior, so deciding when to stop the algorithm can be tricky (see Subsection 3.4.6). However, for a broad class of fairly standard linear hierarchical models, MCMC methods work well and are widely accepted as the "industry standard" for Bayesian modeling.

Example 2.4 Consider once again the normal/normal setting of Example 2.3, but where we now wish to use WinBUGS to sample from the posterior of θ. As of the current writing, the latest version (1.4.3) of the program may be downloaded from www.mrc-bsu.cam.ac.uk/bugs/welcome.shtml. Once installed, a good way to learn the basics of the language is to follow the tutorial: click on Help, pull down to User Manual, and then click on Tutorial. Perhaps even more easily, one can watch "WinBUGS – The Movie," a delightful Flash introduction to running the software available at www.statslab.cam.ac.uk/~krice/winbugsthemovie.html. Finally, a large collection of worked examples are available within WinBUGS by clicking on Help and pulling down to Examples Vol I or Examples Vol II. These examples enable the familiar statistical computing strategy of attempting to find a piece of code that does something *similar* to what you want to do, modifying it to fit your particular model and data set, and then hoping the code still runs once the modifications are made.

WinBUGS solutions to Bayesian hierarchical modeling problems require three basic elements: (1) some WinBUGS code to specify the statistical model, (2) a file containing the data, and (3) a short file giving starting values for the MCMC sampling algorithm. For our normal/normal problem, the first element looks like this:

BUGS code
```
model{
    prec.ybar <- n/sigma2
    prec.theta <- 1/tau2
    ybar   ~ dnorm(theta, prec.ybar)
    theta ~ dnorm(mu, prec.theta)
    }
```

Notice the use of <- (arrow) for assignment and ~ (tilde) for "is distributed as," consistent with usual notation in R and statistical practice generally. However, note that the second parameter in the normal distribution expression dnorm is the *precision* (reciprocal variance), *not* the variance itself, a convention that many students find confusing initially. To be consistent with our notation from Examples 2.2 and 2.3, the code above includes assignment statements for the precision in the data (prec.ybar) and the prior (prec.theta) that give the appropriate conversions from the corresponding variances, σ^2/n and τ^2.

The data file in this simple problem consists of the one line

BUGS code
```
list(ybar=6, mu=2, sigma2=1, tau2=1, n=1)
```

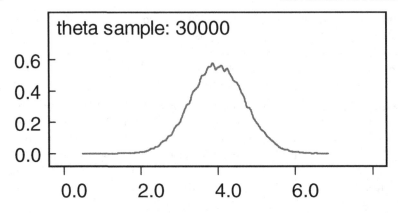

Figure 2.4 *Estimated posterior density based on 30,000 Gibbs samples (histogram not shown), computed by WinBUGS using the "density" command within the* Sample Monitor *tool.*

Thus the data file includes not only the data $\bar{y}$, but also the sample size n and the hyperparameters μ, σ^2, and τ^2. Finally, the initial values file takes the form

BUGS code   `list(theta = 0)`

which simply initializes the sampler at 0 (though in this univariate problem the starting place actually turns out to be arbitrary).

Running the sampler for 30,000 iterations, we obtained a smoothed kernel density estimate plotted in Figure 2.4, which looks very similar to the histogram in Figure 2.3. The sample mean and standard deviation of the 30,000 draws were 3.997 and 0.704, very close to the true values of 4.0 and 0.707.

Again, we do not need all this MCMC power to solve this tiny little problem, but the benefits of WinBUGS will become very apparent as this chapter wears on and we begin tackling models with many more unknown parameters and more complicated distributional structures. ∎

Models with three or more stages

The basic Bayesian model considered in equation (2.1) has two stages, one for $f(\mathbf{y}|\boldsymbol{\theta})$, the likelihood of the data $\mathbf{y}$ given the parameters $\boldsymbol{\theta}$, and one for $\pi(\boldsymbol{\theta}|\boldsymbol{\eta})$, the prior distribution of the model parameters $\boldsymbol{\theta}$ given a vector of hyperparameters $\boldsymbol{\eta}$. In many cases, however, we may need to use a model with more than two stages. For instance, suppose we were unsure as to the proper value for the $\boldsymbol{\eta}$ vector. The proper Bayesian solution would be to quantify this uncertainty in a second-stage prior distribution (sometimes called a *hyperprior*). Denoting this distribution by $h(\boldsymbol{\eta})$, the desired posterior for $\boldsymbol{\theta}$ is now obtained by also marginalizing over $\boldsymbol{\eta}$,

$$
\begin{aligned}
p(\boldsymbol{\theta}|\mathbf{y}) &= \frac{p(\mathbf{y},\boldsymbol{\theta})}{p(\mathbf{y})} = \frac{\int p(\mathbf{y},\boldsymbol{\theta},\boldsymbol{\eta})\,d\boldsymbol{\eta}}{\int\int p(\mathbf{y},\mathbf{u},\boldsymbol{\eta})\,d\boldsymbol{\eta}\,d\mathbf{u}} \\
&= \frac{\int f(\mathbf{y}|\boldsymbol{\theta})\pi(\boldsymbol{\theta}|\boldsymbol{\eta})h(\boldsymbol{\eta})\,d\boldsymbol{\eta}}{\int\int f(\mathbf{y}|\mathbf{u})\pi(\mathbf{u}|\boldsymbol{\eta})h(\boldsymbol{\eta})\,d\boldsymbol{\eta}\,d\mathbf{u}} .
\end{aligned} \tag{2.5}
$$

Alternatively, we might simply replace $\boldsymbol{\eta}$ by an estimate $\hat{\boldsymbol{\eta}}$ obtained as the value that maximizes the marginal distribution $m(\mathbf{y}|\boldsymbol{\eta})$ viewed as a function of $\boldsymbol{\eta}$. Inference is now based on the *estimated posterior* distribution $p(\boldsymbol{\theta}|\mathbf{y},\hat{\boldsymbol{\eta}})$, obtained by plugging $\hat{\boldsymbol{\eta}}$ into equation (2.1) above. Such an approach is often referred to as *empirical Bayes* analysis, since we are using the data to estimate the prior parameter $\boldsymbol{\eta}$. This is the subject of Chapter 5, where we also draw comparisons with the straight Bayesian approach using hyperpriors (c.f. Meeden, 1972; Deely and Lindley, 1981).

Of course, in principle, there is no reason why the hyperprior for $\boldsymbol{\eta}$ cannot itself depend on a collection of unknown parameters $\boldsymbol{\lambda}$, resulting in a generalization of (2.5) featuring a second-stage prior $h(\boldsymbol{\eta}|\boldsymbol{\lambda})$ and a third-stage prior $g(\boldsymbol{\lambda})$. This enterprise of specifying a model over several levels is called *hierarchical modeling*, with each new distribution forming a new level in the hierarchy. The proper number of levels varies with the problem; see the discussions in Sections 2.4 and 4.1 below. Because we are continually adding randomness as we move down the hierarchy, subtle changes to levels near the top are not likely to have much of an impact on the one at the bottom (the data level), which is typically the only one for which we actually have observations.

In order to concentrate on more foundational issues, in the remainder of this section we will typically limit ourselves to the simplest model having only two levels (likelihood and prior), again suppressing the dependence of the prior on the known value of $\boldsymbol{\eta}$.

Bayesian estimation and prediction

Observe that equation (2.1) may be expressed in the convenient shorthand

$$
p(\boldsymbol{\theta}|\mathbf{y}) \propto f(\mathbf{y}|\boldsymbol{\theta})\pi(\boldsymbol{\theta}) ,
$$

or in words, "the posterior is proportional to the likelihood times the prior." Clearly the likelihood may be multiplied by any constant (or even any function of $\mathbf{y}$ alone) without altering the posterior.

Bayes' Theorem may also be used *sequentially*: suppose we have two independently collected samples of data, $\mathbf{y}_1$ and $\mathbf{y}_2$. Then

$$
\begin{aligned}
p(\boldsymbol{\theta}|\mathbf{y}_1,\mathbf{y}_2) &\propto f(\mathbf{y}_1,\mathbf{y}_2|\boldsymbol{\theta})\pi(\boldsymbol{\theta}) \\
&= f_2(\mathbf{y}_2|\boldsymbol{\theta})f_1(\mathbf{y}_1|\boldsymbol{\theta})\pi(\boldsymbol{\theta}) \\
&\propto f_2(\mathbf{y}_2|\boldsymbol{\theta})p(\boldsymbol{\theta}|\mathbf{y}_1) .
\end{aligned} \tag{2.6}
$$

That is, we can obtain the posterior for the full dataset $(\mathbf{y}_1, \mathbf{y}_2)$ by first finding $p(\boldsymbol{\theta}|\mathbf{y}_1)$ and then treating it as the prior for the second portion of the data $\mathbf{y}_2$. This easy algorithm for updating the posterior is quite natural when the data arrive sequentially over time, as in a clinical trial or perhaps a business or economic setting.

Many authors (notably Geisser, 1993) have argued that concentrating on inference for the model parameters is misguided, since $\boldsymbol{\theta}$ is merely an unobservable, theoretical quantity. Switching to a different model for the data may result in an entirely different $\boldsymbol{\theta}$ vector. Moreover, even a perfect understanding of the model does not constitute a direct attack on the problem of *predicting* how the system under study will behave in the future – often the real goal of a statistical analysis. To this end, suppose that y_{n+1} is a future observation, independent of $\mathbf{y}$ given the underlying $\boldsymbol{\theta}$. Then the *predictive distribution* for y_{n+1} is given by

$$
\begin{aligned}
p(y_{n+1}|\mathbf{y}) &= \int p(y_{n+1}, \boldsymbol{\theta}|\mathbf{y})d\boldsymbol{\theta} \\
&= \int f(y_{n+1}|\boldsymbol{\theta}, \mathbf{y})p(\boldsymbol{\theta}|\mathbf{y})d\boldsymbol{\theta} \\
&= \int f(y_{n+1}|\boldsymbol{\theta})p(\boldsymbol{\theta}|\mathbf{y})d\boldsymbol{\theta} \ ,
\end{aligned} \tag{2.7}
$$

the last equality holding thanks to the conditional independence of y_{n+1} and $\mathbf{y}$ given the parameters $\boldsymbol{\theta}$ (i.e., the usual independence of observations often assumed in the likelihood model). The predictive distribution summarizes the information concerning the likely value of a new observation, given the likelihood, the prior, and the data we have observed so far. (We remark that some authors refer to this distribution as the *posterior predictive*, and refer to the marginal distribution $m(y_{n+1}) = \int f(y_{n+1}|\boldsymbol{\theta})\pi(\boldsymbol{\theta})d\boldsymbol{\theta}$ as the *prior predictive*, since the latter summarizes our information concerning y_{n+1} before having seen the data.)

The Bayesian decision-making paradigm improves on the traditional, frequentist approach to statistical analysis in its more philosophically sound foundation, its unified and streamlined approach to data analysis, and its ability to formally incorporate the prior opinion of one or more experimenters into the results via the prior distribution π. Practicing statisticians and biostatisticians, once reluctant to adopt the Bayesian approach due to general skepticism concerning its philosophy and a lack of necessary computational tools, are now turning to it with increasing regularity as traditional analytic approaches emerge as both theoretically and practically inadequate. For example, hierarchical Bayes methods form an ideal setting for combining information from several published studies of the same research area, an emerging scientific discipline commonly referred to as meta-analysis (DuMouchel and Harris, 1983; Cooper and Hedges, 1994). More generally, the hierarchical structure allows for honest assessment of heterogeneity both within and between groups, such as laboratories or census areas.

In the remainder of this chapter, we discuss the steps necessary to conduct a fully Bayesian data analysis, including prior specification, posterior and predictive inference, hierarchical modeling, model selection, and diagnosis of departures from the assumed prior and likelihood form. Throughout our discussion, we illustrate with examples from the realms of biomedical science, public health, and environmental risk assessment as appropriate.

2.2 Prior distributions

Implementation of the Bayesian approach as indicated in the previous subsection depends on a willingness to assign probability distributions not only to data variables like $\mathbf{y}$, but also to parameters like $\boldsymbol{\theta}$. Such a requirement may or may not be consistent with the usual long-run frequency notion of probability. For example, if

$$\theta = \text{true probability of success for a new surgical procedure,}$$

then it is possible (at least conceptually) to think of θ as the limiting value of the observed success rate as the procedure is independently repeated again and again. But if

$$\theta = \text{true proportion of U.S. men who are HIV-positive,}$$

the long-term frequency notion does not apply; it is not possible to even imagine "running the HIV epidemic over again" and reobserving θ. Moreover, the randomness in θ does not arise from any real-world mechanism; if an accurate census of all men and their HIV status were available, θ could be computed exactly. Rather, here θ is random only because it is unknown *to us*, though we may have some feelings about it (say, that $\theta = .05$ is more likely than $\theta = .50$). Bayesian analysis is predicated on such a belief in *subjective probability*, wherein we quantify whatever feelings (however vague) we may have about θ *before* we look at the data $\mathbf{y}$ in a distribution π. This distribution is then updated by the data via Bayes' Theorem (as in (2.1) or (2.5)) with the resulting posterior distribution reflecting a blend of the information in the data and the prior.

Historically, a major impediment to widespread use of the Bayesian paradigm has been that determination of the appropriate form of the prior π (and perhaps the hyperprior h) is often an arduous task. Typically, these distributions are specified based on information accumulated from past studies or from the opinions of subject-area experts. In order to streamline the elicitation process, as well as simplify the subsequent computational burden, experimenters often limit this choice somewhat by restricting π to some familiar distributional family. An even simpler alternative, available in some cases, is to endow the prior distribution with little informative content, so that the data from the current study will be the dominant force in determining the posterior distribution. We address each of these approaches in turn.

2.2.1 Elicited priors

Suppose for the moment that $\boldsymbol{\theta}$ is univariate. Perhaps the simplest approach to specifying $\pi(\theta)$ is first to limit consideration to a manageable collection of θ values deemed "possible," and subsequently to assign probability masses to these values in such a way that their sum is 1, their relative contributions reflecting the experimenter's prior beliefs as closely as possible. If θ is discrete-valued, such an approach may be quite natural, though perhaps time-consuming. If θ is continuous, we must instead assign the masses to intervals on the real line, rather than to single points, resulting in a histogram prior for θ. Such a histogram (necessarily over a bounded region) may seem inappropriate, especially in concert with a continuous likelihood $f(\mathbf{y}|\theta)$, but can perhaps be thought of as just another discrete finite approximation to a continuous underlying "truth," similar to those employed by many numerical integration routines. Moreover, a histogram prior may have as many bins as the patience of the elicitee and the accuracy of his prior opinion will allow. It is vitally important, however, that the range of the histogram be sufficiently wide, since, as can be seen from (2.1), the support of the posterior will necessarily be a subset of that of the prior. That is, the data cannot possibly lend credence to intervals for θ deemed "impossible" in the prior.

Alternatively, we might simply assume that the prior for $\boldsymbol{\theta}$ belongs to a parametric distributional family $\pi(\boldsymbol{\theta}|\boldsymbol{\eta})$, choosing $\boldsymbol{\eta}$ so that the result matches the elicitee's true prior beliefs as nearly as possible. For example, if $\boldsymbol{\eta}$ were two-dimensional, then specification of two moments (say, the mean and the variance) or two quantiles (say, the 50th and 75th) would be sufficient to determine its exact value. This approach limits the effort required of the elicitee, and also overcomes the finite support problem inherent in the histogram approach. It may also lead to simplifications in the posterior computation, as we shall see in Subsection 2.2.2.

A limitation of this approach is of course that it may not be possible for the elicitee to "shoehorn" his or her prior beliefs into any of the standard parametric forms. In addition, two distributions that look virtually identical may in fact have quite different properties. For example, Berger (1985, p. 79) points out that the Cauchy(0,1) and Normal(0, 2.19) distributions have identical 25th, 50th, and 75th percentiles (−1, 0, and 1, respectively) and density functions that appear very similar when plotted, yet may lead to quite different posterior distributions.

Early statistical methods for eliciting priors on an unknown proportion were given by Winkler (1967), and by Kadane et al. (1980) for normal linear models with unknown variance parameters. More general reviews of the area have been published by Kadane and Wolfson (1996, 1998), Chaloner (1996), Garthwaite et al. (1995), and, most recently, in the fine book by O'Hagan et al. (2006). These references provide overviews of the various philosophies of elicitation, as well as reviews of the methods proposed for various models and distributional settings.

In particular, O'Hagan et al. (2006) provide an extensive review of both the statistical and psychological literature, the latter of which has much to say about how the ways humans think about and update probabilistic statements. For instance, many previous psychological studies have shown that humans tend to be *overconfident* in their probability assessments: in situations where the "correct answer" can be ascertained after the fact (e.g., weather forecasting), the 95% subjective confidence intervals tend to include these answers far less than 95% of the time. This appears to be the result of the inherent difficulty in considering the unlikely events that lead to observations in the tails of a distribution. Overconfidence may also result from elicitees failing to condition on events outside their own personal range of experience. A doctor's opinion about the likely success of a new AIDS drug may reflect his own experience successfully treating affluent gay men, but not include the likely outcomes when the drug is given to innercity intravenous drug-using women, an important but very different component of the drug's target population. As a remedy, O'Hagan et al. (2006) recommend avoiding elictation of extreme quantiles (say, the 95th), and instead focus on quantiles closer to the middle of the distribution (e.g., the 50th, 25th, and 75th). A general strategy they recommend is to first elicit the prior median of θ with a question like

> Can you determine a value (your median) such that θ is equally likely to be less than or greater than this point?

and then follow up with a question about the 25th percentile, say

> Suppose you were told that θ is below your assessed median. Can you now determine a new value (the 25th percentile) such that it is equally likely that θ is less than or greater than this value?

Repeating this latter question with "below" replaced by "above" yields an elicited value for the 75th percentile, which can then be used to assess symmetry of the elicitee's opinion, as well as overall consistency with the first two answers.

An opportunity to experiment with both histogram and functional form-matching prior elicitation in a simple setting is afforded below in Exercise 4. We now illustrate multivariate elicitation in a challenging, real-life setting.

Example 2.5 In attempting to model the probability p_{ijkl} of an incorrect response by person j in cue group i on a working memory exam question having k "simple conditions" and "query complexity" level l, Carlin et al.

(1992) consider the model

$$logit(p_{ijkl}) = \begin{cases} \theta_j^{(i)} + \gamma k, & l = 0 \\ \theta_j^{(i)} + \gamma k + \alpha, & l = 1 \\ \theta_j^{(i)} + \gamma k + \alpha + \beta, & l = 2 \end{cases} .$$

In this model, $\theta_j^{(i)}$ denotes the effect for subject ij, γ denotes the effect due to the number of simple conditions in the question, α measures the marginal increase in the logit error rate in moving from complexity level 0 to 1, and β does the same in moving from complexity level 1 to 2. Prior information concerning the parameters was based primarily on rough impressions, precluding precise prior elicitation. As such, $\pi(\{\theta_j^{(i)}\}, \gamma, \alpha, \beta)$ was defined as a product of independent, normally distributed components, namely,

$$\prod_{i=1}^{I} \prod_{j=1}^{J} N(\theta_j^{(i)}|\mu_\theta, \sigma_\theta^2) \times N(\gamma|\mu_\gamma, \sigma_\gamma^2) \times N_2(\alpha, \beta|\mu_\alpha, \mu_\beta, \sigma_\alpha^2, \sigma_\beta^2, \rho) , \quad (2.8)$$

where N represents the univariate normal distribution and N_2 represents the bivariate normal distribution. Hyperparameter mean parameters were chosen by eliciting "most likely" error rates, and subsequently converting to the logit scale. Next, variances were determined by considering the effect on the error rate scale of a postulated standard deviation (and resulting 95% confidence set) on the logit scale. Note that the symmetry imposed by the normal priors on the logit scale does not carry over to the error rate (probability) scale. While neither normality nor prior independence was deemed totally realistic, prior (2.8) was able to provide a suitable starting place, subsequently updated by the information in the data. ■

Even when scientifically relevant prior information is available, elicitation of the precise forms for the prior distribution from the experimenters can be a long and difficult process, with several simplifying assumptions required to complete the task in reasonable time. For instance, in the previous example, the assumption of prior independence across parameters was probably unrealistic, but was necessary to avoid elicitation of manifold covariances as well as variances, say in a multivariate normal prior (see Chapter 6 of O'Hagan et al., 2006, for more on this subject). In addition, the way that questions are presented to the elicitee can have a significant impact on the results. In the previous example, questions asked about 95% confidence intervals probably led to overconfident priors; tail elicitation should have focused on 25th and 75th (or perhaps even the 33rd and 66th) percentiles.

Greenland (2007a) argues for "data priors," which seek the data equivalent of the level of information desired in a prior. Such an approach is useful for elucidating parallels between frequentist and Bayesian inference, and may also allow Bayesian computation to be carried out in standard, frequentist software packages (because now both the prior and likelihood information can be input as data records in the program). In the main, however, prior elicitation issues tend to be application- and elicitee-specific, meaning that general purpose algorithms are typically unavailable; see e.g. Greenland (2006, 2007b) for illustration of his prior data approach in the 2×2 table and linear regression settings, respectively.

As with many other areas of Bayesian endeavor, the difficulty of prior elicitation has been ameliorated somewhat through the addition of interactive computing, especially dynamic graphics and object-oriented computer languages such as R. As of the current writing, the best source of information on up-to-date elicitation software may be www.shef.ac.uk/beep/, the website of the BEEP *(Bayesian Elicitation of Expert Probabilities)* project. BEEP is the grant-funded research effort associated with the O'Hagan et al. (2006) book. Currently available are fairly high-level programs by Paul Garthwaite and David Jenkinson to elicit beliefs about relationships assuming a particular kind of linear model. Another, more low-level approach called SHELF, developed by Tony O'Hagan and Jeremy Oakley, is aimed at eliciting a single distribution in as rigorous and defensible a framework as possible. This R software includes templates for elicitors to follow that will both guide them through the process and also provide a record of the elicitation. Software by John Paul Gosling called ROBEO is also available for implementing the nonparametric elicitation method of Oakley and O'Hagan (2007) and its extensions. The group also plans extensions to multivariate settings and to allow for uncertainty in the elicited probabilities or quantiles.

Example 2.6 In the arena of monitoring clinical trials, Chaloner et al. (1993) show how to combine histogram elicitation, matching a functional form, and interactive graphical methods. Following the advice of Kadane et al. (1980), these authors elicit a prior not on θ (in this case, an unobservable proportional hazards regression parameter), but on corresponding observable quantities familiar to their medically-oriented elicitees, namely the proportion of individuals failing within two years in a population of control patients, p_0, and the corresponding two-year failure proportion in a population of treated patients, p_1. Writing the survivor function at time t for the controls as $S(t)$, under the proportional hazards model we then have that $p_0 = 1 - S(2)$ and $p_1 = 1 - S(2)^{\exp(\theta)}$. Hence, the equation

$$\log[-\log(1 - p_1)] = \theta + \log[-\log(1 - p_0)] \tag{2.9}$$

gives the relationship between p_0, p_1, and θ.

Since the effect of a treatment is typically thought of by clinicians as being relative to the baseline (control) failure probability p_0, they first elicit a best guess for this rate, $\hat{p}_0$. Conditional on this modal value, they then elicit an entire distribution for p_1, beginning with initial guesses for the upper and lower quartiles of p_1's distribution. These values determine initial guesses for the parameters μ and σ of a smooth density function that corresponds on the θ scale to an extreme value distribution. This smooth functional form is then displayed on the computer screen, and the elicitee is allowed to experiment with new values for μ and σ in order to obtain an even better fit to his or her true prior beliefs. Finally, fine tuning of the density is allowed via mouse input directly onto the screen. The density is restandardized after each such change, with updated quantiles computed for the elicitee's approval. At the conclusion of this process, the final prior distribution is discretized onto a suitably fine grid of points $(p_{11}, \ldots, p_{1K})$, and finally converted to a histogram-type prior on θ via equation (2.9), for use in computing its posterior distribution. ■

In this example, the interactive nature of the computer program allows continual reassessment and checking of the elicited prior against the expert's opinion, a good guard against bias due to overconfidence or other sources. However, the inherent difficulty of the elicitation task makes us lean toward its use only in situations where anticipated data sample sizes are small, and the experts possess good reliable prior information (often in the form of intimate knowledge of previous data) on the subject at hand. The only context in which we shall view the use of elicited priors as *essential* will be in the area of *experimental design*, where some idea of the nature of the system being studied must be input in order to plan efficient experiments and accurately predict their operating characteristics. We will return to this subject in detail in Chapter 6.

2.2.2 Conjugate priors

In choosing a prior belonging to a specific distributional family $\pi(\boldsymbol{\theta}|\boldsymbol{\eta})$, some choices may be more computationally convenient than others. In particular, it may be possible to select a member of that family that is *conjugate* with the likelihood $f(\mathbf{y}|\boldsymbol{\theta})$, that is, one that leads to a posterior distribution $p(\boldsymbol{\theta}|\mathbf{y})$ belonging to the same distributional family as the prior. The computational advantages are best illustrated through an example.

Example 2.7 Suppose that X is the number of pregnant women arriving at a particular hospital to deliver their babies during a given month. The discrete count nature of the data plus its natural interpretation as an arrival rate suggest adopting a Poisson likelihood,

$$f(x|\theta) = \frac{e^{-\theta}\theta^x}{x!}, \; x \in \{0, 1, 2, \ldots\}, \; \theta > 0.$$

To effect a Bayesian analysis, we require a prior distribution for θ having support on the positive real line. A reasonably flexible choice is provided by the gamma distribution,

$$\pi(\theta) = \frac{\theta^{\alpha-1}e^{-\theta/\beta}}{\Gamma(\alpha)\beta^\alpha}, \ \theta > 0, \alpha > 0, \ \beta > 0,$$

or $\theta \sim G(\alpha, \beta)$ in distributional shorthand. Note that we have suppressed π's dependence on $\eta = (\alpha, \beta)$ since we assume it to be known. The gamma distribution has mean $\alpha\beta$, variance $\alpha\beta^2$, and can have a shape that is either one-tailed ($\alpha \leq 1$) or two-tailed ($\alpha > 1$); for large α the distribution resembles a normal distribution. The β parameter is a *scale* parameter, stretching or shrinking the distribution relative to 0, but not changing its shape. Using Bayes' Theorem (2.1) to obtain the posterior density, we have

$$\begin{aligned} p(\theta|x) &\propto f(x|\theta)\pi(\theta) \\ &\propto \left(e^{-\theta}\theta^x\right)\left(\theta^{\alpha-1}e^{-\theta/\beta}\right) \\ &= \theta^{x+\alpha-1}e^{-\theta(1+1/\beta)} . \end{aligned} \tag{2.10}$$

Notice that since our intended result is a normalized function of θ, we are able to drop any multiplicative functions that do not depend on θ. (For example, in the first line we have dropped the marginal distribution $m(x)$ in the denominator, since it is free of θ.) But now looking at (2.10), we see that it is proportional to a gamma distribution with parameters $\alpha' = x+\alpha$ and $\beta' = (1+1/\beta)^{-1}$. Note this is the *only* function proportional to (2.10) that still integrates to 1. Because density functions uniquely determine distributions, we know that the posterior distribution for θ is indeed $G(\alpha', \beta')$, and that the gamma is the conjugate family for the Poisson likelihood.

As a concrete illustration, suppose we observe $x = 42$ moms arriving at our hospital to deliver babies during December 2007. Suppose we adopt a $G(5, 6)$ prior, which has mean $5(6) = 30$ and variance $5(6^2) = 180$ (see dotted line in Figure 2.5), reflecting the hospital's totals for the past 24 months, which have been slightly less busy on average. The R code to set up a grid for θ and compute the gamma prior and posterior is

R code
```
alpha<-5
beta<-6
theta<-seq(0, 100, length.out=101)
prior <- dgamma(theta, shape=alpha, scale=beta)
x<-42
posterior <- dgamma(theta, shape=x+alpha, scale=1/(1+1/beta) )
```

using the intrinsic **dgamma** (density of a gamma) function. We may now plot these two densities by typing:

R code
```
plot(theta, posterior, xlab=expression(theta), ylab="density")
lines(theta, prior, lty=3)
```

As in Figure 2.3, we may also sample directly from the gamma posterior, this time using the **rgamma** function:

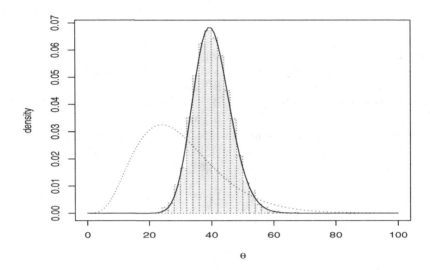

Figure 2.5 *Posterior distribution for θ in the Poisson-gamma example for x = 42: prior (dotted line), posterior (solid line), and posterior MCMC samples (histogram).*

R code
```
postdraw <- rgamma(2000, shape=x+alpha, scale=1/(1+1/beta))
r1<-hist(postdraw, freq=F, breaks=20, plot=F)
lines(r1, lty=3, freq=F, col="gray90")
```

This last command adds the histogram to our existing plot, completing Figure 2.5. ∎

In this example, we might have guessed the identity of the conjugate prior by looking at the Poisson likelihood as a function of θ, instead of x as we typically do. Such a viewpoint reveals the $\theta^a e^{-b\theta}$ structure that is also the basis of the gamma density. This technique is useful for determining conjugate priors in many exponential families, as we shall see as we continue. Since conjugate priors can permit posterior distributions to emerge without numerical integration, we shall use them in many of our subsequent examples.

For multiparameter models, independent conjugate priors may often be specified for each parameter, leading to corresponding conjugate forms for each *conditional* posterior distribution.

Example 2.8 Consider once again the normal likelihood of Example 2.2, where now both θ and σ^2 are assumed unknown, so that $\boldsymbol{\theta} = (\theta, \sigma^2)$. Looking at f as a function of θ alone, we see an expression proportional to $\exp(-(\theta - a)^2/b)$, so that the conjugate prior is again given by the normal distribution. Thus we assume that $\pi_1(\theta) = N(\theta|\mu, \tau^2)$, as before. Next, looking at f as a function of σ, we see a form proportional to $(\sigma^2)^{-a}e^{-b/\sigma^2}$, which is vaguely reminiscent of the gamma distribution. In fact, this is actually the reciprocal of a gamma: if $X \sim G(\alpha, \beta)$, then $Y = 1/X \sim$ Inverse Gamma$(\alpha, \beta) \equiv IG(\alpha, \beta)$, with density function (see Appendix Section A.2)

$$g(y|\alpha, \beta) = \frac{e^{-1/\beta y}}{\Gamma(\alpha)\beta^\alpha y^{\alpha+1}}, \quad y > 0, \ \alpha > 0, \ \beta > 0.$$

Thus, we assume that $\pi_2(\sigma^2) = IG(\sigma^2|a, b)$, where a and b are the shape and scale parameters, respectively. Finally, we assume that θ and σ^2 are *a priori* independent, so that $\pi(\boldsymbol{\theta}) = \pi_1(\theta)\pi_2(\sigma^2)$. With this component-wise conjugate prior specification, the two conditional posterior distributions emerge as expression (2.3) and

$$p(\sigma^2|x, \theta) = IG\left(\sigma^2 \ \middle| \ \frac{1}{2} + a, \ \left[\frac{1}{2}(x - \theta)^2 + \frac{1}{b}\right]^{-1}\right). \qquad (2.11)$$

■

We remark that in the normal setting, closed forms may also emerge for the *marginal* posterior densities $p(\theta|x)$ and $p(\sigma^2|x)$; we leave the details as an exercise. The ability of conjugate priors to produce at least unidimensional conditional posteriors in closed form enables them to retain their importance even in very high dimensional settings analyzed via MCMC.

Finally, while a single conjugate prior may be inadequate to accurately reflect available prior knowledge, a finite mixture of conjugate priors may be sufficiently flexible (allowing multimodality, heavier tails, etc.) while still enabling simplified posterior calculations. For example, suppose we are given the likelihood $f(x|\theta)$ for the univariate parameter θ, and adopt the two-component mixture prior $\pi(\theta) = \alpha\pi_1(\theta) + (1 - \alpha)\pi_2(\theta)$, $0 \le \alpha \le 1$, where both π_1 and π_2 belong to the family of priors conjugate for f. Then

$$\begin{aligned}
p(\theta|x) &= \frac{f(x|\theta)\pi_1(\theta)\alpha + f(x|\theta)\pi_2(\theta)(1 - \alpha)}{\int[f(x|\theta)\pi_1(\theta)\alpha + f(x|\theta)\pi_2(\theta)(1 - \alpha)]d\theta} \\[2mm]
&= \frac{\frac{f(x|\theta)\pi_1(\theta)}{m_1(x)}m_1(x)\alpha + \frac{f(x|\theta)\pi_2(\theta)}{m_2(x)}m_2(x)(1 - \alpha)}{m_1(x)\alpha + m_2(x)(1 - \alpha)} \\[2mm]
&= \frac{f(x|\theta)\pi_1(\theta)}{m_1(x)}w_1 + \frac{f(x|\theta)\pi_2(\theta)}{m_2(x)}w_2 \\[2mm]
&= p_1(\theta|x)w_1 + p_2(\theta|x)w_2 ,
\end{aligned}$$

where $w_1 = m_1(x)\alpha/[m_1(x)\alpha + m_2(x)(1 - \alpha)]$ and $w_2 = 1 - w_1$. Hence, a mixture of conjugate priors leads to a mixture of conjugate posteriors.

2.2.3 Noninformative priors

As alluded to earlier, often no reliable prior information concerning $\boldsymbol{\theta}$ exists, or an inference based solely on the data is desired. At first, it might appear that Bayesian inference would be inappropriate in such settings, but this conclusion is a bit hasty. Suppose we could find a distribution $\pi(\boldsymbol{\theta})$ that contained "no information" about $\boldsymbol{\theta}$ in the sense that it did not favor one $\boldsymbol{\theta}$ value over another (provided both values were logically possible). We might refer to such a distribution as a *noninformative prior* for $\boldsymbol{\theta}$, and argue that all of the information resulting in the posterior $p(\boldsymbol{\theta}|\mathbf{x})$ arose from the data, and hence all resulting inferences were completely *objective*, rather than subjective. Such an approach is likely to be important if Bayesian methods are to compete successfully in practice with their popular likelihood-based counterparts (e.g., maximum likelihood estimation). But is such a "noninformative approach" possible?

In some cases, the answer is an unambiguous "yes." For example, suppose the parameter space is discrete and finite, i.e., $\boldsymbol{\Theta} = \{\theta_1, \ldots, \theta_n\}$. Then clearly the distribution

$$p(\theta_i) = 1/n, \ i = 1, \ldots, n,$$

does not favor any one of the candidate θ values over any other and, as such, is noninformative for θ. If instead we have a bounded continuous parameter space, say $\boldsymbol{\Theta} = [a, b]$, $-\infty < a < b < \infty$, then the uniform distribution

$$p(\theta) = 1/(b - a), \ a < \theta < b,$$

is arguably noninformative for θ (though this conclusion can be questioned, as we shall see later in this subsection and in Example 2.11).

When we move to unbounded parameter spaces, the situation is even less clear. Suppose that $\boldsymbol{\Theta} = (-\infty, \infty)$. Then the appropriate uniform prior would appear to be

$$p(\theta) = c, \ \text{any } c > 0.$$

But this distribution is *improper*, in that $\int p(\theta)d\theta = \infty$, and hence does not appear appropriate for use as a prior. But even here, Bayesian inference is still possible *if* the integral with respect to θ of the likelihood $f(\mathbf{x}|\theta)$ equals some finite value K. (Note that this will not necessarily be the case; the definition of f as a joint probability density requires only the finiteness of its integral with respect to $\mathbf{x}$.) Then

$$p(\theta|\mathbf{x}) = \frac{f(\mathbf{x}|\theta) \cdot c}{\int f(\mathbf{x}|\theta) \cdot c \, d\theta} = \frac{f(\mathbf{x}|\theta)}{K}.$$

Because the integral of this function with respect to θ is 1, it is indeed a proper posterior density, and hence Bayesian inference may proceed as usual. Still, it is worth reemphasizing that care must be taken when using improper priors, since proper posteriors will not always result. A good example (to which we return in Section 7.3) is given by random effects models for longitudinal data, where the dimension of the parameter space increases with the sample size. In such models, the information in the data may be insufficient to identify all the parameters, so at least some of the prior distributions on the individual parameters must be informative.

Example 2.9 Consider again the simple Poisson/gamma setting of Example 2.7. Suppose that past monthly data on the number of pregnant moms arriving at the hospital is not available, so that there is no basis for the $Gamma(5, 6)$ prior, or any other informative prior for that matter. As such, suppose we instead use a uniform prior for the mean arrival count θ. Since θ can theoretically assume any positive value, we could choose a $U(0, \infty)$ prior, but suppose we simply use a $U(0, 1000)$ prior (since this rules out no values that could possibly occur in real life). The posterior will then be a limiting case of that given in (2.10) (i.e., with $\alpha = 1$ and $\beta = \infty$) truncated to the interval $(0, 1000)$. Thus, we lack a strictly conjugate model here, but we may again resort to MCMC methods implemented in WinBUGS to sample from the true posterior. The necessary code is very simple:

BUGS code
```
model{
        x ~ dpois (theta)
        theta ~ dunif(0, 1000)
        }
```

as are the data file,

BUGS code `list(x=42)`

and the initial values file,

BUGS code `list(theta = 10)`

which simply initializes the sampler at 10.

Running the sampler for 30,000 iterations, we obtained a smoothed kernel density estimate plotted in Figure 2.6. Notwithstanding a few "wiggles" near $\theta = 40$ and 50 (which could be smoothed out by taking a Monte Carlo sample size even larger than 30,000), the estimated posterior looks very similar to the gamma prior-based posterior in Figure 2.5. Apparently our switch to the truncated uniform prior has had very little impact on the posterior; that is, our results appear *robust* to the choice of prior distribution in this case. The sample mean of the 30,000 draws was 41.97, very close to the data value of 42; the informative $G(5, 6)$ prior used in Example 2.7 led to a slightly smaller posterior mean of $(42 + 5)(1 + 1/6)^{-1} = 40.3$. ∎

Noninformative priors are closely related to the notion of a *reference prior*, though precisely what is meant by this term depends on the author. Kass and Wasserman (1996) define such a prior as a conventional or "default" choice – not necessarily noninformative, but a convenient place to

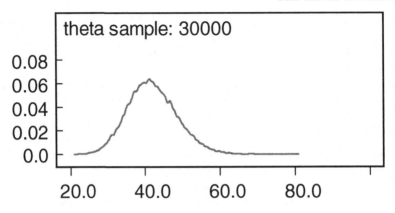

Figure 2.6 *Estimated posterior density based on 30,000 Gibbs samples for the Poisson/uniform example, computed by WinBUGS using the "density" command within the* Sample Monitor *tool.*

begin the analysis. The definition of Bernardo (1979) stays closer to non-informativity, using an expected utility approach to obtain a prior that represents ignorance about θ in some formal sense. Unfortunately, using this approach (and others that are similar), a unique representation for the reference prior is not always possible. Box and Tiao (1973, p.23) suggest that all that is important is that the data dominate whatever information is contained in the prior, since as long as this happens, the precise form of the prior is unimportant. They use the analogy of a properly run jury trial, wherein the weight of evidence presented in court dominates whatever prior ideas or biases may be held by the members of the jury.

Putting aside these subtle differences in definition for a moment, we ask the question of whether a uniform prior is always a good "reference." Sadly, the answer is no, since it is not invariant under reparametrization. As a simple example, suppose we claim ignorance concerning a univariate parameter θ defined on the positive real line, and adopt the prior $p(\theta) = 1$, $\theta > 0$. A sensible reparametrization is given by $\gamma = g(\theta) = \log(\theta)$, since this converts the support of the parameter to the whole real line. Then the prior on γ is given by $p'(\gamma) = |J| \, p(e^\gamma)$, where $J = d\theta/d\gamma$, the Jacobian of the inverse transformation. Here we have $\theta = g^{-1}(\gamma) = e^\gamma$, so that $J = e^\gamma$ and hence

$$p'(\gamma) = e^\gamma, \ -\infty < \gamma < \infty \,,$$

a prior that is clearly *not* uniform. Therefore, using uniformity as a universal definition of prior ignorance, it is possible that "ignorance about θ" does *not* imply "ignorance about γ." A possible remedy to this problem is to rely on the particular modeling context to provide the most reasonable parametrization and, subsequently, apply the uniform prior on this scale.

The *Jeffreys prior* (Jeffreys, 1961, p.181) offers a fairly easy-to-compute alternative that *is* invariant under transformation. In the univariate case, this prior is given by

$$p(\theta) \propto [I(\theta)]^{1/2} , \qquad (2.12)$$

where $I(\theta)$ is the expected Fisher information in the model, namely,

$$I(\theta) = -E_{\mathbf{x}|\theta} \left[\frac{\partial^2}{\partial \theta^2} \log f(\mathbf{x}|\theta) \right] . \qquad (2.13)$$

Notice that the *form* of the likelihood helps to determine the prior (2.12), but not the actual observed *data* since we are averaging over $\mathbf{x}$ in equation (2.13). It is left as an exercise to show that the Jeffreys prior is invariant to 1-1 transformation. That is, in the notation of the previous paragraph, it is true that

$$[I(\gamma)]^{1/2} = [I(\theta)]^{1/2} \left| \frac{d\theta}{d\gamma} \right| , \qquad (2.14)$$

so that computing the Jeffreys prior for γ directly produces the same answer as computing the Jeffreys prior for θ and subsequently performing the usual Jacobian transformation to the γ scale.

In the multiparameter case, the Jeffreys prior is given by

$$p(\boldsymbol{\theta}) \propto |I(\boldsymbol{\theta})|^{1/2} , \qquad (2.15)$$

where $|\cdot|$ now denotes the determinant, and $I(\boldsymbol{\theta})$ is the expected Fisher information *matrix*, having ij-element

$$I_{ij}(\boldsymbol{\theta}) = -E_{\mathbf{x}|\theta} \left[\frac{\partial^2}{\partial \theta_i \partial \theta_j} \log f(\mathbf{x}|\boldsymbol{\theta}) \right] .$$

While (2.15) provides a general recipe for obtaining noninformative priors, it is cumbersome to use in high dimensions. A more common approach is to obtain a noninformative prior for each parameter individually, and then form the joint prior simply as the product of these individual priors. This action is often justified on the grounds that "ignorance" is consistent with "independence," although since noninformative priors are often improper, the formal notion of independence does not really apply.

We close this subsection with two important special cases often encountered in Bayesian modeling. Suppose the density of X is such that $f(x|\theta) = f(x - \theta)$. That is, the density involves θ only through the term $(x - \theta)$. Then θ is called a *location parameter*, and f is referred to as a *location parameter family*. Examples of location parameters include the normal mean and Cauchy median parameters. Berger (1985, pp.83–84) shows that, for a prior on θ to be invariant under location transformations (i.e., transformations of the form $Y = X + c$), it must be *uniform* over the range of θ. Hence the noninformative prior for a location parameter is given by

$$p(\theta) = 1, \ \theta \in \Re .$$

Next, assume instead that the density of X is of the form $f(x|\sigma) = \frac{1}{\sigma}f(\frac{x}{\sigma})$, where $\sigma > 0$. Then σ is referred to as a *scale parameter*, and f is a *scale parameter family*. Examples of scale parameters include the normal standard deviation (as our notation suggests), as well as the σ parameters in the Cauchy and t distributions and the β parameter in the gamma distribution. If we wish to specify a prior for σ that is invariant under scale transformations (i.e., transformations of the form $Y = cX$, for $c > 0$), then Berger shows we need only consider priors of the form k/σ, for $k > 0$. Hence the noninformative prior for a scale parameter is given by

$$p(\sigma) = \frac{1}{\sigma}, \ \sigma > 0 .$$

If we prefer to work on the variance (σ^2) scale, we obtain $p(\sigma^2) = 1/\sigma^2$. Note that, like the uniform prior for the location parameter, both of these priors are also improper because $\int_0^\infty d\sigma/\sigma = \int_0^\infty d\sigma^2/\sigma^2 = \infty$.

Finally, many of the most popular distributional forms (including the normal, t, and Cauchy) combine both location and scale features into the form $f(x|\theta, \sigma) = \frac{1}{\sigma}f(\frac{x-\theta}{\sigma})$, the so-called *location-scale family*. The most common approach in this case is to use the two noninformative priors given above in concert with the notion of prior "independence," obtaining

$$p(\theta, \sigma) = \frac{1}{\sigma}, \ \theta \in \Re, \ \sigma > 0 .$$

As discussed in the exercises, the Jeffreys prior for this situation takes a slightly different form.

2.2.4 Other prior construction methods

There are a multitude of other methods for constructing priors, both informative and noninformative; see Berger (1985, Chapter 3) for an overview. The only one we shall mention is that of using the marginal distribution $m(\mathbf{x}) = \int f(\mathbf{x}|\theta)p(\theta)d\theta$. Here, we choose the prior $p(\theta)$ based on its ability to preserve consistency with the marginal distribution of the observed data. Unlike the previous approaches in this section, this approach uses not only the form of the likelihood, but also the actual observed data values to help determine the prior. We might refer to this method as *empirical* estimation of the prior, and this is, in fact, the method that is used in empirical Bayes (EB) analysis.

Strictly speaking, empirical estimation of the prior is a violation of Bayesian philosophy: the subsequent prior-to-posterior updating in equation (2.1) would "use the data twice" (first in the prior, and again in the likelihood). The resulting inferences from this posterior would thus be "overconfident." Indeed, EB methods that ignore this fact are often referred to as *naive* EB methods. However, much recent work focuses on ways to correct these methods for the effect of their estimation of the prior – many with a good deal of success; see Chapter 5.

2.3 Bayesian inference

Having specified the prior distribution using one of the techniques in the previous section, we may now use Bayes' Theorem (2.1) to obtain the posterior distribution of the model parameters, or perhaps equation (2.7) to obtain the predictive density of a future observation. Since these distributions summarize our current state of knowledge (arising from both the observed data and our prior opinion, if any), we might simply graph the corresponding density (or cumulative distribution) functions and report them as the basis for all posterior inference. However, these functions can be difficult to interpret and, in many cases, may tell us more than we want to know. Hence in this section, we discuss common approaches for summarizing such distributions. In particular, we develop Bayesian analogues for the common frequentist techniques of point estimation, interval estimation, and hypothesis testing. Throughout, we work with the posterior distribution of the model parameters, though most of our methods apply equally well to the predictive distribution – after all, future data values and parameters are merely different types of unknown quantities. We illustrate how Bayesian data analysis is typically practiced, and, in the cases of interval estimation and hypothesis testing, the philosophical advantages inherent in the Bayesian approach.

2.3.1 Point estimation

For ease of notation, consider first the univariate case. To obtain a point estimate $\hat{\theta}(\mathbf{y})$ of θ, we simply need to select a summary feature of $p(\theta|\mathbf{y})$, such as its mean, median, or mode. In principle, the mode is the easiest to compute, since no standardization of the posterior is then required; we may work directly with the numerator of (2.1). Also, note that when the prior $\pi(\theta)$ is flat, the posterior mode will be equal to the maximum likelihood estimate of θ. For this reason, the posterior mode is sometimes referred to as the *generalized maximum likelihood estimate* of θ.

For symmetric posterior densities, the mean and the median will of course be identical; for symmetric unimodal posteriors, all three measures will coincide. For asymmetric posteriors, the choice is less clear, though the median is often preferred since it is intermediate to the mode (which considers only the value corresponding to the maximum value of the density) and the mean (which often gives too much weight to extreme outliers). These differences are especially acute for one-tailed densities. Figure 2.7 provides an illustration using a $G(.5, 1)$ distribution. Here, the modal value (0) seems

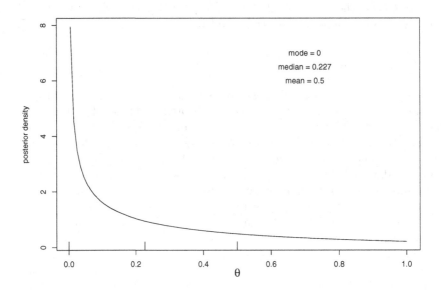

Figure 2.7 *Three point estimates arising from a Gamma(.5, 1) posterior.*

a particularly poor choice of summary statistic, since it is far from the "middle" of the density; in fact, the density is not even finite at this value.

In order to obtain a measure of the accuracy of a point estimate $\hat{\theta}(\mathbf{y})$ of θ, we might use the *posterior variance with respect to* $\hat{\theta}(\mathbf{y})$,

$$E_{\theta|\mathbf{y}}(\theta - \hat{\theta}(\mathbf{y}))^2 \ .$$

Writing the posterior mean $E_{\theta|\mathbf{y}}(\theta) \equiv \mu(\mathbf{y})$ simply as μ for the moment, we have

$$
\begin{aligned}
E_{\theta|\mathbf{y}}(\theta - \hat{\theta}(\mathbf{y}))^2 &= E_{\theta|\mathbf{y}}(\theta - \mu + \mu - \hat{\theta}(\mathbf{y}))^2 \\
&= E_{\theta|\mathbf{y}}[(\theta - \mu)^2 + 2(\theta - \mu)(\mu - \hat{\theta}(\mathbf{y})) + (\mu - \hat{\theta}(\mathbf{y}))^2] \\
&= Var_{\theta|\mathbf{y}}(\theta) + 2[\mu - \hat{\theta}(\mathbf{y})][E_{\theta|\mathbf{y}}(\theta) - \mu] + (\mu - \hat{\theta}(\mathbf{y}))^2 \\
&= Var_{\theta|\mathbf{y}}(\theta) + (\mu - \hat{\theta}(\mathbf{y}))^2 \ ,
\end{aligned}
$$

since the middle term on the third line is identically zero. Hence we have shown that the posterior mean μ minimizes the posterior variance with respect to $\hat{\theta}(\mathbf{y})$ over *all* point estimators $\hat{\theta}(\mathbf{y})$. Furthermore, this minimum value is $Var_{\theta|\mathbf{y}}(\theta) = E_{\theta|\mathbf{y}}[\theta - E_{\theta|\mathbf{y}}(\theta)]^2$, the *posterior variance of θ* (usually referred to simply as the *posterior variance*). This is a possible argument for preferring the posterior mean as a point estimate, and also partially explains why historically only the posterior mean and variance were reported as the results of a Bayesian analysis.

Turning briefly to the multivariate case, we might again take the posterior mode as our point estimate $\hat{\boldsymbol{\theta}} = (\hat{\theta}_1, \ldots, \hat{\theta}_k)$, though it will now be numerically harder to find. When the mode exists, a traditional maximization method such as a grid search, golden section search, or Newton-type method will often succeed in locating it (see e.g., Thisted, 1988, Chapter 4 for a description of these and other maximization methods). The posterior mean $\boldsymbol{\mu} \equiv E_{\boldsymbol{\theta}|\mathbf{y}}(\boldsymbol{\theta})$ is also a possibility, because it is still well-defined and its accuracy is captured nicely by the *posterior covariance matrix*,

$$V = E_{\boldsymbol{\theta}|\mathbf{y}} \left[(\boldsymbol{\theta} - \boldsymbol{\mu})(\boldsymbol{\theta} - \boldsymbol{\mu})' \right] .$$

Similar to the univariate case, one can show that

$$E_{\boldsymbol{\theta}|\mathbf{y}} \left[(\boldsymbol{\theta} - \hat{\boldsymbol{\theta}}(\mathbf{y}))(\boldsymbol{\theta} - \hat{\boldsymbol{\theta}}(\mathbf{y}))' \right] = V + (\boldsymbol{\mu} - \hat{\boldsymbol{\theta}}(\mathbf{y}))(\boldsymbol{\mu} - \hat{\boldsymbol{\theta}}(\mathbf{y}))' ,$$

so that the posterior mean "minimizes" the posterior covariance matrix with respect to $\hat{\boldsymbol{\theta}}(\mathbf{y})$. The posterior median is more difficult to define in multiple dimensions, though it is still often used in one-dimensional subspaces of such problems, i.e., as a summary statistic for $p(\theta_1|\mathbf{y}) = \int \cdots \int p(\boldsymbol{\theta}|\mathbf{y}) \, d\theta_2, \ldots, d\theta_k$.

Example 2.10 For $n = 27$ captured samples of the sirenian species *dugong* (sea cow), Ratkowsky (1983) considers a particular nonlinear growth model to relate an animal's length in meters, Y_i, to its age in years, x_i. In order to avoid a nonlinear model for now, suppose we transform x_i to the log scale. The plot of the transformed data in Figure 2.8 can be created in R by typing

R code
```
x <- c( 1.0,    1.5,   1.5,   1.5, 2.5,    4.0,   5.0,   5.0,   7.0,
        8.0,    8.5,   9.0,   9.5, 9.5,   10.0,  12.0,  12.0,  13.0,
       13.0,  14.5,  15.5,  15.5, 16.5,  17.0,  22.5,  29.0,  31.5)
Y <- c(1.80, 1.85, 1.87, 1.77, 2.02, 2.27, 2.15, 2.26, 2.47,
       2.19, 2.26, 2.40, 2.39, 2.41, 2.50, 2.32, 2.32, 2.43,
       2.47, 2.56, 2.65, 2.47, 2.64, 2.56, 2.70, 2.72, 2.57)
lgage <- log(x)
plot(lgage, Y, xlab="log(age)", ylab="length", pch=20)
```

(The reader need not copy these data values into R by hand, since these data are available along with all our other data and code on the web at www.biostat.umn.edu/~brad/data.html. This example is also available in WinBUGS as the first item in the list that appears after clicking on "Help" and pulling down to "Examples Vol II".) Figure 2.8 reveals an acceptable amount of linearity to make plausible the standard linear regression model,

$$Y_i = \beta_0 + \beta_1 \log(x_i) + \epsilon_i, \ i = 1, \ldots, n$$

where $\epsilon_i \overset{iid}{\sim} N(0, \sigma^2)$, and where once again we define $\tau = 1/\sigma^2$, the precision (reciprocal variance) in each data point. A standard frequentist analysis of this model is easily available in R using the lm command:

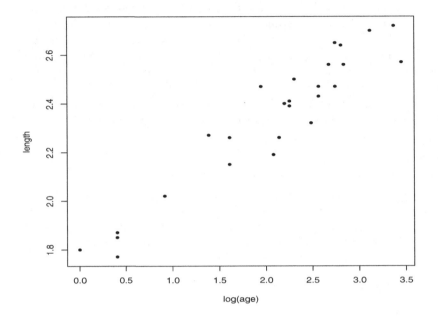

Figure 2.8 *Plot of the linearized dugong data (length versus log age).*

R code
```
reg.out <- summary(lm(Y ~ lgage))
beta1hat <- reg.out$coefficients[2,1]
beta1.se <- reg.out$coefficients[2,2]
```

The usual least squares estimate $\hat{\beta}_1$ and its standard error turn out to be 0.277 and 0.0188, respectively. Recalling that $\hat{\beta}_1$ has a t distribution with $n - 2 = 25$ degrees of freedom, a classical 95% confidence interval for β can be obtained as

R code
```
n <- 27
beta1.CI <- c(beta1hat+qt(0.025, df=n-2)*beta1.se ,
              beta1hat+qt(0.975, df=n-2)*beta1.se)
```

Typing `beta1.CI` reveals this interval to be $(0.239, 0.316)$.

Now suppose we want to compare these classical results to those from a Bayesian formulation of the problem. In an attempt at maximum comparability with our frequentist results above, suppose we place flat priors on both β_0 and β_1. Note that this prior allows the data to determine the appropriate amount of correlation in the $\boldsymbol{\beta} = (\beta_0, \beta_1)'$ joint posterior. If we further assume the data variance σ^2 to be known, the posterior for $\boldsymbol{\beta}$ emerges in closed form as a bivariate normal distribution. This proof is somewhat technical, so we prefer to leave it until we handle the general normal linear model in Subsection 4.1.1.

Instead, for now we use a sampling-based approach, again implemented in WinBUGS. A standard approach in models like ours is to place a vague inverse gamma prior on σ^2 (or equivalently, a gamma prior on τ). The reason for this is that the posterior for β can now can be shown to be a multivariate Student's t distribution – another technical result we defer for now. The Gibbs sampling algorithms in WinBUGS do not require this conjugate prior for τ, but it is often retained since conditional conjugacy still leads to closed forms the software can utilize, perhaps leading to runs that are quicker and/or easier. The WinBUGS code for this model is

BUGS code
```
model{
    for( i in 1:n) {
        logage[i] <- log(x[i])
        Y[i] ~ dnorm(mu[i] , tau)
        mu[i] <- beta0+ beta1*logage[i]
    }

    beta0 ~ dflat()
    beta1 ~ dflat()

    tau ~ dgamma(0.1, 0.1)
    sigma <- 1/sqrt(tau)
}
```

Note here we have used a $Gamma(\epsilon, \epsilon)$ prior with $\epsilon = 0.1$ since, using WinBUGS' parametrization of the gamma distribution (whose second parameter is the reciprocal of ours), this distribution has mean $\epsilon/\epsilon = 1$, but variance $\epsilon/\epsilon^2 = 1/\epsilon = 10$. We hasten to point out that, despite its history, this is a somewhat controversial prior. Its shape for any reasonable ϵ (less than 1) is rather odd, with its infinite peak at 0^+ and infinitely heavy right tail. The corresponding posterior (for τ or σ^2) may be sensitive to the particular choice of ϵ. Still, in many cases the posteriors of the "parameters of interest" (here, the β's) do not experience such sensitivity, so the $Gamma(\epsilon, \epsilon)$ prior remains widely used.

The data are easily read in as

BUGS code
```
list(x = c( 1.0,   1.5,   1.5,   1.5, 2.5,    4.0,   5.0,   5.0,   7.0,
            8.0,   8.5,   9.0,   9.5, 9.5,   10.0, 12.0, 12.0, 13.0,
           13.0, 14.5, 15.5, 15.5, 16.5, 17.0, 22.5, 29.0, 31.5),
      Y = c(1.80, 1.85, 1.87, 1.77, 2.02, 2.27, 2.15, 2.26, 2.47,
            2.19, 2.26, 2.40, 2.39, 2.41, 2.50, 2.32, 2.32, 2.43,
            2.47, 2.56, 2.65, 2.47, 2.64, 2.56, 2.70, 2.72, 2.57),
      n = 27)
```

while the initial value file takes the form

BUGS code
```
list(beta0 = 0, beta1 = 1, tau = 1)   # for gamma prior on tau
```

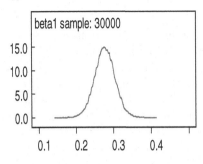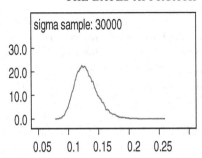

Figure 2.9 *Estimated posterior densities for β_1 and σ based on 30,000 Gibbs samples for the dugong example, Gamma(0.1, 0.1) prior on $\tau \equiv 1/\sigma^2$.*

We then ran our Gibbs sampler for 1000 "burn-in" (pre-convergence) iterations, followed by 30,000 "production" iterations (samples retained for subsequent posterior estimation). We do this in WinBUGS by running 1000 iterations (using the Update tool on the Model menu)) *before* saving the samples for β_1 or σ (using the Samples tool on the Inference menu).

Figure 2.9 shows the posterior density estimates for the slope β_1 and the error standard deviation σ based on our 30,000 post-burn-in MCMC samples. The β_1 samples yield a posterior mean (and median) of 0.277, identical to the classical point estimate obtained above. However, the posterior 2.5 and 97.5 percentiles are 0.222 and 0.332, respectively. Comparing these to the classical 95% confidence interval above, we see that the $G(\epsilon, \epsilon)$ prior does exert some pressure on the tails of the posterior even for our relatively modest choice of $\epsilon = 0.1$. (We defer a full discussion of Bayesian confidence intervals to Subsection 2.3.2.)

Since slopes and intercepts are often correlated, we might want to look at the *joint* posterior of the two elements of β. Our bivariate sample from this distribution can be easily viewed in WinBUGS by clicking on Inference and pulling down to Correlations. Provided both β_0 and β_1 are being monitored, this correlation tool can be used to obtain bivariate scatterplots like the one shown in the left panel of Figure 2.10. Clearly the anticipated negative correlation is substantial, and suggests we should have *centered* our log(age) covariate around its own mean, replacing the definition of mu[i] in our WinBUGS code above with

BUGS code   `mu[i] <- beta0 + beta1*(logage[i] - mean(logage[]))`

The resulting β_0 and β_1 samples are indeed now virtually uncorrelated, as seen in the right panel of Figure 2.10.

As mentioned above, the $G(\epsilon, \epsilon)$ prior is now viewed by many applied Bayesians as suboptimal. Gelman (2006) suggests placing a *uniform* prior on σ, simply bounding the prior away from 0 and ∞ in some sensible way. While such bounds may not be initially obvious, in our setting we start

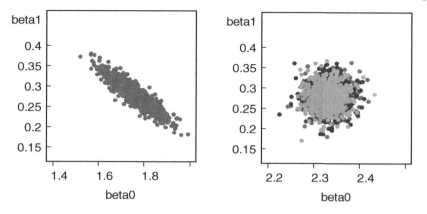

Figure 2.10 *Scatterplot of 2000 bivariate samples of $\boldsymbol{\beta} = (\beta_0, \beta_1)'$, for the dugong example, Gamma(0.1, 0.1) prior on $\tau \equiv 1/\sigma^2$: left panel, uncentered covariate; right panel, centered covariate.*

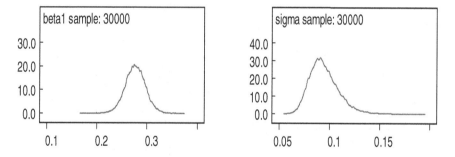

Figure 2.11 *Estimated posterior densities for β_1 and σ based on 30,000 Gibbs samples for the dugong example, Uniform(.01, 100) prior on σ.*

with a $U(.01, 100)$ prior. In WinBUGS, we simply comment out the gamma prior on τ and definition of σ, replacing them with our new prior on σ and definition of τ:

BUGS code
```
#    tau ~ dgamma(0.1, 0.1)
#    sigma <- 1/sqrt(tau)
     tau <- 1/(sigma*sigma)
     sigma ~ dunif(0.01, 100)
```

This switch requires a corresponding small change to our initial values file:

BUGS code
```
list(beta0 = 0, beta1 = 1, sigma = 1)  # for unif prior on sigma
```

Figure 2.11 shows the posterior density estimates for the slope β_1 and the error standard deviation σ based on 30,000 MCMC samples from our new model. The β_1 samples yield a posterior mean and median of 0.277, and upper and lower 2.5 percentiles of 0.238 and 0.318, values extremely similar to the classical point and interval estimates above. A more formal comparison of the $G(.1, .1)$ and $U(.01, 100)$ priors using the DIC criterion (see Subsection 2.4.2 below) is the subject of Exercise 14. ∎

2.3.2 Interval estimation

The Bayesian analogue of a frequentist confidence interval (CI) is usually referred to as a *credible set*, though we will often use the term "Bayesian confidence interval" or simply "confidence interval" in univariate cases for consistency and clarity. More formally,

Definition 2.1 A $100 \times (1 - \alpha)\%$ credible set for $\boldsymbol{\theta}$ is a subset C of $\boldsymbol{\Theta}$ such that

$$1 - \alpha \leq P(C|\mathbf{y}) = \int_C p(\boldsymbol{\theta}|\mathbf{y})d\boldsymbol{\theta} ,$$

where integration is replaced by summation for discrete components of $\boldsymbol{\theta}$.

This definition enables direct probability statements about the likelihood of $\boldsymbol{\theta}$ falling in C, i.e.,

"The probability that $\boldsymbol{\theta}$ lies in C given the observed data $\mathbf{y}$ is at least $(1 - \alpha)$."

This is in stark contrast to the usual frequentist CI, for which the corresponding statement would be something like,

"If we could recompute C for a large number of datasets collected in the same way as ours, about $(1 - \alpha) \times 100\%$ of them would contain the true value of $\boldsymbol{\theta}$."

This is not a very comforting statement, since we may not be able to even imagine repeating our experiment a large number of times (e.g., consider an interval estimate for the 1993 U.S. unemployment rate). In any event, we are in physical possession of only one dataset; our computed C will either contain $\boldsymbol{\theta}$ or it won't, so the actual coverage probability will be either 1 or 0. Thus for the frequentist, the confidence level $(1 - \alpha)$ is only a "tag" that indicates the quality of the procedure (i.e., a narrow 95% CI should make us feel better than an equally narrow 90% one). But for the Bayesian, the credible set provides an actual probability statement, based only on the observed data and whatever prior opinion we have added.

We used "$\leq$" instead of "$=$" in Definition 2.1 in order to accommodate discrete settings, where obtaining an interval with coverage probability *exactly* $(1 - \alpha)$ may not be possible. Indeed, in continuous settings we would like credible sets that do have exactly the right coverage, in order to minimize their size and thus obtain a more precise estimate. More generally, a technique for doing this is given by the *highest posterior density*, or HPD, credible set, defined as the set

$$C = \{\boldsymbol{\theta} \in \boldsymbol{\Theta} : p(\boldsymbol{\theta}|\mathbf{y}) \geq k(\alpha)\} ,$$

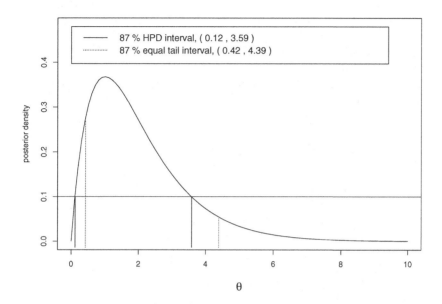

Figure 2.12 *HPD and equal-tail credible intervals arising from a Gamma(2,1) posterior using* $k(\alpha) = 0.1$.

where $k(\alpha)$ is the largest constant satisfying

$$P(C|\mathbf{y}) \geq 1 - \alpha \, .$$

A good way to visualize such a set C is provided by Figure 2.12, which illustrates obtaining an HPD credible interval for θ assumed to follow a $G(2,1)$ posterior distribution. Here, drawing a horizontal line across the graph at $p(\theta|\mathbf{y}) = 0.1$ results in a 87% HPD interval of (0.12, 3.59); see the solid line segments in the figure. That is, $P(0.12 < \theta < 3.59|\mathbf{y}) = .87$ when $\theta|\mathbf{y} \sim G(2,1)$. We could obtain a 90% or 95% HPD interval simply by "pushing down" the horizontal line until the corresponding values on the θ-axis trapped the appropriate, larger probability.

Such a credible set is very intuitively appealing because it groups together the "most likely" θ values and is always computable from the posterior density, unlike classical CIs which frequently rely on "tricks," such as pivotal quantities. However, the graphical "line pushing" mentioned in the previous paragraph translates into iteratively solving the nonlinear equation

$$P(C(k(\alpha))|\mathbf{y}) = 1 - \alpha \, ,$$

in order to find C for a prespecified α level. This calculation is such a nuisance that instead we often simply take the $\alpha/2$- and $(1-\alpha/2)$-quantiles of $p(\theta|\mathbf{y})$ as our $100 \times (1 - \alpha)\%$ credible set for θ. This *equal tail* credible set will be equal to the HPD credible set if the posterior is symmetric and unimodal, but will be a bit wider otherwise.

In the context of our Figure 2.12 example, we obtain an 87% equal tail credible set of $(0.42, 4.39)$; see the dashed line segments in the figure. By construction, we have

$$
\begin{aligned}
P(0.42 < \theta < 4.39|\mathbf{y}) &= 1 - P(\theta < 0.42|\mathbf{y}) - P(\theta > 4.39|\mathbf{y}) \\
&= 1 - \alpha/2 - \alpha/2 \\
&= 1 - \alpha = 0.87 .
\end{aligned}
$$

Due to the asymmetry of our assumed posterior distribution, this interval is 0.5 units wider than the corresponding HPD interval, and includes some larger θ values that are "less likely" while omitting some smaller θ values that are "more likely." Still, this interval is trivially computed for any α using any of several statistical software packages and seems to satisfactorily capture the "middle" of the posterior distribution. Moreover, this interval is invariant to 1-1 transformations of the parameter; for example, the logs of the interval endpoints above provide an 87% equal tail interval for $\eta \equiv \log \theta$.

For a typical Bayesian data analysis, we might summarize our findings by reporting (a) the posterior mean, (b) several important posterior percentiles (corresponding to, say, probability levels .025, .25, .50, .75, and .975), (c) a plot of the posterior itself if it is multimodal, highly skewed, or otherwise badly behaved, and possibly (d) posterior probabilities of the form $P(\theta > c|\mathbf{y})$ or $P(\theta < c|\mathbf{y})$, where c is some important reference point (e.g., 0) that arises from the context of the problem. We defer an illustration of these principles until after introducing the basic concepts of Bayesian hypothesis testing.

2.3.3 Hypothesis testing and Bayes factors

The comparison of predictions made by alternative scientific explanations is, of course, a mainstay of current statistical practice. The traditional approach is to use the ideas of Fisher, Neyman, and Pearson, wherein one states a primary, or *null*, hypothesis H_0, and an *alternative* hypothesis H_a. After determining an appropriate test statistic $T(\mathbf{Y})$, one then computes the *observed significance*, or *p-value*, of the test as

$$
\text{p-value} = P\{T(\mathbf{Y}) \text{ more "extreme" than } T(\mathbf{y}_{obs}) \mid \boldsymbol{\theta}, H_0\} , \qquad (2.16)
$$

where "extremeness" is in the direction of the alternative hypothesis. If the p-value is less than some prespecified Type I error rate, H_0 is rejected; otherwise, it is not.

While classical hypothesis testing has a long and celebrated history in the statistical literature and continues to be a favorite of practitioners, several fairly substantial criticisms of it may be made. First, the approach can be applied straightforwardly only when the two hypotheses in question are *nested*, one within the other. That is, H_0 must be a simplification of H_a (say, by setting one of the model parameters in H_a equal to a constant – usually zero). But many practical hypothesis testing problems involve a choice between two (or more) models that are not nested (e.g., choosing between quadratic and exponential growth models, or between exponential and lognormal error distributions).

A second difficulty is that tests of this type can only offer evidence *against* the null hypothesis. A small p-value indicates that the larger, alternative model has significantly more explanatory power. However, a large p-value does not suggest that the two models are equivalent, but only that we lack evidence that they are not. This difficulty is often swept under the rug in introductory statistics courses, the technically correct phrase "fail to reject [the null hypothesis]" being replaced by "accept."

Third, the p-value itself offers no direct interpretation as a "weight of evidence," but only as a long-term probability (in a hypothetical repetition of the same experiment) of obtaining data at least as unusual as what was actually observed. Unfortunately, the fact that small p-values imply rejection of H_0 causes many consumers of statistical analyses to assume that the p-value is "the probability that H_0 is true," even though (2.16) shows that it is nothing of the sort.

A final, somewhat more philosophical criticism of p-values is that they depend not only on the observed data, but also the total sampling probability of certain *unobserved* datapoints; namely, the "more extreme" $T(\mathbf{Y})$ values in equation (2.16). As a result, two experiments with identical likelihoods could result in different p-values if the two experiments were *designed* differently. As mentioned previously in conjunction with Example 1.2, this fact violates a proposition known as the *Likelihood Principle* (Birnbaum, 1962), which can be stated briefly as follows:

> **The Likelihood Principle:** In making inferences or decisions about θ after $\mathbf{y}$ is observed, all relevant experimental information is contained in the likelihood function for the observed $\mathbf{y}$.

By taking into account not only the observed data $\mathbf{y}$, but also the unobserved but more extreme values of $\mathbf{Y}$, classical hypothesis testing violates the Likelihood Principle.

This final criticism, while theoretically based, has practical import as well. For instance, it is not uncommon for unforeseen circumstances to arise during the course of an experiment, resulting in data that are not exactly what was intended in the experimental design. Technically speaking, in this case a frequentist analysis of the data is not possible, since p-values are only computable for data arising from the original design. Problems like this are especially common in the analysis of clinical trials data, where unplanned interim analyses and unexpected drug toxicities can wreak havoc with the trial's design. Berger and Berry (1988) provide a lucid and nontechnical discussion of these issues, including a clever example showing that a classically significant trial finding can be negated merely by *contemplating* additional trial stages that did not occur.

By contrast, the Bayesian approach to hypothesis testing, due primarily to Jeffreys (1961), is much simpler and more sensible in principle, and avoids all four of the aforementioned difficulties with the traditional approach. As in the case of interval estimation, however, it requires the notion of subjective probability to facilitate its probability calculus. Put succinctly, based on the data that each of the hypotheses is supposed to predict, one applies Bayes' Theorem and computes the posterior probability that the first hypothesis is correct. There is no limit on the number of hypotheses that may be simultaneously considered, nor do any need to be nested within any of the others. As such, in what follows, we switch notation from "hypotheses" H_0 and H_a to "models" M_i, $i = 1, \ldots, m$.

To lay out the mechanics more specifically, suppose we have two candidate parametric models M_1 and M_2 for data $\mathbf{Y}$, and the two models have respective parameter vectors $\boldsymbol{\theta}_1$ and $\boldsymbol{\theta}_2$. Under prior densities $\pi_i(\boldsymbol{\theta}_i)$, $i = 1, 2$, the marginal distributions of $\mathbf{Y}$ are found by integrating out the parameters,

$$p(\mathbf{y}|M_i) = \int f(\mathbf{y}|\boldsymbol{\theta}_i, M_i)\pi_i(\boldsymbol{\theta}_i)d\boldsymbol{\theta}_i \ , \ i = 1, 2 \ . \qquad (2.17)$$

Bayes' Theorem (2.1) may then be applied to obtain the posterior probabilities $P(M_1|\mathbf{y})$ and $P(M_2|\mathbf{y}) = 1 - P(M_1|\mathbf{y})$ for the two models. The quantity commonly used to summarize these results is the *Bayes factor*, BF, which is the ratio of the posterior odds of M_1 to the prior odds of M_1, given by Bayes' Theorem as

$$BF = \frac{P(M_1|\mathbf{y})/P(M_2|\mathbf{y})}{P(M_1)/P(M_2)} \qquad (2.18)$$

$$= \frac{\left[\frac{p(\mathbf{y}|M_1)P(M_1)}{p(\mathbf{y})}\right] / \left[\frac{p(\mathbf{y}|M_2)P(M_2)}{p(\mathbf{y})}\right]}{P(M_1)/P(M_2)}$$

$$= \frac{p(\mathbf{y} \mid M_1)}{p(\mathbf{y} \mid M_2)} \ , \qquad (2.19)$$

the ratio of the observed marginal densities for the two models. Assuming the two models are *a priori* equally probable (i.e., $P(M_1) = P(M_2) = 0.5$), we have that $BF = P(M_1|\mathbf{y})/P(M_2|\mathbf{y})$, the posterior odds of M_1.

Consider the case where both models share the same parametrization (i.e., $\boldsymbol{\theta}_1 = \boldsymbol{\theta}_2 = \boldsymbol{\theta}$), and both hypotheses are simple (i.e., $M_1 : \boldsymbol{\theta} = \boldsymbol{\theta}^{(1)}$

and $M_2 : \boldsymbol{\theta} = \boldsymbol{\theta}^{(2)}$). Then $\pi_i(\boldsymbol{\theta})$ consists of a point mass at $\boldsymbol{\theta}^{(i)}$ for $i = 1, 2$, and so from (2.17) and (2.19) we have

$$BF = \frac{f(\mathbf{y}|\boldsymbol{\theta}^{(1)})}{f(\mathbf{y}|\boldsymbol{\theta}^{(2)})} ,$$

which is nothing but the likelihood ratio between the two models. Hence, in the simple-versus-simple setting, the Bayes factor is precisely the odds in favor of M_1 over M_2 *given solely by the data*. For more general hypotheses, this same "evidence given by the data" interpretation of BF is often used, though Lavine and Schervish (1999) show that a more accurate interpretation is that BF captures the *change* in the odds in favor of model 1 as we move from prior to posterior. In any event, in such cases, BF does depend on the prior densities for the θ_i, which must be specified either as convenient conjugate forms or by more careful elicitation methods. In this case, it becomes natural to ask whether "shortcut" methods exist that provide a rough measure of the evidence in favor of one model over another without reference to any prior distributions.

One answer to this question is given by Schwarz (1978), who showed that for large sample sizes n, an approximation to $-2 \log BF$ is given by

$$\Delta BIC = W - (p_2 - p_1) \log n , \qquad (2.20)$$

where p_i is the number of parameters in model $M_i, i = 1, 2$, and

$$W = -2 \log \left[\frac{max_{M_1} f(\mathbf{y}|\boldsymbol{\theta})}{max_{M_2} f(\mathbf{y}|\boldsymbol{\theta})} \right] ,$$

the usual likelihood ratio test statistic. BIC stands for *Bayesian Information Criterion* (though it is also known as the *Schwarz Criterion*), and Δ denotes the change from Model 1 to Model 2. The second term in ΔBIC acts as a penalty term which corrects for differences in size between the models (to see this, think of M_2 as the "full" model and M_1 as the "reduced" model). Thus $\exp(-\frac{1}{2}\Delta BIC)$ provides a rough approximation to the Bayes factor that is independent of the priors on the θ_i.

There are many other penalized likelihood ratio model choice criteria, the most famous of which is the *Akaike Information Criterion*,

$$\Delta AIC = W - 2(p_2 - p_1) . \qquad (2.21)$$

This criterion is derived from frequentist asymptotic efficiency considerations (where both n and the p_i approach infinity). This statistic is fairly popular in practice, but approaches $-2 \log BF$ asymptotically only if the information in the prior increases at the same rate as the information in the likelihood, an unrealistic assumption in practice. Practitioners have also noticed that AIC tends to "keep too many terms in the model," as would be expected from its smaller penalty term.

While Bayes factors were viewed for many years as the only correct way to carry out Bayesian model comparison, they have come under increasing criticism of late. The most serious difficulty in using them is quite easy to see: if $\pi_i(\boldsymbol{\theta}_i)$ is improper (as is typically the case when using noninformative priors), then $p(\mathbf{y}|M_i) = \int f(\mathbf{y}|\boldsymbol{\theta}_i, M_i)\pi_i(\boldsymbol{\theta}_i)d\boldsymbol{\theta}_i$ necessarily is as well, and so BF as given in (2.19) is not well-defined. Several authors have attempted to modify the definition of BF to repair this deficiency. One approach is to divide the data into two parts, $\mathbf{y} = (\mathbf{y}_1, \mathbf{y}_2)$ and use the first portion to compute model-specific posterior densities $p(\boldsymbol{\theta}_i|\mathbf{y}_1), i = 1, 2$. Using these two posteriors as the priors in equation (2.19), the remaining data $\mathbf{y}_2$ then produce the *partial Bayes factor*,

$$BF(\mathbf{y}_2|\mathbf{y}_1) = \frac{p(\mathbf{y}_2|\mathbf{y}_1, M_1)}{p(\mathbf{y}_2|\mathbf{y}_1, M_2)} .$$

While this solves the nonexistence problem in the case of improper priors, there remains the issue of how many datapoints to allocate to the "training sample" $\mathbf{y}_1$, and how to select them. Berger and Pericchi (1996) propose using an arithmetic or geometric mean of the partial Bayes factors obtained using all possible minimal training samples (i.e., where n_1 is taken to be the smallest sample size leading to proper posteriors $p(\boldsymbol{\theta}_i|\mathbf{y}_1)$). Because the result corresponds to an actual Bayes factor under a reasonable intrinsic prior, they refer to it as an *intrinsic Bayes factor*. O'Hagan (1995) instead recommends a *fractional Bayes factor*, defined as

$$BF_b(\mathbf{y}) = \frac{p(\mathbf{y}, b|M_1)}{p(\mathbf{y}, b|M_2)} ,$$

where $b = n_1/n$, n_1 is again a "minimal" sample size, and

$$p(\mathbf{y}, b|M_i) = \frac{\int f(\mathbf{y}|\boldsymbol{\theta}_i, M_i)\pi_i(\boldsymbol{\theta}_i)d\boldsymbol{\theta}_i}{\int f(\mathbf{y}|\boldsymbol{\theta}_i, M_i)^b\pi_i(\boldsymbol{\theta}_i)d\boldsymbol{\theta}_i} .$$

O'Hagan (1995, 1997) also suggests other, somewhat larger values for b, such as $b = \max(n_1, \log n)/n$ and $b = \max(n_1, \sqrt{n})/n$, on the grounds that this should reduce sensitivity to the prior.

Note that while the intrinsic Bayes factor uses part of the data to rectify the impropriety in the priors, the fractional Bayes factor instead uses a fraction b of the *likelihood function* for this purpose. O'Hagan (1997) offers asymptotic justifications for $BF_b(\mathbf{y})$ as well as comparison with various forms of the intrinsic BF; see also DeSantis and Spezzaferri (1999) for further comparisons, finite-sample motivations for using the fractional BF, and open research problems in this area.

Example 2.11 *Consumer preference data.* We summarize our presentation on Bayesian inference with a simple example, designed to illustrate key concepts in the absence of any serious computational challenges. Suppose 16 consumers have been recruited by a fast food chain to compare two types of ground beef patty on the basis of flavor. All of the patties to be evaluated have been kept frozen for eight months, but one set of 16 has been stored in a high-quality freezer that maintains a temperature that is consistently within one degree of 0 degrees Fahrenheit. The other set of 16 patties has been stored in a freezer wherein the temperature varies anywhere between 0 and 15 degrees Fahrenheit. The food chain executives are interested in whether storage in the higher-quality freezer translates into a substantial improvement in taste (as judged by the consumers), thus justifying the extra effort and cost associated with equipping all of their stores with these freezers.

In a test kitchen, the patties are defrosted and prepared by a single chef. To guard against any possible order bias (e.g., it might be that a consumer will always prefer the first patty he or she tastes), each consumer is served the patties in a random order, as determined by the chef (who has been equipped with a fair coin for this purpose). Also, the study is *double-blind*: neither the consumers nor the servers know which patty is which; only the chef (who is not present at serving time) knows and records this information. This is also a measure to guard against possible bias on the part of the consumers or the servers (e.g., it might be that the server talks or acts differently when serving the cheaper patty). After allowing sufficient time for the consumers to make up their minds, the result is that 13 of the 16 consumers state a preference for the more expensive patty.

In order to analyze this data, we need a statistical model. In Bayesian analysis, this consists of two things: a likelihood and a prior distribution. Suppose we let θ be the probability that a consumer prefers the more expensive patty. Let $Y_i = 1$ if consumer i states a preference for the more expensive patty, and $Y_i = 0$ otherwise. If we assume that the consumers are independent and that θ is indeed constant over the consumers, then their decisions form a sequence of Bernoulli trials. Defining $X = \sum_{i=1}^{16} Y_i$, we have

$$X|\theta \sim Binomial(16, \theta) .$$

That is, the sampling density for x is $f(x|\theta) = \binom{16}{x}\theta^x(1-\theta)^{16-x}$.

Turning to the selection of a prior distribution, looking at the likelihood as a function of θ, we see that the beta distribution offers a conjugate family, since its density function is given by

$$\pi(\theta) = \frac{\Gamma(\alpha+\beta)}{\Gamma(\alpha)\Gamma(\beta)}\theta^{\alpha-1}(1-\theta)^{\beta-1} .$$

Before adopting this family, however, we must ask ourselves if it is sufficiently broad to include our true prior feelings in this problem. Fortunately for us, the answer in this case seems to be yes, as evidenced by Figure 2.13. This figure shows three different members of the beta family, all of which might be appropriate. The solid line corresponds to the $Beta(.5, .5)$ distribution, which is the Jeffreys prior (2.12) for this problem; see Exercise 9. We know this prior is noninformative in a transformation-invariate sense,

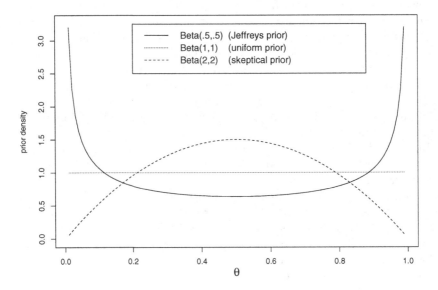

Figure 2.13 *Prior distributions used in analyzing the consumer preference data.*

but it seems to provide extra weight to extreme values of θ. The dotted line displays what we might more naturally think of as a noninformative prior, namely, a flat (uniform) prior, obtained by setting $\alpha = \beta = 1$. Finally, the dashed line follows a $Beta(2, 2)$ prior, which we might interpret as reflecting mild skepticism of an improvement in taste by the more expensive patty (i.e., it favors values of θ near .5). The R code to draw Figure 2.13 is

R code
```
x <- seq(.01,.99,length=99);   top <- 3.25
plot(x,dbeta(x,.5,.5),type="l",xlab=expression(theta),
  ylab="prior density",ylim=c(0,top))
lines(x,dbeta(x,1,1),lty=2)
lines(x,dbeta(x,2,2),lty=3)
legend(.2,top,cex=1.2,legend =c("Beta(.5,.5)  (Jeffreys prior)",
  "Beta(1,1)  (uniform prior)",
  "Beta(2,2)  (skeptical prior)"),lty=1:3)
```

The beta family could also provide priors that are not symmetric about .5 (by taking $\alpha \neq \beta$), if the fast food chain executives had strong feelings or the results of previous studies to support such a conclusion. However, we shall restrict our attention to the three choices in Figure 2.13, since they seem to cover an adequate range of "minimally informative" priors that would not be controversial regardless of who is the ultimate decision-maker (e.g., the corporation's board of directors).

Thanks to the conjugacy of our prior family with the likelihood, the posterior distribution for θ is easily calculated as

$$
\begin{aligned}
p(\theta|x) &\propto f(x|\theta)\pi(\theta) \\
&\propto \theta^{x+\alpha-1}(1-\theta)^{16-x+\beta-1} \\
&\propto Beta(x+\alpha, 16-x+\beta) \ .
\end{aligned}
$$

The three beta posteriors corresponding to our three candidate priors are plotted in Figure 2.14, with corresponding posterior quantile information summarized in Table 2.1. The R code to generate the figure is

R code
```
x <- seq(0,1,length=100);
plot(x,dbeta(x,13.5,3.5),type="l",lty=1,xlab=expression(theta),
  ylab="posterior density")
lines(x,dbeta(x,14,4),lty=2)
lines(x,dbeta(x,15,5),lty=3)
legend(0.1 ,3.5,cex=1.2,legend =c(
  "Beta(13.5,3.5)", "Beta(14,4)", "Beta(15,5)"),lty=1:3)
```

while the R code to create the entries in summary Table 2.1 below is

R code
```
a <- 0.025 ; b <- 0.5; d <- 0.975
a1 <- qbeta(a,13.5,3.5); a2 <- qbeta(b,13.5,3.5)
a3 <- qbeta(d,13.5,3.5); a4 <- pbeta(0.6,13.5,3.5); a5 <- 1-a4
b1 <- qbeta(a,14,4); b2 <- qbeta(b,14,4)
b3 <- qbeta(d,14,4); b4 <- pbeta(0.6,14,4); b5 <- 1-b4
e1 <- qbeta(a,15,5); e2 <- qbeta(b,15,5)
e3 <- qbeta(d,15,5); e4 <- pbeta(0.6,15,5); e5 <- 1-e4

cat("        q_.025     q_.50     q_.975     P(theta >.6)\n")
prob <- c( a1,a2,a3,a5,b1,b2,b3,b5,e1,e2,e3,e5)
p <- t(matrix(prob,4,3))
round(p,3)
```

The three posteriors are ordered in the way we would expect, i.e., the evidence that θ is large decreases as the prior becomes more "skeptical" (concentrated near $1/2$). Still, the differences among them are fairly minor, suggesting that despite the small sample size ($n = 16$), the results are robust to subtle changes in the prior. The posterior medians (which

Prior	Posterior quantile			
distribution	.025	.500	.975	$P(\theta > .6|x)$
$Beta(.5, .5)$	0.579	0.806	0.944	0.964
$Beta(1, 1)$	0.566 .	0.788	0.932	0.954
$Beta(2, 2)$	0.544	0.758	0.909	0.930

Table 2.1 *Posterior summaries, consumer preference data.*

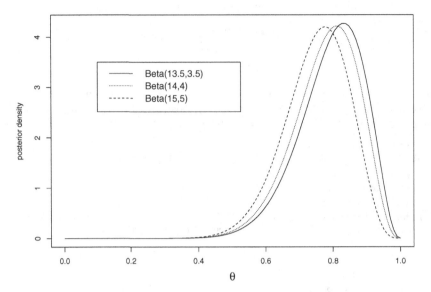

Figure 2.14 *Posterior distributions arising from the three prior distributions chosen for the consumer preference data.*

we might use as a point estimate of θ) are all near 0.8; posterior modes (which would not account for the left skew in the distributions) would be even larger. In particular, the posterior mode under the flat prior is $13/16$ = 0.8125, the maximum likelihood estimate of θ. Also note that all three posteriors produce 95% equal-tail credible intervals that exclude 0.5, implying that we can be quite confident that the more expensive patty offers an improvement in taste.

To return to the executives' question concerning a *substantial* improvement in taste, we must first define "substantial." We (somewhat arbitrarily) select 0.6 as the critical value that θ must exceed in order for the improvement to be regarded as "substantial." Given this cutoff value, it is natural to compare the hypotheses $M_1 : \theta \geq 0.6$ and $M_2 : \theta < 0.6$. Using the uniform prior, the Bayes factor in favor of M_1 is easily computed using equation (2.18) and Table 2.1 as

$$BF = \frac{0.954/0.046}{0.4/0.6} = 31.1 \, ,$$

or adjusted odds of 31.1 to 1 in favor of a substantial improvement in taste. Jeffreys suggested judging BF in units of 0.5 on the common-log-odds scale. Because $\log_{10} 31.1 \approx 1.5$, this rule implies a reasonably strong preference for M_1. Still, we caution that Bayes factors can be much larger than this (e.g., in the millions) – say, when key predictors are left out of a standard linear regression model whose light-tailed normal errors assumption leads to extremely small marginal likelihoods for outlying observations. ∎

2.4 Hierarchical modeling

2.4.1 Normal linear models

As mentioned above, the term "hierarchical model" refers to the extension of the usual Bayesian structure given in equation (2.1) to a hierarchy of l levels, where the joint distribution of the data and the parameters is given by

$$f(\mathbf{y}|\boldsymbol{\theta}_1)\pi_1(\boldsymbol{\theta}_1|\boldsymbol{\theta}_2)\pi_2(\boldsymbol{\theta}_2|\boldsymbol{\theta}_3)\cdots\pi_l(\boldsymbol{\theta}_l|\boldsymbol{\lambda})$$

and we typically seek the marginal posterior of the first stage parameters, $p(\boldsymbol{\theta}_1|\mathbf{y})$. In the case where f and the π_i are normal (Gaussian) distributions with known variance matrices, this marginal posterior is readily available using the results of Lindley and Smith (1972), which we will detail carefully in expressions (4.1)–(4.4) and Example 4.1. However, more complicated settings generally require Monte Carlo methods for their solution. Gelfand et al. (1990) develop such techniques for arbitrary hierarchical normal linear models using the Gibbs sampler. A great many applications of such models have appeared in print. For example, Lange, Carlin, and Gelfand (1992) employed a hierarchical model for the longitudinal CD4 T-cell numbers of a cohort of San Francisco men infected with the HIV virus (see Sections 7.3 and 8.1 for an in-depth discussion of longitudinal data models). Their model allows for incomplete and unbalanced data, population covariates, heterogeneous variances, and errors-in-variables. We now illustrate the fitting of a common normal-errors linear hierarchical model in WinBUGS.

Example 2.12 We consider the two-way analysis of variance (ANOVA) setting in the context of the data in Table 2.2, originally reported and analyzed by Johnson et al. (1999). These are in fact summaries of many time-to-event datasets, namely the estimated log relative hazards $Y_{ij} = \hat{\beta}_{ij}$ obtained by fitting separate Cox proportional hazards regressions to the data from each of $J = 18$ clinical units participating in $I = 6$ different AIDS protocols (studies). Anecdotal evidence collected by the principal investigator (PI) of the data coordinating center administering these trials suggests not all units may be equally adept at carrying out these protocols. Since unit-level effects are difficult to see from the raw data, we wish to simultaneously analyze data from all units across all studies using the *cross-protocol* model,

$$Y_{ij} = a_i + b_j + s_{ij} + \epsilon_{ij}, \ i = 1,\ldots,I, \ j = 1,\ldots,J, \qquad (2.22)$$

where a_i is the main effect for study i, b_j is the main effect for unit j, s_{ij} is a study-unit interaction term, and $\epsilon_{ij} \overset{iid}{\sim} N(0,\sigma_{ij}^2)$. We take the estimated

Estimated Unit-Specific Log Relative Hazards					
Unit (j)	Toxo	ddI/ddC	NuCombo ZDV+ddI	NuCombo ZDV+ddC	Fungal	CMV
A (1)	0.814	NA	-0.406	0.298	0.094	NA
B (3)	-0.203	NA	NA	NA	NA	NA
C (3)	-0.133	NA	0.218	-2.206	0.435	0.145
D (4)	NA	NA	NA	NA	NA	NA
E (5)	-0.715	-0.242	-0.544	-0.731	0.600	0.041
F (6)	0.739	0.009	NA	NA	NA	0.222
G (7)	0.118	0.807	-0.047	0.913	-0.091	0.099
H (8)	NA	-0.511	0.233	0.131	NA	0.017
I (9)	NA	1.939	0.218	-0.066	NA	0.355
J (10)	0.271	1.079	-0.277	-0.232	0.752	0.203
K (11)	NA	NA	0.792	1.264	-0.357	0.807
L (12)	-0.002	0.300	-0.103	-0.431	0.837	0.373
M (13)	-0.076	1.413	0.658	-0.022	-0.164	-0.64
N (14)	0.651	-0.470	0.060	0.421	-0.112	-0.010
O (15)	-0.249	0.098	-0.272	-0.163	0.860	0.081
P (16)	0.003	0.292	0.705	0.608	-0.327	1.044
Q (17)	NA	0.195	0.605	0.187	NA	-0.201
R (18)	1.217	0.165	0.385	0.172	-0.022	0.203

Table 2.2 *Estimated unit-specific log relative hazards and standard errors, cross protocol data set.*

standard errors from the Cox regressions as (known) values of the data standard deviations σ_{ij}. Note that some values are missing ("NA") since not all 18 units participated in all 6 studies, and the Cox estimation procedure did not converge for some units that *did* participate in a study, but experienced few deaths. In particular, note that Unit D ($j = 4$) contributes no data at all (all values missing).

The primary goal of our analysis is to identify which clinics are *opinion leaders* (strongly agree with overall result across studies) and which are *dissenters* (strongly disagree). Overall results from every study favor the treatment over the placebo (i.e., mostly negative Ys) except in Trial 1, the toxoplasmosis prophylaxis trial (Toxo). We thus multiply all the Y_{ij}'s by –1 for $i \neq 1$, so that larger Y_{ij} correspond in all cases to stronger agreement with the overall.

Figure 2.15 shows a plot of the Y_{ij} values and associated approximate 95% frequentist confidence intervals. All of the units in Trial 3 (NuCombo–

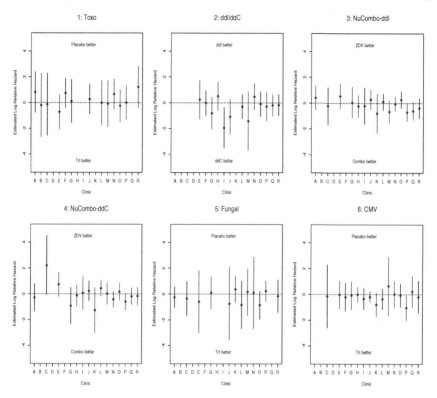

Figure 2.15 *Approximate 95% confidence sets for the log relative hazard for each unit and study.*

ddI) have relatively tight confidence sets because of high event rates, while the units in Trials 1 and 5 (Toxo and Fungal) mostly have wide confidence sets due to their low event rates.

At the second stage of our model, we suppose

$$a_i \overset{iid}{\sim} N(0, 100^2), \ b_j \overset{iid}{\sim} N(0, \sigma_b^2), \ \text{and} \ s_{ij} \overset{iid}{\sim} N(0, \sigma_s^2),$$

while at the third stage we assume

$$\sigma_b \ \sim \ Unif(0.01, 100)$$
$$\text{and} \ \sigma_s \ \sim \ Unif(0.01, 100).$$

That is, we *preclude* borrowing of strength across studies, (since the a_i deal with different endpoints and there is no reason not to think of them as anything but fixed effects). But we also *encourage* borrowing of strength across units and their interactions (if any) with the study effects, since the units are all supposed to be running the same protocols. This outlook is statistically necessary as well, since with $I + J + IJ$ parameters but fewer than IJ data points, there must be shrinkage among *some* of the effects in order for the joint posterior to emerge as proper.

Under Gelman-style (vague uniform) priors on the standard deviation parameters, the WinBUGS code for this model is:

BUGS code
```
model
{
   for(i in 1:I) {
       for(j in 1:J) {
           Y[i,j] ~ dnorm(theta[i,j],P[i,j])
           theta[i,j] <- a[i]+b[j]+s[i,j]        # full model
           s[i,j] ~ dnorm(0.0,C)
           }
       a[i] ~ dnorm(0, 0.0001)
   }
   for (j in 1:J) {
       b[j] ~ dnorm(0.0,B)
   }

   b_sigma ~ dunif(0.01, 100)   # clinic-level s.d.
   s_sigma ~ dunif(0.01,100)    # s.d. of s_ij's
   B <- 1/(b_sigma*b_sigma)
   C <- 1/(s_sigma*s_sigma)
}
```

Using list format, a partial version of the data (with full version available as usual at www.biostat.umn.edu/~brad/data.html) looks like

BUGS code
```
list(I = 6,      # number of studies
         J = 18,      # number of clinics

   Y = structure(
     .Data = c(0.814, -0.203, -0.133, NA, -0.715, 0.739,
               0.118, NA, NA, 0.271,
               ...........
               1.044, -0.201, 0.203), .Dim = c(6, 18)),

   P = structure(
     .Data = c(1.55472, 0.63796, 0.66422, 0.0001, 2.22103, 3.02457,
               1.38408, 0.0001, 0.0001, 2.93207,
               ...........
               4.17327, 2.72863, 2.50757), .Dim = c(6, 18)))
```

To investigate convergence of this model, we experiment with three different sets of initial values,

BUGS code
```
list(b_sigma=5, s_sigma=5, a=c(3,3,3,3,3,3),
     b=c(2,2,2,2,2,2,2,2,2,2,2,2,2,2,2,2,2,2))
```

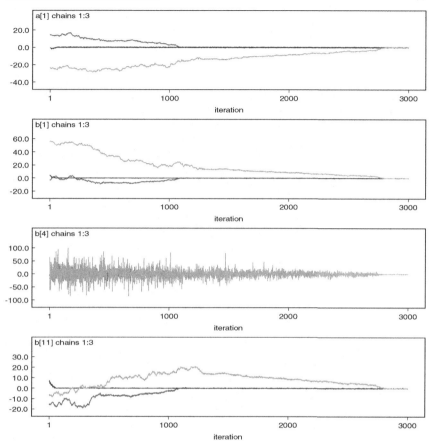

Figure 2.16 *Trace plots for a_1, b_1, b_4, and b_{11}, full model for cross protocol data.*

```
list(b_sigma=25, s_sigma=25, a=c(3,3,3,3,3,3),
b=c(-10,-10,-10,-10,-10,-10,-10,-10,-10,-10,-10,-10,-10,-10,-10,
   -10,-10,-10))

list(b_sigma=50, s_sigma=50, a=c(-3,-3,-3,-3,-3,-3),
b=c(20,20,20,20,20,20,20,20,20,20,20,20,20,20,20,20,20,20))
```

The first of these is fairly close to where we think the support of the true posterior for the a_i and b_j is located (though still far from a "best guess"; remember the b's are centered around zero). The other two sets are very far from the prior means and can reasonably be expected to be "overdispersed" with respect to the true joint posterior distribution ("bad starting values").

Figure 2.16 shows traceplots for four parameters (a_1, b_1, b_4, and b_{11}) under the full model. Note that convergence is essentially immediate from

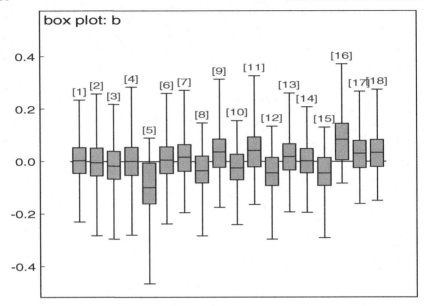

Figure 2.17 *Boxplots of the b_j posteriors, full model for cross protocol data.*

the good starting values, while the other two chains "wander" for a bit before locating the bulk of the posterior mass. This manifests for b_4 only as a gradual decrease in posterior variance; there are no observations from Unit 4 so the prior mean of 0 determines this unit's posterior (since it equals the prior). Note also that even though our Gibbs algorithm uses univariate updating, the samples are all interconnected via the parameters' full conditional distributions. This means that when one parameter in a chain at last begins sampling from the stationary distribution, all parameters do (though there can be preconvergence "crossing" of the stationary region, as seen in the b_1 and b_{11} chains). We hasten to add, however, that the issue of MCMC convergence can be much more subtle than this for more complicated mean structures; see for example the model setting and solution to Chapter 3, Exercise 21. We also obtained much longer waits for the "bad" chains to coalesce simply by replacing the vague normal prior for the a_i with

BUGS code   `a[i] ~ dflat()`

This may provide a practical reason to prefer the vague normal specification, because the impact of this change on the ultimate estimate of the posterior itself is negligible.

Figure 2.17 shows boxplots of the posterior samples of the unit effects b_j. Note that no unit's effect is significantly different from 0 at the customary 0.05 level. However, using a 50% confidence level (the width of the boxes),

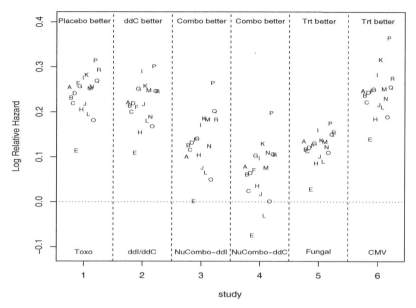

Figure 2.18 *Unit- and study-specific posterior means θ_{ij}, cross protocol data.*

Unit P ($j = 16$) is an opinion leader, while Unit E ($j = 5$) is a dissenter. Having done a little, shall we say, expert opinion-informed pre-analysis of the data, the PI of the data coordinating center was not surprised at E's dissent, but P's leadership role was interesting news to him.

Finally, Figure 2.18 gives the posterior means of the true log-relative hazards θ_{ij} for each study i and unit j. While there is some indication of interaction (e.g., the tighter clustering of the results for the Fungal study, Study 5), there is some indication that a purely additive model (no s_{ij} parameters) would fit nearly as well here. The substantial shrinkage of the b_j towards 0 seen in the previous figure appears to have impacted this plot as well: while we still see mostly positive values, no estimated θ_{ij} emerges as greater than 0.4. ∎

In many applications, a linear model is inappropriate for the underlying mean structure. Our reasons for rejecting the linear model may be theoretical (e.g., an underlying physical process) or practical (e.g., poor model fit). Fortunately, the Bayesian framework requires no major alterations to accommodate nonlinear mean structures. That is, our analytic framework is unchanged; only our computational algorithms must become more sophisticated (typically via Monte Carlo methods). Still, there is often value in investigating an approximating linear model on some appropriate

Patient	Number of Hours Following Drug Administration, x							
	2	4	6	8	10	24	28	32
1	1.09	0.75	0.53	0.34	0.23	0.02	–	–
2	2.03	1.28	1.20	1.02	0.83	0.28	–	–
3	1.44	1.30	0.95	0.68	0.52	0.06	–	–
4	1.55	0.96	0.80	0.62	0.46	0.08	–	–
5	1.35	0.78	0.50	0.33	0.18	0.02	–	–
6	1.08	0.59	0.37	0.23	0.17	–	–	–
7	1.32	0.74	0.46	0.28	0.27	0.03	0.02	–
8	1.63	1.01	0.73	0.55	0.41	0.01	0.06	0.02
9	1.26	0.73	0.40	0.30	0.21	–	–	–
10	1.30	0.70	0.40	0.25	0.14	–	–	–

Table 2.3 *Cadralazine concentration data.*

scale, since MCMC algorithms operating on linear models are often better-behaved and thus more reliable and convenient in practice.

The field of pharmacokinetics offers a large class of nonlinear models, since the underlying process of how a drug is processed by the body is known to follow a complex pattern that is nonlinear in the parameters (see e.g. Wakefield, 1996). While these models are inherently nonlinear, in their simplest forms they can be approximately linearized at little loss in fit or interpretability, as we now demonstrate.

Example 2.13 Consider the data in Table 2.3, which are also plotted in the top panel of Figure 2.19. Here Y_{ij} is the concentration of a particular drug (cadralazine) x_{ij} hours following administration, for $i = 1, \ldots, N = 10$ men observed at times $j = 1, \ldots, T$. Most of the men do not have the full complement of $T = 8$ observations; fortunately, we may just code any missing values as NA in R or WinBUGS. Note also that Patient 2 appears to be an outlier in this data set, with a drug clearance trajectory that is much slower than the rest and almost linear in appearance.

Wakefield et al. (1994) fit a nonlinear "one-compartment" pharmacokinetic model to these data, which will be the subject of Chapter 4, Exercise 2. For now, however, we simplify matters by instead modeling $Z_{ij} = \log Y_{ij}$, since the progression in log-concentration is roughly linear in time, as shown in the bottom panel of Figure 2.19. Note that this figure also reveals the sixth ($x = 24$) observation for Patient 8 to be an apparent data entry error, since a patient's drug concentration must be monotonically decreasing. Replacing this value with NA in the dataset, we proceed with the hierarchical linear model

$$Z_{ij} = \beta_{0i} + \beta_{1i}x_{ij} + \epsilon_{ij} , \qquad (2.23)$$

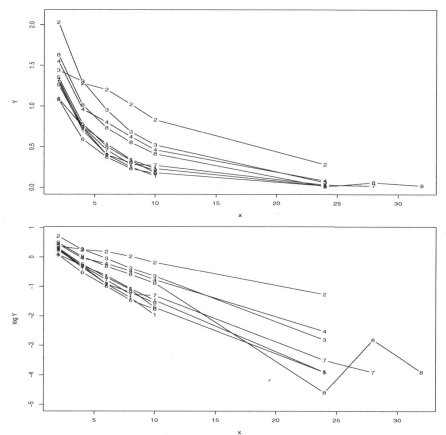

Figure 2.19 *Cadralazine data, where plotting character indicates patient number. Top panel: original scale; bottom panel, log scale. Note apparent data entry error for Patient 8, observation at $x = 24$.*

where $Z_{ij} = \log Y_{ij}$ and $\epsilon_{ij} \stackrel{iid}{\sim} N(0, \sigma^2)$, for $j = 1, \ldots, T$ and $i = 1, \ldots, N$. At the second stage, we choose the conjugate prior

$$\boldsymbol{\beta}_i \stackrel{iid}{\sim} N_2(\boldsymbol{\mu}, \boldsymbol{\Upsilon}) \,,$$

where $\boldsymbol{\beta}_i = (\beta_{0i}, \beta_{1i})'$. Vague hyperpriors on $\boldsymbol{\mu}$ and $\boldsymbol{\Upsilon}$ complete the hierarchical specification.

WinBUGS code for this model is as follows:

BUGS code
```
model{
    for(i in 1:N){
        for(j in 1:T){
            Z[i,j] ~ dnorm (mu[i,j] , tauZ)
            Y[i,j] <- exp(Z[i,j])
            mu[i,j] <- beta[i, 1] + beta[i, 2]* X[j]
```

```
          }
          beta[i , 1:2] ~ dmnorm (betamu[1:2] , betaUpsilon[1:2, 1:2])
          }
          betamu[1:2] ~ dmnorm(meen[1:2] , prec[1:2 ,1:2])
          betaUpsilon[1:2 , 1:2] ~ dwish(R[1:2, 1:2], 2)
          beta.s2[1:2, 1:2] <- inverse(betaUpsilon[1:2, 1:2])
          for (i in 1 : 2) {beta.s[i] <- sqrt(beta.s2[i, i]) }

          tauZ <- 1/(sigmaZ * sigmaZ)
          sigmaZ ~ dunif(0.01, 100)
          }
```

Using list format, the logged observations Z_{ij} and other data can be read in as

```
BUGS code   list(X=c(2,4,6,8,10,24,28,32), N=10, T=8,
              Z = structure(
            .Data = c(0.086, -0.288, -0.635, -1.079, -1.47, -3.912, NA, NA,
                      0.708, 0.247, 0.182, 0.02, -0.186, -1.273, NA, NA,
                      0.365, 0.262, -0.051, -0.386, -0.654, -2.813, NA, NA,
                      0.438, -0.041, -0.223, -0.478, -0.777, -2.526, NA, NA,
                      0.3, -0.248, -0.693, -1.109, -1.715, -3.912, NA, NA,
                      0.077, -0.528, -0.994, -1.47, -1.772, NA, NA, NA,
                      0.278, -0.301, -0.777, -1.273, -1.309, -3.507, -3.912, NA,
                      0.489, 0.01, -0.315, -0.598, -0.892, NA, -2.813, -3.912,
                      0.231, -0.315, -0.916, -1.204, -1.561, NA, NA, NA,
                      0.262, -0.357, -0.916, -1.386, -1.966, NA, NA, NA),
                      .Dim = c(10,8)),
              meen = c(0,0),
              prec = structure(.Data = c(1.0E-6, 0, 0, 1.0E-6), .Dim = c(2,2)),
              R = structure(.Data = c(0.1, 0, 0, 0.1), .Dim = c(2, 2)))
```

Note the hyperprior for μ is made very vague through specification of its precision matrix prec, which has values near 0 on the diagonal and exactly 0 off of it (i.e., its two components are uncorrelated *a priori*). Similarly, the Wishart prior for Υ is made vague by choosing a very rough guess for R and then setting the degrees of freedom parameter equal to 2, the smallest value for which this Wishart prior is proper.

Convergence for this model is rapid; obvious starting places for the fixed effects might be

```
BUGS code   list(sigmaZ =1, betaUpsilon=structure(.Data = c(0.1,0,0,0.1),
              .Dim = c(2, 2)), betamu=c(0,0))
```

with the random effect starting values generated at random using gen inits. The resulting posterior means and 95% equal-tail confidence intervals for the two components of μ (grand intercept and grand slope) are 0.55, (0.41, 0.69) and –0.17, (–0.25, –0.09), respectively. Thus log-concentration experiences a significant decrease over time, but this of course was never in

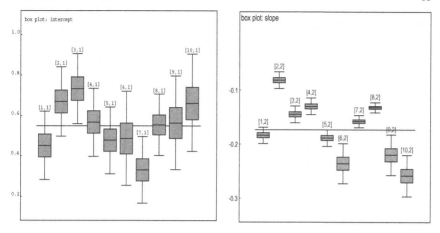

Figure 2.20 *Boxplots of individual-specific intercepts β_{0i} (left) and slopes β_{1i} (right), linearized PK model, cadralazine data.*

doubt. Perhaps more interesting are the boxplots of the β_{0i} and β_{1i} samples across the patients, shown in Figure 2.20. The former samples do not suggest large differences across patients, while the latter do, with Patient 2 again emerging with a significantly higher slope (slower drug clearance) than the other patients.

We close this example by illustrating a useful capability of WinBUGS with respect to missing data. Suppose we assume that the missing Z_{ij} are, in the nomenclature of Little and Rubin (1987), "missing at random." That is, we assume that whether a Z_{ij} is missing or not at a given time point has no connection to its magnitude at that time, or to other confounding patient characteristics (e.g., the patient's health status at that time). If this is the case, the full conditional distribution for any missing Z_{ij} is trivial to obtain: it is simply the normal distribution implied by (2.23), with mean $\beta_{0i} + \beta_{1i}x_{ij}$ and variance σ^2. Even though much of the **Z** vector is not missing, WinBUGS will permit us to add this variable to the others being monitored, resulting in posterior samples being drawn for this (partially stochastic) node. These samples can then be summarized in the usual way. For example, the missing Z_{ij} may be summarized using the Comparison tool on the Inference menu, the same one used to create the box plots of the random effects in Figure 2.20. Since such a large collection of boxplots would be rather hard to read here, Figure 2.21 instead shows a "caterpillar plot," which is nothing but the boxplot display stacked vertically instead of horizontally, and with the boxes indicating the interquartile ranges removed to save space (so that only the posterior means and 95% CIs remain visible). Observations 7 and 8 for Patient 2 emerge as much larger than

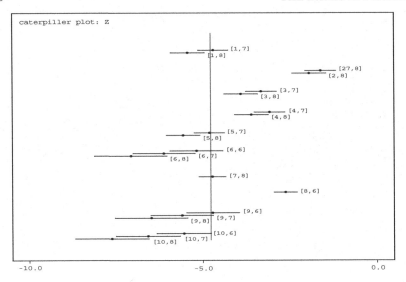

Figure 2.21 *Caterpillar plot of posterior imputations for missing Z_{ij} values, cadralazine data.*

the rest, followed by observation 6 for Patient 8 (the apparently erroneous observation we replaced with NA).

Finally, such imputations are also readily available on the original (Y) scale, since we have added its definition as $\exp(Z)$ into our WinBUGS code. Figure 2.22 shows the posterior distributions of the predicted cadralazine concentrations for Patient 2 had they been observed at times 7 and 8. Using 30,000 post-burn-in iterations, we obtain estimated posterior means of .203 and .148, respectively. Note that these point estimates and indeed the entire density estimates remain significantly above 0, consistent with the plot of the raw data in Figure 2.19. ∎

2.4.2 Effective model size and the DIC criterion

Appropriate statistical selection of the best among a collection of hierarchical models is problematic, due to the ambiguity in the "size" of such models arising from the posterior shrinkage of their random effects toward a common value. To address this problem, Spiegelhalter et al. (2002) suggest a generalization of the Akaike information criterion (AIC) that is based on the posterior distribution of the *deviance* statistic,

$$D(\boldsymbol{\theta}) = -2 \log f(\mathbf{y}|\boldsymbol{\theta}) + 2 \log h(\mathbf{y}) \,, \qquad (2.24)$$

where $f(\mathbf{y}|\boldsymbol{\theta})$ is the likelihood function for the observed data vector $\mathbf{y}$ given the parameter vector $\boldsymbol{\theta}$, and $h(\mathbf{y})$ is some standardizing function of the data

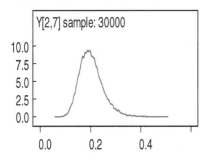 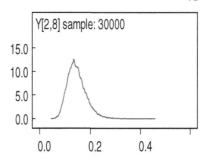

Figure 2.22 *Posterior predictive distributions for $Y_{2,7}$ and $Y_{2,8}$ assuming them to be missing at random, cadralazine data.*

alone (which thus has no impact on model selection). In this approach, the *fit* of a model is summarized by the posterior expectation of the deviance, $\overline{D} = E_{\theta|y}[D]$, while the *complexity* of a model is captured by the effective number of parameters p_D, which is typically less than the total number of model parameters due to the borrowing of strength across individual-level parameters in hierarchical models. In Subsection 4.6.1, we show that a reasonable definition of p_D is the expected deviance minus the deviance evaluated at the posterior expectations,

$$p_D = E_{\theta|y}[D] - D(E_{\theta|y}[\boldsymbol{\theta}]) = \overline{D} - D(\bar{\boldsymbol{\theta}}) . \tag{2.25}$$

The *Deviance Information Criterion* (DIC) is then defined as

$$DIC = \overline{D} + p_D = 2\overline{D} - D(\bar{\boldsymbol{\theta}}) , \tag{2.26}$$

with smaller values of DIC indicating a better-fitting model. Note that DIC is scale-free; the choice of standardizing function $h(\mathbf{y})$ in (2.24) is arbitrary. Thus values of DIC have no intrinsic meaning; as with AIC, only *differences* in DIC across models are meaningful, with differences of 3 to 5 normally being thought of as the smallest that are interesting. Besides its generality, another attractive aspect of DIC is that it may be readily calculated during an MCMC run by monitoring both $\boldsymbol{\theta}$ and $D(\boldsymbol{\theta})$, and at the end of the run simply taking the sample mean of the simulated values of D, minus the plug-in estimate of the deviance using the sample means of the simulated values of $\boldsymbol{\theta}$.

While DIC's generality and easy availability within WinBUGS (as an option on the Inference menu) has led to its broad use in applied Bayesian work, many practical issues have led some to question the appropriateness of DIC for arbitrarily general Bayesian models. For example, DIC is not invariant to parametrization, so (as with prior elicitation) the most plausible parametrization must be carefully chosen beforehand. Unknown scale parameters and other innocuous restructuring of the model can also affect

Model	$\overline{D}$	p_D	DIC
Full model	122.0	12.8	134.8
Drop interactions	123.4	9.7	133.1
No unit effect	123.8	10.0	133.8
No study effect	121.4	9.7	131.1
Unit + intercept	120.3	4.6	124.9
Unit effect only	122.9	6.2	129.1
Study effect only	126.0	6.0	132.0

Table 2.4 *Hierarchical model selection table, cross-protocol example.*

DIC. Finally, p_D can occasionally emerge as *negative* in certain advanced modeling settings where the posterior departs markedly from normality; see Example 4.8 in Subsection 4.6.1.

Example 2.14 In our cross-protocol setting of Example 2.12, we considered only the full model (2.22) for the mean log-relative hazards θ_{ij}. However, because we lack replications for each study-unit (i-j) combination, the interactions s_{ij} in this model were only weakly identified, and the model might well be better off without them (or even without the unit effects b_j). To address this, we compare a variety of reduced models for this dataset, as summarized in the following WinBUGS code:

```
BUGS code   Y[i,j] ~ dnorm(theta[i,j],P[i,j])
       #    theta[i,j] <- a[i]+b[j]+s[i,j]   # full model
       #    theta[i,j] <- a[i] + b[j]        # drop interactions
       #    theta[i,j] <- a[i] + s[i,j]      # no unit effect
       #    theta[i,j] <- b[j] + s[i,j]      # no study effect
       #    theta[i,j] <- a[1] + b[j]        # unit + intercept
       #    theta[i,j] <- b[j]               # unit effect only
            theta[i,j] <- a[i]               # study effect only
```

That is, in addition to the full model, we consider dropping the interactions s_{ij}, the unit effects b_j, and the (fixed) study effects a_i, either alone or in combination. Initializing a single MCMC chain using our "good" starting values, we ran each model for a 20,000-iteration production period following a 1000-iteration burn-in (which Figure 2.16 suggests should be more than sufficient for these starting values).

The results for $\overline{D}$, p_D, and DIC are given in Table 2.4. (We typically include these values after the model name in the comments in the WinBUGS code itself, though there is not room to show this here; see the web version for details.) The best-fitting (smallest $\overline{D}$) model is the one with only an intercept (a role played here by a_1) and the unit effects b_j. The extreme shrinkage of the 18 b_j (evidenced earlier for the full model in Figure 2.17) is apparent from the very low p_D score of 4.6, which includes 1 for the intercept. This dual dominance (good fit plus low effective size) means this model also enjoys the best overall model choice (smallest DIC) score. However, this model is difficult to justify theoretically, since it was only by random chance that all the treatment effects a_i emerged as similar. The DIC-best model that includes these effects is the study effects-only model (DIC=132.0), closely followed by the no-interaction model (DIC = 133.1). The difference in these DIC scores is not too much larger than their possible Monte Carlo errors, so choice of either model could be justified here. Indeed, DIC differences of less than 5 or even 10 are often thought of by practitioners as hardly worth mentioning, again supporting selection of virtually any of these models. Note also the very small p_D values across all the models, which again contributes to the similarity of the DIC scores. ■

We close this subsection with a nonlinear hierarchical modeling example that cannot be readily approximated by a linear model. Our context is that of survival (time-to-event) data, which are very common in biostatistics and engineering. Such data are of course positive and often missing, since we may not observe the event (say, death) for every individual in the study. Thus we need a distribution for a positive random variable that is often *left-censored* (i.e., we know only that the failure time is greater than some observed survival time t) within a model that can accommodate covariates, such as the age, gender and treatment assignment for each individual.

Survival models come in two types: *parametric* models, such as Weibull or gamma, and *nonparametric* (or *semiparametric*) models, such as the celebrated Cox model (see e.g. Cox and Oakes, 1984). Since the WinBUGS language is essentially a framework for parametric sampling, we illustrate only this case here, but emphasize that we will return to the nonparametric case in Sections 2.6 and 7.5 (where WinBUGS code will still be possible in some cases, but more complex).

Example 2.15 The data in Table 2.5 are presented by Carlin and Hodges (1999), and arise from a clinical trial comparing two treatments for *Mycobacterium avium* complex (MAC), a disease common in late stage HIV-infected persons. Eleven clinical centers ("units") have enrolled a total of 69 patients in the trial, of which 18 have died. Suppose for $j = 1, \ldots, n_i$ and $i = 1, \ldots, k$, we let

$$t_{ij} = \text{time to death or censoring}$$
$$\text{and } x_{ij} = \text{treatment indicator for subject } j \text{ in unit } i.$$

We also let γ_{ij} be a death indicator (0 if alive, 1 if dead), where we use $x_{ij} = -1$ for one treatment and 1 for the other, for numerical reasons to be clarified later. The table gives survival times in half-days from the MAC treatment trial, where "+" indicates a censored observation (i.e., the patient was still alive at the end of the observation period).

Unit	Drug	Time	Unit	Drug	Time	Unit	Drug	Time
A	1	74+	E	1	214	H	1	74+
A	2	248	E	2	228+	H	1	88+
A	1	272+	E	2	262	H	1	148+
A	2	344				H	2	162
			F	1	6			
B	2	4+	F	2	16+	I	2	8
B	1	156+	F	1	76	I	2	16+
			F	2	80	I	2	40
C	2	100+	F	2	202	I	1	120+
			F	1	258+	I	1	168+
D	2	20+	F	1	268+	I	2	174+
D	2	64	F	2	368+	I	1	268+
D	2	88	F	1	380+	I	2	276
D	2	148+	F	1	424+	I	1	286+
D	1	162+	F	2	428+	I	1	366
D	1	184+	F	2	436+	I	2	396+
D	1	188+				I	2	466+
D	1	198+	G	2	32+	I	1	468+
D	1	382+	G	1	64+			
D	1	436+	G	1	102	J	1	18+
			G	2	162+	J	1	36+
E	1	50+	G	2	182+	J	2	160+
E	2	64+	G	1	364+	J	2	254
E	2	82						
E	1	186+	H	2	22+	K	1	28+
E	1	214+	H	1	22+	K	1	70+
						K	2	106+

Table 2.5 *Survival times (in half-days) from the MAC treatment trial, from Carlin and Hodges (1999). Here, "+" indicates a censored observation.*

We adopt the usual *proportional hazards* model, where the hazard (instantaneous risk of death at any moment t) for an individual having covariate value x_{ij} is proportional as a function of t to some *baseline hazard* $h_0(t)$. If we then assume a Weibull model for this baseline hazard, $h_0(t_{ij}) = \rho_i t_{ij}^{\rho_i - 1}$, then the hazard for an individual in the ith unit can be modeled as

$$
\begin{aligned}
h(t_{ij}; x_{ij}) &= h_0(t_{ij})\omega_i \exp(\beta_0 + \beta_1 x_{ij}) \\
&= \rho_i t_{ij}^{\rho_i - 1} \exp(\beta_0 + \beta_1 x_{ij} + W_i) ,
\end{aligned}
$$

where $\rho_i > 0$, $\boldsymbol{\beta} = (\beta_0, \beta_1)' \in \Re^2$, and $W_i = \log \omega_i$ is a unit-specific *frailty* term. That is, the W_i capture overall differences among the clinics, while the ρ_i allow differing baseline hazards that either increase ($\rho_i > 1$) or decrease ($\rho_i < 1$) over time. We assume i.i.d. specifications for these random effects,

$$W_i \overset{iid}{\sim} N(0, 1/\tau) \quad \text{and} \quad \rho_i \overset{iid}{\sim} G(\alpha, \alpha) \,,$$

though in Chapter 7 we will extend to frailties having a spatial structure (needed when spatially closer units exhibit similar survival patterns).

Following the `mice` example in the `WinBUGS` manual (click on `Help` and pull down to `Examples Vol I`), if we write

$$\mu_{ij} = \exp(\beta_0 + \beta_1 x_{ij} + W_i) \,,$$

then

$$t_{ij} \sim Weibull(\rho_i, \mu_{ij}) \,.$$

The data as given in Table 2.5 are highly unbalanced, with n_i values ranging from 1 to 13. As such, a bit of preliminary rearrangement of the data in R can streamline our subsequent BUGS code. Specifically, we follow the approach BUGS uses for spatial adjacency modeling and create an "offset" vector that holds the position in the overall data vector where each unit's information starts. The code we use for this is

R code
```
nsub<-c(4,2,1,10,8,12,6,6,13,4,3)
temp<-1
offset<-rep(0,(length(nsub)+1))
for(i in 1:length(nsub)){offset[i]<-temp; temp<-temp+nsub[i]}
offset[ (length(nsub)+1)] <-sum(nsub)+1
```

See www.biostat.umn.edu/~brad/data/MAC_data.txt for the reformatted data. We also remark that, while the drug assignment is coded as 1 or 2 in the raw data, we recode the drug covariate to $(-1,1)$ (i.e., set $x_{ij} = 2drug_{ij} - 3$) to ease collinearity between the slope β_1 and the intercept β_0 in our sampling algorithm. This simple change can be done within the WinBUGS code, which follows:

BUGS code
```
model{
    for(i in 1:nsites){
      for(j in offset[i] : (offset[i+1]-1) ){
        t[j]  ~ dweib (r[i], mu[j] ) I(t.cen[j],)
        X[j] <- 2*drug[j] - 3
        mu[j] <- exp(beta0 + beta1 * X[j] + W[i])
        }
      }
    for (i in 1:nsites) {
      b0star[i] <- beta0+ W[i]
      r[i]  ~ dgamma(alpha, alpha)
      W[i]  ~ dnorm(0, tau.W)
      }
```

```
    alpha <- 10
    tau.W ~ dgamma(1,1);  sigma.W<- 1/sqrt(tau.W)
    beta0 ~ dnorm(0 , 0.001)
    beta1 ~ dnorm(0,  0.001)
    relhaz <- exp(2 * beta1)
}
```

This is what we might call the "basic model," which features two sets of clinic-specific random effects: the Weibull shape parameters r_i and the frailties W_i. Notice here we have placed vague priors on β_0 and β_1, a moderately informative $G(1,1)$ prior on τ, and set $\alpha = 10$ (so that the distribution of the r_i is rather tightly concentrated around 1).

The clever use of the t.cen variable here is also worth special mention. Note that the time-to-event variable t is defined as a *censored* Weibull, accomplished by multiplying the Weibull distribution by the indicator function I(t.cen[j],) where the missing upper bound indicates the bound is infinity. Thus, we are saying that we know t lies between t.cen and infinity. The data set takes the form

BUGS code
```
list(nsites=11,
     offset=c(1,5,7,8,18,26,38,44,50,63,67,70),
      t = c( NA,248,NA,344,
NA,NA,
NA,
NA,64,88,NA,NA,NA,NA,NA,NA,NA,
NA,NA,82,NA,NA,214,NA,262,
6,NA,76,80,202,NA,NA,NA,NA,NA,NA,
NA,NA,102,NA,NA,NA,
NA,NA,NA,NA,NA,162,
8,NA,40,NA,NA,NA,NA,276,NA,366,NA,NA,NA,
NA,NA,NA,254,
NA,NA,NA),
      t.cen= c(74,0,272,0,
4,156,
100,
20,0,0,148,162,184,188,198,382,436,
50,64,0,186,214,0,228,0,
0,16,0,0,0,258,268,368,380,424,428,436,
32,64,0,162,182,364,
22,22,74,88,148,0,
0,16,0,120,168,174,268,0,286,0,396,466,468,
18,36,160,0,
28,70,106),
      drug= c(1,2,1,2,
2,1,
2,
2,2,2,2,1,1,1,1,1,1,
1,2,2,1,1,1,2,2,
```

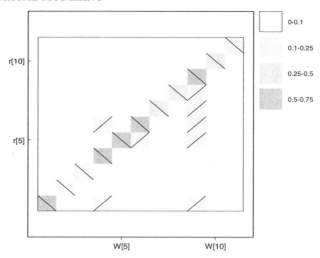

Figure 2.23 *Cross-correlations between the clinic-specific shape parameters r_i and frailties W_i, MAC treatment trial data.*

```
1,2,1,2,2,1,1,2,1,1,2,2,
2,1,1,2,2,1,
2,1,1,1,1,2,
2,2,2,1,1,2,1,2,1,1,2,2,1,
1,1,2,2,
1,1,2
))
```

(The placing of each clinic's data on a separate line is purely for our benefit; WinBUGS decodes these three vectors using the offset variable, not the carriage returns.) Notice that in the dataset, every observed death has t as its death time and t.cen=0, indicating no censoring. But every censored observation has "NA" (missing value) for t and its censoring time as t.cen. Thus these terms contribute only the fact that, had we seen the death, it would have been bigger than t.cen, but otherwise followed the Weibull distribution. We remark that the DIC tool in WinBUGS does incorporate these terms into the deviance here, even though missing observations are not normally included since they normally contain no information.

Initial values for the Gibbs sampler may be specified as

BUGS code
```
list(r=c(1, 1, 1, 1, 1, 1, 1, 1, 1, 1, 1), tau.W=1.0,
     beta1=1.0, beta0=-10.0, W=c(0,0,0,0,0,0,0,0,0,0,0))
```

Despite these relatively benign starting values, the three Gibbs chains we ran were slow to mix, with fairly high autocorrelations (positive out to lag 40 for some parameters). There is also evidence of high negative cross-correlations among the two random effects for the same clinic; see the crude

Node (Unit)	mean	sd	MC error	2.5%	median	97.5%
W_1 (A)	−0.0491	0.835	0.0210	−1.775	−0.0460	1.639
W_3 (C)	−0.1829	0.9173	0.0178	−2.2	−0.1358	1.520
W_5 (E)	−0.0320	0.8107	0.0319	−1.682	−0.0265	1.572
W_6 (F)	0.4173	0.8277	0.0407	−1.066	0.3593	2.227
W_9 (I)	0.2546	0.7969	0.0369	−1.241	0.2164	1.968
W_{11} (K)	−0.1945	0.9093	0.0209	−2.139	−0.1638	1.502
ρ_1 (A)	1.086	0.1922	0.0072	0.7044	1.083	1.474
ρ_3 (C)	0.9008	0.2487	0.0063	0.4663	0.8824	1.431
ρ_5 (E)	1.143	0.1887	0.0096	0.7904	1.139	1.521
ρ_6 (F)	0.935	0.1597	0.0084	0.6321	0.931	1.265
ρ_9 (I)	0.9788	0.1683	0.0087	0.6652	0.9705	1.339
ρ_{11} (K)	0.8807	0.2392	0.0103	0.4558	0.8612	1.394
τ	1.733	1.181	0.0372	0.3042	1.468	4.819
β_0	−7.111	0.689	0.0447	−8.552	−7.073	−5.874
β_1	0.596	0.2964	0.0105	0.06099	0.5783	1.245
relhaz	3.98	2.951	0.1122	1.13	3.179	12.05

Table 2.6 *Posterior summaries, MAC survival model (10,000 samples, after a burn-in of 1000).*

graphical summary in Figure 2.23, obtained using the `matrix` button on the `Correlation` tool. Apparently there is not much information in our small dataset to separate these two components, even using our rather informative priors.

Denoting vectors with boldface, the joint posterior distribution for $\beta, \rho,$ and α is

$$p(\boldsymbol{\beta}, \boldsymbol{\rho}, \alpha \mid \mathbf{t}, \mathbf{x}, \boldsymbol{\gamma}) \propto L(\boldsymbol{\beta}, \boldsymbol{\rho} ; \mathbf{t}, \mathbf{x}, \boldsymbol{\gamma}) \left\{ \prod_{i=1}^{k} p(\rho_i | \alpha) \right\} p(\alpha)$$
$$\propto \left[\prod_{i=1}^{k} \prod_{j=1}^{n_i} \{ \rho_i t_{ij}^{\rho_i - 1} \exp(\rho_i \beta_0 + \beta_1 x_{ij}) \}^{\gamma_{ij}} \exp\{ -t_{ij}^{\rho_i} \exp(\rho_i \beta_0 + \beta_1 x_{ij}) \} \right]$$
$$\times \left\{ \prod_{i=1}^{k} \rho_i^{\alpha-1} \exp(-\alpha \rho_i) \alpha^\alpha / \Gamma(\alpha) \right\} \alpha^{c-1} \exp(-\alpha/d) .$$

`WinBUGS` samples from the relevant full conditionals, obtaining the results shown in Table 2.6 from 20,000 post-burn-in iterations. Units A and E have moderate overall risk ($W_i \approx 0$), but increasing hazards ($\rho > 1$). This is sensible, because these units experienced few deaths, but these occur late. By contrast, Units F and I have high overall risk ($W_i > 0$), but decreasing hazards ($\rho < 1$). These units experienced several early deaths, but also many long-term survivors. Finally, Units C and K have low overall risk ($W_i < 0$) and decreasing hazards ($\rho < 1$). Once again this is plausible, because these units saw no deaths at all, and included just a few survivors. The table also reveals that the two MAC treatments differ significantly, with the 95% equal-tail Bayesian CI for the treatment effect β_1 excluding 0 and an estimated relative hazard ($e^{2\beta_1}$, the ratio of $e^{\beta_0+\beta_1}$ and $e^{\beta_0-\beta_1}$) of nearly 4 between the two groups. The overall fit score $\overline{D}$ is 272.2 while the effective model size p_D is 8.5, meaning that DIC = 280.8. Again the very high degree of shrinkage in the random effects is evident; note there are 22 random effects plus fixed effects in the model.

A great many alternate models (with either more, fewer, or no random effects) could be attempted here and compared to these "basic" results. This is the subject of Exercise 21. ∎

2.5 Model assessment

2.5.1 Diagnostic measures

Having successfully fit a model to a given dataset, the statistician must be concerned with whether or not the fit is adequate and the assumptions made by the model are justified. For example, in standard linear regression, the assumptions of normality, independence, linearity, and homogeneity of variance must all be investigated. Several authors have suggested using the marginal distribution of the data, $m(\mathbf{y})$, in this regard. Observed y_i values for which $m(y_i)$ is small are "unlikely" and, therefore, may be considered outliers under the assumed model. Too many small values of $m(y_i)$ suggest the model itself is inadequate and should be modified or expanded.

A problem with this approach is the difficulty in defining how small is "small," and how many outliers are "too many." In addition, we have the problem of the possible impropriety of $m(\mathbf{y})$ under noninformative priors, already mentioned in the context of Bayes factors. As such, we might instead work with *predictive* distributions, since they will be proper whenever the posterior is. Specifically, suppose we fit our Bayesian model using a *fitting sample* of data $\mathbf{z} = (z_1, \ldots, z_m)'$ and wish to check it using an independent *validation sample* $\mathbf{y} = (y_1, \ldots, y_n)'$. By analogy with classical model checking, we could begin by analyzing the collection of *residuals*, the difference between the observed and fitted values for each point in the validation sample. In this context, we define a Bayesian residual as

$$r_i = y_i - E(Y_i|\mathbf{z}), \; i = 1, \ldots, n. \tag{2.27}$$

Plotting these residuals versus fitted values might reveal failure in a normality or homogeneity of variance assumption; plotting them versus time could reveal a failure of independence. Summing their squares or absolute values could provide an overall measure of fit, but, to eliminate the dependence on the scale of the data, we might first *standardize* the residuals, computing

$$d_i = \frac{y_i - E(Y_i|\mathbf{z})}{\sqrt{Var(Y_i|\mathbf{z})}}, \; i = 1, \ldots, n. \tag{2.28}$$

The above discussion assumes the existence of two independent data samples, which may well be unavailable in many problems. Of course we could simply agree to always reserve some portion (say, 20-30%) of our data at the outset for subsequent model validation, but we may be loathe to do this when data are scarce. As such, Gelfand, Dey, and Chang (1992) suggest a *cross-validatory* (or "leave one out") approach, wherein the fitted value for y_i is computed conditional on all the data *except* y_i, namely, $\mathbf{y}_{(i)} \equiv (y_1, \ldots y_{i-1}, y_{i+1}, \ldots, y_n)'$. That is, we replace (2.27) by

$$r_i' = y_i - E(Y_i|\mathbf{y}_{(i)}) \, ,$$

and (2.28) by

$$d_i' = \frac{y_i - E(Y_i|\mathbf{y}_{(i)})}{\sqrt{Var(Y_i|\mathbf{y}_{(i)})}} \, .$$

Gelfand et al. (1992) provide a collection of alternate definitions for d_i, with associated theoretical support presented in Gelfand and Dey (1994).

Note that in this cross-validatory approach we compute the posterior mean and variance with respect to the *conditional predictive distribution*,

$$f(y_i|\mathbf{y}_{(i)}) = \frac{f(\mathbf{y})}{f(\mathbf{y}_{(i)})} = \int f(y_i|\boldsymbol{\theta}, \mathbf{y}_{(i)})p(\boldsymbol{\theta}|\mathbf{y}_{(i)})d\boldsymbol{\theta} \, , \tag{2.29}$$

which gives the likelihood of each point given the remainder of the data. The actual values of $f(y_i|\mathbf{y}_{(i)})$ (sometimes referred to as the *conditional predictive ordinate*, or CPO) can be plotted versus i as an outlier diagnostic, since data values having low CPO are poorly fit by the model. Notice that, unlike the marginal distribution, $f(y_i|\mathbf{y}_{(i)})$ will be proper if $p(\boldsymbol{\theta}|\mathbf{y}_{(i)})$ is. In addition, the collection of conditional predictive densities $\{f(y_i|\mathbf{y}_{(i)}), \; i = 1, \ldots, n\}$ is equivalent to $m(\mathbf{y})$ when both exist (Besag, 1974), encouraging the use of the former even when the latter is undefined. One might even view the product of the CPO values as a "pseudo marginal likelihood," with large values indicating good overall model fit. For computational convenience, the log of this product (the *log pseudo marginal likelihood*, or LPML), is more often used in this regard; see Subsection 4.6.2.

Example 2.16 We investigate the calculation of Bayesian residuals and CPO statistics in the context of the oft-analyzed stack loss data of Brownlee (1965, p. 454). These data give the stack loss Y, the amount of ammonia escaping in an industrial setting, along with the three covariates air flow (X_1), temperature (X_2), and acid concentration (X_3). Observations 21, 1, 3, and 4 have been previously identified by several analyses as potential outliers. We assume an ordinary normal-errors regression model for Y_i with mean structure

$$E(Y_i) = \beta_0 + \beta_1 z_{i1} + \beta_2 z_{i2} + \beta_3 z_{i3} , \qquad (2.30)$$

where z_{ij} is the centered and scaled version of x_{ij}, i.e., the covariate minus its own mean and divided by its own standard deviation. We also place very vague normal priors on the β_j coefficients, $j = 0, 1, 2, 3$.

We being with the following WinBUGS code:

BUGS code
```
model
{
# standardize x's and coefficients
for (j in 1:p) {
  b[j] <- beta[j] / sd(x[,j ])
  for (i in 1:N) {z[i,j] <- (x[i,j]-mean(x[,j]))/sd(x[,j])}
  }
b0 <- beta0-b[1]*mean(x[,1])-b[2]*mean(x[,2])-b[3]*mean(x[,3])

# statistical model
for (i in 1 : N) {
   Y[i] ~ dnorm(mu[i], tau)
   mu[i] <- beta0 +beta[1]*z[i,1] +beta[2]*z[i,2] +beta[3]*z[i,3]

   sresid_apr[i] <- (Y[i]-mu[i])/sigma
   outlier_apr[i] <- step(sresid_apr[i]-1.5)
                                + step(-(sresid_apr[i]+1.5))
   CPO_apr[i] <-sqrt(tau)* exp(-tau/2*(Y[i]-mu[i])*(Y[i]-mu[i]))
   }

# priors
beta0 ~  dnorm(0, 0.00001)
for (j in 1 : p) {beta[j] ~ dnorm(0, 0.00001)}
sigma ~ dunif(0.01,100)      #  vague Gelman prior for sigma
tau<- 1/(sigma*sigma)
}  # end of stacks model
```

This code computes standardized residuals, an outlier indicator (activated if a standardized residual is greater than 1.5 in absolute value), and CPO values up to a convenient normalizing constant. Note that all three of these calculations are approximate, in that this code does *not* leave out the current datapoint y_i when computing any of these summaries; all are based on the full posterior distribution. This is computationally convenient, but the impact of this approximation is not yet clear.

The data is the same as that of the stacks example in the WinBUGS examples manual, namely

BUGS code
```
list(p = 3, N = 21,
     Y = c(42,37,37,28,18,18,19,20,15,14,14,13,11,12,8,7,8,8,9,15,15),
     x = structure(.Data = c(  80, 27, 89,
```

```
                        80, 27, 88,
                        75, 25, 90,
                        62, 24, 87,
                        62, 22, 87,
                        62, 23, 87,
                        62, 24, 93,
                        62, 24, 93,
                        58, 23, 87,
                        58, 18, 80,
                        58, 18, 89,
                        58, 17, 88,
                        58, 18, 82,
                        58, 19, 93,
                        50, 18, 89,
                        50, 18, 86,
                        50, 19, 72,
                        50, 19, 79,
                        50, 20, 80,
                        56, 20, 82,
                        70, 20, 91), .Dim = c(21, 3)))
```

After inputting two sets of starting values,

```
BUGS code    list(beta0 = 10, beta = c(0,0,0), sigma = 0.1)
             list(beta0 = -10, beta=c(10,10,10), sigma = 10)
```

we ran two Gibbs chains for 20,000 iterations each, following a 1000-iteration burn-in period. Posterior boxplots for the sresid_apr and CPO_apr are shown in Figure 2.24. The standardized residuals indicate observation 21 is an outlier due to overprediction by the model (negative residual), while observations 1, 3, and 4 may be outliers due to underprediction (positive residuals). The approximate CPO values tell a similar story, with observation 21 being the most egregious outlier (smallest CPO), followed by observations 4, 3, and 1, in that order. The individual mean values of outlier_apr were 0.17, 0.37, 0.64, and 0.83 for observations 1, 3, 4, and 21, respectively; no other observation had an estimated outlier probability greater than 0.08. Thus observations 4 and 21 are flagged by our approximate method as more likely than not to be outliers, according to our ± 1.5 standardized residual cutoff.

Obtaining exact cross-validatory results here would appear to require 21 runs of the Gibbs sampler, one for each partial, "leave one out" dataset. A way to "trick" WinBUGS into doing all 21 of these runs simultaneously is to replace the data vector $\mathbf{Y}$ with a 21×21 data *matrix*, having one row for each of the 21 partial datasets. That is, the data Y now takes the form

```
BUGS code    Y = structure(.Data =
                 c(NA, 37, 37, 28, 18, 18, ... 15,
                   42, NA, 37, 28, 18, 18, ... 15,
```

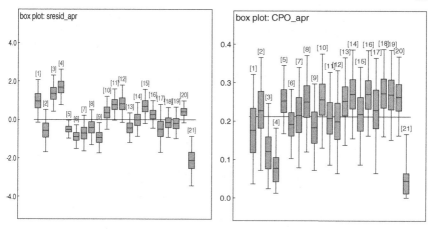

Figure 2.24 *Boxplots of approximate standardized residuals (left) and CPO values (right), stack loss data.*

```
42, 37, NA, 28, 18, 18, ... 15,
42, 37, 37, NA, 18, 18, ... 15,
..............
42, 37, 37, 28, 18, 18, ... NA,
42, 37, 37, 28, 18, 18, ... 15),
.Dim = c(22, 21)),
```

Note that, in fact, we use 22 rows, with the additional final row containing the full dataset; calculations with this row alone should reproduce our approximate results above.

This new data format requires double subscripting for Y and the model parameters, but otherwise adds little additional complexity to our code:

BUGS code
```
model
{
# Standardise x's and coefficients
for(i in 1:(N+1)){for (j in 1:p) {b[i,j] <- beta[i,j]/sd(x[,j]) }}
for (j in 1:p) {for (i in 1:N) {
    z[i,j]<-(x[i,j]-mean(x[,j]))/sd(x[,j])}}
for(i in 1:(N+1)){ b0[i]<-beta0[i]-b[i,1]*mean(x[,1])
                -b[i,2]*mean(x[,2])-b[i,3]*mean(x[,3])}

# statistical model:
for (i in 1:N) {     # replicated dataset with NA in ith location
    for(j in 1:N){   # datapoint in each replicate set
        Y[i,j] ~ dnorm(mu[i,j],tau[i])
        mu[i,j] <- beta0[i]+beta[i,1]*z[j,1]
                +beta[i,2]*z[j,2]+beta[i,3]*z[j,3]}
```

```
# exact CPO:
   CPO[i]<-sqrt(tau[i])* exp(-tau[i]/2*(Y[(N+1),i]-mu[i,i])
                                    *(Y[(N+1),i]-mu[i,i]))
# exact standardized residuals:
   sresid[i] <- (Y[(N+1),i]-mu[i,i])/sigma[i]
   outlier[i] <- step(sresid[i]-1.5) + step(-(sresid[i]+1.5))
   }  # end of i loop

# full data (approximate) model:
for(j in 1:N){
   Y[(N+1),j] ~ dnorm(mu[(N+1),j], tau[(N+1)])
   mu[(N+1),j]<-beta0[(N+1)] + beta[(N+1),1]*z[j,1]
               +beta[(N+1),2]*z[j,2] + beta[(N+1),3]*z[j,3]

# approximate CPO:
   CPO_apr[j]<-sqrt(tau[(N+1)])*exp(-tau[(N+1)]/2
           *(Y[(N+1),j]-mu[(N+1),j])*(Y[(N+1),j]-mu[(N+1),j]))
# approximate standardized residuals:
   sresid_apr[j] <- (Y[(N+1),j]-mu[(N+1),j])/sigma[(N+1)]
   outlier_apr[j]<-step(sresid_apr[j]-1.5)
                              + step(-(sresid_apr[j]+1.5))
   }  # end of j loop

# Priors
   for(i in 1:(N+1)) { beta0[i] ~  dnorm(0, 0.00001)
      sigma[i] ~ dunif(0.01,100)      # vague Gelman prior for sigma
      tau[i]<- 1/(sigma[i]*sigma[i])
      for (j in 1:p) {beta[i,j] ~ dnorm(0, 0.00001)}  # coeffs indep
      }  # end of i loop
   }  # end of WinBUGS code
```

In the exact sresid and CPO calculations, we use the full dataset Y[N+1,]
when computing the corresponding summary, but average over samples
from the appropriate leave-one-out dataset when we do this. Note the 22
sets of posterior samples are obtained independently, because there are
no links between parameters across the 22 corresponding models. This is
perhaps the key feature of our "trick"; we are running 22 Gibbs samplers
in *parallel*, rather than sequentially.

Our new data format also requires a new format for the inits; two
possible choices are as follows:

BUGS code
```
list(beta0 =c(10,10,10,10,10,10,10,10,10,10,10,
              10,10,10,10,10,10,10,10,10,10,10),
   beta=structure(.Data=c(0,0,0,0,0,0,0,0,0,0,0,0,0,0,0,0,0,0,0,
   0,0,0,0,0,0,0,0,0,0,0,0,0,0,0,0,0,0,0,0,0,0,0,0,
   0,0,0,0,0,0,0,0,0,0,0,0,0,0,0,0,0,0,0,0,0,
   0,0,0,0), .Dim = c(22,3)),
   sigma =c(0.1,0.1,0.1,0.1,0.1,0.1,0.1,0.1,0.1,0.1,0.1,0.1,
```

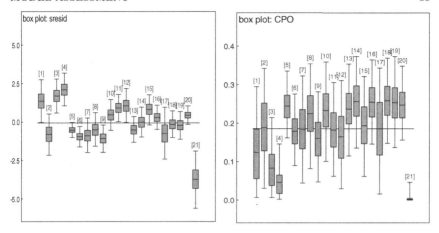

Figure 2.25 *Boxplots of exact standardized residuals (left) and CPO values (right), stack loss data.*

```
0.1,0.1,0.1,0.1,0.1,0.1,0.1,0.1,0.1,0.1))

list(beta0 =c(-10,-10,-10,-10,-10,-10,-10,-10,-10,-10,-10,
             -10,-10,-10,-10,-10,-10,-10,-10,-10,-10,-10),
beta=structure(.Data=c(10,10,10,10,10,10,10,10,10,10,10,10,
10,10,10,10,10,10,10,10,10,10,10,10,10,10,10,10,10,10,
10,10,10,10,10,10,10,10,10,10,10,10,10,10,10,10,
10,10,10,10,10,10,10,10,10,10,10,10,10,10,10,10,10,10,10),
.Dim = c(22, 3)),
sigma =c(1,1,1,1,1,1,1,1,1,1,1,1,1,1,1,1,1,1,1,1,1,1))
```

The code runs well in WinBUGS, albeit 22 times slower for obvious reasons. Using similar burn-in and production runs as above, Figure 2.25 shows the exact standardized residuals and CPOs for this problem. Note the overall appearance of this figure is similar to that of the approximate version in Figure 2.24, but on a dramatically different scale. For example, the exact standardized residual for observation 21 is estimated to be –3.73; it was only –2.14 under the approximate method. The exact results also indicate overwhelming support for this observation being an outlier (posterior mean 0.99), much stronger than previously suggested (0.83) by the approximate method. Clearly the approximation leads to standardized residuals that are too small, a natural outcome of this method's "double usage" of each data point. Finally, the CPO values for observation 21 are also dramatically different, now clustered tightly near 0.

It thus appears that precise "leave one out" results require more than the quick approximation corresponding to Figure 2.24. A slightly more direct solution that uses BRugs, the version of BUGS that can be downloaded and called from within R, is considered below in Exercise 22. Still another possibility arises from an importance sampling approximation we will entertain in equation (3.4). ∎

An alternative to cross-validatory approaches is due to Rubin (1984), who proposes instead *posterior predictive* model checks. The idea here is to construct test statistics or other "discrepancy measures" $D(\mathbf{y})$ that attempt to measure departures of the observed data from the assumed model (likelihood and prior distribution). For example, suppose we have fitted a normal distribution to a sample of univariate data, and wish to investigate the model's fit in the lower tail. We might compare the observed value of the discrepancy measure

$$D(\mathbf{y}) = y_{min}$$

with its posterior predictive distribution, $p(D(\mathbf{y}^*)|\mathbf{y})$, where $\mathbf{y}^*$ denotes a hypothetical future value of $\mathbf{y}$. If the observed value is extreme relative to this reference distribution, doubt is cast on some aspect of the model. Gelman et al. (2004) stress that the choice of $D(\mathbf{y})$ can (and should) be tuned to the aspect of the model whose fit is in question. However, $D(\mathbf{y})$ should not focus on aspects that are parametrized by the model (e.g., taking $D(\mathbf{y})$ equal to the sample mean or variance in the above example), because these would be fit automatically in the posterior distribution. Geweke (2007) also appreciates this fact and uses it to argue in favor of replacing the posterior predictive with the *prior* predictive, which typically (but not always) leads to more conservative checks (see the "using the data twice" discussion below). Still, the basic idea remains comparing some function of the observed data to replicates of this function based on draws from some predictive distribution, with doubt cast on the model if the two disagree in some meaningful way. The idea of frequentist checks using such functions dates to Box (1980); see also Little (2006) for a more modern reinterpretation and development.

In order to be computable in the classical framework, test statistics must be functions of the observed data alone. But, as pointed out by Gelman, Meng, and Stern (1996), basing Bayesian model checking on the posterior predictive distribution allows generalized test statistics $D(\mathbf{y}, \boldsymbol{\theta})$ that depend on the parameters as well as the data. For example, as an omnibus goodness-of-fit measure, Gelman et al. (2004, p.175) recommend

$$D(\mathbf{y}, \boldsymbol{\theta}) = \sum_{i=1}^{n} \frac{[y_i - E(Y_i|\boldsymbol{\theta})]^2}{Var(Y_i|\boldsymbol{\theta})} . \qquad (2.31)$$

With $\boldsymbol{\theta}$ varying according to its posterior distribution, we would now compare the distribution of $D(\mathbf{y}, \boldsymbol{\theta})$ for the observed $\mathbf{y}$ with that of $D(\mathbf{y}^*, \boldsymbol{\theta})$ for a future observation $\mathbf{y}^*$. A convenient summary measure of the extremeness of the former with respect to the latter is the tail area,

$$p_D \equiv P[D(\mathbf{y}^*, \boldsymbol{\theta}) > D(\mathbf{y}, \boldsymbol{\theta})|\mathbf{y}]$$
$$= \int P[D(\mathbf{y}^*, \boldsymbol{\theta}) > D(\mathbf{y}, \boldsymbol{\theta})|\boldsymbol{\theta}] \, p(\boldsymbol{\theta}|\mathbf{y}) d\boldsymbol{\theta} \,. \qquad (2.32)$$

In the case where $D(\mathbf{y}^*, \boldsymbol{\theta})$'s distribution is free of $\boldsymbol{\theta}$ (in frequentist parlance, when $D(\mathbf{y}^*, \boldsymbol{\theta})$ is a *pivotal quantity*), p_D is exactly equal to the frequentist p-value, or the probability of seeing a statistic as extreme as the one actually observed. As such, (2.32) is sometimes referred to as the *Bayesian p-value*. Gelman et al. (2005) extend this thinking to missing and latent data settings (e.g., survival data), constructing plots and diagnostics based on the complete (observed plus imputed unobserved) dataset that have advantages over those based on the observed data alone.

Given our earlier rejection of the p-value (in Example 1.2 and elsewhere), primarily for its failure to obey the Likelihood Principle, our current interest in a quantity that may be identical to it may seem incongruous. But as stressed by Meng (1994), it is important to remember that we are not advocating the use of (2.32) in model choice; p_D is *not* the probability that the model is "correct," and Bayesian p-values should *not* be compared across models. Rather, they serve only as measures of discrepancy between the assumed model and the observed data, and therefore provide information concerning model adequacy. But even on this point there is controversy, since, strictly speaking, this model checking strategy uses the data twice: once to compute the observed statistic $D(\mathbf{y}, \boldsymbol{\theta})$, and again to obtain the posterior predictive reference distribution. It is for this reason that, when sufficient data exist, we prefer the approach that clearly separates the fitting sample $\mathbf{z}$ from the validation sample $\mathbf{y}$. Then we could find

$$p'_D \equiv P[D(\mathbf{y}^*, \boldsymbol{\theta}) > D(\mathbf{y}, \boldsymbol{\theta})|\mathbf{z}]$$
$$= \int P[D(\mathbf{y}^*, \boldsymbol{\theta}) > D(\mathbf{y}, \boldsymbol{\theta})|\boldsymbol{\theta}] \, p(\boldsymbol{\theta}|\mathbf{z}) d\boldsymbol{\theta} \,. \qquad (2.33)$$

Still, (2.32) could be viewed merely as an approximation of the type found in empirical Bayes analyses; see Chapter 5. Moreover, the generalized form in (2.32) accommodates nuisance parameters and provides a Bayesian probability interpretation in terms of future datapoints, thus somewhat rehabilitating the p-value.

Example 2.17 In this example we reconsider the stack loss data of Example 2.16, in order to compute and evaluate Bayesian p-values in this context. Recall these data give the stack loss Y and three covariates air flow (X_1), temperature (X_2), and acid concentration (X_3). Here we will investigate the Bayesian p-value's ability to identify observations 21, 1, 3, and 4, which we previously identified as potential outliers. We again assume the usual normal-errors regression model for Y_i with mean structure given in (2.30), and again place essentially flat normal priors on the β_j coefficients, $j = 0, 1, 2, 3$.

The WinBUGS code we use is as follows:

BUGS code
```
model
{
# standardize x's and coefficients
for (j in 1 : p) {
    b[j] <- beta[j] / sd(x[ , j ])
    for (i in 1 : N) {z[i,j] <- (x[i,j]-mean(x[,j]))/sd(x[,j])}
    }
b0 <- beta0 - b[1]* mean(x[,1])-b[2]*mean(x[,2])-b[3]*mean(x[,3])

# statistical model
for (i in 1 : N) {
    Y[i] ~ dnorm(mu[i], tau)
    mu[i] <- beta0 +beta[1]*z[i,1] +beta[2]*z[i,2] +beta[3]*z[i,3]

    Ystar[i] ~ dnorm(mu[i], tau)
    D[i]<- (Y[i]-mu[i])* (Y[i]-mu[i]) *tau
    Dstar[i] <-   (Ystar[i]-mu[i])* (Ystar[i]-mu[i])*tau
    p_val[i] <- step(Dstar[i] - D[i])
    }

sumDstar <- sum(Dstar[])
sumD <- sum(D[])
Dp <- step(sumDstar-sumD)
Sp <- step(ranked(Y[], 1) - ranked(Ystar[], 1))    # min(Y)= 7
Bp <- step(ranked(Ystar[], N) - ranked(Y[], N))    # max(Y)=42

# priors
beta0 ~   dnorm(0, 0.00001)
for (j in 1 : p) {beta[j] ~ dnorm(0, 0.00001)}
sigma ~ dunif(0.01,100)      #  vague Gelman prior for sigma
tau<- 1/(sigma*sigma)
} # end of stacks model
```

The data is the same as in the first part of Example 2.16. Note that Ystar is being defined as another "copy" of the data vector Y, but because it is not observed, it will be generated as part of the MCMC sampling order. The components of (2.31) are computed as D[i] and Dstar[i] for the actual and artificial data, respectively, with the relative positions of their sums determining the omnibus goodness-of-fit Bayesian p-value Dp; the differences in D[i] and Dstar[i] for a given i provide information on the outlyingness of individual observations. The code also computes two other Bayesian p-values, Sp and Bp, which measure the agreement of the smallest and largest data values in the true and artificial datasets, respectively.

After inputting the starting values,

BUGS code   `list(beta0 = 10, beta = c(0,0,0), sigma = 0.1)`

we ran the Gibbs sampler for 20,000 iterations, following a 1000-iteration burn-in period. We obtained a posterior means of 0.54 for Dp, indicating no overall lack of fit using our omnibus statistic. Similarly, the Bayesian p-values Sp and Bp were 0.96 and 0.35, indicating no problem with fit in either tail of the predictive distribution. The individual lack-of-fit values p_val were 0.395, 0.223, 0.126, and 0.074 for observations 1, 3, 4, and 21, respectively. Thus, observation #21 is flagged as a potential outlier at the 10% level, though its Bayesian p-value seems to reflect the "using the data twice" issue mentioned above, since even this large outlier has p-value greater than the traditional 0.05 cutoff. ■

2.5.2 Model averaging

The discussion so far in this section suggests that the line between model adequacy and model choice is often a blurry one. Indeed, some authors (most notably, Gelfand, Dey, and Chang, 1992) have suggested cross-validatory model adequacy checks as an informal alternative to the more formal approach to model choice offered by Bayes factors and their variants. We defer details until Subsection 4.6.

In some problem settings there may be underlying scientific reasons why one model or another should be preferred (for example, a spatial distribution of observed disease cases, reflecting uncertainty as to the precise number of disease "clusters" or sources of contamination). But in other cases, the process of model selection may be somewhat arbitrary in that a number of models may fit equally well and simultaneously provide plausible explanations of the data (for example, a large linear regression problem with 20 possible predictor variables). In such settings, choosing a single model as the "correct" one may actually be suboptimal, since further inference conditional on the chosen model would of necessity ignore the uncertainty in the model selection process itself. This, in turn, would lead to overconfidence in our estimates of the parameters of interest.

Example 2.18 In this example, we revisit the multiple regression setting of Example 2.17, and consider variable selection for the stack loss data. Recall we have three predictors X_1, X_2, and X_3; we now wish to ascertain which of these three are worth including in the model. To do this, we create a new variable π_j, $j = 1, 2, 3$, that gives the probability of including X_j in the model, and redefine the model's mean structure by altering (2.30) to

$$E(Y_i) = \beta_0 + \pi_1\beta_1 z_{i1} + \pi_2\beta_2 z_{i2} + \pi_3\beta_3 z_{i3} ,$$

where z_{ij} is again the centered and scaled version of x_{ij}. By assigning independent $Bernoulli(1/2)$ priors to the π_j and adding these parameters into the sampling order, we obtain posterior summaries of whether these

Variable	mean	sd	MC error	2.5%	median	97.5%
π_1	0.9993	0.02607	4.53E-4	1.0	1.0	1.0
π_2	0.2914	0.4544	0.01986	0.0	0.0	1.0
π_3	0.00476	0.06883	5.818E-4	0.0	0.0	0.0

Table 2.7 *Posterior summaries, covariate inclusion probabilities π_j, stack loss data.*

predictors should be included in the mean structure or not, assuming we are indifferent as to their inclusion before having seen the data.

The WinBUGS code to implement this trick is

BUGS code
```
model
{
# standardize x's and coefficients
for (j in 1 : p) {
   pi[j] ~ dbern(0.5)
   b[j] <- beta[j] / sd(x[ , j ])
   for (i in 1 : N){z[i,j] <- (x[i,j]- mean(x[,j]))/sd(x[,j])}
   }

# statistical model
for (i in 1 : N) {
   Y[i] ~ dnorm(mu[i], tau)
   mu[i] <- beta0 + pi[1]*beta[1]*z[i,1] + pi[2]*beta[2]*z[i,2]
                  + pi[3]*beta[3]*z[i,3]
   }

# priors
beta0 ~ dnorm(0, 0.00001)
for (j in 1 : p) {beta[j] ~ dnorm(0, 0.00001) }
sigma ~ dunif(0.01,100)     # vague Gelman prior for sigma
tau<- 1/(sigma*sigma)
}
```

Adding the data as given in Example 2.17, and the initial values

BUGS code   `list(beta0 = 10, beta = c(0,0,0), sigma = 1, pi = c(1,1,1))`

we ran 50,000 iterations following a 1000-iteration burn-in period. Longer run times are required here since the binary variable inclusion probabilities π are by their nature difficult to estimate. The results for these parameters are shown in Table 2.7. Clearly X_1 has a great deal of importance and is included in almost every sampled model, while the situation is the opposite for X_3.

We hasten to add that the interpretations of many of the parameters

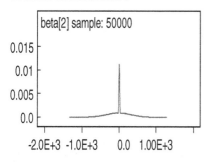 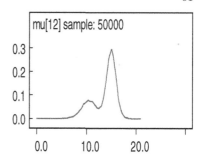

Figure 2.26 *Marginal posterior distributions for β_2 and μ_{12}, stack loss data with stochastic covariate inclusion probability.*

change from the fixed mean structure case in Example 2.17. For example, the posterior of β_2 (left panel of Figure 2.26) is not meaningful, since it reflects the roughly 70%/30% mixture of this parameter's prior (which is a very vague zero-centered normal) and its posterior given its inclusion in the model (which is fairly tightly concentrated near 0). On the other hand, the marginal posteriors of the μ_i remain meaningful, since they represent our best guess as to the mean structure for each parameter *regardless* of whether the β_j are included or not. Thus bimodality is possible for the μ_i, and occurs for the roughly half of them (observations 4, 7–13, and 21) that have influential X_{i2} values. For example, the posterior for μ_{12} (right panel of Figure 2.26) is strongly bimodal, with the heights of the modes reflecting whether β_2 is (minor mode) or is not (major mode) in the model. ∎

The previous example demonstrates an approach to handling the uncertainty regarding model selection; namely, by including this uncertainty in the model, sampling over any unknown parameters, and obtaining the posteriors of quantities whose interpretations do not change across models (e.g., μ_{12}). More broadly, this is an illustration of a Bayesian alternative to model choice known as Bayesian *model averaging.* Suppose $(M_1, \ldots, M_m)$ again index our collection of candidate models and γ is a quantity of interest, assumed to be well-defined for every model. Given a set of prior model probabilities $\{p(M_1), \ldots, p(M_m)\}$, the posterior distribution of γ is

$$p(\gamma|\mathbf{y}) = \sum_{i=1}^{m} p(\gamma|M_i, \mathbf{y})p(M_i|\mathbf{y}) , \qquad (2.34)$$

where $p(\gamma|M_i, \mathbf{y})$ is the posterior for γ under the ith model, and $p(M_i|\mathbf{y})$ is the posterior probability of this model, computable as

$$p(M_i|\mathbf{y}) = \frac{p(\mathbf{y}|M_i)p(M_i)}{\sum_{j=1}^{m} p(\mathbf{y}|M_j)p(M_j)} ,$$

where $p(\mathbf{y}|M_i)$ is the marginal distribution of the data under the ith model,

given previously in equation (2.17). Suppose we rank models based on
the log of their conditional predictive ordinate evaluated at γ, namely,
$\log p(\gamma|M_i, \mathbf{y})$. Using the nonnegativity of the Kullback-Leibler divergence,
Madigan and Raftery (1994) show that for each model $i = 1, \ldots, m$,

$$E\left[\log\left(\sum_{j=1}^{m} p(\gamma|M_j, \mathbf{y})p(M_j|\mathbf{y})\right)\right] \geq E\left[\log p(\gamma|M_i, \mathbf{y})\right] ,$$

where the expectations are with respect to the joint posterior distribution
of γ and the M_is. That is, averaging over all the models results in a better
log predictive score than using any one of the models individually.

Despite this advantage, there are several difficulties in implementing a
solution via equation (2.34). All of them basically stem from the fact that
the number of potential models m is often extremely large. For example, in
our 20-predictor regression setting, $m = 2^{20}$ (since each variable can be in
or out of the model), and this does not count variants of each model arising
from outlier deletion or variable transformation. Specifying the prior probabilities for this number of models is probably infeasible unless some fairly
automatic rule is implemented (say, $p(M_i) = 1/m$ for all i). In addition,
direct computation of the sum in (2.34) may not be possible, forcing us to
look for a smaller summation with virtually the same result. One possibility
(also due to Madigan and Raftery, 1994) is to limit the class of models to
those which meet some minimum level of posterior support, e.g., the class

$$\mathcal{A}' = \left\{ M_i : \frac{\max_j p(M_j|\mathbf{y})}{p(M_i|\mathbf{y})} \leq C \right\} ,$$

say, for $C = 20$. They further eliminate the set $\mathcal{B}$ of models that have
less posterior support than their own submodels, and then perform the
summation in (2.34) over the set $\mathcal{A} = \mathcal{A}'\backslash\mathcal{B}$. Madigan and Raftery (1994)
give a greedy-search algorithm for locating the elements of $\mathcal{A}$, and offer
several examples to show that their procedure loses little of the predictive
power available by using the entire model collection.

An alternative would be to search the entire sample space using MCMC
methods, locating the high-probability models and including them in the
summation. We defer further discussion of these issues until Chapter 4,
after we have fully developed the computational machinery to enable routine calculation of the residuals and predictive summaries in Chapter 3.
We do however point out that all of the model averaging approaches discussed so far assume that the search may be done in "model only" space
(i.e., the parameters $\boldsymbol{\theta}$ can be integrated out analytically prior to beginning
the algorithm). For searches in "model-parameter" space, matters are more
complicated; algorithms for this situation are provided by George and Mc-
Culloch (1993), Carlin and Chib (1995), Chib (1995), Green (1995), Phillips

and Smith (1996), Dellaportas et al. (2002), and Godsill (2001); again see Chapter 4.

2.6 Nonparametric methods

The previous section describes how we might mix over various models M_i in obtaining a posterior or predictive distribution that more accurately reflects the underlying uncertainty in the precise choice of model. In a similar spirit, it might be that no single parametric model $f(y|\theta)$ could fully capture the variability inherent in y's conditional distribution, suggesting that we average over a collection of such densities. For example, for the sake of convenience we might average over normal densities,

$$f(y|\boldsymbol{\mu}, \boldsymbol{\sigma^2}, \mathbf{q}) = \sum_{k=1}^{K} q_k \, \phi(y|\mu_k, \sigma_k^2) \,, \tag{2.35}$$

where $\boldsymbol{\mu} = (\mu_1, \ldots, \mu_K)$, $\boldsymbol{\sigma^2} = (\sigma_1^2, \ldots, \sigma_K^2)$, $\mathbf{q} = (q_1, \ldots, q_K)$, and $\phi(.|.,.)$ denotes the normal density function. Equation (2.35) is usually referred to as a *mixture* density, which has mixing probabilities q_k satisfying $\sum_{k=1}^{K} q_k = 1$. Mixture models are attractive since they can accommodate an arbitrarily large range of model anomalies (multiple modes, heavy tails, and so on) simply by increasing the number of components in the mixture. They are also quite natural if the observed population is more realistically thought of as a combination of several distinct subpopulations (say, a collection of ethnic groups). An excellent reference on the Bayesian analysis of models of this type is the book by Titterington, Makov, and Smith (1985).

Unfortunately, inference using mixture distributions like (2.35) is often difficult. Because increasing one of the σ_k^2 and increasing the number of mixture components K can have a similar effect on the shape of the mixture, the parameters of such models are often not fully identified by the data. This forces the adoption of certain restrictions, such as requiring $\sigma_k^2 = \sigma^2$ for all k, or fixing the dimension K ahead of time. These restrictions in turn lead to difficulties in computing the relevant posterior and predictive densities; in this regard see West (1992) and Diebolt and Robert (1994) for a list of potential problem areas and associated remedies.

Example 2.19 Volume II of the `WinBUGS` Examples manual contains a mixture model called `eyes`. The data consist of measurements on the peak sensitivity wavelengths for individual microspectrophotometric records on a small collection of monkey's eyes. The observations from one particular monkey (see the `data` step below) appear bimodal, suggesting the data may arise as a two-component normal mixture model with common variance, i.e.,

$$Y_i \quad \sim \quad N(\lambda_{T_i}, \sigma^2), \quad i = 1, \ldots, N = 48 \,,$$

where T_i (1 or 2) gives the group identifier of the ith observation, and we assume $T_i = a_i + 1$ where $a_i \sim Bernoulli(p)$. A slight reparametrization to $\lambda_2 = \lambda_1 + \theta$ for $\theta > 0$ is needed to better identify the components of the mixture. We also sort the data into ascending order, and fix $T_1 = 1$ and $T_N = 2$, so that the labels on the two mixture components will not keep switching back and forth as the MCMC sampler progresses.

The WinBUGS code is:

BUGS code
```
model
{
for( i in 1 : N ) {
   y[i] ~ dnorm(mu[i], tau)
   mu[i] <- lambda[T[i]]
   T[i] ~ dcat(P[])
   }
P[1:2] ~ ddirch(alpha[])
theta ~ dnorm(0.0, 1.0E-6) I(0.0, )
lambda[2] <- lambda[1] + theta
lambda[1] ~ dnorm(0.0, 1.0E-6)

Ystar1 ~ dnorm(lambda[1], tau)
Ystar2 ~ dnorm(lambda[2], tau)
Tstar ~ dcat(P[]);   astar <- Tstar-1
Ystar <- Ystar1*(1-astar) + Ystar2*astar

sigma ~ dunif(0.01,100)   #   vague Gelman prior for sigma
tau<- 1/(sigma*sigma)
}
```

Note the nested subscripting in the definition of mu[i], and the use of truncation in the vague normal prior for θ, in order to ensure its positivity. The code also uses dcat and ddirch (instead of dbern and dbeta) in order to place the distribution directly on the T_i. We also add a predictive step for T^*, the group membership of a future observation, and Y^*, the value of that future observation.

The data are

BUGS code
```
list(y = c(529.0, 530.0, 532.0, 533.1, 533.4,
           533.6, 533.7, 534.1, 534.8, 535.3,
           535.4, 535.9, 536.1, 536.3, 536.4,
           536.6, 537.0, 537.4, 537.5, 538.3,
           538.5, 538.6, 539.4, 539.6, 540.4,
           540.8, 542.0, 542.8, 543.0, 543.5,
           543.8, 543.9, 545.3, 546.2, 548.8,
           548.7, 548.9, 549.0, 549.4, 549.9,
           550.6, 551.2, 551.4, 551.5, 551.6,
           552.8, 552.9,553.2), N = 48, alpha = c(1, 1),
     T = c(1, NA, NA, NA, NA, NA, NA, NA, NA, NA,
```

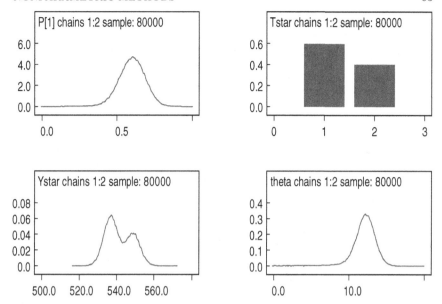

Figure 2.27 *Marginal posterior distributions for P, T^*, Y^*, and θ, eyes data with two-component mixture model.*

```
         NA, NA, NA, NA, NA, NA, NA, NA, NA, NA,
         NA, NA, NA, NA, NA, NA, NA, NA, NA, NA,
         NA, NA, NA, NA, NA, NA, NA, NA, NA, NA,
         NA, NA, NA, NA, NA, NA, NA, 2))
```

while two sets of `inits` are

BUGS code   `list(lambda = c(535, NA), theta = 5, sigma = 1.0, P=c(0.5, 0.5))`
               `list(lambda = c(535, NA), theta =15, sigma = 5.0, P=c(0.2, 0.8))`

Figure 2.27 shows the estimated posterior densities of p, T^*, Y^*, and θ based on two post-burn-in chains of 40,000 iterations each. Since p is `P[1]` in the code, the fact that its posterior is centered slightly above 0.5 suggests that $T = 1$ is the dominant component in the mixture. This is borne out by the posteriors of T^* and Y^*, the latter of which reflects the data's bimodality.

Finally, the posterior for θ enables a precise way to estimate the difference in the means of the two mixture components. Note that it is well separated from 0, confirming that the truncation of this parameter is really only part of avoiding the "label switching" problem, not a real limitation on the MCMC algorithm. However, we remark that, had the data not suggested such a clear separation of the two modes, or had we used a higher dimensional (e.g., three mode) likelihood, or mode-specific precisions τ_i, or seen any other change that *weakens the identifiability of the parameters by the data*, the MCMC algorithm might well have struggled. Mixture models are notorious for their complicated and weak identifiability patterns and, as a result, are often difficult nuts for MCMC algorithms to crack. ∎

A potentially appealing alternative to discrete finite mixtures like (2.35) would be *continuous* mixtures of the form

$$f(y|g) = \int f(y|\boldsymbol{\theta})g(\boldsymbol{\theta})d\boldsymbol{\theta} .$$

However, this creates the problem of choosing an appropriate mixing distribution g. Of course we could select a member of some parametric family $g(\boldsymbol{\theta}|\eta)$, but the appropriate choice might not be apparent before seeing the data. As such, we might wish to specify g *nonparametrically*, in effect, allowing the data to suggest the appropriate mixing distribution.

The most popular method for doing this is via the *Dirichlet process* (DP) prior, introduced by Ferguson (1973). This prior is a distribution on the set of *all* probability distributions $\mathcal{P}$. Suppose we are given a base distribution G_0 and a scalar precision parameter ν. Then, for any finite partition of the parameter space $\{B_1, \ldots, B_m\}$, the DP prior is a random probability measure $G \in \mathcal{P}$ that assigns a Dirichlet distribution with m-dimensional parameter vector $(\nu G_0(B_1), \ldots, \nu G_0(B_m))$ to the vector of probabilities $(G(B_1), \ldots, G(B_m))$. (The Dirichlet distribution is a multivariate generalization of the beta distribution; see Appendix A.) A surprising consequence of this formulation is that the DP assigns probability 1 to the subset of *discrete* probability measures. We use the standard notation $G \sim DP(\nu G_0)$ to denote this Dirichlet process prior for G.

The use of the Dirichlet process in the context of mixture distributions essentially began with Antoniak (1974). Here, we work with the set of probability distributions $\mathcal{F}$ for y having densities of the form

$$f(y|G) = \int f(y|\boldsymbol{\theta})dG(\boldsymbol{\theta}) ,$$

where $G \sim DP(\nu G_0)$. Hence $\mathcal{F}$ is a nonparametric mixture family, with $G(\boldsymbol{\theta})$ playing the role of the conditional distribution of $\boldsymbol{\theta}$ given G. Note that we may partition $\boldsymbol{\theta}$ as $(\boldsymbol{\theta}_1, \boldsymbol{\theta}_2)$ and DP mix on only the subvector $\boldsymbol{\theta}_1$, placing a traditional parametric prior on $\boldsymbol{\theta}_2$; many authors refer to such models as *semiparametric*.

In practice, we will usually have independent data values $y_i, i = 1, \ldots, n$. Inserting a latent parameter $\boldsymbol{\theta}_i$ for each y_i, we obtain the hierarchical model

$$
\begin{aligned}
Y_i|\boldsymbol{\theta}_i &\overset{ind}{\sim} f_i(y_i|\boldsymbol{\theta}_i) \\
\boldsymbol{\theta}_i|G &\overset{iid}{\sim} G(\boldsymbol{\theta}) \\
G|G_0, \nu &\sim DP(\nu G_0)
\end{aligned}
\tag{2.36}
$$

where here we DP mix over the entire $\boldsymbol{\theta}$ vector.

Computational difficulties precluded the widespread use of DP mixture models until recently, when a series of papers (notably Escobar, 1994, and Escobar and West, 1992, 1995) showed how MCMC methods could be used to obtain the necessary posterior and predictive distributions. These methods avoid the infinite dimension of G by analytically marginalizing over this dimension in the model, obtaining the likelihood

$$\left[\prod_{i=1}^{n} f_i(y_i|\boldsymbol{\theta}_i)\right] p(\boldsymbol{\theta}_1, \ldots, \boldsymbol{\theta}_n|G_0, \nu) . \qquad (2.37)$$

Note that this marginalization implies that the $\boldsymbol{\theta}_i$ are no longer conditionally independent as in (2.36). Specifying a prior distribution on ν and the parameters of the base distribution G_0 completes the Bayesian model specification. The basic computational algorithm requires that G_0 be conjugate with the f_i, but a recent extension due to MacEachern and Müller (1994) eliminates this restriction (though, of course, not without additional algorithmic complexity). Mukhopadhyay and Gelfand (1997) combine DP mixtures with overdispersed exponential family models (Gelfand and Dalal, 1990), producing an extremely broad extension of the generalized linear model setting. (See also Gelfand and Kottas (2002) for a recent overview of modern computational approaches for Bayesian DP mixture modeling.)

DP mixing is a relatively recent addition to the applied Bayesian toolkit, and many practical issues remain to be addressed. For example, note that large values of the DP precision parameter ν favor adoption of the base measure G_0 as the posterior for the $\boldsymbol{\theta}_i$, while small values will produce the empirical cdf of the data as this posterior. As such, the precise choice of the prior distribution for ν (and the sensitivity of the results to this decision) is still an area of substantial debate. Other subjects of ongoing research include the proper selection of computational algorithm and associated convergence checks, and the suitability and feasibility of the models in real data settings. Finally, since WinBUGS is essentially a parametric tool, non-parametric Bayes MCMC algorithms must still often be coded "by hand" in C, Fortran, or some other fairly low-level language. Identifiability is always a concern, and MCMC convergence is often difficult to obtain. Still, the substantial current research activity in nonparametric Bayesian methods and data analysis suggests a bright future in practice. The paper by Gelfand (1998) offers a review of various semiparametric models, including DP mixing, Polya trees and mixtures of Polya trees (c.f. Hanson and Johnson, 2002; Hanson, 2006) and gamma and beta process modeling. Basu and Chib (2003) show how to compute Bayes factors for DP mixture models.

2.7 Exercises

1. Let $Y_1, Y_2 \ldots$ be a sequence of random variables (not necessarily independent) taking on the values 0 and 1. Let $S_n = Y_1 + \ldots Y_n$.

   (a) Represent $P(Y_{n+1} = 1 | S_n = s)$ in terms of the joint distribution of the Y_i.

   (b) Further refine the representation when all permutations of the Y_i such that $S_n = s$ have equal probability (i.e., the Y_i are exchangeable).

   (c) What is the conditional probability if the Y_i are independent?

2. Confirm expression (2.3) for the conditional posterior distribution of θ given σ^2 in the conjugate normal model, and show that the precision in this posterior is the sum of the precisions in the likelihood and the normal prior, where *precision* is defined to be the reciprocal of the variance.

3. Confirm expression (2.11) for the conditional posterior distribution of σ^2 given θ in the conjugate normal model. If your random number generator could produce only χ^2 random variates, how could you obtain samples from (2.11)?

4. Let θ be the height of the tallest individual currently residing in the city where you live. Elicit your own prior density for θ by

   (a) a histogram approach

   (b) matching an appropriate distributional form from Appendix A.

   What difficulties did you encounter under each approach?

5. For each of the following densities from Appendix A, provide a conjugate prior distribution for the unknown parameter(s), if one exists:

   (a) $X \sim Bin(n, \theta)$, n known

   (b) $X \sim NegBin(r, \theta)$, r known

   (c) $\mathbf{X} \sim Mult(n, \boldsymbol{\theta})$, n known

   (d) $X \sim G(\alpha, \beta)$, α known

   (e) $X \sim G(\alpha, \beta)$, β known

   (f) $\mathbf{X} \sim N_k(\boldsymbol{\theta}, \boldsymbol{\Sigma})$, $\boldsymbol{\Sigma}$ known

   (g) $\mathbf{X} \sim N_k(\boldsymbol{\theta}, \boldsymbol{\Sigma})$, $\boldsymbol{\theta}$ known

6. For the following densities from Appendix A, state whether each represents a location family, scale family, location-scale family, or none of these. For any of the first three situations, suggest a simple noninformative prior for the unknown parameter(s):

   (a) $X \sim DE(1, \sigma^2)$

   (b) $X \sim Unif(\theta - a, \theta + a)$

   (c) $X \sim Logistic(\mu, 1)$

(d) $X \sim G(\alpha, \beta)$, α known

(e) $X \sim G(\alpha, \beta)$, β known

7. Let θ be a univariate parameter of interest, and let $\gamma = g(\theta)$ be a 1-1 transformation. Use (2.12) and (2.13) to show that (2.14) holds, i.e., that the Jeffreys prior is invariant under reparametrization. (*Hint:* What is the expectation of the so-called *score statistic*, $d \log f(\mathbf{x}|\theta)/d\theta$?)

8. Suppose that $f(x|\theta, \sigma)$ is normal with mean θ and standard deviation σ.

(a) Suppose that σ is known, and show that the Jeffreys prior (2.12) for θ is given by

$$p(\theta) = 1, \ \theta \in \Re .$$

(b) Next, suppose that θ is known, and show that the Jeffreys prior for σ is given by

$$p(\sigma) = \frac{1}{\sigma}, \ \sigma > 0 .$$

(c) Finally, assume that θ and σ are both unknown, and show that the bivariate Jeffreys prior (2.15) is given by

$$p(\theta, \sigma) = \frac{1}{\sigma^2}, \ \theta \in \Re, \ \sigma > 0 ,$$

a slightly different form than that obtained by simply multiplying the two individual Jeffreys priors obtained above.

9. Show that the Jeffreys prior based on the binomial likelihood $f(x|\theta) = \binom{n}{x}\theta^x (1 - \theta)^{n-x}$ is given by the $Beta(.5, .5)$ distribution.

10. Suppose that in the situation of Example 2.3 we adopt the noninformative prior $p(\theta, \sigma) = \frac{1}{\sigma}$, $\theta \in \Re$, $\sigma > 0$ (i.e., $p(\theta, \sigma^2) = \frac{1}{\sigma^2}$, $\theta \in \Re$, $\sigma^2 > 0$).

(a) Show that the marginal posterior for $t = \sqrt{n}(\theta - \bar{y})/s$, where $s^2 = \sum_{i=1}^{n}(y_i - \bar{y})^2/(n-1)$, is proportional to

$$[1 + t^2/(n - 1)]^{-n/2} ,$$

a Student's t distribution with $(n-1)$ degrees of freedom.

(b) Show that the marginal posterior for σ^2 is proportional to

$$(\sigma^2)^{-[(n-1)/2+1]} \exp\left[-\frac{1}{2\sigma^2} \sum_{i=1}^{n}(y_i - \bar{y})^2\right] ,$$

an $IG\left(\frac{n-1}{2}, \frac{2}{\sum_{i=1}^{n}(y_i - \bar{y})^2}\right)$ density.

11. In a pediatric clinical study carried out to see how effective aspirin is in reducing temperature, $n = 12$ five-year-old children suffering from influenza had their temperature taken immediately before (X) and 1 hour after (Y) administration of aspirin. The data are in Table 2.8, and

Patient	Before (X)	After (Y)	Patient	Before (X)	After (Y)
1	102.4	99.6	7	102.5	101.0
2	103.2	100.1	8	103.1	100.1
3	101.9	100.2	9	102.8	100.7
4	103.0	101.1	10	102.3	101.1
5	101.2	99.8	11	101.9	101.3
6	100.7	100.2	12	101.4	100.2

Table 2.8 *Temperatures of $n = 12$ children before and after taking aspirin.*

also online at www.biostat.umn.edu/~brad/data/aspirin_data.txt. Let $Z = X - Y$, the observed reduction in temperature. The study investigators suspect that the reduction in temperature Z may be related to the initial temperature X. Thus, they would like to fit the simple linear regression (SLR) model,

$$Z_i = \beta_0 + \beta_1(x_i - \bar{x}) + \epsilon_i, \ \epsilon_i \overset{iid}{\sim} N(0, \sigma^2), \ i = 1, \ldots, n.$$

(a) Use R to plot the raw data. Does the linear model appear plausible?

(b) Following the model of Example 2.10, fit the SLR model in WinBUGS, using vague priors for all parameters. Find a posterior density estimate and a 95% credible interval for the initial temperature effect β_1. Do your results confirm the investigators' suspicion? Also check the effective model size p_D and DIC score for the fitted SLR.

(c) Suppose a 13th patient arrives with an initial temperature of 100.0 degrees. Find an estimate of the predictive density of Z_{13}, the predicted reduction in temperature for this patient, and a point estimate for $P(Z_{13} > 0|Z_1, \ldots, Z_{12})$, the predictive probability that this new patient's temperature is reduced by aspirin.

(d) Follow the model of Example 2.16 to investigate approximate residual and CPO values for your model (do not bother with exact cross-validatory results here). Are any data points outliers in any sense? Also check effective model size p_D and DIC score for the fitted SLR.

(e) Suppose we attempt to improve model fit by adding a quadratic term, $\beta_2(x_i - \bar{x})^2$, to our model. Make this change and recheck p_D and DIC, as well as the posterior of β_2. Is this model enhancement well-justified by the data? Also check this change's impact on your answer to part (c) above. Is either Z_{13} prediction well-justified by the data?

12. Consider www.biostat.umn.edu/~brad/data/land_data.txt, for which a few records are shown in Table 2.9. Here, Y represents log-transformed

Property	Y	X_1	X_2	X_3
1	6.907755	0.881749	0.9475592	0.9177695
2	6.55108	0.890181	0.9524228	0.9268535
3	6.55108	0.872585	0.9460401	0.9245727
4	6.684612	0.8628099	0.9418296	0.9207009
...	...	...	...	...
389	5.298317	0.9350303	0.8371726	0.9082236

Table 2.9 *Land values near Chicago.*

land values for $n = 389$ pieces of real estate in and around Chicago, Illinois, while the covariates X_1, X_2, and X_3 are distances from each property to Lake Michigan, Midway airport, and O'Hare airport, respectively. We wish to fit the normal errors linear regression model,

$$Y_i = \beta_1 + X_{1i}\beta_2 + X_{2i}\beta_3 + X_{3i}\beta_4 + \epsilon_i, \quad \epsilon_i \sim Normal(0, \sigma^2),$$

for $i = 1, \ldots, n$, where the β_k are unknown regression coefficients, and σ^2 is the error variance.

(a) Use WinBUGS to fit this model to the data using vague priors – say, independent flat or $N(0, 10^{-8})$ priors for the β_k, and an $IG(3, 1/2\mu)$ for σ^2 where μ is a best guess for this parameter (e.g., $\hat{\sigma}^2$ from OLS regression). Run multiple chains to help assess MCMC convergence. Interpret the resulting posterior estimates of the β_k. Proximities to which Chicago landmarks are desirable from the perspective of real estate value? Which are detrimental? (*Hint:* An $IG(3, 1/2\mu)$ prior for σ^2 is equivalent to a $G(3, 1/2\mu)$ prior for $\tau = 1/\sigma^2$.)

(b) Next, suppose we wish to use the joint improper prior $p(\boldsymbol{\beta}, \sigma^2) \propto \frac{1}{\sigma^2}$, i.e., the product of a flat prior on $\boldsymbol{\beta}$ and the limit of an IG prior on σ^2. Such a prior is not possible in WinBUGS, but a closed form solution can be obtained by generalizing the results of the previous exercise to the regression setting (i.e., replacing θ by $\mathbf{x}'\boldsymbol{\beta}$). That is, derive the conditional posterior distribution $p(\boldsymbol{\beta}, |\sigma^2, \mathbf{y})$ and the marginal posterior $p(\sigma^2|\mathbf{y})$, and then write an R program to sample directly from the joint posterior. Compare your results to those above, and note any significant discrepancies.

(c) Compute a Bayesian p-value for the fit of the model in part (a), following the approach of Example 2.17.

13. The classic Behrens-Fisher problem is that of estimating normal means from independent populations with unknown (and thus possibly unequal) variances. Here we consider this problem in the context of an

Group	n	$\bar{y}$	$\hat{\sigma}$
Exposed (A)	36	1.173	0.20
Control (B)	32	1.013	0.24

Table 2.10 *Calcium flow summary statistics, chicken data.*

experiment on the effects of magnetic fields on the flow of calcium out of chicken brains. The experiment comprised measurements on two groups of chickens, an exposed group (A) and a control group (B), with each chicken measured once. Summary statistics are shown in Table 2.10.

(a) Use classical methods to test the hypothesis $H_0 : \mu_A = \mu_B$ versus $H_a : \mu_A \neq \mu_B$.

(b) Compare your answer in part (a) to those of a Bayesian analysis under the improper joint priors $p(\mu_A, \sigma_A^2) \propto 1/\sigma_A^2$ and $p(\mu_B, \sigma_B^2) \propto 1/\sigma_B^2$. (*Hint:* Composition sampling in R may be easier and better here than Gibbs sampling in WinBUGS; see part (b) of the previous problem.)

(c) Use DIC to compare the models under the two competing hypotheses $H_0 : \mu_A = \mu_B, \ \sigma_A^2 = \sigma_B^2$ and $H_a : \mu_A \neq \mu_B, \ \sigma_A^2 \neq \sigma_B^2$, and again comment on how these results compare to those in the previous two parts.

14. Refer to Example 2.10. Compare p_D and DIC for the two prior distributions considered for the error terms, namely, the $Gamma(.1, .1)$ prior on the error precision τ, and the $Unif(.01, 100)$ prior on the error standard deviation σ. Which prior emerges as "best"? In your opinion, is it sensible to use a model choice criterion like DIC to choose between *prior* distributions?

15. Suppose that $Y|\theta \sim G(1, \theta)$ (i.e., the *exponential* distribution with mean θ), and that $\theta \sim IG(\alpha, \beta)$.

(a) Find the posterior distribution of θ.

(b) Find the posterior mean and variance of θ.

(c) Find the posterior mode of θ.

(d) Write down two integral equations that could be solved to find the 95% equal-tail credible interval for θ.

16. Suppose $X_1, \ldots, X_n \overset{iid}{\sim} NegBin(r, \theta)$, r known, and that θ is assigned a $Beta(\alpha, \beta)$ prior distribution.

(a) Find the posterior distribution of θ.

(b) Suppose further that $r = 5$, $n = 10$, and we observe $\sum_{i=1}^{n} x_i = 70$. Of the two hypotheses $H_1 : \theta \leq .5$ and $H_2 : \theta > .5$, which has greater posterior probability under the mildly informative $Beta(2, 2)$ prior?

(c) What is the Bayes factor in favor of H_1? Does it suggest strong evidence in favor of this hypothesis?

17. Consider the two-stage binomial problem wherein $\theta \sim G$, and $X_i|\theta \overset{iid}{\sim}$ $Bernoulli(\theta)$, $i = 1, \ldots, n$.

(a) Find the joint marginal distribution of $\mathbf{X}_n = (X_1, \ldots, X_n)$.

(b) Are the components independent? In any case, describe features of the marginal distribution.

(c) Let $S_n = X_1 + \cdots + X_n$. Under what conditions does S_n have a binomial distribution? Describe features of its distribution.

(d) If $dG(\theta) = d\theta$, find the distribution of S_n and discuss its features.

18. For the Dirichlet process prior described in Section 2.6, use the properties of the Dirichlet distribution to show that for any measurable set B,

(a) $E[G(B)] = G_0(B)$

(b) $Var[G(B)] = G_0(B)(1 - G_0(B))/(\nu + 1)$

What might be an appropriate choice of vague prior distribution on ν?

19. The data in Table 2.11 come from a 1980 court case (*Reynolds v. Sheet Metal Workers*) involving alleged applicant discrimination. Perform a Bayesian analysis of these data. Is there evidence of discrimination?

	Selected	Rejected	Total
Black	14	30	44
White	41	39	80
Total	55	69	124

Table 2.11 *Summary table from* Reynolds v. Sheet Metal Workers.

20. The following are increased hours of sleep for 10 patients treated with soporific B compared with soporific A (Cushny and Peebles, presented by Fisher in *Statistical Methods for Research Workers*):

$$1.2, 2.4, 1.3, 1.3, 0.0, 1.0, 1.8, 0.8, 4.6, 1.4$$

Perform a Bayesian analysis of this data and draw conclusions, assuming each component of the likelihood to be

(a) normal

(b) Student's t with 3 degrees of freedom (t_3)

(c) Cauchy (t_1)

(d) Bernoulli (i.e., positive with some probability θ), dealing with the 0 somehow (for example, by ignoring it).

Don't forget to comment on relevant issues in model assessment and model selection.

21. Refer to Example 2.15. Repeat the analysis of the MAC treatment data assuming a model having

(a) three sets of random effects r_i, W_i, and β_{1i}, adding clinic-specific treatment effect parameters.

(b) two sets of random effects W_i and β_{1i}, but setting $r_1 = 1$ for all i.

(c) one set of random effects W_i, but setting $r_1 = 1$, $\beta_{1i} = \beta_1$ for all i.

(d) no random effects, setting $r_1 = 1$, $\beta_{1i} = \beta_1$, and $W_i = 0$ for all i.

Compare convergence behavior, boxplots of the random effects (where applicable), and posterior distribution(s) for the drug effect(s) across models, noting any interesting discrepancies. Also compute the fit $(\overline{D})$, effective size (p_D), and overall model choice score (DIC) across models. Which appears the most sensible for these data?

22. Refer to Example 2.16. Repeat the exact "leave one out" analysis of the stack loss data using the BRugs language. Here we would avoid the "trickery" of the fully WinBUGS solution given in the example, and instead simply format the data using R to leave out the ith observation. We then call BUGS from R for this dataset, and loop through the $n = 21$ leave-one-out datasets within R. (*Hint:* The BRugs manual available online at

cran.r − project.org/web/packages/BRugs/BRugs.pdf

will likely be of great assistance.)

23. Refer to Example 2.18.

(a) Repeat the analysis of the stack loss data assuming a slightly generalized model where we let

$$\pi_j \overset{iid}{\sim} Bernoulli(\theta) \, ,$$

where $\theta \sim Unif(0, 1)$. That is, allow prior variable inclusion probabilities other than $1/2$. Do you find any substantial changes to the posteriors for the π_j or the μ_j?

(b) Repeat your previous solution, but now using dcat and ddirch instead of dbern and dbeta. Is your model more general than that in part (a), in any sense?

(c) Now include prior dependence among the selection probabilities – say, by using a trivariate normal prior on the logit-transformed vector of π_j. Now do your answers change at all?

## CHAPTER 3

Bayesian computation

3.1 Introduction

As discussed in Chapter 2, determination of posterior distributions comes down to the evaluation of complex, often high-dimensional integrals (i.e., the denominator of expression (2.1)). In addition, posterior summarization often involves computing moments or quantiles, which leads to more integration (i.e., now integrating the numerator of expression (2.1)). A solution to an important special case of the problem was provided in Subsection 2.2.2, which described how *conjugate* prior forms may often be found that enable at least partial analytic evaluation of these integrals. Still, in all but the simplest model settings (typically linear models with normal likelihoods), some intractable integrations remain. In the last chapter, we were content to let WinBUGS handle implementation of the necessary computational methods. In this chapter, we present a full description of the most commonly used such methods, for readers whose MCMC needs may exceed the capabilities of WinBUGS and other standard software packages.

The earliest solution to this problem involved using asymptotic methods to obtain analytic approximations to the posterior density. The simplest such result is to use a normal approximation to the posterior, essentially a Bayesian version of the Central Limit Theorem. More complicated asymptotic techniques, such as Laplace's method (Tierney and Kadane, 1986), enable more accurate, possibly asymmetric posterior approximations. Asymptotic methods are the subject of Section 3.2.

When approximate methods are intractable or result in insufficient accuracy, we must resort to numerical integration. For many years, traditional methods such as Gaussian quadrature were the methods of choice for this purpose; see Thisted (1988, Chapter 5) or Givens and Hoeting (2005, also Chapter 5) for a general discussion of these methods. But their application was limited to models of low dimension (say, up to 10 parameters), because the necessary formulae become intractable for larger models. They also suffer from the so-called "curse of dimensionality," namely, the fact that the number of function evaluations required to maintain a given accuracy level depends exponentially on the number of dimensions.

An alternative is provided by the expectation-maximization (EM) algorithm of Dempster, Laird, and Rubin (1977); again see Givens and Hoeting (2005, Chapter 4) or Subsection 5.2.2 below. Originally developed as a tool for finding MLEs in missing data problems, it applies equally well to finding the mode of a posterior distribution in many Bayesian settings. Indeed, much of the subsequent work broadening the scope of the algorithm to provide standard error estimates and other relevant summary information was motivated by Bayesian applications; see e.g. Meng and Rubin (1991, 1992, 1993), and Meng and Van Dyk (1997).

While the EM algorithm has seen enormous practical application, it is a fairly slow algorithm, and remains primarily geared toward finding the posterior mode, rather than estimation of the whole posterior distribution. As a result, most applied Bayesians have turned to Monte Carlo integration methods, which provide more complete information and are comparatively easy to program, even for very high-dimensional models. These methods include traditional noniterative methods such as importance sampling (Geweke, 1989) and simple rejection sampling, and are the subject of Section 3.3. Even more powerful are the iterative Monte Carlo methods, including the Metropolis-Hastings algorithm (Hastings, 1970) and the Gibbs sampler (Geman and Geman, 1984; Gelfand and Smith, 1990). These methods produce a Markov chain, the output of which corresponds to a (correlated) sample from the joint posterior distribution. These methods, already used but only cursorily explained in Chapter 2, are presented along with methods for assessing their convergence in Section 3.4.

While our discussion related to MCMC methods is by far the longest in this chapter, space does not permit a full description of this enormous, rapidly growing area. A wealth of additional material may be found in the burgeoning collection of textbooks on the subject. Among Bayesian statisticians, perhaps the most celebrated of these is the book by Gilks et al. (1996). While this book appeared more than a decade ago (making it relatively old for a computing book), it was so far ahead of its time when published that it remains an excellent source of good, practical advice on MCMC and its implementation. Chen et al. (2000) and Gamerman and Lopes (2006) are somewhat more recent books that explicate MCMC in the specific context of Bayesian modeling and data analysis, similar to our own focus. By contrast, the books by Robert and Casella (1999) and Liu (2001) are more explicitly focused on computing itself, and are somewhat broader in that they attempt to cover all Monte Carlo methods (not just MCMC). Of these two, Robert and Casella (1999) is somewhat lower-level, offering nice reviews of random variate generation, Monte Carlo optimization, and algorithms for missing data models. By contrast, Liu (2001) is more of a reference book, and contains a lengthy discussion of *sequential Monte Carlo* methods (a generalization of importance sampling in which the importance weights are calculated recursively), as well as a wealth of specialized algorithms useful in particular application areas, such as DNA sequencing and image analysis. In general, the goal of each of these specialized algorithms is more efficient sampling from a particular probability distribution than would be possible using the extremely general (but occasionally slow) Metropolis-Hastings method.

Before proceeding on with our presentation, two very recent Bayesian computing books by Albert (2007) and Marin and Robert (2007) are worthy of special mention. Both of these books adopt the R language as their sole computing platform; indeed, both include R tutorials in their first chapters. As already discussed at some length, statisticians (especially academics and students) have grown very fond of R in the last few years, thanks to its free availability, its ability to combine programmability with advanced graphics and statistical package features, and the ready availability of built-in and user-contributed extensions from the Comprehensive R Archive Network (CRAN). The books reflect the distinct and well-established tastes of their authors. Albert (2007) aims at North American first-year graduate or perhaps advanced undergraduate students, building carefully from first principles and including an R package, LearnBayes, for implementing a great many standard models. The writing is in a clear, teach-by-example style, though this of necessity limits the book's coverage somewhat; the author kindly mentions the book you are reading as one of those that could be used as a methodological companion to his. By contrast, the level of formality and mathematical rigor in Marin and Robert (2007) is at least that of its fairly mature stated audience of second-year master's students. Demanding homework problems, including both mathematical derivations and computer programming problems (but only occasionally a real dataset) are interspersed throughout the presentation.

We have great admiration for what the authors of both of these books are trying to do, and have ourselves at times considered sole use of R as our MCMC engine (for example, when we're struggling to decipher WinBUGS' often unhelpful error messages). But ultimately we could not turn our backs on the power and easy-to-learn syntax offered by WinBUGS. And while it does still crash too often and its error messages can be cryptic, to recreate all its power in R requires either a willingness to use other people's functions (as Albert suggests), or write them all yourself (as Marin and Robert suggest). Neither approach seems likely to be successful with the diverse audience for good applied Bayesian work that we envision, an audience that needs a platform that will work over a very broad class of models with no need for sophisticated coding. As such, we will continue to recommend use of both WinBUGS and R (and perhaps BRugs), though this chapter will cover several advanced or specialized tools (e.g., slice sampling, Langevin-Hastings algorithms, blocking via structured MCMC, etc.) whose current availability in WinBUGS is either limited or nonexistent. No doubt many of these tools will emerge in `OpenBUGS` (the still-developing open-source version of `WinBUGS`) over time.

3.2 Asymptotic methods

3.2.1 Normal approximation

When the number of datapoints is fairly large, the likelihood will be quite peaked, and small changes in the prior will have little effect on the resulting posterior distribution. Some authors refer to this as "stable estimation," with the likelihood concentrated in a small region $\Omega \subset \Theta$ where $\pi(\theta)$ is nearly constant. In this situation, the following theorem shows that $p(\theta|\mathbf{x})$ will be approximately normal.

Theorem 3.1 Suppose that $X_1, \ldots, X_n \overset{iid}{\sim} f_i(x_i|\boldsymbol{\theta})$, and thus $f(\mathbf{x}|\boldsymbol{\theta}) = \prod_{i=1}^n f_i(x_i|\boldsymbol{\theta})$. Suppose the prior $\pi(\boldsymbol{\theta})$ and $f(\mathbf{x}|\boldsymbol{\theta})$ are positive and twice differentiable near $\hat{\boldsymbol{\theta}}^\pi$, the posterior mode (or "generalized MLE") of $\boldsymbol{\theta}$, assumed to exist. Then under suitable regularity conditions, the posterior distribution $p(\boldsymbol{\theta}|\mathbf{x})$ for large n can be approximated by a normal distribution having mean equal to the posterior mode, and covariance matrix equal to minus the inverse Hessian (second derivative matrix) of the log posterior evaluated at the mode. This matrix is sometimes notated as $[I^\pi(\mathbf{x})]^{-1}$, since it is the "generalized" observed Fisher information matrix for $\boldsymbol{\theta}$. More specifically,

$$I_{ij}^\pi(\mathbf{x}) = - \left[\frac{\partial^2}{\partial \theta_i \partial \theta_j} \log\left(f(\mathbf{x}|\boldsymbol{\theta})\pi(\boldsymbol{\theta})\right) \right]_{\boldsymbol{\theta}=\hat{\boldsymbol{\theta}}^\pi} .$$

Other forms of the normal approximation are occasionally used. For instance, if the prior is reasonably flat, we might ignore it in the above calculations. This in effect replaces the posterior mode $\hat{\theta}^\pi$ by the MLE $\hat{\theta}$, and the generalized observed Fisher information matrix by the usual observed Fisher information matrix, $\widehat{I}(\mathbf{x})$, where

$$\begin{aligned} \widehat{I}_{ij}(x) &= -\left[\frac{\partial^2}{\partial \theta_i \partial \theta_j} \log f(\mathbf{x}|\boldsymbol{\theta}) \right]_{\boldsymbol{\theta}=\widehat{\boldsymbol{\theta}}} \\ &= -\sum_{l=1}^n \left[\frac{\partial^2}{\partial \theta_i \partial \theta_j} \log f_l(x_l|\boldsymbol{\theta}) \right]_{\boldsymbol{\theta}=\widehat{\boldsymbol{\theta}}} . \end{aligned}$$

Alternatively, we might replace the posterior mode by the posterior mean $\boldsymbol{\mu}^\pi(\mathbf{x})$, and replace the variance estimate based on observed Fisher information with the posterior covariance matrix $V^\pi(\mathbf{x})$), or even the *expected* Fisher information matrix $I(\hat{\boldsymbol{\theta}})$, which in the case of i.i.d. sampling is given by

$$I_{ij}(\boldsymbol{\theta}) = -nE_{X_1|\boldsymbol{\theta}} \left[\frac{\partial^2}{\partial \theta_i \partial \theta_j} \log f(X_1|\boldsymbol{\theta}) \right] .$$

The first options listed above are the most commonly used; however, since adding the prior distribution $\pi(\boldsymbol{\theta})$ complicates matters little, and the remaining options (e.g., computing a mean instead of a mode) involve integrations – the very act the normal approximation was designed to avoid.

Just as the well-known Central Limit Theorem enables a broad range of frequentist inference by showing that the sampling distribution of the sample mean $\bar{X}$ is asymptotically normal, Theorem 3.1 enables a broad range of Bayesian inference by showing that the posterior distribution of a continuous parameter $\boldsymbol{\theta}$ is also asymptotically normal. For this reason, Theorem 3.1 is sometimes referred to as the *Bayesian Central Limit Theorem*. We do not attempt a general proof of the theorem here, but provide an outline in the unidimensional case. Write $l(\theta) = \log f(\mathbf{x}|\theta)\pi(\theta)$ and use a Taylor expansion of $l(\theta)$ as follows:

$$
\begin{aligned}
p(\theta|\mathbf{x}) &\propto f(\mathbf{x}|\theta)\pi(\theta) = \exp\{\log f(\mathbf{x}|\theta)\pi(\theta)\} = \exp\{l(\theta)\} \\
&\approx \exp\left\{\left[l(\hat{\theta}) + \frac{\partial}{\partial\theta}l(\theta)\Big|_{\theta=\hat{\theta}^{\pi}}(\theta - \hat{\theta}) + \frac{\partial^2}{\partial\theta^2}l(\theta)\Big|_{\theta=\hat{\theta}^{\pi}} \cdot \frac{(\theta - \hat{\theta})^2}{2}\right]\right\} \\
&= \exp\left[l(\hat{\theta}) - \frac{1}{2}I^{\pi}(\mathbf{x})(\theta - \hat{\theta})^2\right] \\
&\propto N(\hat{\theta}^{\pi}, [I^{\pi}(\mathbf{x})]^{-1}) .
\end{aligned}
$$

The second term in the Taylor expansion is zero because the log posterior has slope zero at its mode, while the third term gives rise to the generalized Fisher information matrix, completing the proof. ∎

Example 3.1 Consider again the beta/binomial example of Section 2.11. Using a flat prior on θ, we have $f(x|\theta)\pi(\theta) \propto \theta^x(1-\theta)^{n-x}$, so that $l(\theta) = x\log\theta + (n-x)\log(1-\theta)$. Taking the derivative of $l(\theta)$ and equating to zero, we obtain $\hat{\theta}^{\pi} = x/n$, the familiar binomial proportion. The second derivative is

$$
\frac{\partial^2 l(\theta)}{\partial\theta^2} = \frac{-x}{\theta^2} - \frac{n-x}{(1-\theta)^2} ,
$$

so that

$$
\frac{\partial^2 l(\theta)}{\partial\theta^2}\Big|_{\theta=\hat{\theta}^{\pi}} = -\frac{x}{\hat{\theta}^2} - \frac{n-x}{(1-\hat{\theta})^2} = -\frac{n}{\hat{\theta}} - \frac{n}{1-\hat{\theta}} .
$$

Thus $[I^{\pi}(x)]^{-1} = \left(\frac{n}{\hat{\theta}} + \frac{n}{1-\hat{\theta}}\right)^{-1} = \left(\frac{n}{\hat{\theta}(1-\hat{\theta})}\right)^{-1} = \frac{\hat{\theta}(1-\hat{\theta})}{n}$, which is the usual frequentist expression for the variance of $\hat{\theta} = x/n$. Thus,

$$
p(\theta|x) \,\dot\sim\, N\left(\hat{\theta}^{\pi}, \frac{\hat{\theta}(1-\hat{\theta})}{n}\right) .
$$

Figure 3.1 shows this normal approximation to the posterior of θ for the consumer preference data, previously analyzed in Example 2.11. Also

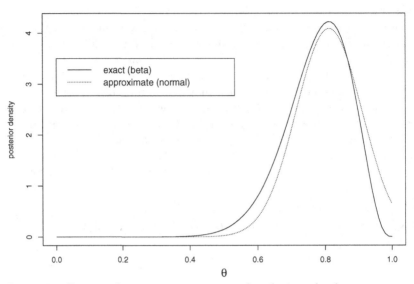

Figure 3.1 *Exact and approximate posterior distributions for θ, consumer preference data example.*

shown is the exact posterior distribution, which we recall under the flat prior on θ is a $Beta(14, 4)$ distribution. The R code to create this figure is:

R code
```
x <- seq(0,1,length=100);
plot(x,dbeta(x,14,4),type="l",lty=1,xlab=expression(theta),
  ylab="posterior density")
phat <- 13/16;  var <- phat*(1-phat)/16
lines(x,dnorm(x,phat,sqrt(var)),lty=2)
legend(0 ,3.5,cex=1.2,legend =c(
  "exact (beta)", "approximate (normal)"),lty=1:2)
```

Notice that the two distributions have very similar modes, but that the approximate posterior density fails to capture the left skew in the exact posterior; it does not go to 0 at $\theta = 1$, and drops off too quickly as θ decreases from $\hat{\theta} = 13/16 = .8125$. This inaccuracy is also reflected in the corresponding equal tail credible sets: while the exact interval is $(.57, .93)$, the one based on the normal distribution is $(.62, 1.0)$. Had the observed x been even larger, the upper limit for the latter interval would have been greater than 1, a logical impossibility in this problem. Still, the difference is fairly subtle, especially given the small sample size involved (only 16). ∎

3.2.2 Laplace's method

The posterior approximation techniques of the previous subsection are often referred to as *modal approximations* because they estimate $\boldsymbol{\theta}$ by the mode of the posterior. They are also called *first order approximations*, since one can show the error in using such approximations goes to 0 at a rate proportional to $1/n$ as $n \to \infty$. Estimates of posterior moments and quantiles may be obtained simply as the corresponding features of the approximating normal density. But as Figure 3.1 shows, these estimates may be poor if the true posterior differs significantly from normality. This raises the question of whether we can obtain more accurate posterior estimates without significantly more effort in terms of higher order derivatives or complicated transformations.

An affirmative answer to this question is provided by clever application of an expansion technique known as *Laplace's Method*. This technique draws its inspiration from work originally published in 1774 by Laplace (reprinted 1986); the most frequently cited paper on the subject is the rather more recent one by Tierney and Kadane (1986). These authors' clever approach provides a *second* order approximation to the posterior mean of $g(\boldsymbol{\theta})$, i.e., one for which the error vanishes at a rate proportional to $1/n^2$. Remarkably, the approach still requires the calculation of only first and second derivatives of the log-posterior (although it turns out we do need exactly twice as many of these calculations as we would to produce the simple normal approximation). By contrast, most other second-order expansion methods require computation of third- and perhaps higher-order derivatives.

There are several advantages to asymptotic methods. First, they replace numerical integration with numerical differentiation, which has historically been thought of as more stable numerically. They are deterministic algorithms (i.e., that do not rely on random numbers), so two different analysts should be able to produce a common answer given the same dataset, model, and prior distribution. Finally, they greatly reduce the computational complexity in any study of *robustness*, i.e., an investigation of how sensitive our conclusions are to modest changes in the prior or likelihood function. For example, suppose we wish to find the posterior expectation of a function of interest $g(\boldsymbol{\theta})$ under a new prior distribution, $\pi_{NEW}(\boldsymbol{\theta})$. We can write

$$E_{NEW}[g(\boldsymbol{\theta})|\mathbf{y}] = \frac{\int g(\boldsymbol{\theta})f(\mathbf{y}|\boldsymbol{\theta})\pi(\boldsymbol{\theta})b(\boldsymbol{\theta})d\boldsymbol{\theta}}{\int f(\mathbf{y}|\boldsymbol{\theta})\pi(\boldsymbol{\theta})b(\boldsymbol{\theta})d\boldsymbol{\theta}} \; ,$$

where $b(\cdot)$ is the appropriate *perturbation function*, which in this case is $b(\boldsymbol{\theta}) = \pi_{NEW}(\boldsymbol{\theta})/\pi(\boldsymbol{\theta})$. We can avoid starting the posterior calculations from scratch by using a result due to Kass, Tierney, and Kadane (1989), namely,

$$E_{NEW}[g(\boldsymbol{\theta})|\mathbf{y}] \approx \frac{b(\boldsymbol{\theta}^*)}{b(\hat{\boldsymbol{\theta}})} E[g(\boldsymbol{\theta})|\mathbf{y}], \tag{3.1}$$

where $\boldsymbol{\theta}^*$ and $\hat{\boldsymbol{\theta}}$ maximize $\log[g(\boldsymbol{\theta})f(\mathbf{y}|\boldsymbol{\theta})\pi(\boldsymbol{\theta})]$ and $\log[f(\mathbf{y}|\boldsymbol{\theta})\pi(\boldsymbol{\theta})]$, respectively. Note that if an asymptotic approximation was used to compute the original posterior expectation $E[g(\boldsymbol{\theta})|\mathbf{y}]$, then $\hat{\boldsymbol{\theta}}$ and perhaps $\boldsymbol{\theta}^*$ will already be available. Hence any number of alternate priors may be investigated simply by evaluating the ratios $b(\theta^*)/b(\hat{\theta})$; no new derivative calculations are required. A somewhat more refined alternative is to obtain the original posterior expectation precisely using a more costly numerical quadrature or Monte Carlo integration method, and then use equation (3.1) to carry out a quick sensitivity analysis. We shall return to this issue in Subsection 4.2.1.

An asymptotic approximation also has several limitations, however. For the approximation to be valid, the posterior distribution must be unimodal, or nearly so. Its accuracy also depends on the parameterization used (e.g., θ versus $\log(\theta)$), and the correct one may be difficult to ascertain. Similarly, the size of the dataset, n, must be fairly large; moreover, it is hard to judge how large is "large enough." That is, the second-order accuracy provided by Laplace's method is comforting, and is surely better than the first-order accuracy provided by the normal approximation, but for any given dataset we will still lack a numerical measurement of how far our approximate posterior expectations are from their exact values. Worse yet, since the asymptotics are in n, the size of the dataset, we will not be able to improve the accuracy of our approximations without collecting additional data! Finally, for moderate- to high-dimensional θ (say, bigger than 10), Laplace's method will rarely be of sufficient accuracy, and numerical computation of the associated Hessian matrices will be prohibitively difficult anyway. Unfortunately, high-dimensional models (such as random effects models that feature a parameter for every subject in the study) are fast becoming the norm in many branches of applied statistics.

For these reasons, practitioners now often turn to iterative methods, especially those involving Monte Carlo sampling, for posterior calculation. These methods typically involve longer runtimes, but can be applied more generally and are often easier to program. As such, they are the subject of the remaining sections in this chapter.

3.3 Noniterative Monte Carlo methods

3.3.1 Direct sampling

We begin with the most basic definition of Monte Carlo integration, found in many calculus texts. Suppose $\theta \sim h(\theta)$ and we seek $\gamma \equiv E[c(\theta)] = \int c(\theta)h(\theta)d\theta$. Then if $\theta_1, \ldots, \theta_N \overset{iid}{\sim} h(\theta)$, we have

$$\hat{\gamma} = \frac{1}{N} \sum_{j=1}^{N} c(\theta_j) , \qquad (3.2)$$

which converges to $E[c(\theta)]$ with probability 1 as $N \to \infty$, by the Strong Law of Large Numbers. In our case, $h(\theta)$ is a posterior distribution and γ is the posterior mean of $c(\boldsymbol{\theta})$. Hence the computation of posterior expectations requires only a sample of size N from the posterior distribution.

Notice that, in contrast to the methods of Section 3.2, the quality of the approximation in (3.2) improves as we increase N, the Monte Carlo sample size (which we choose), rather than n, the size of the dataset (which is typically beyond our control). Also note that increasing the dimension of $\boldsymbol{\theta}$ does not alter the basic structure of (3.2), although more samples will be needed to produce sufficiently precise estimates. A final contrast with asymptotic methods is that the structure of (3.2) also allows us to evaluate its accuracy for any fixed N. Because $\hat{\gamma}$ is itself a sample mean of independent observations, we have that $Var(\hat{\gamma}) = Var[c(\boldsymbol{\theta})]/N$. But $Var[c(\boldsymbol{\theta})]$ can be estimated by the sample variance of the $c(\boldsymbol{\theta}_i)$ values, so that a standard error estimate for $\hat{\gamma}$ is given by

$$\hat{se}(\hat{\gamma}) = \sqrt{\frac{1}{N(N-1)} \sum_{j=1}^{N} [c(\boldsymbol{\theta}_j) - \hat{\gamma}]^2} \ . \tag{3.3}$$

Finally, the Central Limit Theorem implies that $\hat{\gamma} \pm 2\,\hat{se}(\hat{\gamma})$ provides an approximate 95% confidence interval for the true value of the posterior mean γ. Again, N may be chosen as large as necessary to provide any preset level of narrowness to this interval. While it may seem strange for a practical Bayesian textbook to recommend use of a frequentist interval estimation procedure, Monte Carlo simulations provide one (and perhaps the only) example where they are clearly appropriate!

In addition, letting $I_{(a,b)}$ denote the indicator function of the set (a, b), note that

$$p \equiv P\{a < c(\boldsymbol{\theta}) < b|\mathbf{y}\} = E\{I_{(a,b)}[c(\boldsymbol{\theta})]|\mathbf{y}\} \ ,$$

so that an estimate of p is available simply as

$$\hat{p} = \frac{\text{number of } c(\boldsymbol{\theta}_j)\text{s} \in (a, b)}{N} \ ,$$

with associated binomial standard error estimate $\sqrt{\hat{p}(1-\hat{p})/N}$. In fact, in the univariate case, this suggests that a histogram of the sampled θ_is would estimate the posterior itself, since the probability in each histogram bin converges to the true bin probability. Alternatively, we could use a kernel density estimate to "smooth" the histogram,

$$\hat{p}(\theta|\mathbf{y}) = \frac{1}{Nh_N} \sum_{j=1}^{N} K\left(\frac{\theta - \theta_j}{h_N}\right) \ ,$$

where K is a "kernel" density (typically a normal or rectangular distribution) and h_N is a window width satisfying $h_N \to 0$ and $Nh_N \to \infty$ as $N \to \infty$. The interested reader is referred to the excellent book by Silverman (1986) for more on this and other methods of density estimation.

Example 3.2 Let $Y_i \overset{\text{iid}}{\sim} N(\mu, \sigma^2)$, $i = 1, \ldots, n$, and suppose we adopt the reference prior $\pi(\mu, \sigma) = \frac{1}{\sigma}$. Then one can show (see Berger (1985, problem 4.12), or Lee (1997, Section 2.12)) that the joint posterior of μ and σ^2 is given by

$$\mu|\sigma^2, \mathbf{y} \quad \sim \quad N(\bar{y}, \sigma^2/n) \,,$$

$$\text{and} \quad \sigma^2|\mathbf{y} \quad \sim \quad K\chi_{n-1}^{-2} = K \cdot IG\left(\frac{n-1}{2}, 2\right) \,,$$

where $K = \sum_{i=1}^{n}(y_i - \bar{y})^2$. We may thus generate samples from the joint posterior quite easily as follows. First sample $\sigma_j^2 \sim K\chi_{n-1}^{-2}$, and then sample $\mu_j \sim N(\bar{y}, \sigma_j^2/n)$, $j = 1, \ldots, N$. This then creates the set $\{(\mu_j, \sigma_j^2), j = 1, \ldots, N\}$ from $p(\mu, \sigma^2|\mathbf{y})$. To estimate the posterior mean of μ, we would use

$$\hat{E}(\mu|\mathbf{y}) = \frac{1}{N}\sum_{j=1}^{N}\mu_j \,.$$

To obtain a 95% equal-tail posterior credible set for μ, we might simply use the empirical .025 and .975 quantiles of the sample of μ_j values.

Estimates of functions of the parameters are also easily obtained. For example, suppose we seek an estimate of the distribution of $\gamma = \sigma/\mu$, the coefficient of variation. We simply define the transformed Monte Carlo samples $\gamma_j = \sigma_j/\mu_j$, $j = 1, \ldots, N$, and create a histogram or kernel density estimate based on these values.

As a final illustration, suppose we wish to estimate $P(Y_{n+1} > c|\mathbf{y})$, where Y_{n+1} is a *new* observation, not part of $\mathbf{y}$. Writing $\boldsymbol{\theta} = (\mu, \sigma^2)$, we have

$$\begin{aligned}
P(Y_{n+1} > c|\mathbf{y}) &= \int I_{(c,\infty)}(y_{n+1})f(y_{n+1}|\mathbf{y})dy_{n+1} \\
&= \int \int I_{(c,\infty)}(y_{n+1})f(y_{n+1}|\boldsymbol{\theta})p(\boldsymbol{\theta}|\mathbf{y})d\boldsymbol{\theta}dy_{n+1} \\
&= \int \left[\int I_{(c,\infty)}(y_{n+1})f(y_{n+1}|\boldsymbol{\theta})dy_{n+1}\right]p(\boldsymbol{\theta}|\mathbf{y})d\boldsymbol{\theta},
\end{aligned}$$

the second equality coming from the fact that $f(y_{n+1}|\boldsymbol{\theta}, \mathbf{y}) = f(y_{n+1}|\boldsymbol{\theta})$ because the Y_is are conditionally independent given $\boldsymbol{\theta}$. But now the quantity inside the brackets in the third line is nothing but $P(Y_{n+1} > c|\boldsymbol{\theta}) = 1 - \Phi((c - \mu)/\sigma)$, where Φ is the cdf of a standard normal distribution. Therefore,

$$P(Y_{n+1} > c|\mathbf{y}) \approx \frac{1}{N}\sum_{j=1}^{N}[1 - \Phi((c - \mu_j)/\sigma_j)] \,,$$

since the $\boldsymbol{\theta}_j \overset{\text{iid}}{\sim} p(\boldsymbol{\theta}|\mathbf{y})$. ∎

3.3.2 Indirect methods

Example 3.2 suggests that given a sample from the posterior distribution, almost any quantity of interest can be estimated. But what if we can't directly sample from this distribution? This is an old problem that predates its interest by Bayesian statisticians by many years. As a result, there are several approaches one might try, of which we shall discuss only three: importance sampling, rejection sampling, and the weighted bootstrap.

Importance sampling

This approach is outlined carefully by Hammersley and Handscomb (1964); it has been championed for Bayesian analysis by Geweke (1989). Suppose we wish to approximate a posterior expectation, say

$$E(c(\boldsymbol{\theta})|\mathbf{y}) = \frac{\int c(\boldsymbol{\theta})L(\boldsymbol{\theta})\pi(\boldsymbol{\theta})d\boldsymbol{\theta}}{\int L(\boldsymbol{\theta})\pi(\boldsymbol{\theta})d\boldsymbol{\theta}} \ ,$$

where for notational convenience we again suppress any dependence of the function of interest f and the likelihood L on the data $\mathbf{y}$. Suppose we can roughly approximate the normalized likelihood times prior, $cL(\boldsymbol{\theta})\pi(\boldsymbol{\theta})$, by some density $g(\boldsymbol{\theta})$ from which we can easily sample – say, a multivariate t density, or perhaps a "split-t" (i.e., a t that uses possibly different scale parameters on either side of the mode in each coordinate direction; see Geweke, 1989, for details). Then defining the *weight function* $w(\boldsymbol{\theta}) = L(\boldsymbol{\theta})\pi(\boldsymbol{\theta})/g(\boldsymbol{\theta})$, we have

$$
\begin{aligned}
E(c(\boldsymbol{\theta})|\mathbf{y}) &= \frac{\int c(\boldsymbol{\theta})w(\boldsymbol{\theta})g(\boldsymbol{\theta})d\boldsymbol{\theta}}{\int w(\boldsymbol{\theta})g(\boldsymbol{\theta})d\boldsymbol{\theta}} \\
&\approx \frac{\frac{1}{N}\sum_{j=1}^{N} c(\boldsymbol{\theta}_j)w(\boldsymbol{\theta}_j)}{\frac{1}{N}\sum_{j=1}^{N} w(\boldsymbol{\theta}_j)} \ ,
\end{aligned}
\tag{3.4}
$$

where $\boldsymbol{\theta}_j \overset{iid}{\sim} g(\boldsymbol{\theta})$. Here, $g(\boldsymbol{\theta})$ is called the *importance function*; how closely it resembles $cL(\boldsymbol{\theta})\pi(\boldsymbol{\theta})$ controls how good the approximation in (3.4) is. To see this, note that if $g(\boldsymbol{\theta})$ is a good approximation, the weights will all be roughly equal, which in turn will minimize the variance of the numerator and denominator (see Ripley, 1987, Exercise 5.3). If, on the other hand, $g(\boldsymbol{\theta})$ is a poor approximation, many of the weights will be close to zero, and thus a few $\boldsymbol{\theta}_j$s will dominate the sums, producing an inaccurate approximation.

Example 3.3 Suppose $g(\theta)$ is taken to be the relatively light-tailed normal distribution, but $cL(\theta)\pi(\theta)$ has much heavier, Cauchy-like tails. Then it will take many draws from g to obtain a few samples in these tails, and these points will have disproportionately large weights (since g will be small relative to $cL\pi$ for these points), thus destabilizing the estimate (3.4). As a result, a very large N will be required to obtain an approximation of acceptable accuracy. ∎

We may check the accuracy of approximation (3.4) using the following formula:

$$\text{Var}\left(\frac{\bar{x}}{\bar{y}}\right) \approx \frac{\hat{\sigma}_x^2}{\bar{y}^2} + \frac{\bar{x}\hat{\sigma}_y^2}{\bar{y}^2} - \frac{\bar{x}\hat{\sigma}_{xy}}{\bar{y}^3}$$

where $\hat{\sigma}_x^2 = \frac{1}{N-1}\sum(x_j - \bar{x})^2$, $\hat{\sigma}_y^2 = \frac{1}{N-1}\sum(y_j - \bar{y})^2$, and $\hat{\sigma}_{xy} = \frac{1}{N-1}\sum(x_j - \bar{x})(y_j - \bar{y})$. (We would of course plug in $x_j = c(\boldsymbol{\theta}_j)w(\boldsymbol{\theta}_j)$ and $y_j = w(\boldsymbol{\theta}_j)$.)

Finally, West (1992) recommends adaptive approximation of posterior densities using mixtures of multivariate t distributions. That is, after drawing a sample of size N_0 from an initial importance sampling density g_0, we compute the weighted kernel density estimate

$$g_1(\boldsymbol{\theta}) = \sum_{j=1}^{N_0} w_j K(\boldsymbol{\theta}|\boldsymbol{\theta}_j, Vh_{N_0}^2)\,.$$

Here, K is the density function of a multivariate t density with mode $\boldsymbol{\theta}_j$ and scale matrix $Vh_{N_0}^2$, where V is an estimate of the posterior covariance matrix and h_{N_0} is a kernel window width. We then iterate the procedure, drawing N_1 importance samples from g_1 and revising the mixture density to g_2, and so on until a suitably accurate estimate is obtained.

Rejection sampling

This is an extremely general and quite common method of random generation; excellent summaries are given in the books by Ripley (1987, Section 3.2) and Devroye (1986, Section II.3). In this method, instead of trying to approximate the normalized posterior

$$h(\boldsymbol{\theta}) = \frac{L(\boldsymbol{\theta})\pi(\boldsymbol{\theta})}{\int L(\boldsymbol{\theta})\pi(\boldsymbol{\theta})d\boldsymbol{\theta}}\,,$$

we try to "blanket" it. That is, suppose there exists an identifiable constant $M > 0$ and a smooth density $g(\boldsymbol{\theta})$, called the *envelope function*, such that $L(\boldsymbol{\theta})\pi(\boldsymbol{\theta}) < Mg(\boldsymbol{\theta})$ for all $\boldsymbol{\theta}$ (this situation is illustrated for the one-dimensional case in Figure 3.2). The rejection method proceeds as follows:

(i) Generate $\boldsymbol{\theta}_j \sim g(\boldsymbol{\theta})$.

(ii) Generate $U \sim \text{Uniform}(0, 1)$.

(iii) If $MUg(\boldsymbol{\theta}_j) < L(\boldsymbol{\theta}_j)\pi(\boldsymbol{\theta}_j)$, accept $\boldsymbol{\theta}_j$; otherwise, reject $\boldsymbol{\theta}_j$.

(iv) Return to step (i) and repeat, until the desired sample $\{\boldsymbol{\theta}_j,\ j = 1, \dots, N\}$ is obtained. The members of this sample will then be random variables from $h(\boldsymbol{\theta})$.

A formal proof of this result is available in Devroye (1986, pp. 40–42) or Ripley (1987, pp. 60–62); we provide only a heuristic justification in the

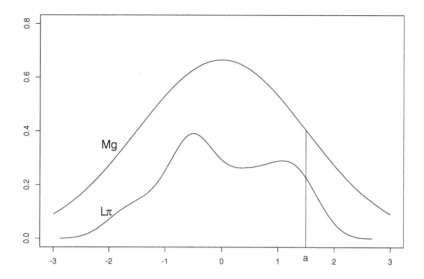

Figure 3.2 *Unstandardized posterior distribution and proper rejection envelope.*

case of a univariate θ. Consider a fairly large sample of points generated from $g(\theta)$. An appropriately scaled histogram of these points would have roughly the same shape as the curve labeled "Mg" in Figure 3.2. Now consider the histogram bar centered at the point labeled "a" in the figure. The rejection step in the above algorithm has the effect of slicing off the top portion of the bar (i.e., the portion between the two curves), since only those points having $MUg(\theta_j)$ values below the lower curve are retained. But this is true for every potential value of "a" along the horizontal axis, so a histogram of the accepted θ_j values would mimic the shape of the lower curve, which is proportional to the posterior distribution $h(\theta)$, as desired.

Intuition suggests that M should be chosen as small as possible, so as not to unnecessarily waste samples. This is easy to confirm, since if K denotes the number of iterations required to get one accepted candidate θ_j, then K is a *geometric* random variable, i.e.,

$$P(K = i) = (1 - p)^{i-1}p \,, \tag{3.5}$$

where p is the probability of acceptance. So $P(K = i)$ decreases monotonically, and at an exponential rate. It is left as an exercise to show that $p = c/M$, where c is the normalizing constant for the posterior $h(\theta)$. So our geometric distribution has mean $E(K) = p^{-1} = M/c$, and thus we do indeed want to minimize M. Note that if h were available for selection as the g function, we would choose the minimal acceptable value $M = c$, obtaining an acceptance probability of 1.

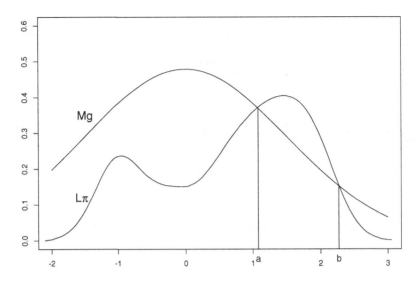

Figure 3.3 *Unstandardized posterior distribution and deficient rejection envelope.*

Like an importance sampling density, the envelope density g should be similar to the posterior in general appearance, but with heavier tails and sharper infinite peaks, in order to assure that there are sufficiently many rejection candidates available across its entire domain. One also has to be careful that Mg is actually an "envelope" for the unnormalized posterior $L\pi$. To see what happens if this condition is not met, suppose

$$S_M = \{\theta : L(\theta)\pi(\theta) > Mg(\theta)\} \ ,$$

the situation illustrated in Figure 3.3 with $S_M = (a, b)$. Then the distribution of the accepted θs is *not* $h(\theta)$, but really

$$h^*(\theta) = \begin{cases} \dfrac{L(\theta)\pi(\theta)}{\int_{S_M^C} L(\theta)\pi(\theta)d\theta + MP_g(S_M)} \ , & \theta \in S_M^C \\[4mm] \dfrac{Mg(\theta)}{\int_{S_M^C} L(\theta)\pi(\theta)d\theta + MP_g(S_M)} \ , & \theta \in S_M \end{cases} . \tag{3.6}$$

Unfortunately, $\int_{S_M} L(\theta)\pi(\theta)d\theta > MP_g(S_M)$, so even if $P_g(S_M)$ is small, there is no guarantee that (3.6) is close to $h(\theta)$. That is, only a few observed envelope violations do not necessarily imply a small inaccuracy in the posterior sample.

To address this, when we find a θ_j such that $L(\theta_j)\pi(\theta_j) > Mg(\theta_j)$, we may do a local search in the neighborhood of θ_j, and increase M accordingly. Of course, we should really go back and recheck all of the previously accepted θ_js because some may no longer be acceptable with the new, larger M.

Weighted bootstrap

This method was presented by Smith and Gelfand (1992), and is very similar to the *sampling-importance resampling* algorithm of Rubin (1988). Suppose an M appropriate for the rejection method is not readily available, but that we do have a sample $\boldsymbol{\theta}_1, \ldots \boldsymbol{\theta}_N$ from some approximating density $g(\boldsymbol{\theta})$. Define

$$w_i = \frac{L(\boldsymbol{\theta}_i)\pi(\boldsymbol{\theta}_i)}{g(\boldsymbol{\theta}_i)} \quad \text{and} \quad q_i = \frac{w_i}{\sum_{j=1}^{N} w_j} .$$

Now draw $\boldsymbol{\theta}^*$ from the *discrete* distribution over $\{\boldsymbol{\theta}_1 \ldots \boldsymbol{\theta}_N\}$ which places mass q_i at $\boldsymbol{\theta}_i$. Then

$$\boldsymbol{\theta}^* \overset{\cdot}{\sim} h(\boldsymbol{\theta}) = \frac{L(\boldsymbol{\theta})\pi(\boldsymbol{\theta})}{\int L(\boldsymbol{\theta})\pi(\boldsymbol{\theta})d\boldsymbol{\theta}} ,$$

with the approximation improving as $N \to \infty$. This is a *weighted* bootstrap, since instead of resampling from the set $\{\boldsymbol{\theta}_1, \ldots \boldsymbol{\theta}_N\}$ with equally likely probabilities of selection, we are resampling some points more often than others due to the unequal weighting.

To see that the method does perform as advertised, notice that for the standard bootstrap,

$$
\begin{aligned}
P(\boldsymbol{\theta}^* \le a) \quad &= \quad \sum_{i=1}^{N} \frac{1}{N} I_{(-\infty,a]}(\boldsymbol{\theta}_i) \\
&\overset{N\to\infty}{\longrightarrow} \quad E_g I_{(-\infty,a]}(\boldsymbol{\theta}) \quad = \quad \int_{-\infty}^{a} g(\boldsymbol{\theta})d\boldsymbol{\theta}
\end{aligned}
$$

so that $\boldsymbol{\theta}^*$ is approximately distributed as $g(\boldsymbol{\theta})$. For the weighted bootstrap,

$$
\begin{aligned}
P(\boldsymbol{\theta}^* \le a) \quad &= \quad \sum_{i=1}^{N} q_i I_{(-\infty,a]}(\boldsymbol{\theta}_i) \quad = \quad \frac{\frac{1}{N}\sum_{i=1}^{N} w_i I_{(-\infty,a]}(\boldsymbol{\theta}_i)}{\frac{1}{N}\sum_{i=1}^{N} w_i} \\
&\overset{N\to\infty}{\longrightarrow} \quad \frac{E_g \frac{L(\boldsymbol{\theta})\pi(\boldsymbol{\theta})}{g(\boldsymbol{\theta})} \cdot I_{(-\infty,a]}(\boldsymbol{\theta})}{E_g \frac{L(\boldsymbol{\theta})\pi(\boldsymbol{\theta})}{g(\boldsymbol{\theta})}} \quad = \quad \frac{\int_{-\infty}^{a} L(\boldsymbol{\theta})\pi(\boldsymbol{\theta})d\boldsymbol{\theta}}{\int_{-\infty}^{\infty} L(\boldsymbol{\theta})\pi(\boldsymbol{\theta})d\boldsymbol{\theta}} \\
&= \quad \int_{-\infty}^{a} h(\boldsymbol{\theta})d\boldsymbol{\theta}
\end{aligned}
$$

so that $\boldsymbol{\theta}^*$ is now approximately distributed as $h(\boldsymbol{\theta})$, as desired.

Note that, similar to the previous two indirect sampling methods, we need $g(\boldsymbol{\theta}) \approx h(\boldsymbol{\theta})$, or else a very large N will be required to obtain acceptable accuracy. In particular, the "tail problem" mentioned before is potentially even more harmful here, since if there are no candidate $\boldsymbol{\theta}_i$s located in the tails of $h(\boldsymbol{\theta})$, there is of course no way to resample them!

In any of the three methods discussed above, if the prior $\pi(\boldsymbol{\theta})$ is proper, it can play the role of $g(\boldsymbol{\theta})$, as shown in the following example.

Example 3.4 Suppose $Y_1, \ldots, Y_n \overset{\text{iid}}{\sim} N(\theta, \sigma^2)$ and $\pi(\theta) = \text{Cauchy}(\mu, \tau)$, with σ^2, μ, and τ known. The likelihood $L(\theta) \propto \exp\{-\frac{1}{2\sigma^2}\sum_{i=1}^n (y_i - \theta)^2\}$ is maximized at $\hat{\theta} = \bar{y}$. Let $M = L(\hat{\theta})$ in the rejection method, and let $g(\theta) = \pi(\theta)$. Then

$$L(\theta)\pi(\theta) \leq Mg(\theta) \quad \Longleftrightarrow \quad L(\theta)\pi(\theta) \leq L(\hat{\theta})\pi(\theta)$$
$$\Longleftrightarrow \quad L(\theta) \leq L(\hat{\theta}) .$$

So we simply generate $\theta_j \sim \pi(\theta)$, $U \sim \text{Uniform}(0,1)$, and accept θ_j if $MUg(\theta_j) < L(\theta_j)\pi(\theta_j)$, i.e., if

$$U < \frac{L(\theta_j)}{L(\hat{\theta})} .$$

Clearly this ratio is also the probability of accepting a θ_j candidate. Hence this approach will be quite inefficient unless $\pi(\theta)$ is not too flat relative to the likelihood $L(\theta)$, so that most of the candidates have a reasonable chance of being accepted. Unfortunately, this will not normally be the case, because in most applications the data will carry much more information about θ than the prior. In these instances, setting $g = \pi$ will also result in poor performance by importance sampling or the weighted bootstrap, since g will be a poor approximation to $L\pi$. As a result, g should be chosen as the prior π as a last resort, when methods of finding a better approximation have failed. ∎

3.4 Markov chain Monte Carlo methods

Importance sampling, rejection sampling, and the weighted bootstrap are all "one-off," or noniterative, methods: they draw a sample of size N, and stop. Hence there is no notion of the algorithm "converging" – we simply require N large enough for sufficiently precise estimates. But for many problems, especially high-dimensional ones, it may be quite difficult or impossible to find an importance sampling density (or envelope function) that is an acceptably accurate approximation to the log posterior, but still easy to sample from.

In such cases, it is now standard Bayesian practice to turn to *Markov chain Monte Carlo (MCMC)* methods. As seen in Chapter 2, these methods operate by sequentially sampling parameter values from a Markov chain whose stationary distribution is exactly the desired joint posterior distribution of interest. The great increase in generality of these methods comes at the price of requiring an assessment of *convergence* of the Markov chain to its stationary distribution, something that can sometimes be shown theoretically, but more typically is judged using plots or numerical summaries of the sampled output from the chain (see Subsection 3.4.6 below).

As already mentioned, the majority of Bayesian MCMC computing is accomplished using one of two basic algorithms, the *Metropolis-Hastings (M-H)* algorithm (Metropolis et al., 1953; Hastings, 1970) and the *Gibbs sampler* (Geman and Geman, 1984; Gelfand and Smith, 1990). Among Bayes and MCMC textbook authors, it has become somewhat fashionable to present the M-H algorithm first, since it is older (dating at least to 1953) and since the Gibbs sampler is essentially a special case of M-H (or more precisely, a hybrid of several M-H algorithms; see Subsection 3.4.4 below). However, we present the Gibbs sampler first, mostly because its reduction of a hard multivariate problem (sampling from a posterior of arbitrarily large dimension) to a series of simple lower-dimensional (and often one-dimensional) problems is the key idea that at long last made Bayesian analysis feasible in practice. This key idea is also at the core of WinBUGS, our MCMC software of choice.

3.4.1 Gibbs sampler

Suppose our model features k parameters, $\boldsymbol{\theta} = (\theta_1, \ldots, \theta_k)'$. To implement the Gibbs sampler, we must assume that samples can be generated from each of the *full* (or *complete*) *conditional* distributions $\{p(\theta_i \mid \boldsymbol{\theta}_{j \neq i}, \mathbf{y}), i = 1, \ldots, k\}$ in the model. Such samples might be available directly (say, if the full conditionals were familiar forms, like normals and gammas) or indirectly (say, via a rejection sampling approach). In this latter case two popular alternatives are the adaptive rejection sampling (ARS) algorithm of Gilks and Wild (1992), and the Metropolis algorithm described in Subsection 3.4.2. In either case, under mild conditions (see e.g. Besag, 1974), the collection of full conditional distributions uniquely determine the joint posterior distribution, $p(\boldsymbol{\theta}|\mathbf{y})$, and hence all marginal posterior distributions $p(\theta_i|\mathbf{y})$, $i = 1, \ldots, k$.

Given an arbitrary set of starting values $\{\theta_2^{(0)}, \ldots, \theta_k^{(0)}\}$, the algorithm proceeds as follows:

Gibbs Sampler: For $(t = 1, \ldots, T)$, repeat:

   Step 1: Draw $\theta_1^{(t)}$ from $p\left(\theta_1 \mid \theta_2^{(t-1)}, \theta_3^{(t-1)}, \ldots, \theta_k^{(t-1)}, \mathbf{y}\right)$

   Step 2: Draw $\theta_2^{(t)}$ from $p\left(\theta_2 \mid \theta_1^{(t)}, \theta_3^{(t-1)}, \ldots, \theta_k^{(t-1)}, \mathbf{y}\right)$

$$\vdots$$

   Step k: Draw $\theta_k^{(t)}$ from $p\left(\theta_k \mid \theta_1^{(t)}, \theta_2^{(t)}, \ldots, \theta_{k-1}^{(t)}, \mathbf{y}\right)$

Under mild regulatory conditions that are generally satisified for most statistical models (see, e.g., Geman and Geman, 1984; Schervish and Carlin, 1992; Roberts and Smith, 1993), one can show that the k-tuple obtained at iteration t, $(\theta_1^{(t)}, \ldots, \theta_k^{(t)})$, converges in distribution to a draw from the true joint posterior distribution $p(\theta_1, \ldots, \theta_k|\mathbf{y})$. This means that for t sufficiently large (say, bigger than t_0), $\{\boldsymbol{\theta}^{(t)}, t = t_0 + 1, \ldots, T\}$ is a (correlated) sample from the true posterior, from which any posterior quantities of interest may be estimated. For example, a histogram of the $\{\theta_i^{(t)}, t = t_0+1, \ldots, T\}$ themselves provides a simulation-consistent estimator of the marginal posterior distribution for θ_i, $p(\theta_i \mid \mathbf{y})$. We might also use a sample mean to estimate the posterior mean, i.e.,

$$\widehat{E}(\theta_i|\mathbf{y}) = \frac{1}{T - t_0} \sum_{t=t_0+1}^{T} \theta_i^{(t)} . \tag{3.7}$$

The time from $t = 0$ to $t = t_0$ is commonly known as the *burn-in* period; popular methods for selection of an appropriate t_0 are discussed below.

In practice, we may actually run m *parallel* Gibbs sampling chains, instead of only 1, for some modest m (say, $m = 3$ or 5). We will see below that such parallel chains may be useful in assessing sampler convergence, and can be produced with no extra time on a multiprocessor computer. In this case, we would again discard all samples from the burn-in period, obtaining the posterior mean estimate,

$$\widehat{E}(\theta_i|\mathbf{y}) = \frac{1}{m(T - t_0)} \sum_{j=1}^{m} \sum_{t=t_0+1}^{T} \theta_{i,j}^{(t)} , \tag{3.8}$$

where now the second subscript on $\theta_{i,j}$ indicates chain number. If instead our goal is estimation of the entire marginal posterior density of θ_i, note that a mixture density estimate is available as

$$
\begin{aligned}
p(\theta_i|\mathbf{y}) &= \int \int p(\theta_i|\boldsymbol{\theta}_{\ell \neq i}, \mathbf{y}) p(\boldsymbol{\theta}_{\ell \neq i}|\mathbf{y}) d\boldsymbol{\theta}_{\ell \neq i} \\
&\approx \frac{1}{m(T - t_0)} \sum_{j=1}^{m} \sum_{t=t_0+1}^{T} p(\theta_i|\boldsymbol{\theta}_{\ell \neq i,j}^{(t)}, \mathbf{y}) , \tag{3.9}
\end{aligned}
$$

where $\boldsymbol{\theta}_{\ell \neq i} = (\theta_1, \ldots, \theta_{i-1}, \theta_{i+1}, \ldots, \theta_k)'$. This formula makes use of the fact that, at convergence, the $(k-1)$-tuples $\boldsymbol{\theta}_{\ell \neq i,j}^{(t)}$ are distributed according to the joint posterior distribution $p(\boldsymbol{\theta}_{\ell \neq i}|\mathbf{y})$. We remark that the smoothing inherent in formula (3.9) leads to a less variable estimate of $p(\theta_i|\mathbf{y})$ than could be obtained by kernel density estimation, which would ignore the known shape of θ_i's full conditional distribution. Gelfand and Smith (1990) refer to this process as "Rao-Blackwellization," since the conditioning and resulting variance reduction is reminiscent of the well-known Rao-Blackwell Theorem. It does require knowledge of the *normalized* form of the full conditional for θ_i; if we lack this, we can resort to an importance-weighted marginal density estimate (see Chen, 1994) or simply a kernel density estimate (Silverman, 1986, Chapter 3). Again, we defer comment on how the issues how to choose t_0 and how to assess the quality of (3.8), (3.9), and related estimators until later in this section.

As a historical footnote, we add that Geman and Geman (1984) apparently chose the name "Gibbs sampler" because the distributions used in their context (image restoration, where the parameters were actually the colors of pixels on a screen) were Gibbs distributions (see e.g. Banerjee et al., 2004, Chapter 3). These were in turn named after J.W. Gibbs, a nineteenth-century American physicist and mathematician generally regarded as one of the founders of modern thermodynamics and statistical mechanics. While the family of Gibbs distributions can be viewed as including most standard statistical models as special cases, most Bayesian applications do not require anywhere near this level of generality, typically dealing solely with standard statistical distributions (normal, gamma, etc.). Yet, despite a few attempts by some Bayesians to choose a more descriptive name (e.g., the "successive substitution sampling" (SSS) moniker due to Schervish and Carlin, 1992), the "Gibbs sampler" name has stuck.

Example 3.5 The conditioning inherent in the full conditional distributions implies that for a very large class of hierarchical models with conjugate priors and hyperpriors, the distributions required by the Gibbs sampler will be available in closed form. For instance, consider the three-stage hierarchical model, where, using p as a generic symbol for a density function, we have a likelihood $p(y|\theta)$, a prior $p(\theta|\eta)$, and a hyperprior $p(\eta)$. (Any or all of y, θ, and η may be vectors, but for convenience we suppress this in the notation.) The most common example of a model of this type would be a hierarchical normal means model, where $y|\theta \sim N(y|\theta, \Sigma_y)$, $\theta|\eta \sim N(\theta|\eta, \Sigma_\theta)$, and $\eta \sim N(\eta|\mu_\eta, \Sigma_\eta)$, with μ_η and the three variances assumed known for the moment. Assuming (as in this normal setting) that the prior is conjugate with the likelihood, we can obtain a closed form for the marginal distribution of the data,

$$p(y|\eta) = \int p(y|\theta)p(\theta|\eta)d\theta$$

so that the conditional posterior of θ is also available in closed form as

$$p(\theta|y, \eta) = \frac{p(\theta, y, \eta)}{p(y, \eta)} = \frac{p(y|\theta)p(\theta|\eta)}{p(y|\eta)} .$$

Next, notice that

$$
\begin{aligned}
p(\eta|\theta, y) &= \frac{p(\eta, \theta, y)}{p(\theta, y)} = \frac{p(\theta|\eta)p(\eta)}{\int p(\theta|\eta)p(\eta)d\eta} \\
&= \frac{p(\theta|\eta)p(\eta)}{p(\theta)} = p(\eta|\theta) .
\end{aligned}
$$

This distribution will also be available in closed form if the hyperprior and the prior are a conjugate pair. Thus both full conditional distributions are available in closed form, and the Gibbs sampler can be readily used to estimate the two marginal posteriors,

$$p(\theta|y) = \int p(\theta|y, \eta)p(\eta|y)d\eta \quad \text{and} \quad p(\eta|y) = \int p(\eta|\theta)p(\theta|y)d\theta .$$

■

The Gibbs algorithm in this case of $k = 2$ full conditionals is essentially the "data augmentation" algorithm of Tanner and Wong (1987), developed for model settings where θ plays the role of the (possibly vector) parameter of interest, and η plays the role of a vector of missing data values. However, in order to obtain an estimate of the marginal density $p(\theta|\mathbf{y})$ (instead of merely a single sample from it), these authors complicate the algorithm slightly. At the ith stage of the algorithm, they generate $\eta_1^{(i)}, \ldots, \eta_m^{(i)} \overset{iid}{\sim} p(\eta|\theta^{(i-1)}, \mathbf{y})$, and then generate a single θ value as

$$\theta^{(i)} \sim \hat{p}_i(\theta|\mathbf{y}) = \frac{1}{m} \sum_{j=1}^{m} p(\theta|\eta_j^{(i)}, \mathbf{y}) .$$

Here, $\hat{p}_i(\theta|\mathbf{y})$ is a Monte Carlo estimate of $p_i(\theta|\mathbf{y}) = \int p(\theta|\eta, \mathbf{y})p_i(\eta|\mathbf{y})d\eta$. This produces a less variable $\theta^{(i)}$ sample at each iteration, but, more importantly, it automatically produces a smooth estimate of the marginal density of interest $p(\theta|\mathbf{y})$ at convergence of the algorithm. Tanner (1993) observes that $m = 1$ is sufficient for convergence of the algorithm, and refers to this approach as "chained data augmentation."

A full justification of why the Gibbs sampler works requires Markov chain theory that is beyond the scope of this book. In a nutshell, we need the Markov chain to be *irreducible* (i.e., for every set A with positive posterior probability, the probability of the chain ever entering A is positive for every starting point $\theta^{(0)}$) and *aperiodic* (i.e., the chain can move from any state to any other; there can be no "absorbing" states from which escape is impossible). Essentially, aperiodicity ensures convergence of the chain to its stationary distribution (the true joint posterior), while irreducibility ensures this stationary distribution is unique. The more mathematically inclined reader is referred to the widely-referenced paper by Tierney (1994) or the textbook by Robert and Casella (1999) for full details.

Thinking more practically, a minimal requirement for Gibbs convergence related to aperiodicity is that the parameter space be fully *connected*, in the sense that it have no "holes" in it that might prevent the chain from moving from one part of the space to another. To illustrate, consider an idealized joint posterior distribution for two univariate parameters θ and η that has two disconnected regions of support: one lying in the first quadrant (i.e., $\theta > 0$ and $\eta > 0$) and one in the third quadrant (i.e., $\theta < 0$ and $\eta < 0$). Suppose we start a Gibbs chain at $\theta^{(0)} > 0$. Then clearly this chain can never escape the first quadrant, because conditional on $\theta^{(0)} > 0$ we must get $\eta^{(1)} > 0$, then conditional on $\eta^{(1)} > 0$ we must get $\theta^{(1)} > 0$, and so on. Similarly, a chain started with a negative value of either θ or η can never escape the third quadrant. Fortunately, the vast majority of statistical models we will work with (normal, binomial, Poisson, etc.) feature continuous and fully connected joint parameter spaces, but this requirement is worth noting. We will return to the issue of MCMC convergence and how to diagnose it in Subsection 3.4.6.

Even if the space is fully connected, note that the quality of our estimates in (3.8) or (3.9) will be poor if θ and η are highly correlated because this will lead to high autocorrelation ("slow mixing") in the resulting $\eta^{(i)}$ sequence. That is, the chain might get "stuck" in one part of the joint distribution, leading to a density estimate based on samples that represent only a portion of the distribution. To remedy this, some authors (notably Raftery and Lewis, 1992) have recommended retaining only every m^{th} iterate after convergence has obtained, where m is large enough that the retained samples are effectively uncorrelated. This option, known as *thinning*, is available in WinBUGS and does serve to simplify the estimation of the quality of our MCMC estimates, since they will now be built out of nearly independent components. However, in Subsection 3.4.5 we argue that thinning is usually suboptimal, in that one can nearly always reduce the variance by using *all* the post-convergence MCMC output. Therefore, we do not recommend thinning in general.

Chapter 2 already featured several examples using the Gibbs sampler as implemented in WinBUGS. Here we delve a little deeper, showing how to derive the full conditional distributions required in the sampling order.

Example 3.6 Consider again the Poisson/gamma model of Example 2.7, expanded slightly to

$$Y_i|\theta_i \stackrel{ind}{\sim} Poisson(\theta_i t_i), \ \theta_i \stackrel{ind}{\sim} G(\alpha, \beta), \ i = 1, \ldots, k,$$

where the t_i are known constants. For the moment we assume that α is also known, but that β has an $IG(c, d)$ hyperprior, with known hyperparameters c and d. Thus the density functions corresponding to the three stages of the model are

$$f(y_i|\theta_i) = \frac{e^{-(\theta_i t_i)}(\theta_i t_i)^{y_i}}{y_i!}, \ y_i \geq 0, \ \theta_i > 0,$$

$$g(\theta_i|\beta) = \frac{\theta_i^{\alpha-1}e^{-\theta_i/\beta}}{\Gamma(\alpha)\beta^\alpha}, \ \alpha > 0, \ \beta > 0,$$

$$\text{and} \quad h(\beta) = \frac{e^{-1/(\beta d)}}{\Gamma(c)d^c\beta^{c+1}}, \ c > 0, \ d > 0.$$

Our main interest lies in the marginal posterior distributions of the θ_i, $p(\theta_i|\mathbf{y})$. Note that while the gamma prior is conjugate with the Poisson likelihood and the inverse gamma hyperprior is conjugate with the gamma prior, no closed form for $p(\theta_i|\mathbf{y})$ is available. However, the full conditional distributions of β and the θ_i needed to implement the Gibbs sampler *are* readily available. To see this, recall that by Bayes' Rule, each is proportional to the full Bayesian model specification,

$$\left[\prod_{i=1}^{k} f(y_i|\theta_i)g(\theta_i|\beta)\right] h(\beta) . \tag{3.10}$$

We can find the necessary full conditional distributions by dropping irrelevant terms from (3.10) and normalizing. For θ_i we have

$$\begin{aligned}
p(\theta_i|\theta_{j\neq i}, \beta, \mathbf{y}) \;\;&\propto\;\; f(y_i|\theta_i)g(\theta_i|\beta) \\
&\propto\;\; \theta_i^{y_i+\alpha-1} e^{-\theta_i(t_i+1/\beta)} \\
&\propto\;\; G\left(\theta_i \mid y_i + \alpha, \, (t_i + 1/\beta)^{-1}\right) ,
\end{aligned} \tag{3.11}$$

while for β, we have

$$\begin{aligned}
p(\beta|\{\theta_i\}, \mathbf{y}) \;\;&\propto\;\; \left[\prod_{i=1}^{k} g(\theta_i|\beta)\right] h(\beta) \\
&\propto\;\; \left[\prod_{i=1}^{k} \frac{e^{-\theta_i/\beta}}{\beta^\alpha}\right] \frac{e^{-1/(\beta d)}}{\beta^{c+1}} \\
&\propto\;\; \frac{e^{-\frac{1}{\beta}\left(\sum_{i=1}^{k}\theta_i + \frac{1}{d}\right)}}{\beta^{k\alpha+c+1}} \\
&\propto\;\; IG\left(\beta \;\Big|\; k\alpha + c, \left(\sum_{i=1}^{k}\theta_i + 1/d\right)^{-1}\right) .
\end{aligned} \tag{3.12}$$

Thus our *conditionally conjugate* prior specification (stage 2 with stage 1, and stage 3 with stage 2) has allowed both of these distributions to emerge as standard forms, and the $\{\theta_i^{(t)}\}$ and $\beta^{(t)}$ may be sampled directly from (3.11) and (3.12), respectively.

As a specific illustration, we apply this approach to the pump data in Table 3.1, which is also available in the WinBUGS Examples Volume I manual as the Pump example. These are the numbers of pump failures, Y_i, observed in t_i thousands of hours for $k = 10$ different systems of a certain nuclear power plant. We complicate matters somewhat by also treating α as an unknown parameter to be estimated. The complication is that its full conditional,

$$p(\alpha|\{\theta_i\}, \beta, \mathbf{y}) \propto \left[\prod_{i=1}^{k} g(\theta_i|\alpha, \beta)\right] h(\alpha) ,$$

is not proportional to any standard family for any choice of hyperprior

i	Y_i	t_i	r_i
1	5	94.320	.053
2	1	15.720	.064
3	5	62.880	.080
4	14	125.760	.111
5	3	5.240	.573
6	19	31.440	.604
7	1	1.048	.954
8	1	1.048	.954
9	4	2.096	1.910
10	22	10.480	2.099

Table 3.1 *Pump failure data (from Gaver and O'Muircheartaigh, 1987).*

$h(\alpha)$. However, if we select an $Expo(\mu)$ hyperprior for α, then

$$p(\alpha|\{\theta_i\},\beta,\mathbf{y}) \propto \left[\prod_{i=1}^{k} \frac{\theta_i^{\alpha-1}}{\Gamma(\alpha)\beta^\alpha}\right] e^{-\alpha/\mu} ,$$

which can be shown to be log-concave in α. Thus we may use the adaptive rejection sampling (ARS) algorithm of Gilks and Wild (1992) to update this parameter.

The WinBUGS code to fit this model is as follows:

BUGS code
```
model
{
    for (i in 1:k) {
            theta[i] ~ dgamma(alpha,beta)
            lambda[i] <- theta[i]*t[i]
            Y[i] ~ dpois(lambda[i])
    }
    alpha ~ dexp(1.0)
    beta ~ dgamma(0.1, 1.0)
}

#  DATA:
list(k = 10, Y = c(5, 1, 5, 14, 3, 19, 1, 1, 4, 22),
    t = c(94.320, 15.72, 62.88, 125.76, 5.24, 31.44, 1.048,
            1.048, 2.096, 10.48))

#  INITIAL VALUES:
list(theta=c(1,1,1,1,1,1,1,1,1,1), alpha=1, beta=1)
```

We chose the values $\mu = 1$, $c = 0.1$, and $d = 1.0$, resulting in reasonably

Parameter	mean	sd	2.5%	median	97.5%
α	0.7001	0.2699	0.2851	0.6634	1.338
β	0.9290	0.5325	0.1938	0.8315	2.205
θ_1	0.0598	0.02542	0.02128	0.05627	0.1195
θ_5	0.6056	0.3150	0.1529	0.5529	1.359
θ_6	0.6105	0.1393	0.3668	0.5996	0.9096
θ_{10}	1.993	0.4251	1.264	1.958	2.916

Table 3.2 *Posterior summaries, pump data problem.*

vague hyperpriors for α and β. WinBUGS recognizes the conjugate forms
(3.11) and (3.12) and samples directly from them; it also recognizes the
log-concavity for α and uses ARS for this parameter.

The results from running 1000 burn-in samples, followed by a "production" run of 10,000 samples (single chain) are given in Table 3.2. Note that
while θ_5 and θ_6 have very similar posterior means, the latter posterior is
much narrower (i.e., smaller posterior standard deviation). This is because,
while the crude failure rates for the two pumps are similar, the latter is
based on a far greater number of hours of observation ($t_6 = 31.44$, while
$t_5 = 5.24$). Therefore, we "know" more about pump 6, and this is properly
reflected in its posterior distribution. ∎

In many examples, the prior distribution of each parameter can be chosen
to be conjugate with the corresponding likelihood term that preceded it in
the hierarchy. This enables all of the full conditional distributions necessary
for implementing the Gibbs sampler to emerge in closed form as members
of familiar distributional families, so that sample generation is straightforward. But for many parameters, we may want to select a prior that is not
conjugate; for others (e.g., α in the previous example), no conjugate prior
will exist. How does one perform the required sampling in such cases?

Provided that the model specification (likelihood and priors) are available
in closed form, note that *any* given full conditional will be available at least
up to a constant of proportionality, because

$$p(\theta_i|\boldsymbol{\theta}_{\ell \neq i}, \mathbf{y}) \propto f(\mathbf{y}|\boldsymbol{\theta})\pi(\boldsymbol{\theta}) \equiv L(\boldsymbol{\theta})\pi(\boldsymbol{\theta}) \ . \tag{3.13}$$

Hence the "nonconjugacy problem" for Gibbs sampling outlined in the
previous paragraph is nothing but a univariate version of the basic computation problem of sampling from nonstandardized densities. This suggests
using one of the indirect methods given in Subsection 3.3.2 to sample from
these full conditionals. Unfortunately, the fact that this problem is embed-
ded within a Gibbs sampler makes some of these techniques prohibitively

expensive to implement. For example, recall that the weighted bootstrap operates by resampling from a finite sample from an approximation to the target distribution with appropriately determined unequal probabilities of selection. Sampling from (3.13) at the tth iteration of the Gibbs algorithm would thus require drawing a large approximating sample, of which only one would be retained. Worse yet, this initial sample could not typically be reused, since the values of the conditioning random variables $\boldsymbol{\theta}_{\ell \neq i}$ will have changed by iteration $(t + 1)$.

Rejection sampling will have similarly high start-up costs when applied in this context, though several ingenious modifications to reduce them have been proposed in the MCMC Bayesian literature. Wakefield et al. (1991) developed a generalized version of the ratio-of-uniforms method (see e.g. Devroye, 1986, p.194), a tailored rejection method that is particularly useful for sampling from nonstandardized densities with heavy tails. Gilks and Wild (1992) instead use traditional rejection, but restrict attention to log-concave densities (i.e., ones for which the second derivative of the log of (3.13) with respect to θ_i is nonpositive everywhere). For these densities, an envelope on the log scale can be formed by two lines, one of which is tangent to $\log p$ left of the mode, and the other tangent right of the mode. The exponential of this piecewise linear function is then guaranteed to envelope $\exp(\log p) = p$. In a subsequent paper, Gilks (1992) developed a derivative-free version of this algorithm based on secant (instead of tangent) lines. The log-concavity property can be a nuisance to check, but holds for most likelihoods (with the Student's t being a notable exception).

Carlin and Gelfand (1991b) recommended a *multivariate* rejection algorithm, wherein a single multivariate split-normal or split-t envelope function is obtained for $L(\boldsymbol{\theta})\pi(\boldsymbol{\theta})$ at the algorithm's outset. Cross-sectional slices of this envelope in the appropriate coordinate direction may then serve as univariate envelopes for the full conditional distributions regardless of the values of the conditioning random variables at a particular iteration. Unfortunately, while this approach greatly reduces the amount of computational overhead in the algorithm, the difficulty in finding an envelope whose shape even vaguely resembles that of $L(\boldsymbol{\theta})\pi(\boldsymbol{\theta})$ across $\boldsymbol{\theta}$ limits its applicability to fairly regular surfaces of moderate dimension.

When WinBUGS encounters a nonconjugacy problem for a given parameter, it first performs a check for log concavity. If this passes, it samples this full conditional using ARS; if not, it resorts to one of the methods in the next two sections, depending on whether the support of the parameter is bounded or unbounded. In the former case, it uses slice sampling (Subsection 3.4.3), while in the latter it uses Metropolis sampling (Subsection 3.4.2). Strictly speaking, WinBUGS uses a *hybrid* of these three approaches; Subsection 3.4.4 offers a more formal definition of this term and why such combinations can still lead to a convergent MCMC algorithm.

3.4.2 Metropolis-Hastings algorithm

The Gibbs sampler is easy to understand and implement, but requires the ability to readily sample from each of the full conditional distributions, $p(\theta_i \mid \theta_{\ell \neq i}, \mathbf{y})$. Unfortunately, when the prior $\pi(\theta)$ and the likelihood $f(\mathbf{y}|\theta)$ are not a conjugate pair, one or more of these full conditionals may not be available in closed form. Even in this setting, however, $p(\theta_i \mid \theta_{\ell \neq i}, \mathbf{y})$ *will* be available up to a proportionality constant, because as already mentioned it is proportional to the portion of $f(\mathbf{y}|\theta)\pi(\theta)$ that involves θ_i.

The *Metropolis algorithm* (and its *Metropolis-Hastings* extension) is a rejection algorithm that attacks precisely this problem, since it requires only a function proportional to the distribution to be sampled, at the cost of requiring a rejection step from a particular *candidate* density. Like the Gibbs sampler, this algorithm was not developed by statisticians for the purpose of estimating posterior distributions. In this case, the development was by nuclear physicists working on the Manhattan Project in the 1940s seeking to understand the particle movement theory underlying the first atomic bomb; interestingly, one of the coauthors on the original Metropolis et al. (1953) paper was Edward Teller, who is often referred to as "the father of the hydrogen bomb."

While as mentioned above our main interest in the algorithm is for generation from (typically univariate) full conditionals, it is most easily described (and theoretically supported) for the full multivariate θ vector. Thus, suppose for now that we wish to generate from a joint posterior distribution $p(\theta|\mathbf{y}) \propto h(\theta) \equiv f(\mathbf{y}|\theta)\pi(\theta)$. (Note that, in contrast to Section 3.3, we are now using h to denote the *un*normalized posterior.) We begin by specifying a *candidate* (or *proposal*) density $q(\theta^*|\theta^{(t-1)})$ that is a valid density function for every possible value of the conditioning variable $\theta^{(t-1)}$, and satisfies $q(\theta^*|\theta^{(t-1)}) = q(\theta^{(t-1)}|\theta^*)$, i.e., q is *symmetric* in its arguments. Given a starting value $\theta^{(0)}$ at iteration $t = 0$, the algorithm proceeds as follows:

Metropolis Algorithm: For $(t = 1, \ldots, T)$, repeat:

1. Draw θ^* from $q(\cdot|\theta^{(t-1)})$

2. Compute the ratio $r = h(\theta^*)/h(\theta^{(t-1)}) = \exp[\log h(\theta^*) - \log h(\theta^{(t-1)})]$

3. If $r \geq 1$, set $\theta^{(t)} = \theta^*$;

   if $r < 1$, set $\theta^{(t)} = \begin{cases} \theta^* & \text{with probability } r \\ \theta^{(t-1)} & \text{with probability } 1 - r \end{cases}$.

Then under similarly mild conditions as those supporting the Gibbs sampler, a draw $\theta^{(t)}$ converges in distribution to a draw from the true posterior density $p(\theta|\mathbf{y})$. Note however that the shape of h is not used in the candidate generation, unlike the Gibbs sampler, which draws from full conditionals *derived* from h. This suggests the Metropolis algorithm may be less efficient if not properly tuned.

Indeed, the Metropolis algorithm affords substantial flexibility through the selection of the candidate density q, but this flexibility can be a blessing and a curse. While theoretically we are free to pick almost anything, in practice only a "good" choice will result in sufficiently many candidate acceptances. The usual approach (after $\boldsymbol{\theta}$ has been transformed to have support $\Re^k$, if necessary) is to set

$$q(\boldsymbol{\theta}^*|\boldsymbol{\theta}^{(t-1)}) = N(\boldsymbol{\theta}^*|\boldsymbol{\theta}^{(t-1)}, \widetilde{\Sigma}) \,, \tag{3.14}$$

because this distribution obviously satisfies the symmetry property, and is "self correcting" (candidates are always centered around the current value of the chain). This approach of using Gaussian proposals centered around the current value of the chain is often called *random walk Metropolis*, since each new candidate is obtained as a single random walk step away from the current value. Specification of q in (3.14) thus comes down to specification of $\widetilde{\Sigma}$. Here we might try to mimic the posterior variance by setting $\widetilde{\Sigma}$ equal to an empirical estimate of the true posterior variance, derived from a preliminary sampling run.

The reader might well imagine an optimal choice of q would produce an empirical acceptance ratio of 1, the same as the Gibbs sampler (and with no apparent "waste" of candidates). However, the issue is rather more subtle than this: accepting all or nearly all of the candidates is often the result of an overly narrow candidate density. Such a density will "baby-step" around the parameter space, leading to high acceptance, but also high autocorrelation in the sampled chain. An overly wide candidate density will also struggle, proposing leaps to places far from the bulk of the posterior's support, leading to high rejection and, again, high autocorrelation. Thus the "folklore" here is to choose $\widetilde{\Sigma}$ so that roughly 50% of the candidates are accepted. Subsequent theoretical work (e.g., Gelman et al., 1996) indicates even lower acceptance rates are optimal, but this result varies with the dimension and true posterior correlation structure of $\boldsymbol{\theta}$. Specifically, when the target density is univariate normal, the first order autocorrelation in the chain is minimized by taking $\sqrt{\widetilde{\Sigma}}$ equal to 2.4 times the true posterior standard deviation, and this candidate density leads to a 44.1% acceptance rate. However, for the multivariate target density having product form

$$p(\boldsymbol{\theta}) = \prod_{i=1}^{K} g(\theta_i)$$

for some one-dimensional density g, the optimal acceptance rate approaches 23.4% as the dimension of the parameter space k goes to infinity. This suggests that in high dimensions, it is worth risking an occasional sticking point in the algorithm in order to gain the benefit of an occasional large jump across the parameter space.

In practice, choice of $\widetilde{\Sigma}$ is often done *adaptively*. For instance, in one dimension (setting $\widetilde{\Sigma} = \widetilde{\sigma}$, and thus avoiding the issue of correlations among the elements of $\boldsymbol{\theta}$), a common trick is to simply pick some initial value of $\widetilde{\sigma}$, and then keep track of the empirical proportion of candidates that are accepted. If this fraction is too high (70 to 100%), we simply increase $\widetilde{\sigma}$; if it is too low (0 to 20%), we decrease it. Since certain kinds of adaptation can actually disturb the chain's convergence to its stationary distribution (see Example 3.9 below), the simplest approach is to allow this adaptation only during the burn-in period, a practice sometimes referred to as *pilot adaptation* (Gilks et al., 1998). This is in fact the approach currently used by WinBUGS, where the default pilot period is 4000 iterations. A more involved alternative is to allow adaptation at *regeneration points* which, once defined and identified, break the Markov chain into independent sections. See e.g. Mykland, Tierney, and Yu (1995), Johnson (1998), Mira and Sargent (2003), and Hobert et al. (2002) for discussions of the use of regeneration.

As mentioned above, in practice the Metropolis algorithm is often found as a substep in a larger Gibbs sampling algorithm, used to generate from awkward full conditionals. Such hybrid Gibbs-Metropolis applications are sometimes known as "Metropolis within Gibbs" (or simply "Metropolis substeps,") and users would worry about how many such substeps should be used. Fortunately, it was soon realized that a single substep was sufficient to ensure convergence of the overall algorithm, and so this is now standard practice: when we encounter an awkward full conditional (say, for θ_i), we simply draw one Metropolis candidate, accept or reject it, and move on to θ_{i+1}. See Subsection 3.4.4 for further discussion of convergence properties and implementation of hybrid MCMC algorithms.

We now present the important generalization of the Metropolis algorithm devised by Hastings (1970). In this variant, we drop the requirement that q be symmetric in its arguments, which is often useful for bounded parameter spaces (say, $\theta > 0$) where Gaussian proposals as in (3.14) are not natural.

Metropolis-Hastings Algorithm: When using a candidate density q for which $q(\boldsymbol{\theta}^* | \boldsymbol{\theta}^{(t-1)}) \neq q(\boldsymbol{\theta}^{(t-1)} | \boldsymbol{\theta}^*)$, replace the acceptance ratio r in Step 2 of the Metropolis algorithm above by

$$r = \frac{h(\boldsymbol{\theta}^*)q(\boldsymbol{\theta}^{(t-1)} | \boldsymbol{\theta}^*)}{h(\boldsymbol{\theta}^{(t-1)})q(\boldsymbol{\theta}^* | \boldsymbol{\theta}^{(t-1)})} \ . \tag{3.15}$$

Then again under mild conditions, a draw $\boldsymbol{\theta}^{(t)}$ converges in distribution to a draw from the true posterior density $p(\boldsymbol{\theta}|\mathbf{y})$ as $t \to \infty$. Again, a full justification of why the algorithm works requires Markov chain theory we prefer to avoid, but see Chib and Greenberg (1995a) for a not-too-technical justification of (3.15) using the theory of *reversible* Markov chains.

In practice, we often set $q(\boldsymbol{\theta}^* \mid \boldsymbol{\theta}^{(t-1)}) = q(\boldsymbol{\theta}^*)$, i.e., we use a proposal density that ignores the current value of the variable. This algorithm is sometimes referred to as a *Hastings independence chain*, so named because the proposals (though not the final $\boldsymbol{\theta}^{(t)}$ values) form an independent sequence. Here, using equation (3.15) we obtain the acceptance ratios $r = w(\boldsymbol{\theta}^*)/w(\boldsymbol{\theta}^{(t-1)})$ where $w(\boldsymbol{\theta}) = h(\boldsymbol{\theta})/q(\boldsymbol{\theta})$, the usual weight function used in importance sampling. This suggests using the same guidelines for choosing q as for choosing a good importance sampling density (i.e., making q a good match for h), but perhaps with heavier tails to obtain an acceptance rate closer to 0.5, as encouraged by Gelman, Roberts, and Gilks (1996). While easy to implement, this algorithm can be difficult to tune since it will converge slowly unless the chosen q is rather close to the true posterior (which is of course unknown in advance).

Example 3.7 We illustrate the Metropolis-Hastings algorithm using the data in Table 3.3, which are taken from Bliss (1935). These data record the number of adult flour beetles killed after five hours of exposure to various levels of gaseous carbon disulphide (CS_2). Since the variability in these data cannot be explained by the standard logistic regression model, we attempt to fit the generalized logit model suggested by Prentice (1976),

$$P(\text{death}|w) \equiv g(w) = \{\exp(x)/(1 + \exp(x))\}^{m_1} .$$

Here, w is the predictor variable (dose), and $x = (w - \mu)/\sigma$ where $\mu \in \Re$ and σ^2, $m_1 > 0$. Suppose there are y_i flour beetles dying out of n_i exposed at level w_i, $i = 1, \ldots, N$.

Dosage w_i	# Killed y_i	# Exposed n_i
1.6907	6	59
1.7242	13	60
1.7552	18	62
1.7842	28	56
1.8113	52	63
1.8369	53	59
1.8610	61	62
1.8839	60	60

Table 3.3 *Flour beetle mortality data (from Bliss, 1935).*

For our prior distributions, we assume that $m_1 \sim \text{Gamma}(a_0, b_0)$, $\mu \sim \text{Normal}(c_0, d_0^2)$, and $\sigma^2 \sim \text{Inverse Gamma}(e_0, f_0)$, where $a_0, b_0, \ c_0, d_0, e_0,$ and f_0 are known, and $m_1, \mu,$ and σ^2 are independent. While these common families may appear to have been chosen to preserve some sort of conjugate structure, this is not the case: there is no closed form available for any of the three full conditional distributions needed to implement the Gibbs sampler. Thus we instead resort to the Metropolis algorithm. Our likelihood-prior specification implies the joint posterior distribution

$$p(\mu, \sigma^2, m_1 | \mathbf{y}) \quad \propto \quad f(\mathbf{y}|\mu, \sigma^2, m_1)\pi(\mu, \sigma^2, m_1)$$

$$\propto \quad \left\{ \prod_{i=1}^{N} [g(w_i)]^{y_i} [1 - g(w_i)]^{n_i - y_i} \right\} \frac{m_1^{a_0 - 1}}{\sigma^{2(e_0 + 1)}}$$

$$\times \exp\left[-\frac{1}{2}\left(\frac{\mu - c_0}{d_0}\right)^2 - \frac{m_1}{b_0} - \frac{1}{f_0 \sigma^2} \right].$$

We begin by making a change of variables from (μ, σ^2, m_1) to $\boldsymbol{\theta} = (\theta_1, \theta_2, \theta_3) = (\mu, \frac{1}{2}\log \sigma^2, \log m_1)$. This transforms the parameter space to $\Re^3$ (necessary if we wish to work with Gaussian proposal densities), and also helps to symmetrize the posterior distribution. Accounting for the Jacobian of this transformation, our target density is now

$$p(\boldsymbol{\theta}|\mathbf{y}) \propto h(\boldsymbol{\theta}) \quad = \quad \left\{ \prod_{i=1}^{N} [g(w_i)]^{y_i} [1 - g(w_i)]^{n_i - y_i} \right\} \exp(a_0 \theta_3 - 2e_0 \theta_2)$$

$$\times \exp\left[-\frac{1}{2}\left(\frac{\theta_1 - c_0}{d_0}\right)^2 - \frac{\exp(\theta_3)}{b_0} - \frac{\exp(-2\theta_2)}{f_0} \right].$$

As mentioned above, numerical stability is improved by working on the log scale, i.e., by computing the acceptance ratio as $r = \exp[\log h(\boldsymbol{\theta}^*) - \log h(\boldsymbol{\theta}^{(t-1)})]$.

We complete the prior specification by choosing the same hyperparameter values as in Carlin and Gelfand (1991b). That is, we take $a_0 = .25$ and $b_0 = 4$, so that m_1 has prior mean 1 (corresponding to the standard logit model) and prior standard deviation 2. We then specify rather vague priors for μ and σ^2 by setting $c_0 = 2$, $d_0 = 10$, $e_0 = 2.000004$, and $f_0 = 1000$; the latter two choices imply a prior mean of .001 and a prior standard deviation of .5 for σ^2. Figure 3.4 shows the output from three parallel Metropolis sampling chains, each run for 10,000 iterations using a $N_3(\boldsymbol{\theta}^{(t-1)}, \widetilde{\Sigma})$ proposal density where

$$\widetilde{\Sigma} = D = Diag(.00012, .033, .10). \tag{3.16}$$

The histograms in the figure include all 24,000 samples obtained after a burn-in period of 2000 iterations (reasonable given the appearance of the monitoring plots). The chains mix very slowly, as can be seen from the extremely high lag 1 sample autocorrelations, estimated from the chain #2 output and printed above the monitoring plots. The reason for this slow convergence is the high correlations among the three parameters, estimated

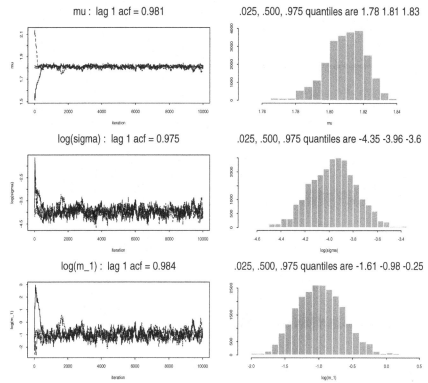

Figure 3.4 *Metropolis analysis of the flour beetle mortality data using a Gaussian proposal density with a diagonal $\widetilde{\Sigma}$ matrix. Monitoring plots use three parallel chains, and histograms use all samples following iteration 2000. Overall Metropolis acceptance rate: 13.5%.*

as $\widehat{Corr}(\theta_1, \theta_2) = -0.78$, $\widehat{Corr}(\theta_1, \theta_3) = -0.94$, and $\widehat{Corr}(\theta_2, \theta_3) = 0.89$ from the chain #2 output. As a result, the proposal acceptance rate is low (13.5%) and convergence is slow, despite the considerable experimentation in arriving at the three variances in our proposal density.

We can accelerate convergence by using a nondiagonal proposal covariance matrix designed to better mimic the posterior surface itself. From the output of the first algorithm, we can obtain an estimate of the posterior covariance matrix in the usual way as $\widehat{\Sigma} = \frac{1}{G} \sum_{g=1}^{G} (\boldsymbol{\theta}_g - \bar{\boldsymbol{\theta}})(\boldsymbol{\theta}_g - \bar{\boldsymbol{\theta}})'$, where $g = 1, \ldots, G$ indexes the post-convergence Monte Carlo samples. We then reran the algorithm with proposal variance matrix

$$\widetilde{\Sigma} = 2\widehat{\Sigma} = \begin{pmatrix} 0.000292 & -0.003546 & -0.007856 \\ -0.003546 & 0.074733 & 0.117809 \\ -0.007856 & 0.117809 & 0.241551 \end{pmatrix}. \qquad (3.17)$$

The results, shown in Figure 3.5, indicate improved convergence, with lower

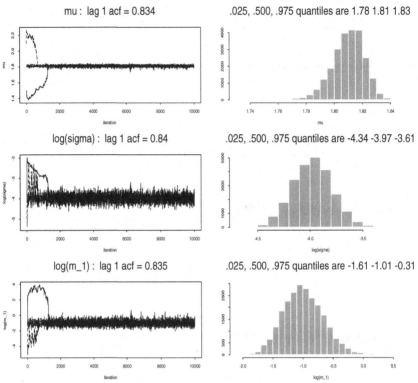

Figure 3.5 *Metropolis analysis of the flour beetle mortality data using a Gaussian proposal density with a nondiagonal $\widetilde{\Sigma}$ matrix. Monitoring plots use three parallel chains, and histograms use all samples following iteration 2000. Overall Metropolis acceptance rate: 27.3%.*

observed autocorrelations and a higher Metropolis acceptance rate (27.3%). This rate is close to the optimal rate derived by Gelman, Roberts, and Gilks (1996) of 31.6%, though this rate applies to target densities having three *independent* components. The marginal posterior results in both figures are consistent with the results obtained for this prior-likelihood combination by Carlin and Gelfand (1991b) and Müller (1991). ∎

Langevin-Hastings algorithm

As mentioned above, the Metropolis algorithm is typically implemented with a multivariate normal proposal density given in (3.14), i.e., the candidate value $\boldsymbol{\theta}^*$ is drawn from $q(\boldsymbol{\theta}^*|\boldsymbol{\theta}^{(t-1)}) = N(\boldsymbol{\theta}^*|\boldsymbol{\theta}^{(t-1)}, \widetilde{\Sigma})$. While these random walk proposals (centered at the current value of the chain, $\boldsymbol{\theta}^{(t-1)}$) can be very effective, in many situations we can do even better by introducing a systematic "drift" in the mean of this density, in order to nudge the chain in the direction of the posterior mode. One has to be very careful, however, to ensure that the Markov chain still converges to the correct posterior distribution.

An increasingly popular approach of this type is based on the theory of *Langevin diffusions* as described by Grenander and Miller (1994). These diffusions are *continuous time* Markov processes based on stochastic differential equation theory that is beyond the scope of this book. However, the key point for us, originally pointed out by Besag (1994) in his discussion of Grenander and Miller (1994), is that these continuous time processes can be approximated in discrete time, and these can, in turn, form the basis of a Metropolis-Hastings algorithm that will often have excellent convergence properties. Our brief outline here is similar to that in Robert and Casella (1999); see also Christensen et al. (2006) for an excellent example in a challenging non-Gaussian spatial modeling setting.

Langevin-Hastings Algorithm: For $(t = 1, \ldots, T)$, repeat:

1. Draw $\boldsymbol{\theta}^*$ from $N_k(\boldsymbol{\theta}^* | \widetilde{\boldsymbol{\mu}}(\boldsymbol{\theta}^{(t-1)}), \widetilde{\Sigma})$, where the proposal mean vector and covariance matrix are given by

$$\widetilde{\boldsymbol{\mu}}(\boldsymbol{\theta}^{(t-1)}) = \boldsymbol{\theta}^{(t-1)} + \frac{\tilde{\sigma}^2}{2} \nabla \log h(\boldsymbol{\theta}^{(t-1)}) \quad \text{and} \quad \widetilde{\Sigma} = \tilde{\sigma}^2 I_k \;, \qquad (3.18)$$

   where $\nabla f(\mathbf{a})$ for any function of a k-vector $\mathbf{a}$ is $\left(\frac{\partial f(\mathbf{a})}{\partial a_1}, \ldots, \frac{\partial f(\mathbf{a})}{\partial a_k}\right)'$, the vector of coordinate-specific derivatives of $f(\mathbf{a})$.

2. Compute the Hastings acceptance ratio (3.15), which in this case becomes

$$r = \frac{h(\boldsymbol{\theta}^*) \exp[-\frac{1}{2\tilde{\sigma}^2} ||\boldsymbol{\theta}^{(t-1)} - \boldsymbol{\theta}^* - \frac{\tilde{\sigma}^2}{2} \nabla \log h(\boldsymbol{\theta}^*)||^2]}{h(\boldsymbol{\theta}^{(t-1)}) \exp[-\frac{1}{2\tilde{\sigma}^2} ||\boldsymbol{\theta}^* - \boldsymbol{\theta}^{(t-1)} - \frac{\tilde{\sigma}^2}{2} \nabla \log h(\boldsymbol{\theta}^{(t-1)})||^2]} \;, \qquad (3.19)$$

   where $||\mathbf{a}||^2 = \sum_{i=1}^k a_i^2$, the squared length of $\mathbf{a}$.

3. Then as with any Metropolis-Hastings sampler,
   if $r \geq 1$, set $\boldsymbol{\theta}^{(t)} = \boldsymbol{\theta}^*$;

   if $r < 1$, set $\boldsymbol{\theta}^{(t)} = \begin{cases} \boldsymbol{\theta}^* & \text{with probability } r \\ \boldsymbol{\theta}^{(t-1)} & \text{with probability } 1 - r \end{cases}$.

As before, under mild conditions the draw $\boldsymbol{\theta}^{(t)}$ converges in distribution to a draw from the true posterior density $p(\boldsymbol{\theta}|\mathbf{y})$ as $t \to \infty$.

The intuition behind the drift term in (3.18) is clear in the case of a unimodal log-target density. Thinking in one dimension ($k = 1$) for simplicity, if $\theta^{(t-1)}$ is left of the mode, then the derivative of the log-target $\log h(\theta^{(t-1)})$ will be positive, encouraging the candidate to be farther to the right; if instead $\theta^{(t-1)}$ is right of the mode, now the negative derivative will push the candidate to the left. Roberts and Tweedie (1996a) show that the theoretical rate of convergence of Langevin-Hastings (L-H) algorithms can be substantially better than that of ordinary random walk Metropolis, and these advantages tend to grow with the size of the parameter space k. Of course, the derivative of $\log h$ must be available in closed form, though as with any Metropolis-type algorithm, we do not need to know the normalizing constant for h. Comparison of L-H with random walk Metropolis for different parameter dimensions k is left to the student as Exercise 18.

L-H algorithms do require some care in their implementation. Their convergence can sometimes break down if the chain is started very far from the posterior mode. An easy fix for this is to occasionally mix in a few ordinary random walk Metropolis iterations; see the discussion of mixtures of MCMC algorithms in Subsection 3.4.4 below. L-H will typically take longer to implement per iteration than straight Metropolis (due to the derivative calculation), but again the more rapid chain mixing speed often compensates for this.

Finally, L-H can perform poorly with heterogeneously scaled posteriors – say, if the true posterior of θ_1 had variance 1, but the true posterior of θ_2 had variance 10 (see Roberts and Stramer, 2003, for more examples and discussion). Note that in such a case, the algorithm's use of a single, univariate proposal variance parameter $\tilde{\sigma}^2$ would indeed seem to offer insufficient flexibility. According to Christensen et al. (2006), the best strategy is to start by making a transformation from $\boldsymbol{\theta}$-space to a new parameter space that standardizes the covariance structure, and perhaps also makes the parameters more nearly independent (i.e., produces bivariate contours that are more nearly circular, instead of elliptical or worse). If this can be done, then results in Roberts and Rosenthal (1998; 2001) for a true posterior consisting of i.i.d. standard normal random variables can be applied to select $\tilde{\sigma}^2$. Specifically, these authors recommend choosing $\tilde{\sigma}^2$ to deliver an acceptance rate of approximately 57.4%. These authors go on to recommend use of $\tilde{\sigma}^2 = \hat{\ell}^2/k^{1/3}$, where $\hat{\ell} = 1.65$, the solution to $2\Phi(-\hat{\ell}^3/8) = 0.574$.

On a related note, the requirement of a spherically symmetric and independent proposal density in (3.18) above can be relaxed with only a small change in the Langevin-Hastings algorithm. Suppose Σ^* is a a full rank square matrix that is a good preliminary estimate of the true posterior covariance matrix of $\boldsymbol{\theta}$. We then replace the proposal mean $\tilde{\boldsymbol{\mu}}$ and covariance matrix $\tilde{\Sigma}$ in (3.18) with

$$\tilde{\boldsymbol{\mu}}(\boldsymbol{\theta}^{(t-1)}) = \boldsymbol{\theta}^{(t-1)} + \frac{\tilde{\sigma}^2}{2}\Sigma^*\nabla \log h(\boldsymbol{\theta}^{(t-1)}) \quad \text{and} \quad \tilde{\Sigma} = \tilde{\sigma}^2\Sigma^* ,$$

and also make the corresponding modifications to the Langevin-Hastings acceptance ratio (3.19). As of the present writing, the use of *adaptively* scaled L-H algorithms of this sort is an area of current research.

3.4.3 Slice sampler

Yet another variant on the basic MCMC approach is to use *auxiliary variables*. In this approach, in order to obtain a sample from a particular posterior distribution $p(\boldsymbol{\theta}|\mathbf{y})$, we *expand* the original parameter space Θ using one or more *auxiliary* (or *latent*) variables. In the spirit of the original data augmentation algorithm, a clever enlargement of the state space can broaden the class of distributions we may sample, and also serve to accelerate convergence to the (marginal) stationary distribution. Use of this idea in MCMC sampling arose in statistical physics (Swendsen and Wang, 1987; Edwards and Sokal, 1988), was brought into the statistical mainstream by Besag and Green (1993), and recently developed further by Higdon (1998), and Damien et al. (1999).

The most popular auxiliary variable algorithm is the *slice sampler* (Neal, 2003). In its most basic form, suppose we seek to sample a univariate θ from $p(\theta|\mathbf{y}) \propto h(\theta) = f(\mathbf{y}|\theta)\pi(\theta)$ where $h(\theta)$ is known. Suppose we add an auxiliary variable U such that $U|\theta \sim Unif(0, h(\theta))$. Then the joint distribution of θ and U is

$$p(\theta, u) \propto h(\theta) \frac{1}{h(\theta)} \cdot I(U < h(\theta)) = 1 \cdot I(U < h(\theta)) , \qquad (3.20)$$

where I denotes the indicator function. This is a joint uniform density over the region $0 < U < h(\theta)$. A Gibbs sampler for this joint distribution would require only two uniform updates: $U|\theta \sim Unif(0, h(\theta))$, and $\theta|U \sim Unif(\theta : U \leq h(\theta))$. The first of these generations is trivial; the second is more complicated since it requires sampling a uniform restricted to the set $S_U = \{\theta : U < h(\theta)\}$. Iterating these two steps in the usual way produces samples from $p(\theta, U)$, and hence from the marginal distribution of θ, $p(\theta|\mathbf{y})$.

Figure 3.6 reveals why this approach is referred to as "slice sampling." U "slices" the nonnormalized density, and the resulting "footprint" on the axis provides S_U. If we can enclose S_U in an interval, we can draw θ uniformly on this interval and simply retain it only if $U < h(\theta)$ (i.e., if $\theta \in S_U$). It is for this reason that WinBUGS uses slice sampling to solve the nonconjugacy problem when the parameter in question has a *bounded* domain (e.g., $0 < \theta < 1$). If $\boldsymbol{\theta}$ is instead multivariate, S_U is more complicated and now we would need a bounding *hyperrectangle*.

Note that if $h(\theta) = h_1(\theta)h_2(\theta)$ where, say, h_1 is a standard density that is easy to sample while h_2 is nonstandard and difficult to sample, then we can introduce an auxiliary variable U such that $U|\theta \sim U(0, h_2(\theta))$. Now $p(\theta, u) = h_1(\theta)I(U < h_2(\theta))$. Again $U|\theta$ is routine to sample, while to sample $\theta|U$ we would now draw θ from $h_1(\theta)$ and retain it only if θ is such that $U < h_2(\theta)$.

Slice sampling incurs problems similar to rejection sampling in that we may have to draw many θs from h_1 before we are able to retain one. On the other hand, it has an advantage over the Metropolis-Hastings algorithm

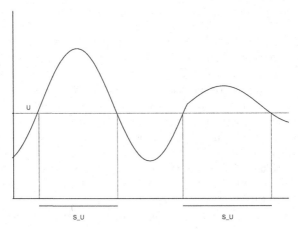

Figure 3.6 *Illustration of slice sampling. For this bimodal distribution, S_U is the union of two disjoint intervals.*

in that it always samples from the exact full conditional $p(\theta|u)$. As noted above, Metropolis-Hastings does not do this, and thus slice sampling would be expected to converge more rapidly. Nonetheless, overall comparison of computation time may make one method a winner for some cases, and the other a winner in other cases. Note also that, like the Gibbs sampler, this algorithm does not require the user to specify any tuning constants or proposal densities, and yet, like the Metropolis-Hastings algorithm, it allows sampling in nonconjugate model settings. Indeed, Higdon (1998) shows that the Metropolis algorithm can be viewed as a special case of the slice sampler.

Mira and Tierney (2001) and Roberts and Rosenthal (1999) give mild and easily verified regularity conditions that ensure rapid convergence of the slice sampler. Future work with slice sampling looks to its "perfect sampling" extension (Mira, Møller, and Roberts, 1999; see also our brief discussion of this area on page 154), as well as its application in challenging applied settings beyond those concerned primarily with image restoration, as in Hurn (1997) and Higdon (1998). An exciting development in this regard is the *polar slice sampler* (Roberts and Rosenthal, 2002), which has excellent convergence properties even for high dimensional target densities.

3.4.4 Hybrid forms, adaptive MCMC, and other algorithms

One of the nice features of MCMC algorithms is that they may be combined in a single problem, in order to take advantage of the strengths of each. To illustrate, suppose that $P_1, \ldots, P_m$ are MCMC samplers all having stationary distribution p. These might correspond to a Gibbs sampler and $(m-1)$ Metropolis-Hastings algorithms, each with a different proposal density. A *mixture* of the algorithms arises when at iteration t, P_i is chosen with probability α_i, where $\sum_{i=1}^m \alpha_i = 1$. Alternatively, we might simply use the algorithms in a *cycle*, wherein each P_i is used in a prespecified order. Clearly, a mixture or cycle of convergent algorithms will also produce a convergent algorithm.

Example 3.8 Let P_1 correspond to a standard Gibbs sampler, and P_2 correspond to an independence chain Hastings algorithm (i.e., one where $q(v, u) = g(v)$, independent of u). Suppose g is overdispersed with respect to the true posterior. For example, g might have roughly the same mode as p, but twice as large a standard deviation. Then a mixture of P_1 and P_2 with $(\alpha_1, \alpha_2) = (.9, .1)$ will serve to occasionally "jump start" the sampler, possibly with a large leap across the parameter space. This will in turn reduce autocorrelations while still preserving the convergence to the stationary distribution p. Perhaps more simply, P_2 could be used once every 10 or 100 iterations in a deterministic cycle. ∎

Another profitable method of combining MCMC algorithms is motivated by the nonconjugacy problem mentioned at the end of Subsection 3.4.1. Suppose that $\boldsymbol{\theta} = (\theta_1, \ldots, \theta_k)$, and that all of the full conditional distributions are available in closed form *except* for $p_i(\theta_i | \theta_{j \neq i}, \mathbf{y})$, perhaps due to nonconjugacy in the prior for θ_i. Since the Metropolis-Hastings algorithm does not require normalized densities from which to sample, we might naturally think of running a Gibbs sampler with a Metropolis *subalgorithm* to obtain the necessary $\theta_i^{(t)}$ sample at iteration t. That is, we would choose a symmetric *conditional* candidate density

$$q\left(v, \theta_i^{(t-1)} \;\middle|\; \theta_{j<i}^{(t)}, \; \theta_{j>i}^{(t-1)}, \mathbf{y}\right) \;,$$

and run a Metropolis subalgorithm for T iterations, accepting or rejecting the candidates as appropriate. We would then set $\theta_i^{(t)}$ equal to the end result of this subalgorithm, and proceed on with the outer Gibbs loop. Note that this same approach could be applied to any parameter for which we lack a closed form full conditional. This approach is often referred to as *Metropolis within Gibbs*, though many authors (notably Besag et al., 1995; Chib and Greenberg, 1995a; and Jain and Neal, 2007) dislike this term, noting that Metropolis et al. (1953) and Hastings (1970) envisioned such algorithms in their original papers – long before the appearance of the Gibbs sampler. As a result, the term *univariate Metropolis steps* might be used to distinguish the approach from univariate Gibbs steps (where the candidate is always accepted) and the fully multivariate Metropolis algorithm introduced in Subsection 3.4.2.

In using such an algorithm, two convergence issues arise. The first is a theoretical one: does this hybrid algorithm actually converge to the correct stationary distribution? The algorithm does not quite meet our definition of a cycle, because it is a deterministic combination of Gibbs and Metropolis steps that would *not* themselves converge if applied individually (since the former does not update θ_i, while the latter does not update the others). However, each of the components *does* preserve the stationary distribution of the chain, so it is easy to show that convergence still obtains.

The second issue is more practical: how should we select T, the number of Metropolis subiterations at each stage of the outer algorithm? Because convergence will occur for *any* value of T, we are free to pick whatever value we like, though clearly some choices will be better than others. $T = 100,000$ would no doubt be a poor choice, since all we would gain from this extensive sampling is high confidence that $\theta_i^{(t)}$ is indeed a sample from the correct full conditional distribution. This would be of little value early on (i.e., when t is small), since the outer Gibbs algorithm would still be far from locating the true stationary distribution, and so the expensive sample would not likely be retained for subsequent inference. On the other hand, $T = 1$ might correspond to such a crude approximation to the true full conditional that convergence of the overall algorithm is retarded. In practice, $T = 1$ is often adopted, effectively substituting one Metropolis rejection step for each Gibbs step lacking an easily sampled complete conditional.

Adaptive MCMC

The difficulty in finding good M-H candidate densities has led to a recent flurry of research into *adaptive* algorithms that attempt to use the early output from a chain to refine and improve the sampling as it progresses. While this is an extremely intuitive and sensible idea, great care is required because such algorithms are no longer Markov (they use more information from the chain that just the immediately previous value $\theta^{(t-1)}$ in determining $\theta^{(t)}$.

We begin with an early and widely-referenced (but sadly never-published) technical report by Müller (1991). This paper suggested using univariate Metropolis substeps for generation from unstandardized complete conditionals, as well as a simple transformation to approximately orthogonalize the parameter space. Specifically, at iteration t let $\widehat{\Sigma} = KK'$ be a current estimate and Cholesky decomposition of the covariance matrix of θ. Let $\eta^{(t)} = K^{-1}\theta^{(t)}$, and run a univariate Metropolis algorithm on the components of η instead of θ. If $p(\theta|\mathbf{y})$ resembles a normal distribution, then the components of η should be approximately uncorrelated, thus speeding convergence of the sampler. Müller recommends running this algorithm in parallel in order to obtain independent samples from which to obtain the covariance estimate $\widehat{\Sigma}$. For those η_is requiring a Metropolis update, the author suggests using a normal proposal density with variance 1, a logical choice if the orthogonalizing transformation has indeed been successful (though recall later work by Gelman, Roberts, and Gilks (1996) showed that the larger variance $(2.4)^2$ gives even better performance).

The adaptive aspect of Müller's idea is his suggestion to update the $\widehat{\Sigma}$ estimate at regular intervals using the current iterates, again in order to improve the orthogonalization and further speed convergence. Note that this continual updating of $\widehat{\Sigma}$ amounts to yet another form of hybrid MCMC algorithm: one that adaptively switches the transition kernel based on the output produced by the algorithm so far. While this seems perfectly sensible in this context, adaptive switching does *not* in general produce a convergent algorithm. The potential danger arises due to the fact that the algorithm is no longer a Markov chain, since it uses more than just the previous iteration to determine the next value (rather, it uses *all* of the previous values to determine the candidate variance $\widehat{\Sigma}$. For a more vivid illustration of the problem, we consider the following example, which was suggested by G.O. Roberts and developed more fully by Gelfand and Sahu (1994).

Example 3.9 Consider the following simple joint posterior distribution:

$$\begin{pmatrix} X \\ Y \end{pmatrix} \sim N \left(\mathbf{0}, \begin{pmatrix} 1 & \rho \\ \rho & 1 \end{pmatrix} \right)$$

Suppose we use the following adaptive algorithm. At iteration t, if $X^{(t)} < 0$ we transform to the new parameters $U = X + Y$ and $V = X - Y$, run one Gibbs step under this new (U, V) parametrization, and finally transform back to (X, Y). Elementary multivariate normal calculations show that the joint distribution of (U, V) is also normal with mean vector zero and covariance matrix

$$\begin{pmatrix} 1 & 1 \\ 1 & -1 \end{pmatrix} \begin{pmatrix} 1 & \rho \\ \rho & 1 \end{pmatrix} \begin{pmatrix} 1 & 1 \\ 1 & -1 \end{pmatrix} = \begin{pmatrix} 2(1+\rho) & 0 \\ 0 & 2(1-\rho) \end{pmatrix}.$$

That is, U and V can be independently generated as $N(0, 2(1 + \rho))$ and $N(0, 2(1 - \rho))$ variates, respectively.

If $X^{(t)} > 0$, we do not transform to (U, V)-space, instead drawing from the Gibbs full conditionals in the original parameter space. Clearly both of these transition kernels have the proper stationary distribution, but consider how the output from the adaptive algorithm will look. When $X^{(t)} < 0$, $X^{(t+1)}$ will be positive with probability $1/2$ and negative with probability $1/2$; its full conditional is symmetric about 0 and is independent of $X^{(t)}$. But when $X^{(t)} > 0$, $X^{(t+1)}$ will be positive with probability greater than $1/2$, since

$$X^{(t+1)} \,\Big|\, X^{(t)} \sim N(\rho^2 X^{(t)}, \, 1 - \rho^4) \,.$$

Taken together, these two results imply that long sequences from this algorithm will tend to have more positive Xs than negative. This is borne out in the simulation study conducted by Gelfand and Sahu (1994). With

$\rho = .9$ and $X^{(0)} \sim N(0,1)$, they obtained $\hat{P}(X^{(100)} > 0) = .784$ and $\hat{E}(X^{(100)}) = .6047.$ ∎

The first provably convergent adaptive MCMC algorithm seems to be the one due to Haario et al. (2001). These authors prove that the basic idea in Müller's approach *is* in fact acceptable *if* implemented with a proposal variance $\widehat{\Sigma}$ that is bounded and whose changes are diminishing over time – say, if the $\widehat{\Sigma}$ used at time t depends on *all* the chain output so far, $\boldsymbol{\theta}^{(0)}, \ldots, \boldsymbol{\theta}^{(t)}$. In particular, they suggest a $N(\boldsymbol{\theta}^{(t-1)}, (2.38)^2 \widehat{\Sigma}/d)$ proposal density, where d is the dimension of $\boldsymbol{\theta}$. A form of this adaptation is available in the OpenBUGS language. Subsequent work by Roberts and Rosenthal (2007) provides three conditions that guarantee convergence of an adaptive MCMC algorithm. While the conditions are technical and require Markov chain theory beyond the scope of this book, the key idea is that either the adaptation needs to be small with high probability (as in the Müller/Haario et al. approach), or else the adaptation needs to take place with probability that decreases to 0 as the number of iterations t grows large. Alternatively, Mira and Sargent (2003) show that adapting every m iterations for m fixed in advance will in most cases still maintain the chain's stationary distribution. Such *systematic adaptation* might be quite useful – say, in improving the current covariance estimate $\widehat{\Sigma}$ in Müller's approach above, or in the structured MCMC approach described in the next subsection.

In summary, adaptive (output-dependent) MCMC algorithms are often acceptable, but should be used with caution. Perhaps the safest approach is to adapt only in the initial, exploratory phase of the algorithm (pilot adaptation), and then stick with a single, nonadaptive approach when saving samples to be used for subsequent inference.

While the Gibbs sampler, the Metropolis-Hastings algorithm, and their hybrids are by far the two most common MCMC methods currently in use, there are a variety of extensions and other MCMC algorithms that are extremely useful in analyzing certain specific model settings or achieving particular methodological goals. For example, Bayesian model choice typically requires specialized algorithms, since the models compared are generally not of identical size (see Sections 4.4 and 4.5 for a discussion of these algorithms). The remainder of this subsection describes a few useful specialized algorithms.

Overrelaxation

As already mentioned, high correlations within the sample space can significantly retard the convergence of both Gibbs and univariate Metropolis-Hastings algorithms. Tranformations to reduce correlations, as in Müller's (1991) approach, offer a direct solution, but will likely be infeasible in high dimensions. An alternative is to use *overrelaxation*, which attempts to speed the sampled chain's progress through the parameter space by ensuring that each new value in the chain is *negatively* correlated with its predecessor. This method is the MCMC analogue of the overrelaxation methods long used to accelerate iterative componentwise maximization algorithms, such as the Gauss-Seidel method (see e.g. Thisted, 1988, Sec. 3.11.3).

In the case where all full conditional distributions $p(\theta_i|\theta_{j\neq i}, \mathbf{y})$ are Gaussian, Adler (1981) provides an effective overrelaxation method. Suppose θ_i has conditional mean μ_i and conditional variance σ_i^2. Then at iteration t, Adler's method draws $\theta_i^{(t)}$ as

$$\theta_i^{(t)} = \mu_i + \alpha(\theta_i^{(t-1)} - \mu_i) + \sigma_i(1 - \alpha^2)^{1/2}Z , \qquad (3.21)$$

where $-1 \leq \alpha \leq 1$, and Z is a standard normal random variate. Choosing $\alpha = 0$ produces ordinary Gibbs sampling; choosing $\alpha < 0$ produces overrelaxation, i.e., a $\theta_i^{(g)}$ sample on the other side of the conditional mean from $\theta_i^{(t-1)}$ ($\alpha > 0$ corresponds to underrelaxation). When $\alpha = -1$, the chain will not be convergent (it will get "stuck" tracing a single contour of the density), so values close to -1 (say, $\alpha = -0.98$) are often recommended.

Several authors generalize this approach beyond Gaussian distributions using transformations or Metropolis accept-reject steps, but these methods are often insufficiently general or suffer from unacceptably high rejection rates. Neal (1998) instead offers a generalization based on order statistics. Here we start by generating K independent values $\{\theta_{i,k}\}_{k=1}^{K}$ from the full conditional $p(\theta_i|\theta_{j\neq i}, \mathbf{y})$. Arranging these values and the old value in nondecreasing order (and breaking any ties randomly), we have

$$\theta_{i,0} \leq \theta_{i,1} \leq \cdots \leq \theta_{i,r} \equiv \theta_i^{(t-1)} \leq \cdots \leq \theta_{i,K} , \qquad (3.22)$$

so that r is the index of the old value in this list. Then we simply take $\theta_{i,K-r}$ as our new, overrelaxed value for $\theta_i^{(t)}$. The parameter K plays the role of α in Adler's method: $K = 1$ produces Gibbs sampling, while $K \to \infty$ is analogous to choosing $\alpha = -1$ in (3.21).

Neal (1998) shows how to implement this approach without explicit generation of the K random variables in (3.22) by transforming the problem to one of overrelaxation for a $Unif(0, 1)$ distribution. This in turn requires the ability to compute the cumulative distribution function and its inverse for each full conditional to be overrelaxed. Still, the method may be worthwhile even in settings where this cannot be done because the increase in runtime created by the K generations is typically offset by the resulting increase in sampling efficiency. Indeed, the WinBUGS package incorporates overrelaxation via direct sampling of the values in (3.22).

Blocking and Structured MCMC

A general approach to accelerating MCMC convergence is *blocking*, i.e., updating multivariate blocks of (typically highly correlated) parameters.

Recent work by Liu (1994) and Liu et al. (1994) confirms its good performance for a broad class of models, though Liu et al. (1994, Sec. 5) and Roberts and Sahu (1997, Sec. 2.4) give examples where blocking actually *slows* a sampler's convergence. Unfortunately, success stories in blocking are often application-specific, and general rules have been hard to find.

Sargent, Hodges, and Carlin (2000) introduce a simple, general, and flexible blocked MCMC method for a large class of hierarchical models called *Structured MCMC* (SMCMC). The approach accelerates convergence for many of these models by blocking parameters and taking full advantage of the posterior correlation structure induced by the model and data. As an elementary illustration, consider the balanced one-way random-effects model with JK observations y_{ij}, $i = 1, \ldots, K$, $j = 1, \ldots, J$. The first stage of the model assumes that $y_{ij} = \theta_i + \epsilon_{ij}$ where $\epsilon_{ij} \sim N(0, \sigma^2)$. At the second stage $\theta_i = \mu + \delta_i$ where $\delta_i \sim N(0, \nu)$, while at the third stage $\mu = M + \xi$ where $\xi \sim N(0, \tau^2)$. Adding prior distributions for σ^2 and ν completes the model specification. Rewriting the second and third stages as

$$0 = -\theta_i + \mu + \delta_i,$$
$$\text{and} \quad M = \mu - \xi,$$

the model can be expressed in the form of a linear model as

$$
\begin{bmatrix} \mathbf{y} \\ \hline \mathbf{0}_K \\ \hline M \end{bmatrix}
=
\left[
\begin{array}{ccc|c}
\mathbf{1}_J & \cdots & \mathbf{0}_J & \\
\vdots & \ddots & \vdots & \mathbf{0}_{JK} \\
\mathbf{0}_J & \cdots & \mathbf{1}_J & \\
\hline
\multicolumn{3}{c|}{-I_{K \times K}} & \mathbf{1}_K \\
\hline
\multicolumn{3}{c|}{\mathbf{0}_K'} & 1
\end{array}
\right]
\begin{bmatrix} \theta_1 \\ \vdots \\ \theta_K \\ \hline \mu \end{bmatrix}
+
\begin{bmatrix} \boldsymbol{\epsilon} \\ \hline \boldsymbol{\delta} \\ \hline -\xi \end{bmatrix}
\tag{3.23}
$$

where $\mathbf{y}' = (y_{11}, \ldots, y_{1J}, \cdots, y_{K1}, \ldots, y_{KJ})'$, $\boldsymbol{\epsilon}$ is partitioned similarly, and $\boldsymbol{\delta}' = (\delta_1, \ldots, \delta_K)'$. Hodges (1998) shows that Bayesian inferences based on (3.23) are identical to those drawn from the standard formulation.

A wide variety of hierarchical models can be reexpressed in a general form of which (3.23) is a special case, namely,

$$
\begin{bmatrix} \mathbf{y} \\ \hline \mathbf{0} \\ \hline M \end{bmatrix}
=
\left[
\begin{array}{c|c}
X_1 & \mathbf{0} \\
\hline
H_1 & H_2 \\
\hline
G_1 & G_2
\end{array}
\right]
\begin{bmatrix} \boldsymbol{\theta}_1 \\ \hline \boldsymbol{\theta}_2 \end{bmatrix}
+
\begin{bmatrix} \boldsymbol{\epsilon} \\ \hline \boldsymbol{\delta} \\ \hline \xi \end{bmatrix} .
\tag{3.24}
$$

In more compact notation, (3.24) can be expressed as

$$Y = X\boldsymbol{\theta} + E, \tag{3.25}$$

where X and Y are known, $\boldsymbol{\theta}$ is unknown, and E is an error term having

$Cov(E) = \Gamma$ with Γ block diagonal having blocks corresponding to the covariance matrices of each of ϵ, δ, and ξ, i.e.,

$$Cov(E) = \Gamma = Diag\left(Cov(\epsilon), Cov(\delta), Cov(\xi)\right) . \qquad (3.26)$$

In our simple example (3.23), Γ is fully (not just block) diagonal.

Hodges (1998) uses the term "data case" to refer to rows of X, Y, and E in (3.25) corresponding to X_1 in (3.24). These are the terms in the joint posterior into which the outcomes $\mathbf{y}$ enter directly. Rows of X, Y, and E corresponding to the H_i are "constraint cases." They place restrictions (stochastic constraints) on possible values of $\boldsymbol{\theta}_1$. Finally, the rows of X, Y, and E corresponding to the G_i are called "prior cases," this label being reserved for cases with known (specified) error variances. We refer to this format as the *constraint case formulation* of a hierarchical model.

For models with normal errors for the data, constraint, and prior cases, Gibbs sampling and (3.25) allow multivariate draws from the marginal posterior of $\boldsymbol{\theta}$. From (3.25), $\boldsymbol{\theta}$ has conditional posterior

$$\boldsymbol{\theta} \mid Y, X, \Gamma \sim N\left((X'\Gamma^{-1}X)^{-1}X'\Gamma^{-1}Y , (X'\Gamma^{-1}X)^{-1}\right) . \qquad (3.27)$$

With conjugate priors for the variance components (i.e., gamma or Wishart priors for their inverses), the full conditional for the variance components is a product of gamma and/or Wishart distributions. Convergence for such samplers is often virtually immediate; see Hodges (1998, Sec. 5.2), and the associated discussion by Wakefield (1998). The constraint case formulation works by using, in each MCMC draw, all the information in the model and data about the posterior correlations among the elements of $\boldsymbol{\theta}$.

The algorithm just outlined has three key features. First, it samples $\boldsymbol{\theta}$ (i.e., all of the mean-structure parameters) as a single block. Second, it does so using the conditional posterior covariance of $\boldsymbol{\theta}$, supplied by both the mean structure of the model, captured by X, and the covariance structure of the model, captured by Γ. Finally, it does so using the conditional distribution in (3.27), for suitable definitions of X and Γ. A SMCMC algorithm for a hierarchical model is any algorithm having these three features.

For linear models with normal errors, the suitable values of X and Γ are those in (3.25) and (3.26). These permit several different SMCMC implementations, the simplest of which is the blocked Gibbs sampler discussed above. However, although this Gibbs implementation may converge in few iterations, computing may be slow due to the need to invert a large matrix, $X'\Gamma^{-1}X$, at each iteration. An alternative would be to use (3.27) as a *candidate* distribution in a Hastings independence chain algorithm. Here we might occasionally update Γ during the algorithm's burn-in period. This pilot adaptation scheme is simple, but forces us to use a single (possibly imperfect) Γ for the post-convergence samples that are summarized for posterior inference. An even better alternative might be to continually update

Γ every m iterations for some preselected m (say, $m = 100$), a *systematic* adaptation scheme.

Example 3.10 A recent clinical trial (Abrams et al., 1994) compared didanosine (ddI) and zalcitabine (ddC) in HIV-infected patients. This dataset is described in detail in Section 8.1; for now, we take the response variable Y_i for patient i to be the change in CD4 count (for which higher levels indicate a healthier immune system) between baseline and the two-month follow-up, for the $K = 367$ patients who had both measurements. Three binary predictor variables and their interactions are of interest: treatment group (x_{1i}: $1 = $ ddC, $0 = $ ddI), reason for eligibility (x_{2i}: $1 = $ failed ZDV, $0 = $ intolerant to ZDV), and baseline Karnofsky score (x_{3i}: $1 = $ score > 85, $0 = $ score ≤ 85; higher scores indicate better health).

Consider the saturated model for this 2^3 factorial design:

$$Y_i = \theta_0 + x_{1i}\theta_1 + x_{2i}\theta_2 + x_{3i}\theta_3$$
$$+ x_{1i}x_{2i}\theta_{12} + x_{1i}x_{3i}\theta_{13} + x_{2i}x_{3i}\theta_{23} + x_{1i}x_{2i}x_{3i}\theta_{123} + \epsilon_i \,, \tag{3.28}$$

where $\epsilon_i \overset{iid}{\sim} N(0, \sigma^2)$, $i = 1, \ldots, 367$. We use flat priors for the intercept θ_0 and main effects ($\theta_1, \theta_2, \theta_3$), but place hierarchical constraints on the two- and three-way interaction terms, namely $\theta_l \sim N(0, \sigma_l^2)$ for $l = 12, 13, 23$, and 123. This linear model is easily written in the form (3.24). Adopting a vague $G(0.0001, 0.0001)$ prior for $\tau \equiv 1/\sigma^2$ (i.e., having mean 1 and variance 10^4) and independent $G(1, 1)$ priors for the $h_l \equiv 1/\sigma_l^2$, $l = 12, 13, 23, 123$, completes the specification.

Equation (3.27) yields a SMCMC implementation that alternately samples from the multivariate normal full conditional $p(\boldsymbol{\theta}|\mathbf{h}, \tau, \mathbf{y})$, and the gamma full conditionals $p(\tau|\boldsymbol{\theta}, \mathbf{h}, \mathbf{y})$ and $p(h_l|\boldsymbol{\theta}, \tau, \mathbf{y})$. As previously mentioned, this Gibbs sampler is a SMCMC implementation that updates the candidate Γ at every iteration. We also consider a pilot-adaptive SMCMC implementation, in which we update Γ at iterations $1, \ldots, 10$, then every 10th iteration until iteration 1000, and then use the value of Γ at iteration 1000 in the "production" run, discarding the 1000 iteration burn-in period. In this chain, $p(\boldsymbol{\theta}|\tilde{\mathbf{h}}, \tilde{\tau}, \mathbf{y})$ is used as a candidate distribution for $\boldsymbol{\theta}$ in a Hastings independence subchain, where $\tilde{\mathbf{h}}$ and $\tilde{\tau}$ are the components of Γ at iteration 1000.

Our model is fully conjugate, so WinBUGS handles it easily. Besides a standard univariate Gibbs sampler, WinBUGS allows only blocking of the fixed effects ($\theta_0, \theta_1, \theta_2, \theta_3$) into a single vector, for which we specified a multivariate normal prior having near-zero precision. We found pilot-adaptive SMCMC dominated the other three in terms of generation rate in "effective samples" per second (see Subsection 3.4.5 for a precise definition of effective sample size); for $\boldsymbol{\theta}$, the advantage is substantial. The WinBUGS chains were fast, but produced highly autocorrelated samples, hurting their effective generation rate. By contrast, SMCMC implemented as a fully blocked

Gibbs sampler produced essentially uncorrelated draws, but at the cost of long runtimes because of repeated matrix inversions. ∎

Next consider hierarchical nonlinear models, i.e., models in which the outcome $\mathbf{y}$ is not linearly related to the parameters in the mean structure. The data cases of such models do not fit the form (3.24), but the constraint and prior cases can often be written as a linear model. Because of the nonlinearity in the data cases, it is now less straightforward to specify the "suitable definitions of X and Γ" needed for a SMCMC algorithm. To do so, we must supply a linear structure approximating the data's contribution to the posterior. We do this by constructing artificial outcome data $\tilde{y}$ with the property that $E(\tilde{y}|\boldsymbol{\theta}_1)$ is roughly equal to $\boldsymbol{\theta}_1$. Rough equality suffices because we use $\tilde{y}$ only in a Metropolis-Hastings implementation to generate candidate draws for $\boldsymbol{\theta}$. Specifically, for a given $\tilde{y}$ with covariance matrix V, we can write the approximate linear model

$$
\begin{bmatrix} \tilde{y} \\ \hline 0 \\ \hline M \end{bmatrix} \approx \begin{bmatrix} I & | & 0 \\ \hline H_1 & | & H_2 \\ \hline G_1 & | & G_2 \end{bmatrix} \begin{bmatrix} \boldsymbol{\theta}_1 \\ \hline \boldsymbol{\theta}_2 \end{bmatrix} + \begin{bmatrix} \epsilon \\ \hline \delta \\ \hline \xi \end{bmatrix} , \tag{3.29}
$$

where $\epsilon \sim N(0, V)$. Equation (3.29) supplies the necessary X matrix for a SMCMC algorithm, and the necessary Γ uses $\text{cov}(\epsilon) = V$ and the appropriate covariance matrices for δ and ξ in (3.26).

The artificial outcome data $\tilde{y}$ can be supplied by two general strategies. The first makes use of crude parameter estimates from the nonlinear part of the model (without the prior or constraint cases), provided they are available. This might occur when the ratio of data elements to parameters is large and the constraint and prior cases are included solely to induce shrinkage in estimation. We call such estimates "unshrunk estimates," and use them and an estimate of their variance as $\tilde{y}$ and V, respectively.

In certain problems, however, such unshrunk estimates may be unstable, or the model may not be identified without constraint and prior cases. Here one can instead run a simple univariate Metropolis algorithm for a small number of iterations, using prior distributions for the variance components that *insist* on little shrinkage. The posterior mean of $\boldsymbol{\theta}$ approximated by this algorithm is a type of artificial data $\tilde{y}$ which, when used with the constraint and prior cases in (3.29), approximates the nonlinear fit well enough for the present purpose. These "low shrinkage" estimates and corresponding estimates of their variance can then be used as $\tilde{y}$ and V, respectively; see Sargent et al. (2000) for implementational details.

Many of the hybrid algorithms presented in this subsection attempt to remedy slow convergence due to high correlations within the parameter space. Examples include MCMC algorithms which employ *blocking* (i.e.,

updating parameters in medium-dimensional groups, as in SMCMC) and
collapsing (i.e., generating from partially marginalized distributions), as
well as overrelaxation and auxiliary variable methods like the slice sampler.
But of course, the problem of estimation in the face of high correlations
is not a new one; it has long been the bane of maximization algorithms
applied to likelihood surfaces. Such algorithms often use a transformation
designed to make the likelihood function more regular (e.g., in 2-space, a
switch from a parametrization wherein the effective support of the likelihood function is oblong to one where the support is more nearly circular). Müller's (1991) "orthogonalizing" transformation applied within his
adaptive univariate Metropolis algorithm is one such approach; see also
Hills and Smith (1992) in this regard. Within the class of hierarchical linear mixed models, Gelfand et al. (1995) discuss simple hierarchical centering reparametrizations that often enable improved algorithm convergence.
See Gelman (2004) for a more recent discussion of the issues surrounding
parametrization in Bayesian models.

3.4.5 Variance estimation

We now turn to the problem of obtaining estimated variances (equivalently,
standard errors) for posterior means obtained from MCMC output. Our
summary here is rather brief; the reader is referred to Ripley (1987, Chapter
6) for further details and alternative approaches.

Suppose that for a given parameter λ we have a single long chain of
MCMC samples $\{\lambda^{(t)}\}_{t=1}^{N}$, which for now we assume come from the stationary distribution of the Markov chain (we attack the convergence problem
in the next subsection). The simplest estimate of $E(\lambda|\mathbf{y})$ is then given by

$$\hat{E}(\lambda|\mathbf{y}) = \hat{\lambda}_N = \frac{1}{N} \sum_{t=1}^{N} \lambda^{(t)} \, ,$$

analogous to the estimator given in (3.2) for the case of i.i.d. sampling.
Continuing this analogy, then, we could attempt to estimate $Var(\hat{\lambda}_N)$ by
following the approach of (3.3). That is, we would simply use the sample
variance, $s_\lambda^2 = \frac{1}{N-1} \sum_{t=1}^{N} (\lambda^{(t)} - \hat{\lambda}_N)^2$, divided by N, obtaining

$$\widehat{Var}_{iid}(\hat{\lambda}_N) = s_\lambda^2/N = \frac{1}{N(N-1)} \sum_{t=1}^{N} (\lambda^{(t)} - \hat{\lambda}_N)^2 \, .$$

While this estimate is easy to compute, it would very likely be an underestimate due to positive autocorrelation in the MCMC samples. This
problem could be ameliorated somewhat by combining the draws from
a collection of initially overdispersed sampling chains. Alternatively, one
could simply use the sample variance of the Nth iteration output from m

independent parallel chains. But this approach is horribly wasteful, discarding an appalling $(N-1)/N$ of the samples. A potentially cheaper but similar alternative would be to subsample a single chain, retaining only every k^{th} sample with k large enough that the retained samples are approximately independent. However, MacEachern and Berliner (1994) give a simple proof using the Cauchy-Schwarz inequality that such systematic subsampling from a stationary Markov chain always increases the variance of sample mean estimators (though more recent work by MacEachern and Peruggia, 2000a, shows that the variance can decrease if the subsampling strategy is tied to the actual updates made). Thus, in the spirit of Fisher, it is better to retain all the samples and use a more sophisticated variance estimate, rather than systematically discard a large portion of them merely to achieve approximate independence.

One such alternative uses the notion of *effective sample size*, or *ESS* (Kass et al., 1998, p. 99). *ESS* is defined as

$$ESS = N/\kappa(\lambda) ,$$

where $\kappa(\lambda)$ is the *autocorrelation time* for λ, given by

$$\kappa(\lambda) = 1 + 2 \sum_{k=1}^{\infty} \rho_k(\lambda) ,$$

where $\rho_k(\lambda)$ is the autocorrelation at lag k for the parameter of interest λ. We may estimate $\kappa(\lambda)$ using sample autocorrelations estimated from the MCMC chain, cutting off the summation when these drop below, say, 0.1 in magnitude (though see Geyer, 1992, or Roberts, 1996, p.50, for a statistically more defensible approach based on an initial convex sequence estimator). The variance estimate for $\hat{\lambda}_N$ is then

$$\widehat{Var}_{ESS}(\hat{\lambda}_N) = s_\lambda^2/ESS(\lambda) = \frac{\kappa(\lambda)}{N(N-1)} \sum_{t=1}^{N} (\lambda^{(t)} - \hat{\lambda}_N)^2 .$$

Note that unless the $\lambda^{(t)}$ are uncorrelated, $\kappa(\lambda) > 1$ and $ESS(\lambda) < N$, so that $\widehat{Var}_{ESS}(\hat{\lambda}_N) > \widehat{Var}_{iid}(\hat{\lambda}_N)$, in concert with intuition. That is, because we have fewer than N effective samples, we expect some inflation in the variance of our estimate.

A final and somewhat easier (though also more naive) method of estimating $Var(\hat{\lambda}_N)$ is through *batching*. Divide our single long run of length N into m successive batches of length k (i.e., $N = mk$), with batch means $B_1, \ldots, B_m$. Clearly $\hat{\lambda}_N = \bar{B} = \frac{1}{m}\sum_{i=1}^{m} B_i$. We then have the variance estimate

$$\widehat{Var}_{batch}(\hat{\lambda}_N) = \frac{1}{m(m-1)} \sum_{i=1}^{m} (B_i - \hat{\lambda}_N)^2 , \qquad (3.30)$$

provided that k is large enough so that the correlation between batches is negligible, and m is large enough to reliably estimate $Var(B_i)$. It is important to verify that the batch means are indeed roughly independent – say, by checking whether the lag 1 autocorrelation of the B_i is less than 0.1. If this is not the case, we must increase k (hence N, unless the current m is already quite large), and repeat the procedure.

Regardless of which of the above estimates $\hat{V}$ is used to approximate $Var(\hat{\lambda}_N)$, a 95% confidence interval for $E(\lambda|\mathbf{y})$ is then given by

$$\hat{\lambda}_N \pm z_{.025}\sqrt{\hat{V}}\,,$$

where $z_{.025} = 1.96$, the upper .025 point of a standard normal distribution. If the batching method is used with fewer than 30 batches, it is a good idea to replace $z_{.025}$ by $t_{m-1,.025}$, the upper .025 point of a t distribution with $m-1$ degrees of freedom.

3.4.6 Convergence monitoring and diagnosis

We conclude our presentation on MCMC algorithms with a discussion of their convergence. Because their output is random and autocorrelated, even the definition of this concept is often misunderstood. When we say that an MCMC algorithm has *converged* at time T, we mean that its output can be safely thought of as coming from the true stationary distribution of the Markov chain for all $t > T$. This definition is not terribly precise since we have not defined "safely," but it is sufficient for practical implementation: no pre-convergence samples should be retained for subsequent inference.

Using a rigorous mathematical framework, many authors have attempted to establish conditions for convergence of various MCMC algorithms in broad classes of problems. For example, Roberts and Smith (1993) provide relatively simple conditions for the convergence of the Gibbs sampler and the Metropolis-Hastings algorithm. Roberts and Tweedie (1996b) show the geometric ergodicity of a broad class of Metropolis-Hastings algorithms, which in turn provides a central limit theorem (i.e., asymptotic normality of suitably standardized ergodic sums of the output from such an algorithm); see also Meyn and Tweedie (1993, Chapter 17) in this regard. Results such as these require elements of advanced probability theory that are well beyond the scope of this book, and hence we do not discuss them further. We do however discuss some of the common causes of convergence failure, and provide a brief review of several diagnostic tools used to make stopping decisions for MCMC algorithms.

Overparametrization and identifiability

Ironically, the most common source of MCMC convergence difficulties is a result of the methodology's own power. The MCMC approach is so generally applicable and easy to use that the class of candidate models for a given dataset now appears limited only by the user's imagination. However, with this generality has come the temptation to fit models so large that their parameters are *unidentified*, or nearly so. To see why this translates into convergence failure, consider the problem of finding the posterior distribution of θ_1 where the likelihood is defined by

$$Y_i|\theta_1, \theta_2 \stackrel{iid}{\sim} N(\theta_1 + \theta_2 , 1) , \quad i = 1, \ldots, n , \tag{3.31}$$

and we adopt independent flat priors for both θ_1 and θ_2. Only the sum of the two parameters is identified by the data, so without proper priors for θ_1 and θ_2, their marginal posterior distributions will be improper as well. Unfortunately, a naive application of the Gibbs sampler in this setting would not reveal this problem. The complete conditionals for θ_1 and θ_2 are both readily available for sampling as normal distributions. And for any starting point, the sampler would remain reasonably stable, due to the ridge in the likelihood surface $L(\theta_1, \theta_2)$ at $\theta_1 + \theta_2 = \bar{y}$. The inexperienced user might be tempted to use a smoothed histogram of the $\theta_1^{(g)}$ samples obtained as a (proper) estimate of the (improper) posterior density $p(\theta_1|\bar{y})$!

An experienced analyst might well argue that Gibbs samplers like the one described in the preceding paragraph are perfectly legitimate, *provided* that their samples are used only to summarize the posterior distributions of identifiable functions of the parameters (in this case, $g(\boldsymbol{\theta}) = \theta_1 + \theta_2$). While the deficiency in model (3.31) is immediately apparent, in more complicated settings (e.g., hierarchical and random effects models) failures in identifiability can be very subtle. Moreover, models that are overparametrized (either deliberately or accidentally) typically lead to high posterior correlations among the parameters (*crosscorrelations*), which will dramatically retard the movement of the Gibbs sampler through the parameter space. Even in models which *are* identified, but "just barely" so (e.g., model (3.31) with a vague but proper prior for θ_1), such high crosscorrelations (and the associated high *autocorrelations* in the realized sample chains) can lead to excruciatingly slow convergence. MCMC algorithms defined on such spaces are thus appropriate only when the model permits a firm understanding of which parametric functions are well-identified and which are not.

For the underidentified Gaussian linear model $\mathbf{y} = X\boldsymbol{\beta} + \boldsymbol{\epsilon}$ with X less than full column rank, Gelfand and Sahu (1999) provide a surprising MCMC convergence result. They show that under a flat prior on $\boldsymbol{\beta}$, the Gibbs sampler for the full parameter vector $\boldsymbol{\beta}$ is divergent, but the samples from the identified subset of parameters (say, $\boldsymbol{\delta} = X_1\boldsymbol{\beta}$) form an *exact* sample from their (unique) posterior density $p(\boldsymbol{\delta}|\mathbf{y})$. That is, such a sampler will produce identically distributed draws from the true posterior for $\boldsymbol{\delta}$, and convergence is immediate. In subsequent work, Gelfand, Carlin, and Trevisani (2000) consider a broad class of Gaussian models with covariates, namely,

$$\mathbf{Y} = (X_0 \ \ X_0\Delta_1 \ \ \cdots \ \ X_0\Delta_r) \, \boldsymbol{\beta} + \boldsymbol{\epsilon}$$

with $\boldsymbol{\beta} = (\beta_0, \beta_1, \ldots, \beta_r)'$. It then turns out $Corr(\boldsymbol{\delta}^{(t)}, \boldsymbol{\delta}^{(t+1)})$ approaches 0 as the prior variance component for β_0 goes to infinity once the chain has converged. This then permits independent sampling from the posterior distributions of estimable parameters. Exact sampling for these parameters is also possible in this case provided that *all* of the $\boldsymbol{\beta}$ prior variance components go to infinity. Simulation work by these authors suggests these results still hold under priors (rather than fixed values) for the variance components that are imprecise but with large means. For more on over-parametrization and posterior impropriety, see Natarajan and McCulloch (1995), Hobert and Casella (1996), and Natarajan and Kass (2000). Carlin and Louis (2000) point out that these dangers might motivate a return to EB analysis in such cases; c.f. Ten Have and Localio (1999) in this regard.

Proving versus diagnosing convergence

Theoretical results often provide convergence rates, which can assist in selecting between competing algorithms (e.g., linear versus quadratic convergence). But because the rates are typically available only up to an arbitrary constant, they are of little use in deciding when to stop a given algorithm. Some authors have made progress in obtaining bounds on the number of iterations T needed to guarantee that distribution being sampled at that time, $p^{(T)}(\boldsymbol{\theta})$, is in some sense within ϵ of the true stationary distribution, $p(\boldsymbol{\theta})$. For example, Rosenthal (1993; 1995a,b) uses Markov minorization conditions to provide bounds in continuous settings involving finite sample spaces and certain hierarchical models. Approaches like these hold great promise, but typically involve sophisticated mathematics in sometimes laborious derivations. Cowles and Rosenthal (1998) ameliorate this problem somewhat by showing how auxiliary simulations may often be used to verify the necessary conditions numerically, and at the same time provide specific values for use in the bound calculations. A second difficulty is that the bounds obtained in many of the examples analyzed to date in this way are fairly loose, suggesting numbers of iterations that are several orders of magnitude beyond what would be reasonable or even feasible in practice (though see Rosenthal, 1996, for some tight bounds in model (5.5), the two-stage normal-normal compound sampling model).

A closely related area that is the subject of intense recent research is *exact* or *perfect sampling*. This refers to MCMC simulation methods that can guarantee that a sample drawn at a given time will be *exactly* distributed according to the chain's stationary distribution. The most popular of these is the "coupling from the past" algorithm, initially outlined for discrete state spaces by Propp and Wilson (1996). Here, a collection of chains from different initial states are run at a sequence of starting times going *back-ward* into the past. When we go far enough back that all the chains have "coupled" by time 0, this sample is guaranteed to be an exact draw from the target distribution. Green and Murdoch (1999) extend this approach to continuous state spaces. The idea has obvious appeal, since it seems to eliminate the convergence problem altogether! However, to date, such algorithms have been made practical only for relatively small problems within fairly well-defined model classes; their extension to high-dimensional models (in which the conditioning offered by the Gibbs sampler would be critical) remains an open problem. Moreover, coupling from the past delivers only a *single* exact sample; to obtain the collection of samples needed for inference, the coupling algorithm would need to be repeated a large number of times. This would produce uncorrelated draws, but at considerably more expense than the usual "forward" MCMC approach, where if we can be sure we are past the burn-in period, every subsequent (correlated) sample can safely be thought of as a draw from the true posterior.

As a result of the practical difficulties associated with theoretical convergence bounds and exact sampling strategies, almost all of the applied work involving MCMC methods has relied on the application of diagnostic tools to output produced by the algorithm. Early attempts by statisticians in this regard (e.g., Gelfand et al., 1990) involved comparing the empirical distributions of output produced at consecutive (or nearly consecutive) iterations, and concluding convergence when the difference between the two was negligible in some sense. This led to wasteful samplers employing a large number of parallel, independent chains. Worse, this statistic was easily fooled into prematurely concluding convergence in slowly mixing samplers. This is because it measures the distance separating the sampled distribution at two different iterations, rather than that separating either from the true stationary distribution. Of course, since the stationary distribution will always be unknown to us in practice, this same basic difficulty will plague *any* convergence diagnostic. Indeed, this is what leads many theoreticians to conclude that all such diagnostics are fundamentally unsound.

Despite the fact that no diagnostic can "prove" convergence of a MCMC algorithm (since it uses only a finite realization from the chain), many statisticians rely heavily on such diagnostics, if for no other reason than "a weak diagnostic is better than no diagnostic at all." We agree that they do have value, though it lies primarily in giving a rough idea as to the nature of the mixing in the chain, and perhaps in keeping the analyst from making a truly embarrassing mistake (like reporting results from an improper posterior). As such, we now present a summary of a few of the methods proposed, along with their goals, advantages, and limitations. Cowles and Carlin (1996) and Mengersen et al. (1999) present much more complete reviews, describing the methodological basis and practical implementation of many such diagnostics that have appeared in the recent literature. Brooks and Roberts (1998) offer a similar review focusing on the underlying mathematics of the various approaches, rather than their implementation.

Convergence diagnostics

MCMC convergence diagnostics have a variety of characteristics, which we now list along with a brief explanation.

1. *Theoretical basis.* Convergence diagnostics have been derived using a broad array of mathematical machinery and sophistication, from simple cusum plots to advanced probability theory. Less sophisticated methods are often more generally applicable and easier to program, but may also be more misleading in complex modeling scenarios.

2. *Diagnostic goal.* Most diagnostics address the issue of *bias*, or the distance of the estimated quantities of interest at a particular iteration from their true values under the target distribution. But a few also consider *variance*, or the precision of those estimates.

3. *Output format.* Some diagnostics are quantitative, producing a single number summary, while others are qualitative, summarized by a graph or other display. The former type is easier to interpret (and perhaps even automate), but the latter can often convey additional, otherwise hidden information to the experienced user.

4. *Replication.* Some diagnostics may be implemented using only a single MCMC chain, while others require a (typically small) number of parallel chains. (This is one of the more controversial points in this area, as we shall illustrate below.)

5. *Dimensionality.* Some diagnostics consider only univariate summaries of the posterior, while others attempt to diagnose convergence of the full joint posterior distribution. Strictly speaking, the latter type is preferable because posterior summaries of an apparently converged parameter may be altered by slow convergence of other parameters in the model. Unfortunately, joint posterior diagnostics are notoriously difficult and likely infeasible in high-dimensional problems, leading us to substitute univariate diagnostics on a representative collection of parameters.

6. *Applicability.* Some diagnostics apply only to output from Gibbs samplers, some to output from any MCMC scheme, and some to a subset of algorithms somewhere between these two extremes. Again, there is often a tradeoff between broad applicability and diagnostic power or interpretability.

7. *Ease of use.* Finally, generic computer code is freely available to implement some convergence diagnostics. Generic code may be written by the user for others, and subsequently applied to the output from any MCMC sampler. Other diagnostics require that problem-specific code be written, while still others require advance analytical work in addition to

problem-specific code. Not surprisingly, the diagnostics that come with their own generic code are the most popular among practitioners.

We now consider several diagnostics, highlighting where they lie with respect to the above characteristics. Perhaps the single most popular approach is due to Gelman and Rubin (1992). Here, we run a small number (m) of parallel chains with different starting points, as in Figures 3.4 and 3.5. These chains must be initially overdispersed with respect to the true posterior; the authors recommend use of a preliminary mode-finding algorithm to assist in this decision. Running the m chains for $2N$ iterations each, we then attempt to check whether the variation within the chains for a given parameter of interest λ approximately equals the total variation across the chains during the latter N iterations. Specifically, we monitor convergence by the estimated *scale reduction factor*,

$$\sqrt{\hat{R}} = \sqrt{\left(\frac{N-1}{N} + \frac{m+1}{mN}\frac{B}{W}\right)\frac{df}{df-2}}\,, \qquad (3.32)$$

where B/N is the variance between the means from the m parallel chains, W is the average of the m within-chain variances, and df is the degrees of freedom of an approximating t density to the posterior distribution. Equation (3.32) is the factor by which the scale parameter of the t density might shrink if sampling were continued indefinitely; the authors show it must approach 1 as $N \rightarrow \infty$. This approach is available in WinBUGS, and an R implementation is also downloadable from CRAN.

Gelman and Rubin's approach can be derived using standard statistical techniques, and is applicable to output from any MCMC algorithm. It is quite easy to use and produces a single summary number on an easily interpreted scale (i.e., values close to 1 suggest good convergence). However, it has been criticized on several counts. First, because it is a univariate diagnostic, it must be applied to each parameter or at least a representative subset of the parameters; the authors also suggest thinking of minus twice the log of the joint posterior density as an overall summary parameter to monitor. The approach focuses solely on the bias component of convergence, providing no information as to the accuracy of the resulting posterior estimates, i.e.,

$$\hat{E}(\lambda|\mathbf{y}) = \frac{1}{mN}\sum_{j=1}^{m}\sum_{t=N+1}^{2N}\lambda_j^{(t)}\,.$$

Finally, the method relies heavily on the user's ability to find a starting distribution that is actually overdispersed with respect to the true posterior of λ, a condition we can't really check without knowledge of the latter!

Brooks and Gelman (1998) extend the Gelman and Rubin approach three important ways, describing an iterated graphical implementation of the original approach, broadening its definition to avoid the original's implicit normality assumption, and developing a multivariate generalization for simultaneous convergence diagnosis for every parameter in a model.

In a companion paper to the original Gelman and Rubin article, Geyer (1992) criticizes the authors' "initial overdispersion" assumption as unverifiable, and the discarding of early samples from several parallel chains as needlessly wasteful. He instead recommends a dramatically different approach wherein we run a *single* chain, and focus not on the bias but on the variance of resulting estimates, as computed for example in Subsection 3.4.5 above. Geyer argues that less than 1% of the run will normally be a sufficient burn-in period whenever the run is long enough to give much precision (recall Gelman and Rubin discard the first 50% of each chain). Still, this approach does not really address the possibility of bias in the posterior estimate, simply starting its sole chain at the MLE or some other "likely" place.

The diagnostic of Raftery and Lewis (1992) begins by retaining only every k^{th} sample after burn-in, with k large enough that the retained samples are approximately independent. More specifically, their procedure supposes our goal is to estimate a posterior quantile $q = P(\lambda < u|\mathbf{y})$ to within $\pm r$ units with probability s. (That is, with $q = .025$, $r = .005$, and $s = .95$, our reported 95% credible sets would have true coverage between .94 and .96.) To accomplish this, they assume $Z_t^{(k)} = Z_{1+(t-1)k}$ is a Markov chain, where $Z_t = I(\lambda^{(t)} < u)$, and use results from Markov chain convergence theory.

The Raftery and Lewis approach addresses both the bias and variance diagnostic goals, is broadly applicable, and is relatively easy to use thanks to computer code provided by the authors (and implemented within the CODA function for analyzing WinBUGS output in R or S-plus). However, Raftery and Lewis' formulation in terms of a specific posterior quantile leads to the rather odd result that the time until convergence for a given parameter differs for different quantiles. Moreover, systematic subsampling typically inflates the variance of the resulting estimates, as pointed out in Subsection 3.4.5.

Summary

Cowles and Carlin (1996) compare the performance of several convergence diagnostics in two relatively simple models, and conclude that while some are often successful, all can also fail to detect the sorts of convergence failures they were designed to identify. As a result, for a generic application of an MCMC method, they recommend a variety of diagnostic tools, rather than any single plot or statistic. Here, then, is a possible diagnostic strategy (and the one we essentially followed in our Chapter 2 WinBUGS investigations):

- Run a few (3 to 5) parallel chains, with starting points drawn (perhaps systematically, rather than at random) from a distribution believed to be overdispersed with respect to the stationary distribution (say, covering ± 3 prior standard deviations from the prior mean).

- Visually inspect these chains by overlaying their sampled values on a common graph (as in Figures 3.4 and 3.5) for each parameter, or, for very high-dimensional models, a representative subset of the parameters. (For a standard hierarchical model, such a subset might include most of the fixed effects, some of the variance components, and a few well-chosen random effects – say, corresponding to two individuals who are at opposite extremes relative to the population.)

- Annotate each graph with the Gelman and Rubin (1992) statistic and lag 1 autocorrelations, since they are easily calculated and the latter helps to interpret the former (large G&R statistics may arise from either slow mixing or multimodality).

- Investigate crosscorrelations among parameters suspected of being confounded, just as one might do regarding collinearity in linear regression.

The use of multiple algorithms with a particular model is often helpful, since each will have its own convergence properties and may reveal different features of the posterior surface. Considering multiple models is also generally a good idea; besides deepening our understanding of the data, this improves our chances of detecting convergence failure or hidden problems with model identifiability.

Note that convergence *acceleration* is a task closely related to, but more tractable than, convergence *diagnosis*. In many problems, the time spent in a dubious convergence diagnosis effort might well be better spent searching for ways to accelerate the MCMC algorithm, since if more nearly uncorrelated samples can be obtained, the diagnosis problem may vanish. We thus remind the reader that reparametrizations can often improve a model's correlation structure, and hence speed convergence; see Hills and Smith (1992) for a general discussion and Gelfand et al. (1995, 1996) for "hierarchical centering" approaches for random effects models. The more sophisticated MCMC algorithms presented in Subsection 3.4.4 (mixtures and cycles, overrelaxation, structured MCMC, slice sampling, etc.) also often lead to substantial reductions in runtime.

3.5 Exercises

1. Use Theorem 3.1 to obtain a first-order approximation to the Bayes factor BF, given in equation (2.19). Is there any computational advantage in the case where the two models are nested, i.e., $\boldsymbol{\theta}_1 = (\boldsymbol{\theta}_2, \boldsymbol{\gamma})$?

2. Suppose that $f(\mathbf{y}|\boldsymbol{\theta}) = \prod_{i=1}^n f(y_i|\boldsymbol{\theta})$. Describe how equation (3.1) could be used to investigate the effect of single case deletion (i.e., removing x_i from the dataset, for $i = 1, \ldots, n$) on the posterior expectation of $g(\boldsymbol{\theta})$.

3. Show that the probability of accepting a given θ_j candidate in the rejection sampling is c/M, where $c = \int L(\theta)\pi(\theta)d\theta$, the normalizing constant for the posterior distribution $h(\theta)$.

4. Suppose we have a sample $\{\theta_1, \ldots, \theta_N\}$ from a posterior $p_1(\theta|\mathbf{y})$, and we want to see the effect of changing the prior from $\pi_1(\theta)$ to $\pi_2(\theta)$. Describe how the weighted bootstrap procedure could be helpful in this regard. Are there any restrictions on the shape of the new prior π_2 for this method to work well?

Y	X	Y	X
925	1	480	20
870	2	486	22
809	4	462	25
720	4	441	25
694	5	426	30
630	8	368	35
626	10	350	40
562	10	348	50
546	12	322	50
523	15		

Table 3.4 *County property tax (Y) and age of home (X).*

5. Use quadratic regression to analyze the data in Table 3.4, which relate property taxes and age for a random sample of 19 homes. The response variable Y is property taxes assessed on each home (in dollars), while the explanatory variable X is age of the home (in years). Assume a normal likelihood and the noninformative prior $\pi(\boldsymbol{\beta}, \sigma) = 1/\sigma$, where $\boldsymbol{\beta}' = (\beta_0, \beta_1, \beta_2)$ and the model is $Y_i = \beta_0 + \beta_1 X_i + \beta_2 X_i^2 + \epsilon_i$. Find the first two moments (hence a point and interval estimate) for the intercept β_0 assuming:

   (a) standard frequentist methods.

   (b) asymptotic posterior normality (first order approximation).

6. In the previous problem, suppose that instead of the intercept β_0, the parameter of interest is the age corresponding to minimum tax under this model. Describe how you would modify the above analysis in this case.

7. Actually perform the point and interval estimation of the age corresponding to minimum property tax in the previous problem using a

noniterative Monte Carlo approach. That is, write an R program to generate $\beta^{(i)}$ values from the joint posterior distribution, and then, using the formula for the minimum tax age in terms of the βs, obtain Monte Carlo values from its posterior distribution.

(*Hint:* $\sigma^2|\mathbf{y} \sim IG(\frac{n-p}{2}, \frac{2}{RSS})$, where IG denotes the inverse gamma distribution, $p = 3$, the number of regressors, and $RSS = ||\mathbf{y} - \mathbf{X}\hat{\beta}||^2$, the residual sum of squares.)

Do you think the quadratic model is appropriate for these data? Why or why not?

8. Repeat the analysis of the previous problem in WinBUGS. Do your answers agree? List two advantages and two disadvantages of MCMC methods relative to noniterative MC methods.

9. Consider the following hierarchical changepoint model for the number of occurrences Y_i of some event during time interval i:

$$Y_i \sim \begin{cases} \text{Poisson}(\theta), & i = 1, \ldots, k \\ \text{Poisson}(\lambda), & i = k+1, \ldots, n \end{cases}$$

$\theta \sim G(a_1, b_1)$, $\lambda \sim G(a_2, b_2)$, θ, and λ independent;

$b_1 \sim IG(c_1, d_1)$, $b_2 \sim IG(c_2, d_2)$, b_1 and b_2 independent,

where G denotes the gamma and IG the inverse gamma distributions.

(a) Apply this model to the data in Table 3.5, which gives counts of coal mining disasters in Great Britain by year from 1851 to 1962. (Here, "disaster" is defined as an accident resulting in the deaths of 10 or more miners.) Set $a_1 = a_2 = .5$, $c_1 = c_2 = 1$, and $d_1 = d_2 = 1$ (a collection of "moderately informative" values). Also assume $k = 40$ (corresponding to the year 1890), and write an R program to obtain marginal posterior density estimates for θ, λ, and $R = \theta/\lambda$ using formula (3.9) with output from the Gibbs sampler.

(b) Re-do this problem in WinBUGS. Are your answers comparable?

10. In the previous problem, assume k is unknown, and adopt the following prior:

$$k \sim \text{Discrete Uniform}(1, \ldots, n), \text{ independent of } \theta \text{ and } \lambda.$$

Add k into the sampling chain, and obtain a marginal posterior density estimate for it. What is the effect on the posterior for R?

11. In the previous problem, replace the third-stage prior given above with

$$b_1 \sim G(c_1, d_1), b_2 \sim G(c_2, d_2), b_1 \text{ and } b_2 \text{ independent},$$

thus destroying the conjugacy for these two complete conditionals. Resort to rejection or Metropolis-Hastings subsampling for these two components instead, choosing appropriate values for c_1, c_2, d_1, and d_2. What is the effect on the posterior for b_1 and b_2? For R? For k?

Year	Count	Year	Count	Year	Count	Year	Count
1851	4	1879	3	1907	0	1935	2
1852	5	1880	4	1908	3	1936	1
1853	4	1881	2	1909	2	1937	1
1854	1	1882	5	1910	2	1938	1
1855	0	1883	2	1911	0	1939	1
1856	4	1884	2	1912	1	1940	2
1857	3	1885	3	1913	1	1941	4
1858	4	1886	4	1914	1	1942	2
1859	0	1887	2	1915	0	1943	0
1860	6	1888	1	1916	1	1944	0
1861	3	1889	3	1917	0	1945	0
1862	3	1890	2	1918	1	1946	1
1863	4	1891	2	1919	0	1947	4
1864	0	1892	1	1920	0	1948	0
1865	2	1893	1	1921	0	1949	0
1866	6	1894	1	1922	2	1950	0
1867	3	1895	1	1923	1	1951	1
1868	3	1896	3	1924	0	1952	0
1869	5	1897	0	1925	0	1953	0
1870	4	1898	0	1926	0	1954	0
1871	5	1899	1	1927	1	1955	0
1872	3	1900	0	1928	1	1956	0
1873	1	1901	1	1929	0	1957	1
1874	4	1902	1	1930	2	1958	0
1875	4	1903	0	1931	3	1959	0
1876	1	1904	0	1932	3	1960	1
1877	5	1905	3	1933	1	1961	0
1878	5	1906	1	1934	1	1962	1

Table 3.5 *Number of British coal mining disasters by year.*

12. Write an R program to reanalyze the flour beetle mortality data in Table 3.3, replacing the multivariate Metropolis algorithm used in Example 3.7 with

   (a) a Hastings algorithm employing independence chains drawn from a $N(\tilde{\boldsymbol{\theta}}, \tilde{\Sigma})$ candidate density, where $\tilde{\boldsymbol{\theta}} = (1.8, -4.0, -1.0)'$ (roughly the true posterior mode) and $\tilde{\Sigma}$ as given in (3.17).

   (b) a univariate Metropolis (Metropolis within Gibbs) algorithm using proposal densities $N(\theta_i^{(t-1)}, D_{ii})$, $i = 1, 2, 3$, with D as given in equation (3.16).

(c) a cycle or mixture of these two algorithms, as suggested in Example 3.8.

Based on your results, what are the advantages and disadvantages of each approach?

13. Returning again to the flour beetle mortality data and model of Example 3.7, note that the decision to use $\widetilde{\Sigma} = 2\widehat{\Sigma}$ in equation (3.17) was rather arbitrary. That is, univariate Metropolis "folklore" suggests a proposal density having variance roughly twice that of the true target should perform well, creating bigger jumps around the parameter space while still mimicking the target's correlation structure. But the *optimal* amount of variance inflation might well depend on the dimension and precise nature of the target distribution, the type of sampler used (multivariate versus univariate, Metropolis versus Hastings, etc.), or any number of other factors.

Explore these issues in the context of the flour beetle mortality data in Table 3.3 by resetting $\widetilde{\Sigma} = c\widehat{\Sigma}$ for $c = 1$ (candidate variance matched to the target) and $c = 4$ (candidate standard deviations twice those in the target) using

(a) the multivariate Hastings algorithm in part (a) of problem 12.
(b) the univariate Metropolis algorithm in part (b) of problem 12.
(c) the multivariate Metropolis algorithm originally used in Example 3.7.

Evaluate and compare performance using acceptance rates and lag 1 sample autocorrelations. Do your results offer any additions (or corrections) to the "folklore"?

14. Consider the generalized Hastings algorithm that uses the following candidate density:

$$q(v, u) = \begin{cases} p(v_i | u_{j \neq i}) & \text{for } v_{j \neq i} = u_{j \neq i} \\ 0 & \text{otherwise} \end{cases}.$$

That is, the algorithm chooses (randomly or deterministically) an index $i \in \{1, \ldots, K\}$, and then uses the full conditional distribution along the ith coordinate as the candidate density. Show that this algorithm has Hastings acceptance ratio (3.15) identically equal to 1 (as called for by the Gibbs algorithm), and hence that the Gibbs sampler is a special case of the Hastings algorithm.

15. Show that Adler's overrelaxation method (3.21) leaves the desired distribution invariant. That is, show that if $\theta_i^{(g-1)}$ is conditionally distributed as $N(\mu_i, \sigma_i^2)$, then so is $\theta_i^{(g)}$.

16. A random variable Z defined on $(0, \infty)$ is said to have a *D-distribution* with parameters $\delta, \beta > 0$ and $k \in \{0, 1, 2, \ldots\}$ if its density function is defined (up to a constant of proportionality) by

$$p_Z(z) \propto z^{\delta-1} e^{-\beta z} (1 - e^{-z})^k . \qquad (3.33)$$

This density emerges in many Bayesian nonparametric problems (e.g., Damien, Laud, and Smith, 1995). We wish to generate observations from this density.

(a) If Z's density is log-concave, we know we can generate the necessary samples using the method of Gilks and Wild (1992). Find a condition (or conditions) under which this density is guaranteed to be log-concave.

(b) When Z's density is not log-concave, we can instead use an auxiliary variable approach (Walker, 1995), adding the new random variable $U = (U_1, \ldots, U_k)$. The U_i are defined on $(0, 1)$ and mutually independent given Z, such that the joint density function of Z and U is defined (up to a constant of proportionality) by

$$p_{Z,U}(z, u) \propto z^{\delta-1} e^{-\beta z} \prod_{i=1}^{k} I_{(e^{-z}, 1)}(u_i) .$$

Show that this joint density function $p_{Z,U}$ does indeed have marginal density function p_Z given in equation (3.33) above.

(c) Find the full conditional distributions for Z and U, and describe a sampling algorithm to obtain the necessary Z samples. What special subroutines would be needed to implement your algorithm?

17. Consider three approaches for sampling from a $N(0, 1)$ target density, $p(x) \propto \exp(-\frac{1}{2}x^2)$:

- a standard Metropolis algorithm using a Gaussian proposal density, $x^* | x^{(t-1)} \sim N(x^{(t-1)}, \sigma^2)$,
- the simple slice sampler (3.20), and
- a Langevin-Hastings algorithm (3.18).

(a) Evaluate and compare the Metropolis and Langevin-Hastings acceptance ratios, the latter of which is given by (3.19).

(b) Write an R program to run single chains of 1000 iterations for each of these samplers, starting both chains at $x^{(0)} = 0$. For which values of the Metropolis proposal variance σ^2 do the various samplers produce the smallest lag 1 autocorrelations (i.e., faster convergence)?

18. Repeat the preceding investigation for a *multivariate* $N_k(\mathbf{0}, I_k)$ target density, $p(\mathbf{x}) \propto \exp(-\frac{1}{2}\mathbf{x}'\mathbf{x})$, where $\mathbf{x}' = (x_1, \ldots, x_k)$. Now we wish to compare these three samplers:

- a univariate Metropolis algorithm, updating one component at a time,
- a multivariate Metropolis algorithm, which updates all k components at one time, and
- a multivariate Langevin-Hastings algorithm (3.18).

(a) Suppose $k = 2$ (not much more challenging than the setting of the previous problem). Which sampler performs better?

(b) Suppose $k = 1000$. Does Langevin-Hastings now offer a noticeable improvement in efficiency, as claimed by Roberts and Rosenthal (1998)?

19. Consider the following two complete conditional distributions, originally analyzed by Casella and George (1992):

$$f(x|y) \propto ye^{-yx}, \quad 0 < x < B < \infty$$
$$f(y|x) \propto xe^{-xy}, \quad 0 < y < B < \infty$$

(a) Obtain an estimate of the marginal distribution of X when $B = 10$ using the Gibbs sampler.

(b) Now suppose $B = \infty$, so that the complete conditional distributions are ordinary (untruncated) exponential distributions. Show *analytically* that $f_x(t) = 1/t$ is a solution to the integral equation

$$f_x(x) = \int \left[\int f_{x|y}(x|y) f_{y|t}(y|t) dy \right] f_x(t) dt$$

in this case. Would a Gibbs sampler converge to this solution? Why or why not?

20. Consider the balanced, additive, one-way ANOVA model,

$$Y_{ij} = \mu + \alpha_i + \epsilon_{ij}, \quad i = 1, \ldots, I, \ j = 1, \ldots, J, \quad (3.34)$$

where $\epsilon_{ij} \overset{iid}{\sim} N(0, \sigma_e^2)$, $\mu \in \Re$, $\alpha_i \in \Re$, and $\sigma_e^2 > 0$. We adopt a prior structure that is a product of independent conjugate priors, wherein μ has a flat prior, $\alpha_i \overset{iid}{\sim} N(0, \sigma_\alpha^2)$, and $\sigma_e^2 \sim IG(a, b)$. Assume that σ_α^2, a, and b are known.

(a) Derive the full conditional distributions for μ, α_i, and σ_e^2, necessary for implementing the Gibbs sampler in this problem.

(b) What is meant by "convergence diagnosis"? Describe some tools you might use to assist in this regard. What might you do to improve a sampler suffering from "slow convergence"?

(c) Suppose for simplicity that σ_e^2 is also known. What conditions on the data or the priors might lead to slow convergence for μ and the α_i? (*Hint:* What conditions weaken the identifiability of the parameters?)

(d) Now let $\eta_i = \mu + \alpha_i$, so that η_i "centers" α_i. Then we can consider two possible parametrizations: (1) (μ, α), and (2) (μ, η). Generate a sample of data from likelihood (3.34), assuming $I = 5$, $J = 1$, and $\sigma_e = \sigma_\alpha = 1$. Write an R or WinBUGS program to investigate the sample crosscorrelations and autocorrelations produced by Gibbs samplers operating on parametrizations (1) and (2) above. Which performs better?

(e) Rerun your program for the case where $\sigma_e = 1$ and $\sigma_\alpha = 10$. Now which parametrization performs better? What does this suggest about the benefits of hierarchical centering reparametrizations?

21. To further study the relationship between identifiability and MCMC convergence, consider again the two-parameter likelihood model

$$Y \sim N(\theta_1 + \theta_2, 1),$$

with prior distributions $\theta_1 \sim N(a_1, b_1^2)$ and $\theta_2 \sim N(a_2, b_2^2)$, θ_1 and θ_2 independent.

(a) Clearly θ_1 and θ_2 are individually identified only by the prior; the likelihood provides information only on $\mu = \theta_1 + \theta_2$. Still, the full conditional distributions $p(\theta_1|\theta_2, y)$ and $p(\theta_2|\theta_1, y)$ are available as normal distributions, thus defining a Gibbs sampler for this problem. Find these two distributions.

(b) In this simple problem, we can also obtain the marginal posterior distributions $p(\theta_1|y)$ and $p(\theta_2|y)$ in closed form. Find these two distributions. Do the data update the prior distributions for these parameters?

(c) Set $a_1 = a_2 = 50$, $b_1 = b_2 = 1000$, and suppose we observe $y = 0$. Run the Gibbs sampler defined in part (a) in R for $t = 100$ iterations, starting each of your sampling chains near the prior mean (say, between 40 and 60), and monitoring the progress of θ_1, θ_2, and μ. Does this algorithm "converge" in any sense? Estimate the posterior mean of μ. Does your answer change using $t = 1000$ iterations?

(d) Now keep the same values for a_1 and a_2, but set $b_1 = b_2 = 10$. Again run 100 iterations using the same starting values as in part (b). What is the effect on convergence? Again repeat your analysis using 1000 iterations; is your estimate of $E(\mu|y)$ unchanged?

(e) Summarize your findings, and make recommendations for running and monitoring convergence of samplers running on "partially unidentified" and "nearly partially unidentified" parameter spaces.

## CHAPTER 4

Model criticism and selection

To this point we have seen the basic elements of Bayesian methods, arguments on behalf of their use from several different philosophical standpoints, and an assortment of computational algorithms for carrying out the analysis. We have observed that the generality of the methodology coupled with the power of modern computing enables consideration of a wide variety of hierarchical models for a given dataset. Given all this, the most natural questions for the reader to ask might be:

1. How can I tell if any of the assumptions I have made (e.g., the specific choice of prior distribution) is having an undue impact on my results?

2. How can I tell if my model is providing adequate fit to the data?

3. Which model (or models) should I ultimately choose for the final presentation of my results?

The first question concerns the *robustness* of the model, the second involves *assessment* of the model, and the third deals with *selection* of a model (or group of models). An enormous amount has been written on these three subjects over the last fifty or so years because they are the same issues faced by classical applied statisticians. The Bayesian literature on these subjects is of course smaller, but still surprisingly large, especially given that truly applied Bayesian work is a relatively recent phenomenon. We group the three areas together here since they all involve criticism of a model that has already been fit to the data.

Subsection 2.3.3 presented the fundamental tool of Bayesian model selection, the Bayes factor, while Section 2.5 outlined the basics of model assessment and model averaging. Armed with the computational techniques presented in Chapter 3, we now revisit and expand on these model building tools. With new MCMC-based model checking ideas arriving all the time (e.g., Dey et al., 1998; O'Hagan, 2003; Bayarri and Castellanos, 2007), we cannot hope to review all of what has been done. Still, in this chapter we will attempt to present the tools most useful for the applied Bayesian, along with sufficient exemplification so that the reader may employ the approaches independently.

4.1 Bayesian modeling

We begin this chapter with a discussion of several broad principles and strategies in Bayesian statistical modeling, intended to build and expand on the introductory hierarchical modeling material in Subsection 2.4. In particular, we wish to illustrate the basics of linear and nonlinear modeling for normal and binary data using `WinBUGS` in both the "flat" (nonhierarchical; no random effects) and hierarchical cases. Readers wishing to learn even more advanced modeling "tricks" (in both the statistical and `WinBUGS` senses of this word) may wish to consult the two books by Congdon (2003; 2007a) or the forthcoming book by Spiegelhalter et al. (2008).

4.1.1 Linear models

We begin then with the general linear model for normally distributed data, arguably the single most important contribution of statistics to the field of scientific inquiry. The bulk of statistical models appearing in print have this basic form, with regression and analysis of variance models being particularly widely used special cases. A Bayesian analysis of this model was first presented in the landmark paper by Lindley and Smith (1972), which we summarize here. Suppose that $\mathbf{Y}|\boldsymbol{\theta}_1 \sim N(A_1\boldsymbol{\theta}_1, C_1)$, where $\mathbf{Y}$ is an $n \times 1$ data vector, $\boldsymbol{\theta}_1$ is a $p_1 \times 1$ parameter vector, A_1 is an $n \times p_1$ known design matrix, and C_1 is an $n \times n$ known covariance matrix. Suppose further that we adopt the prior distribution $\boldsymbol{\theta}_1 \sim N(A_2\boldsymbol{\theta}_2, C_2)$, where $\boldsymbol{\theta}_2$ is a $p_2 \times 1$ parameter vector, A_2 is a $p_1 \times p_2$ design matrix, C_2 is a $p_1 \times p_1$ covariance matrix, and $\boldsymbol{\theta}_2, A_2$, and C_2 are all known. Then the *marginal* distribution of $\mathbf{Y}$ is

$$\mathbf{Y} \sim N(A_1A_2\boldsymbol{\theta}_2, C_1 + A_1C_2A_1'), \tag{4.1}$$

and the *posterior* distribution of $\boldsymbol{\theta}_1$ is

$$\boldsymbol{\theta}_1|\mathbf{y} \sim N(D\mathbf{d}, D), \tag{4.2}$$

where

$$D^{-1} = A_1'C_1^{-1}A_1 + C_2^{-1}, \tag{4.3}$$

and

$$\mathbf{d} = A_1'C_1^{-1}\mathbf{y} + C_2^{-1}A_2\boldsymbol{\theta}_2. \tag{4.4}$$

Thus $E(\boldsymbol{\theta}_1|\mathbf{y}) = D\mathbf{d}$ provides a point estimate for $\boldsymbol{\theta}_1$, with associated variability captured by the posterior covariance matrix $Var(\boldsymbol{\theta}_1|\mathbf{y}) = D$.

Example 4.1 As a concrete illustration, we revisit the "linearized" version of the dugong (sea cow) growth data originally plotted in Figure 2.8 and analyzed in Example 2.10. There we employed a simple linear regression model,

$$Y_i = \beta_0 + \beta_1 \log(x_i) + \epsilon_i,$$

where Y_i is the length of the dugong in meters, x_i is the log of its age in years, and the ϵ_i are i.i.d. normal with mean zero and precision $\tau = 1/\sigma^2$. This model can be cast in our general linear model framework by setting $\theta_1 = \beta = (\beta_0, \beta_1)'$, $C_1 = \sigma^2 I_n$, and

$$A_1 = X = \begin{pmatrix} 1 & \log(x_1) \\ 1 & \log(x_2) \\ \vdots & \vdots \\ 1 & \log(x_n) \end{pmatrix}.$$

A noninformative prior is provided by taking $C_2^{-1} = \mathbf{0}$, i.e., setting the prior precision matrix equal to a $p_1 \times p_1$ matrix of zeroes. Then from equations (4.3) and (4.4), we have

$$D^{-1} = X'(\sigma^2 I_n)^{-1}X + \mathbf{0} = \frac{1}{\sigma^2}(X'X), \tag{4.5}$$

and

$$\mathbf{d} = X'(\sigma^2 I_n)^{-1}\mathbf{y} + \mathbf{0} = \frac{1}{\sigma^2}(X'\mathbf{y}), \tag{4.6}$$

so that the posterior mean is given by

$$D\mathbf{d} = \left[\frac{1}{\sigma^2}(X'X)\right]^{-1} \frac{1}{\sigma^2}(X'\mathbf{y}) = (X'X)^{-1}X'\mathbf{y} = \hat{\beta}_{LS},$$

the usual least squares estimate of β. From (4.2), the posterior distribution of β is

$$\beta|\mathbf{y} \sim N(\hat{\beta}_{LS}, \sigma^2(X'X)^{-1}). \tag{4.7}$$

Recall that the sampling distribution of the least squares estimate is given by $\hat{\beta}_{LS}|\beta \sim N(\beta, \sigma^2(X'X)^{-1})$, so that classical and noninformative Bayesian inferences regarding β will be formally identical in this example.

If we return to a $N_2(\mu, R^{-1})$ prior for β, but continue to (unrealistically) assume for the moment that τ, μ, and R are all fixed and known, then equations (4.5) and (4.6) yield

$$D^{-1} = X'(\sigma^2 I_n)^{-1}X + R = \tau X'X + R,$$

and

$$\mathbf{d} = X'(\sigma^2 I_n)^{-1}\mathbf{y} + R\mu = \tau(X'\mathbf{y}) + R\mu.$$

Taking $R = 0$ delivers the simple posterior distribution for β given in (4.7).

The closed form solutions we have just described are easily studied in the R language. For our dugong data, suppose we make our bivariate normal prior for β virtually "flat" by by setting $R = Diag(0.001, 0.001)$. We also set $\mu = (0,0)'$ and consider two values of τ, 1 and 100. The R code to set up the data and prior values is

R code
```
X <-cbind(rep(1,n),lgage)
mu <- c(0,0)
R <- matrix(c(0.001,0,0,0.001), nrow=2)
tau <- c(1,100)
```

where Y, x and lgage are as previously defined in Example 2.10. A simple function to determine the posterior of β is then

R code
```
postfn <- function(Y, X, mu, R, tau){
    p <- dim(X)[2]
    D <- solve( tau*t(X)%*%X + R)
    d <- tau*t(X)%*%Y + R%*%diag(rep(1,p))%*%mu
    postmean <- D%*%d
    postsd <- sqrt(diag(D))
    return(list(mean= D%*%d, sd=sqrt(diag(D)) ))
    }
```

Note this function makes liberal use of R's matrix multiply function, %*%. We now apply our new postfn function to the dugong data to obtain the posterior mean and variance of the slope β_1, as well as a 95% confidence interval. For $\tau = 1$ we have

R code
```
post1 <- postfn(Y, X, mu, R, tau[1]) # posterior for tau=1
beta <- post1$mean[2]
postsd <- post1$sd[2]
CI1 <- c(beta+qnorm(0.025)*postsd , beta+qnorm(0.975)*postsd)
```

Now typing CI1 reveals the interval to be $(-0.132, 0.687)$. Repeating the above calculation for $\tau = 100$

R code
```
post2 <- postfn(Y, X, mu, R, tau[2]) # posterior for tau=100
beta <- post1$mean[2]
postsd <- post1$sd[2]
CI2 <- c(beta+qnorm(0.025)*postsd , beta+qnorm(0.975)*postsd)
```

This instead delivers a CI of $(0.236, 0.318)$. Note that this second interval is very similar to the frequentist interval we obtained in Example 2.10. The first interval is enormously wider because the far smaller τ value implies far lower faith in the data.

Finally, we can easily compute and plot the two posterior densities:

R code
```
theta <- seq(-0.2,0.6,length.out=100)
dens1<-dnorm(theta, mean=post1$mean[2], sd=post1$sd[2] )
dens2<-dnorm(theta, mean=post2$mean[2], sd=post2$sd[2] )
plot(theta, dens1, xlab=expression(beta), xlim=c(-0.2, 0.6),
    ylim=c(0,20), ylab="posterior density", type="n")
lines(theta, dens1, lty=1)        # posterior density for tau=1
lines(theta, dens2, lty=2)        # posterior density for tau=100
```

Setting $\tau = 1$ (i.e., high data precision) leads to the solid (lty = 1, line type #1) posterior in Figure 4.1. If we instead use $\tau = 100$, we obtain the dashed (lty = 2, line type #2) curve in the figure. Again, the increase in posterior precision arises from the tighter prior's willingness to lend more credence to the data.

Finally, we can also draw a sample from this posterior, and add this histogram to our plot along with a legend:

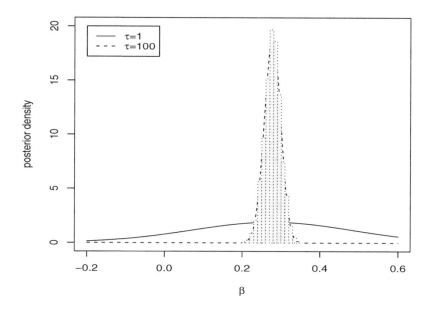

Figure 4.1 *Posterior for β_1, linearized dugong data, for two values of the data precision τ. A histogram of posterior samples is also shown in the $\tau = 100$ case.*

R code
```
postdraw <- rnorm(2000, mean=post2$mean[2], sd=post2$sd[2] )
r1<-hist(postdraw,freq=F,breaks=20, plot=F)
lines(r1,lty=3, freq=F, col="gray90")    #posterior draws, tau=100
legend(-0.2, 20, legend=c(expression(paste(tau,"=1", sep="")),
    expression(paste(tau,"=100", sep=""))), lty=c(1,2), ncol=1)
```

These values are shown as the dotted (`lty = 3`, line type #3) histogram in Figure 4.1. ∎

If $C_1 = Var(\mathbf{Y}|\boldsymbol{\theta}_1)$ is unknown in a general linear model, or if the model has more than two levels (say, a third-stage hyperprior, in addition to the likelihood and the prior), then a closed form analytic solution for the posterior mean of $\boldsymbol{\beta}$ will typically be unavailable. Fortunately, sampling-based computational methods are available to escape this unpleasant situation, as we have already seen in Example 2.10 and elsewhere.

Example 4.2 In the previous example, we adopted a $N_2(\boldsymbol{\mu}, R^{-1})$ prior for $\boldsymbol{\beta}$, but assumed $\boldsymbol{\mu}$ and R were fixed and known. (Our flat priors on β_0 and β_1 are asymptotically equivalent to setting $R^{-1} = 0$.) Now suppose we wish to place third-stage hyperpriors on these parameters; specifically, a flat hyperprior on $\boldsymbol{\mu}$ and a $Wishart(\nu, \Omega)$ hyperprior on the prior precision matrix R. Recall that we can make this latter hyperprior vague (but still proper) by setting $\nu = rank(R) = 2$.

Because we must use sampling-based methods in WinBUGS to handle this setting, we follow our original Example 2.10 dugong model and place a uniform prior on the data standard deviation σ as well. The WinBUGS code for this problem is

BUGS code
```
model{
    for( i in 1:n) {
        logage[i] <- log(x[i])
        Y[i] ~ dnorm(meen[i] , tau)
        meen[i] <- beta[1]+ beta[2]*(logage[i] - mean(logage[]))
        }

    beta[1:2] ~ dmnorm(mu[], R[ , ])
    mu[1] ~ dflat()
    mu[2] ~ dflat()
    R[1:2, 1:2] ~ dwish(Omega[ , ], 2)

    tau <- 1/(sigma*sigma)
    sigma ~ dunif(0.01, 100)
}
```

Note that now β_1 is the intercept and β_2 is the slope, since WinBUGS does not permit 0 subscripts. The full conditional for β remains exactly as in (4.7), since when developing the full conditional for a parameter we are allowed to condition on all other parameters in the model. Thus we would now *label* the full conditional as $p(\beta|y, \mu, \sigma, R)$ instead of $p(\beta|y)$, but its form is completely unchanged. The full conditional for μ is also bivariate normal, though WinBUGS generates its two components separately as univariate normals, since the flat priors on the two μ components are being specified separately in the code. The full conditional for R is Wishart by conditional conjugacy, while σ's nonconjugate specification over a bounded domain causes WinBUGS to select slice sampling as its updating method.

The Data and Inits files are natural extensions of the ones seen in Example 2.10:

BUGS code
```
# Data:
list(x = c( 1.0,   1.5,   1.5,   1.5, 2.5,   4.0,   5.0,   5.0,   7.0,
            8.0,   8.5,   9.0,   9.5, 9.5,   10.0, 12.0, 12.0, 13.0,
           13.0, 14.5, 15.5, 15.5, 16.5, 17.0, 22.5, 29.0, 31.5),
     Y = c(1.80, 1.85, 1.87, 1.77, 2.02, 2.27, 2.15, 2.26, 2.47,
           2.19, 2.26, 2.40, 2.39, 2.41, 2.50, 2.32, 2.32, 2.43,
           2.47, 2.56, 2.65, 2.47, 2.64, 2.56, 2.70, 2.72, 2.57),
     n = 27,
     Omega = structure(.Data = c(0.1, 0, 0, 0.1), .Dim = c(2, 2)))

# Inits:
```

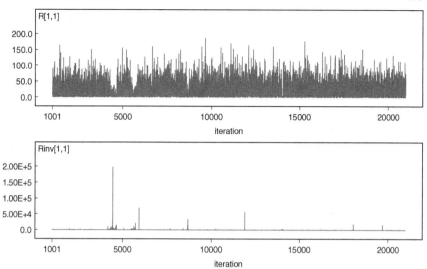

Figure 4.2 *Traceplots for R_{11} and $(R^{-1})_{11}$, linearized dugong model.*

```
list( beta = c(0, 0), sigma=1, mu = c(0,0),
      R = structure(.Data = c(1,0,0,1), .Dim = c(2, 2)))
```

The primary differences here are the initialization of β and μ as vectors, the initialization of R, and the specification of the Ω matrix (a rough guess for R/ν). Here we simply set $\Omega = Diag(0.1, 0.1)$ because, while we expect the βs to be roughly uncorrelated thanks to our centering of the logage covariate, we have little idea as to their scale (at least not without "peeking" at the β posteriors obtained in previous examples). Running this code produces very similar DIC scores and posterior estimates for β and σ as obtained in Example 2.10, or even Example 4.1 when fixing $\tau = 100$.

The sampler converges well, but the hyperparameters remain hard to estimate. To see this, we monitor both R and R^{-1}. To do this we must first explicitly add the latter into our code using the intrinsic **inverse** function:

BUGS code
```
Rinv[1:2, 1:2] <- inverse(R[ , ])
priorsd[1] <- sqrt(Rinv[1,1])
priorsd[2] <- sqrt(Rinv[2,2])
```

Notice we have also defined a bivariate node **priorsd** to capture the prior standard deviations of the intercept and slope, respectively. Traceplots for R_{11} and $(R^{-1})_{11}$ based on 20,000 post-burn-in iterations are shown in Figure 4.2. While the former traces are reasonably stable and well-behaved, the occasional near-singularity of R causes a few extreme values to dominate the $(R^{-1})_{11}$ traces. This instability is inherited by **priorsd**: both of its component distributions are extremely heavy-tailed, with posterior means that fluctuate over time but reasonably stable posterior medians (both around 0.49 after 60,000 iterations).

A related point is that these two estimates are not the same as the posterior standard deviations of the βs themselves (which are about 0.018 and 0.020, respectively – quite a bit smaller). These latter values reflect the bulk of information in the data, while the priorsd values essentially just reflect the minimal contribution of the Wishart hyperprior. This illustrates why a three-stage model is perhaps unnecessary here; there are no "replications" β_k from which to estimate the second-stage variability captured by R^{-1}. If however we had multiple *species* or *herds* of dugong indexed by k, then the third-stage of our current model would make perfect sense, and the posteriors of the priorsd parameters would now likely reflect any cross-species or cross-herd variability in the data, and not just whatever information we placed in the hyperprior. ∎

4.1.2 Nonlinear models

Next we consider the case of nonlinear modeling, where we replace the linear mean structure $Y_i = \mathbf{x}_i'\boldsymbol{\beta} + \epsilon_i$ with a more generic form $Y_i = g(\mathbf{x}_i, \boldsymbol{\beta}) + \epsilon_i$ for some known function g. Here the Lindley and Smith (1972) result (4.2) typically will not apply to the full conditional distribution of the main effects $\boldsymbol{\beta}$, but may still be useful for higher stage parameters having linear model hyperpriors. In any case, posterior samples may still be generated using a mixture of Gibbs and Metropolis steps.

Example 4.3 Carlin and Gelfand (1991b) present a nonconjugate Bayesian analysis of the dugong data set, following the approach of Ratkowsky (1983). The idea is to model the untransformed data (closed circles in Figure 4.4 below) using the nonlinear growth model

$$Y_i = \alpha - \beta\gamma^{x_i} + \epsilon_i, \ i = 1, \ldots, n , \tag{4.8}$$

where $\alpha > 0$, $\beta > 0$, $0 \le \gamma \le 1$ and as usual $\epsilon_i \overset{iid}{\sim} N(0, \sigma^2)$ for $\sigma^2 > 0$. Thus α corresponds to the average length of a fully grown dugong, $(\alpha - \beta)$ is the length of a dugong at birth, and γ determines the growth rate. Specifically, lower values of γ produce an initially steep growth curve (rapid progression to adulthood immediately after birth), while higher γ values lead to much more gradual, almost linear growth.

The nonlinearity of the model eliminates any hope for a closed form full conditional for γ regardless of our choice of prior, so we proceed with a sampling-based solution. The WinBUGS code to fit this model (adapted from that given in the **Examples Vol II** section of the **Help** pull-down) is

BUGS code
```
model{
    for( i in 1 : N){
        Y[i] ~ dnorm(mu[i], tau)
```

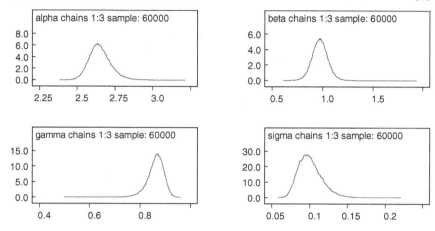

Figure 4.3 *Posterior density estimates, nonlinear dugong model.*

```
    mu[i] <- alpha - beta * pow(gamma,x[i])
    }
alpha ~ dflat()
beta ~ dflat()
gamma ~ dunif(0.5, 1.0)

tau <- 1/(sigma*sigma)
sigma ~ dunif(0.01, 100)
}
```

Note flat priors are perfectly suitable for the two "endpoint" parameters α and β, but the harder-to-estimate growth parameter γ benefits from a tighter (albeit uniform) specification.

We run three parallel Gibbs sampling chains of 20,000 iterations each following a 1000-iteration burn-in, with the chains initialized as

BUGS code   # Inits:
```
list(alpha = 1, beta = 1, sigma = 1, gamma = 0.9)
list(alpha = 10, beta = 10, sigma = 10, gamma = 0.7)
list(alpha = 100, beta = 100, sigma = 100, gamma = 0.5)
```

The resulting Gibbs sampler seems to mix well, as indicated by rapid agreement among the three chains' sample traces (not shown). Figure 4.3 shows the resulting density estimates for all four model parameters. Estimation appears good, with the fitted growth curve (obtained by plugging point estimates for α, β, and γ into equation (4.8)) closely following the observed data; again see Figure 4.4.

Autocorrelation plots (not shown) obtained via the `auto cor` button on the `Sample` monitor tool suggest rather high autocorrelations (i.e., rather different from 0 even at lag 20) only for α and γ. The bivariate scatterplot

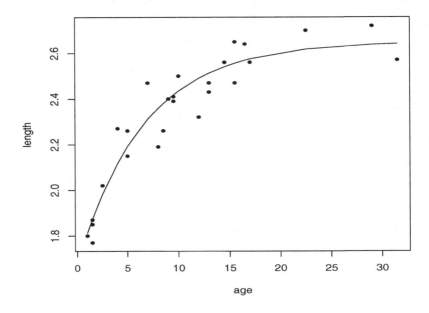

Figure 4.4 *Plot of the untransformed dugong data (length versus age) with fitted nonlinear growth curve superimposed.*

of the post-burn-in sampled values of these two parameters shown in Figure 4.5 reveals the reason for this slow convergence: these two parameters are highly correlated with a posterior that is roughly "boomerang-shaped." Larger values of α (the length of full grown dugongs) are generally associated with larger values of γ (i.e., a more gradual growth curve), but this relationship is complex and highly nonlinear. This complexity does not in and of itself cast doubt on the model, but is certainly typical of the problems we face when fitting nonlinear models: they often produce awkward joint posterior surfaces that will cause almost any MCMC algorithm to struggle. ∎

4.1.3 Binary data models

Having handled the case of continuous response data, the next natural case to address is that of discrete response (binary or count) data. We do this in the context of the dugong data somewhat artificially by discretizing the response, as in the following example.

Example 4.4 Consider the following binary version of the dugong exam-

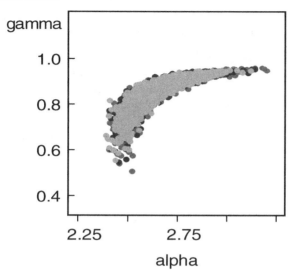

Figure 4.5 *Bivariate sample plot, α versus γ, nonlinear dugong model.*

ple, where we define the response

$$Z_i = \begin{cases} 1 & \text{if } Y_i > 2.4 \text{ (i.e., the dugong is full grown)} \\ 0 & \text{otherwise} \end{cases}$$

That is, $Z_i = 1$ if the dugong is "full grown," and 0 otherwise. We can then model $p_i = P(Z_i = 1)$ as

$$logit(p_i) = \log[p_i/(1 - p_i)] = \beta_0 + \beta_1 log(age) ,$$

a classical, nonhierarchical logistic regression model. In practice, two other commonly used link functions are the *probit*

$$probit(p_i) = \Phi^{-1}(p_i) = \beta_0 + \beta_1 log(age) ,$$

and the *complementary log-log* (cloglog),

$$cloglog(p_i) = \log[-\log(1 - p_i)] = \beta_0 + \beta_1 log(age) .$$

All of these models are easily handled in `WinBUGS` using the code:

BUGS code
```
model{
    for( i in 1:n) {
      logage[i] <- log(x[i])
      z[i] ~ dbern( p[i] )
      logit(p[i]) <- beta0 + beta1*(logage[i] - mean(logage[]))
#     p[i] <- phi(beta0 + beta1*(logage[i] - mean(logage[])))
#     cloglog(p[i]) <- beta0 + beta1*(logage[i] - mean(logage[]))
      }
    beta0 ~ dflat()
```

```
        beta1 ~ dflat()
      } # end of WinBUGS code
```

Notice that the probit and cloglog models are commented out for the moment, using the pound sign (#). We also remark that this formulation of the probit (specifying p_i directly using the standard normal cdf phi) appears to be more numerically stable than using the `probit` function in WinBUGS.

The Data file is the obvious extension of the one used in previous dugong examples, namely

```
BUGS code   # Data:
            list(x = c( 1.0,   1.5,   1.5,   1.5, 2.5,    4.0,   5.0,   5.0,   7.0,
                        8.0,   8.5,   9.0,   9.5, 9.5,   10.0,  12.0,  12.0,  13.0,
                       13.0,  14.5,  15.5,  15.5, 16.5,  17.0,  22.5,  29.0,  31.5),
                    z = c(0, 0, 0, 0, 0, 0, 0, 0, 1, 0, 0, 0, 0, 1, 1, 0, 0, 1,
                          1, 1, 1, 1, 1, 1, 1, 1, 1),
                    n = 27)
```

For our `Inits` values, we run three parallel sampling chains, overdispersed with respect to the true values of β_0 and β_1:

```
BUGS code   # Inits:
            list( beta0 = -10, beta1 = -5)
            list( beta0 =   0, beta1= 10)
            list( beta0 =  10, beta1= 25)
```

We hasten to add that these starting values are too far from the true posterior to enable convergence in the probit case, and so for this model we instead use the "easier" values,

```
BUGS code   list(beta0 = -.8, beta1 = 3.3)
```

Using MCMC production runs of 20,000 post-burn in iterations each, Table 4.1 gives the model choice summaries, with fit captured by $\overline{D}$, effective size by p_D, and their sum providing the overall DIC score. Clearly the fit of all three models is very similar, with the cloglog enjoying a small advantage in $\overline{D}$ score. Since all three agree on p_D score (just under 2.0 "effective parameters"), this produces a slight win for the cloglog on DIC score. It would be a mistake to read too much into this "win," however, because the magnitude of the difference in $\overline{D}$ score is only slightly greater than its own Monte Carlo variability (crudely judged by running a few more MCMC iterations and recomputing the score).

Noting the fitted values (posterior means) of β_0 and β_1 for all three models, we may plot the data and fitted curves using the following R code:

```
R code   xgrid <- seq(0.5,32,length=101)
         Y <- c(1.80, 1.85, 1.87, 1.77, 2.02, 2.27, 2.15, 2.26, 2.47,
                2.19, 2.26, 2.40, 2.39, 2.41, 2.50, 2.32, 2.32, 2.43,
                2.47, 2.56, 2.65, 2.47, 2.64, 2.56, 2.70, 2.72, 2.57)
```

Model	$\overline{D}$	p_D	DIC
logit	19.62	1.85	21.47
probit	19.30	1.87	21.17
cloglog	18.77	1.84	20.61

Table 4.1 *Model selection table, binarized dugong example.*

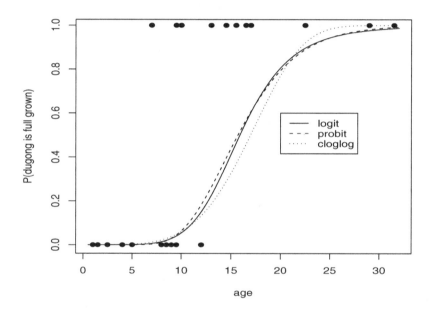

Figure 4.6 *Plot of the fitted logit, probit, and complementary log-log (cloglog) models to the binarized dugong data (shown as filled circles).*

```
lgage <- log(xgrid)
ctlgage <- lgage - mean(lgage)
z <- as.integer(Y>2.4)

beta0 <- -1.52;  beta1 <- 6.19
p_logit <- exp(beta0 + beta1*ctlgage)/(1 + exp(beta0 + beta1*ctlgage))`

beta0 <- -0.79;  beta1 <- 3.39
p_probit <- pnorm(beta0 + beta1*ctlgage)

beta0 <- -1.79;  beta1 <- 4.58
```

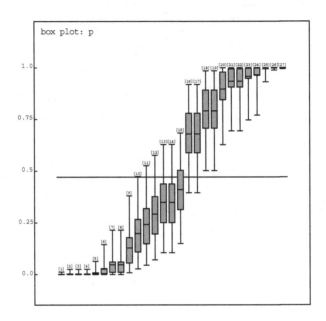

Figure 4.7 *Boxplots of p_i posterior distributions, binarized dugong data with complementary log-log link function.*

```
p_cloglog <- 1 - exp(-exp(beta0 + beta1*ctlgage))

plot(xgrid, p_logit, xlab="age", ylab="P(dugong is full grown)",
    pch=20,type="l")
lines(xgrid, p_probit, lty=2)
lines(xgrid, p_cloglog, lty=3)
points(age, z, pch=19)
legend(20, .6, legend=c("logit", "probit", "cloglog"), lty=1:3,
    ncol=1)
```

The resulting plot is shown in Figure 4.6. The logit and probit fits appear very similar, but the cloglog fitted curve is slightly different, increasing more slowly at first, but then accelerating and reaching its upper asymptote of 1 a bit sooner. Finally, boxplots of the posterior distributions of each of the p_i, induced by the link function and the posteriors for β_0 and β_1, are easily obtained using the **box plot** button on the comparison tool (**Inference** menu) in **WinBUGS**. Figure 4.7 shows the results under our slightly preferred cloglog link function. ∎

We hope the collection of examples in this section (and in particular, the model choice example just completed) gives some idea about Bayesian modeling in practice, and sets up our more expansive discussions of model adequacy and comparison in the remainder of this chapter.

4.2 Bayesian robustness

As we have already mentioned, a commonly voiced concern with Bayesian methods is their dependence on various aspects of the modeling process. Possible sources of uncertainty include the prior distribution, the precise form of the likelihood, and the number of levels in the hierarchical model. Of course, these are concerns for the frequentist as well. For example, the statistics literature is filled with methods for investigating the effect of case deletion, which is a special type of likelihood sensitivity. Still, the problem appears more acute for the Bayesian due to the appearance of the prior distribution (even though it may have been deliberately chosen to be noninformative).

In the next subsection, we investigate the robustness of the conclusions of a Bayesian analysis by checking whether or not they remain essentially unchanged in the presence of perturbations in the prior, likelihood, or some other aspect of the model. Subsection 4.2.2 considers a reverse attack on the problem, wherein we attempt to characterize the class of model characteristics (say, prior distributions) that lead to a certain conclusion given the observed data. Both of these approaches differ slightly from what many authors call the "robust Bayesian viewpoint," which seeks model settings (particularly prior distributions) that are robust from the outset, rather than picking a (perhaps informative) baseline and checking it later, or cataloging assumptions based on a predetermined decision. The case studies in Chapter 8 feature a variety of robustness investigations, while further methodological tools are provided in Chapter 7.

4.2.1 Sensitivity analysis

The most basic tool for investigating model uncertainty is the *sensitivity analysis*. That is, we simply make reasonable modifications to the assumption in question, recompute the posterior quantities of interest, and see whether they have changed in a way that has practical impact on interpretations or decisions. If not, the data are strongly informative with respect to this assumption and we need not worry about it further. If so, then our results are sensitive to this assumption, and we may wish to communicate the sensitivity, think more carefully about it, collect more data, or all of these.

Example 4.5 Suppose our likelihood features a mean parameter θ and a variance parameter σ^2, and in our initial analysis we employed a $N(\mu, \tau^2)$

prior on θ and an $IG(a, b)$ prior on σ^2. To investigate possible sensitivity of our results to the prior for θ, we might first consider shifts in its mean, e.g., by increasing and decreasing μ by one prior standard deviation τ. We also might alter its precision, say, by doubling and halving τ. Alternatively, we could switch to a heavier-tailed t prior with the same moments.

Sensitivity to the prior for σ^2 could be investigated in the same way, although we would now need to solve for the hyperparameter values required to increase and decrease its mean or double and halve its standard deviation. ∎

The increased ease of computing posterior summaries via Monte Carlo methods has led to a corresponding increased ease in performing sensitivity analyses. For example, given a converged MCMC sampler for a posterior distribution $p(\theta|\mathbf{y})$, it should take little time for the sampler to adjust to the new posterior distribution $p_{NEW}(\theta|\mathbf{y})$ arising from some modest change to the prior or likelihood. Thus, the additional sampling could be achieved with a substantial reduction in the burn-in period. Still, for a model having p hyperparameters, simply considering two alternatives (one larger, one smaller) to the baseline value for each results in a total of 3^p possible priors – possibly infeasible even using MCMC methods.

Fortunately, a little algebraic work eliminates the need for further sampling. Suppose we have a sample $\{\theta_1, \dots, \theta_N\}$ from a posterior $p(\theta|\mathbf{y})$, which arises from a likelihood $f(\mathbf{y}|\theta) = \prod_{i=1}^n f(y_i|\theta)$ and a prior $p(\theta)$. To study the impact of deleting case k, we see that the new posterior is

$$p_{NEW}(\theta|\mathbf{y}) \propto \frac{f(\mathbf{y}|\theta)\pi(\theta)}{f(y_k|\theta)} \propto \frac{1}{f(y_k|\theta)} p(\theta|\mathbf{y}) .$$

In the notation of Subsection 3.3.2, we can use $g(\theta) = p(\theta|\mathbf{y})$ as an importance sampling density, so that the weight function is given by

$$w(\theta) = \frac{p_{NEW}(\theta|\mathbf{y})}{p(\theta|\mathbf{y})} = \frac{1}{f(y_k|\theta)} .$$

Thus for any posterior function of interest $h(\theta)$, we have $\hat{E}[h(\theta)|\mathbf{y}] = \sum_{j=1}^N h(\theta_j)/N$, and

$$\hat{E}_{NEW}[h(\theta)|\mathbf{y}] = \frac{\sum_{j=1}^N h(\theta_j)w(\theta_j)}{\sum_{j=1}^N w(\theta_j)} , \qquad (4.9)$$

using equation (3.4). Because p and p_{NEW} should be reasonably similar, the former should be a good importance sampling density for the latter and hence the approximation in equation (4.9) should be good for moderate N. We could also obtain an approximate sample from p_{NEW} via the weighted bootstrap by resampling θ_i values with probability q_i, where $q_i = w(\theta_i)/\sum_{j=1}^N w(\theta_j)$.

In situations where large changes in the prior are investigated, or where outlying or influential cases are deleted, p_{NEW} may differ substantially from p. A substantial difference between the two posteriors can lead to highly variable importance sampling weights and, hence, poor estimation. Importance link function transformations of the Monte Carlo sample (MacEachern and Peruggia, 2000b) can be used to effectively alter the distribution from which the $\{\theta_1, \ldots, \theta_N\}$ were drawn, stablizing the importance sampling weights and improving estimation.

Sensitivity analysis via asymptotic approximation

Notice that importance sampling may be used for sensitivity analysis even if the original posterior sample $\{\theta_1, \ldots, \theta_N\}$ was obtained using some other method (e.g., the Gibbs sampler). The new posterior estimates in (4.9) are not likely to be as accurate as the original, but are probably sufficient for the purpose of a sensitivity analysis. In the same vein, the asymptotic methods of Section 3.2 can be used to obtain approximate sensitivity results without any further sampling *or* summation. For example, we have already seen in equation (3.1) a formula requiring only two maximizations (both with respect to the original posterior) that enables any number of sensitivity investigations. In the realm of model comparison, Kass and Vaidyanathan (1992) use such approximations to advantage in determining the sensitivity of Bayes factors to prior and likelihood input. Let $\pi_{NEW,k}(\theta_k)$ be new priors on θ_k where $k = 1, 2$ indexes the model. Writing BF for the Bayes factor under the original priors and BF_{NEW} for the Bayes factor under the new priors, Kass and Vaidyanathan (1992, equation (2.23)) show that

$$BF_{NEW} = BF \cdot \frac{r_1(\tilde{\theta}_1)}{r_2(\tilde{\theta}_2)} \cdot \left\{ 1 + O\left(\frac{1}{n}\right) \right\} \qquad (4.10)$$

where $r_k(\tilde{\theta}_k) = \pi_{NEW,k}(\tilde{\theta}_k)/\pi_k(\tilde{\theta}_k)$ and $\tilde{\theta}_k$ is the posterior mode using the original prior $\pi_k(\theta_k)$. Thus, one only needs to evaluate r_k at $\tilde{\theta}_k$ for $k = 1, 2$ for each new pair of priors, and these computations are sufficiently easy and rapid that a very large number of new priors can be examined without much difficulty.

A further simplification comes from a decomposition of the parameter vector into two components, $\theta_k = (\theta_k^{(1)}, \theta_k^{(2)})$ with the first component containing all parameters that are common to both models. We define $\theta^{(1)} = \theta_1^{(1)} (= \theta_2^{(1)})$. This parameter vector might be considered a nuisance parameter while the $\theta_k^{(2)}$, $k = 1, 2$ become the parameter vectors of interest. If we take the priors to be such that the two components are *a priori* independent under both models, we may write $\pi_k(\theta_k) = \pi_k^{(1)}(\theta^{(1)}) \cdot \pi_k^{(2)}(\theta_k^{(2)})$. Suppose this is the case, and furthermore $\pi_1^{(1)}(\theta^{(1)}) = \pi_2^{(1)}(\theta^{(1)})$, and that both these conditions hold for the new priors considered. In this situation, if it turns out that the first-component posterior modes under the two models are approximately equal (formally, to order $O(n^{-1})$), then

$$\frac{\pi_{NEW,1}(\tilde{\theta}_1)}{\pi_{NEW,2}(\tilde{\theta}_2)} \doteq \frac{\pi^{(2)}_{NEW,1}(\tilde{\theta}^{(2)}_1)}{\pi^{(2)}_{NEW,2}(\tilde{\theta}^{(2)}_2)}$$

to the same order of accuracy as equation (4.10). The ratio $r_1(\tilde{\theta}_1)/r_2(\tilde{\theta}_2)$ in (4.10) thus involves the new priors only through their second components. In other words, if the modal components $\tilde{\theta}^{(1)}_k$ are nearly equal for $k = 1$ and 2, when we perform the sensitivity analysis we do not have to worry about the effect of modest modifications to the prior on the nuisance-parameter component of θ_k. Carlin, Kass, Lerch, and Huguenard (1992) use this approach for investigating sensitivity of a comparison of two competing models of human working memory load to outliers and changes in the prior distribution.

Sensitivity analysis via scale mixtures of normals

The conditioning feature of MCMC computational methods enables another approach to investigating the sensitivity of distributional specifications in either the likelihood or prior for a broad class of common hierarchical models. Consider the model $y_i = \mu_i + \epsilon_i$, $i = 1, \ldots, n$, where the μ_i are unknown mean structures and the ϵ_i are independent random errors having density f with mean 0. For convenience, one often assumes that the ϵ_i form a series of independent normal errors, i.e., $\epsilon_i | \sigma^2 \sim N(0, \sigma^2)$, $i = 1, \ldots, n$. However, the normal distribution's light tails may well make this an unrealistic assumption, and so we wish to investigate alternative forms that allow greater variability in the observed y_i values. Andrews and Mallows (1974) show that expanding our model to

$$\epsilon_i | \sigma^2, \lambda_i \sim N(0, \lambda_i \sigma^2), \ i = 1, \ldots, n \,,$$

and subsequently placing a prior on λ_i enables a variety of familiar (and more widely dispersed) error densities to emerge. That is, we create f as the *scale mixture of normal distributions*,

$$f(\epsilon_i | \sigma^2) = \int_\Lambda p(\epsilon_i | \sigma^2, \lambda_i) p(\lambda_i) d\lambda_i \,, \ i = 1, \ldots, n \,.$$

The following list identifies the necessary functional forms for $p(\lambda_i)$ to obtain some of the possible departures from normality:

- Student's t errors: If $\nu/\lambda_i \sim \chi^2_\nu$ (i.e., if $\lambda_i \sim IG(\frac{\nu}{2}, \frac{2}{\nu})$), then $\epsilon_i | \sigma \sim t_\nu(0, \sigma)$.

- Double exponential errors: If $\lambda_i \sim Expo(2)$, the exponential distribution having mean 2, then $\epsilon_i | \sigma \sim DE(0, \sigma)$, where DE denotes the double exponential distribution.

- Logistic errors: If $1/\sqrt{\lambda_i}$ has the asymptotic Kolmogorov distance distribution, then $\epsilon_i|\sigma$ is logistic (see Andrews and Mallows, 1974).

Generally one would not view the addition of n parameters to the model as a simplifying device, but would prefer instead to work directly with the nonnormal error density in question. However, Carlin and Polson (1991) point out that the conditioning feature of the Gibbs sampler makes this augmentation of the parameter space quite natural. To see this, note first that under an independence prior on the λ_i, we have $\lambda_i|\{\lambda_{j\neq i}\}, \{\mu_j\}, \sigma^2, \mathbf{y} \sim \lambda_i|\mu_i, \sigma^2, y_i$. But by Bayes theorem, $p(\lambda_i|\mu_i, \sigma^2, y_i) \propto p(y_i|\mu_i, \sigma^2, \lambda_i)p(\lambda_i)$, where the appropriate normalization constant, $p(y_i|\mu_i, \sigma^2)$, is known by construction as the desired nonnormal error density. Hence the complete conditional for λ_i will always be of known functional form. Generation of the required samples may be done directly if this form is a standard density; otherwise, a carefully selected rejection method may be employed. Finally, since the remaining complete conditionals (for σ^2 and the μ_j) are determined given $\boldsymbol{\lambda}$, convenient prior specifications may often be employed with the normal likelihood, again leading to direct sampling.

Example 4.6 Consider the model $y_i = \mu_i + \epsilon_{ij}$, where j indexes the model error distribution (i.e., the mixing distribution for the λ_i), and i indexes the observation, as before. Suppose that $\mu_i = f(\mathbf{x}_i, \theta)\boldsymbol{\beta}$, where $\boldsymbol{\beta}$ is a k-dimensional vector of linear nuisance parameters, and $f(\mathbf{x}_i, \theta) = (f_1(\mathbf{x}_i, \theta), \ldots, f_k(\mathbf{x}_i, \theta))$ is a collection of known functions, possibly nonlinear in θ. Creating the $n \times k$ matrix $F_\theta = (f(\mathbf{x}_i, \theta))$, the log-likelihood $\log p(\mathbf{y} \mid \theta, \boldsymbol{\beta}, \boldsymbol{\lambda}, \sigma^2)$ is

$$-\frac{1}{2\sigma^2}(\mathbf{y} - F_\theta\boldsymbol{\beta})^T \Sigma_i^{-1}(\mathbf{y} - F_\theta\boldsymbol{\beta}) - n\log\sigma - \frac{1}{2}\sum_{i=1}^n \log\lambda_i ,$$

where $\Sigma_i = Diag(\lambda_1, \ldots, \lambda_n)$. Assume that $\boldsymbol{\beta} \sim N(\boldsymbol{\beta}_0, \Sigma_0)$, where $\boldsymbol{\beta}_0$ and Σ_0 are known. In addition, let $\sigma^2 \sim IG(a_0, b_0)$, and let θ have prior distribution $p(\theta)$.

Suppose that, due to uncertainty about the error density and the impact of possible outliers, we wish to compare the three models

$M = 1: \quad \epsilon_i \sim N(0, \sigma^2) ,$

$M = 2: \quad \epsilon_i \sim t(0, \sigma^2, \nu = 2) ,$ and

$M = 3: \quad \epsilon_i \sim DE(0, \sigma) ,$

where M indicates which error distribution (model) we have selected. For $M = 1$, clearly the λ_i are not needed, so their full conditional distributions are degenerate at 1. For $M = 2$, we have the full conditional distribution

$$IG\left(\frac{\nu + 1}{2}, \left\{\frac{1}{2}\left[\frac{(y_i - f(\mathbf{x}_i, \theta)\boldsymbol{\beta})^2}{\sigma^2} + \nu\right]\right\}^{-1}\right), \quad i = 1, \ldots, n ,$$

in a manner very similar to that of Andrews and Mallows (1974). Finally, for $M = 3$ we have a complete conditional proportional to

$$\lambda_i^{-\frac{1}{2}} \exp\left(-\frac{1}{2}\left(\lambda_i + \frac{(y_i - f(\mathbf{x}_i, \theta)\beta)^2}{\lambda_i \sigma^2}\right)\right) ,$$

that is, $\lambda_i|\theta, \beta, \sigma^2, \mathbf{y} \sim GIG\left(\frac{1}{2}, 1, (y_i - f(\mathbf{x}_i, \theta)\beta)^2/\sigma^2\right)$ for $i = 1, \ldots, n$, where GIG denotes the generalized inverse Gaussian distribution (see Devroye, 1986, p. 478). In order to sample from this density, we note that it is the reciprocal of an

$$\text{Inverse Gaussian}\left(\left|\frac{\sigma}{y_i - f(\mathbf{x}_i, \theta)\beta}\right|, 1\right) ,$$

a density from which we may easily sample (see e.g. Devroye, 1986, p. 149). ∎

Thus, the scale mixture of normals approach enables investigation of nonnormal error distributions as a component of the original model, rather than later as part of a sensitivity analysis, and with little additional computational complexity. If the λ_i all have posterior distributions centered tightly about 1, the additional modeling flexibility is unnecessary, and the assumption of normality is appropriate. In this same vein, the λ_i can be thought of as outlier diagnostics, because extreme observations will correspond to extreme fitted values of these scale parameters. Wakefield et al. (1994) use this idea to detect outlying individuals in a longitudinal study by employing the normal scale mixture at the *second* stage of their hierarchical model (i.e., on the prior for the random effect component of μ_i).

Example 4.7 Using the normal scale mixture idea, consider a reanalysis of Fisher's sleep data from Chapter 2, Problem 20. These are the increased hours of sleep for 10 patients treated with soporific B compared with soporific A, given as

$$1.2, 2.4, 1.3, 1.3, 0.0, 1.0, 1.8, 0.8, 4.6, 1.4 .$$

Suppose we use WinBUGS to compare models having mean θ and the three error distributions considered in Example 4.6, namely the normal, the t_2, and the DE. These are readily available as normal scale mixtures using mixing parameters λ_i as follows:

$M = 1 :$ $\lambda_i = 1$

$M = 2 :$ $\lambda_i \sim IG(1, 1)$

$M = 3 :$ $\lambda_i \sim Expo(2)$

These three choices are easily coded in WinBUGS. In fact, in the code that follows we use a "data duplication" trick that enables us to fit all three models *simultaneously*, facilitating model comparison:

BUGS code

```
model
{
    for(i in 1:N) {
      # Duplicate the data
      Y1[i] <- Y[i]
      Y2[i] <- Y[i]
      Y3[i] <- Y[i]

      # Weighted precision parameters
      tau1[i] <- tau0[1]/lambda1[i];
      tau2[i] <- tau0[2]/lambda2[i];
      tau3[i] <- tau0[3]/lambda3[i];

      # Mean structures (all same)
      Y1[i] ~ dnorm(theta[1],tau1[i]);
      Y2[i] ~ dnorm(theta[2],tau2[i]);
      Y3[i] ~ dnorm(theta[3],tau3[i]);

      # Error distributions
      lambda1[i] <- 1;                 # M1 = e_i ~ N(0,tau1)
      lambda2[i] <- 1/inv.lambda2[i];  # So that lambda ~ IG
      inv.lambda2[i] ~ dgamma(1,1);    # M2 = e_i ~ t_2(0, tau2)
      lambda3[i] ~ dexp(0.5);          # M3 = e_i ~ DE(0,tau3)
    }
    # Priors
    for (k in 1:3) {
      theta[k] ~ dnorm(0,0.00001)      # Vague normal on mean
      sigma[k] ~ dunif(0.01,100)       # Uniform on sigma
      tau0[k] <- 1/(sigma[k]*sigma[k])
    }
}
```

Here, $k = 1, \ldots, 3$ indexes the model while $i = 1, \ldots, 10$ indexes the observation. WinBUGS will run all three models simultaneously, but their posteriors will be independent since there is no connection across k anywhere in the code. The Compare, Stats, and DIC tools may now produce convenient displays featuring all 3 models (see below).

Potential outliers can be readily identified by examining the boxplots of the λ_i posterior distributions for Models 2 and 3 in Figure 4.8. Subject 9 (who got 4.6 additional hours of sleep) is easily identified by the posterior of λ_9, which is centered near 10 and has 95th percentile well over 100 for Model 2 (note a log scale is being used on the vertical axis to adjust the plot for the extremely heavy upper tail). The large λ_9 provides the variance inflation needed to accommodate this outlying data point. Subjects 2 (2.4 additional hours) and 5 (0.0 additional hours) are also potential outliers, but to a far lesser degree.

To understand the benefit of nonnormal errors a bit better, consider the

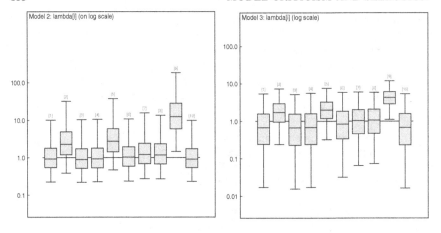

Figure 4.8 *Comparison of outlier diagnostics* (λ_i *posterior boxplots) for the two nonnormal error distributions, Fisher's sleep data.*

Model	mean	SD	2.5%	median	97.5%
1 (normal errors)	1.57	0.48	0.62	1.58	2.51
2 (t_2 errors)	1.31	0.28	0.78	1.29	1.91
3 (DE errors)	1.32	0.27	0.80	1.30	1.91

Table 4.2 *Posterior comparison for the grand mean θ across the three error distributions, Fisher's sleep data.*

fixed effect estimates in Table 4.2, as well as the boxplots of the three posterior distributions for the grand mean θ in Figure 4.9. Notice that the mean is shifted down in Models 2 and 3 (away from the large outlier and back toward the bulk of the data) relative to Model 1. In addition, θ is more precisely estimated in the t_2 and DE errors cases, as indicated by the narrower 95% central confidence intervals in Table 4.2 and the narrower boxplots in Figure 4.9. This is because the heavier tails of the t_2 and DE distributions allow them to accommodate the large outlier more readily, resulting in posterior estimates for the mean sleep increase that are both more accurate and more precise. ∎

4.2.2 Prior partitioning

The method of sensitivity analysis described in the previous subsection is a direct and conceptually simple way of measuring the effect of the prior distributions and other assumptions made in a Bayes or empirical Bayes

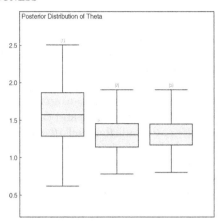

Figure 4.9 *Comparison of posterior boxplots for θ for three error distributions, Fisher's sleep data.*

analysis. However, it does not free us from careful development of the original prior, which must still be regarded as a reasonable baseline. This can be impractical because the prior beliefs and vested interests of the potential consumers of our analysis may be unimaginably broad. For example, if we are analyzing the outcome of a government-sponsored clinical trial to determine the effectiveness of a new drug relative to a standard treatment for a given disease, our results will likely be read by doctors in clinical practice, epidemiologists, regulatory workers (e.g., from the U.S. Food and Drug Administration), legislators, and, of course, patients suffering from or at risk of contracting the disease itself. These groups are likely to have widely divergent opinions as to what constitutes "reasonable" prior opinion; the clinician who developed the drug is likely to be optimistic about its value, while the regulatory worker (who has seen many similar drugs emerge as ineffective) may be more skeptical. What we need is a method for communicating the robustness of our conclusions to *any* prior input the reader deems appropriate.

A potential solution to this problem is to "work the problem backward," as follows. Suppose that, rather than fix the prior and compute the posterior distribution, we fix the posterior (or set of posteriors) that produce a given conclusion, and determine which prior inputs are consistent with this desired result, given the observed data. The reader would then be free to determine whether the outcome was reasonable according to whether the prior class that produced it was consistent with his or her own prior beliefs. We refer to this approach simply as *prior partitioning* since we are subdividing the prior class based on possible outcomes, though it is important to remember that such partitions also depend on the data and the decision to be reached. As such, the approach is not strictly Bayesian (the data are playing a role in determining the prior), but it does provide valuable robustness information while retaining the framework's philosophical, structural, documentary, and communicational advantages.

To illustrate the basic idea, consider the point null testing scenario $H_0 : \theta = \theta_0$ versus $H_1 : \theta \neq \theta_0$. Without loss of generality, set $\theta_0 = 0$. Suppose our data x has density $f(x|\theta)$, where θ is an unknown scalar parameter. Let π represent the prior probability of H_0, and $G(\theta)$ the prior cumulative distribution function (cdf) of θ conditional on $\{\theta \neq 0\}$. The complete prior cdf for θ is then $F(\theta) = \pi I_{[0,\infty)}(\theta) + (1 - \pi)G(\theta)$, where I_S is the indicator function of the set S. Hence, the posterior probability of the null hypothesis is

$$P_G(\theta = 0|x) = \frac{\pi f(x|0)}{\pi f(x|0) + (1 - \pi) \int f(x|\theta)dG(\theta)} \,. \qquad (4.11)$$

Prior partitioning seeks to characterize the G for which this probability is less than or equal to some small probability $p \in (0, 1)$, in which case we reject the null hypothesis. (Similarly, we could also seek the G leading to $P_G(\theta \neq 0|x) \leq p$, in which we would reject H_1.) Elementary calculations show that characterizing this class of priors $\{G\}$ is equivalent to characterizing the set $\mathcal{H}_c$, defined as

$$\mathcal{H}_c = \left\{ G : \int f(x|\theta)dG(\theta) \geq c = \frac{1-p}{p} \frac{\pi}{1-\pi} f(x|0) \right\} \,. \qquad (4.12)$$

Carlin and Louis (1996) establish results regarding the features of $\mathcal{H}_c$, and then use these results to obtain sufficient conditions for $\mathcal{H}_c$ to be nonempty for classes of priors that satisfy various moment and percentile restrictions. The latter are somewhat more useful, since percentiles and tail areas of the conditional prior G are transform-equivariant, and Chaloner et al. (1993) have found that elicitees are most comfortable describing their opinions through a "best guess" (mean, median, or mode) and a few relatively extreme percentiles (say, the 5th and the 95th).

To lay out the technical detail of this percentile restriction approach, let θ_L and θ_U be such that $P_G(\theta \leq \theta_L) = a_L$ and $P_G(\theta > \theta_U) = a_U$, where a_L and a_U lie in the unit interval and sum to less than 1. For fixed values for θ_L and θ_U, we seek the region of (a_L, a_U) values (i.e., the class of priors) that lead to a given decision. To accomplish this we can take a general G, alter it to pass through (θ_L, a_L) and $(\theta_U, 1 - a_U)$, and then search over all G subject only to the constraint that the intervals $(-\infty, a_L]$, $(a_L, a_U]$, and (a_U, ∞) all have positive probability.

Assume that $f(x|\theta)$ is a unimodal function of θ for fixed x that vanishes in both tails, an assumption that will be at least approximately true for large datasets due to the asymptotic normality of the observed likelihood function. Keeping θ_L, θ_U, and $\mathbf{a}$ fixed, we seek the supremum and infimum of $\int f(x|\theta)dG(\theta)$. The infimum will always be given by

$$(1 - a_L - a_U)\min\{f(x|\theta_L), f(x|\theta_U)\}, \qquad (4.13)$$

since a unimodal f must take its minimum over the central support interval at one of its edges. However, depending on the location of the maximum likelihood estimator $\hat{\theta}$, $\sup_G \int f(x|\theta)dG(\theta)$ equals

$$\begin{array}{ll} a_L f(x|\hat{\theta}) + (1 - a_L - a_U)f(x|\theta_L) + a_U f(x|\theta_U), & \hat{\theta} \leq \theta_L \\ a_L f(x|\theta_L) + (1 - a_L - a_U)f(x|\hat{\theta}) + a_U f(x|\theta_U), & \theta_L < \hat{\theta} \leq \theta_U \\ a_L f(x|\theta_L) + (1 - a_L - a_U)f(x|\theta_U) + a_U f(x|\hat{\theta}), & \hat{\theta} > \theta_U \end{array} \quad (4.14)$$

Notice that the infimum is obtained by pushing mass as far away from the MLE as allowed by the constraints, while the supremum is obtained by pushing the mass as close to the MLE as possible. In conjunction with π, the prior probability of H_0, the supremum and infimum may be used to determine the prior percentiles compatible with $P(\theta = 0|x) \leq p$ and $P(\theta \neq 0|x) \leq p$, respectively. Because $\mathcal{H}_c$ is empty if the supremum does not exceed c, we can use the supremum expression to determine whether there are any G that satisfy the inequality in (4.12), i.e., whether any priors G exist that enable stopping to reject the null hypothesis. Similarly, the infimum expression may be useful in determining whether any G enable stopping to reject the alternative hypothesis, H_1.

We may view as fixed either the (θ_L, θ_U) pair, the (a_L, a_U) pair, or both. As an example of the first case, suppose we seek the (a_L, a_U) compatible with a fixed (θ_L, θ_U) pair (an *indifference zone*) for which $\int f(x|\theta)dG(\theta) \geq c$. Then given the location of $\hat{\theta}$ with respect to the indifference zone, equation (4.14) may be easily solved to obtain the half-plane in which acceptable as must lie. When combined with the necessary additional constraints $a_L \geq 0, a_U \geq 0$, and $a_L + a_U \leq 1$, the result is a polygonal region that is easy to graph and interpret. Graphs of acceptable (θ_L, θ_U) pairs for fixed (a_L, a_U) may be obtained similarly, although the solution of equation (4.14) is now more involved and the resulting regions may no longer be compact. We will return to these ideas in the context of our Section 8.2 data example.

Since the ideas behind prior partitioning are somewhat more theoretical and less computational than those driving sensitivity analysis, the approach has been well developed in the literature. Two often-quoted early references are the paper by Edwards, Lindman, and Savage (1963), who explore robust Bayesian methods in the context of psychological models, and the book by Mosteller and Wallace (1964), who discuss bounds on the prior probabilities necessary to choose between two simple hypotheses (authorship of a given disputed *Federalist* paper by either Alexander Hamilton or James Madison).

The subsequent literature in the area is vast; see Berger (1985, Section 4.7) or, more recently, Berger (1994) for a comprehensive review. Here we mention only a few particularly important papers. In the point null setting,

Berger and Sellke (1987) and Berger and Delampady (1987) show that we attain the minimum of $P(\theta = \theta_0|x)$ over all conditional priors G for $\theta \neq \theta_0$ when G places all of its mass at $\hat{\theta}$, the maximum likelihood estimate of θ. Even in this case, where G is working with the data against H_0, these authors showed that the resulting $P(\theta = \theta_0|x)$ values are typically still larger than the corresponding two-sided p-value, suggesting that the standard frequentist approach is biased against H_0 in this case. In the interval null hypothesis setting, prior partitioning is reminiscent of the work of O'Hagan and Berger (1988), who obtain bounds on the posterior probability content of each of a collection of intervals that form the support of a univariate parameter, under the restriction that the prior probability assignment to these intervals is in a certain sense unimodal. In the specific context of clinical trial monitoring, Greenhouse and Wasserman (1995) compute bounds on posterior expectations and tail areas (stopping probabilities) over an ϵ-contaminated class of prior distributions (Berger and Berliner, 1986).

Further restricting the prior class

Sargent and Carlin (1996) extend the above approach to the case of an interval null hypothesis, i.e., $H_0 : \theta \in [\theta_L, \theta_U]$ versus $H_1 : \theta \notin [\theta_L, \theta_U]$. This formulation is useful in the context of clinical trial monitoring, where $[\theta_L, \theta_U]$ is thought of as an *indifference zone*, within which we are indifferent as to the use of treatment or placebo. For example, we might take $\theta_U > 0$ if there were increased costs or toxicities associated with the treatment. Let π again denote the prior probability of H_0, and let $G(\theta)$ now correspond to the prior cdf of θ given $\theta \notin [\theta_L, \theta_U]$. Making the simplifying assumption of a uniform prior over the indifference zone, the complete prior density for θ may be written as

$$p(\theta) = \frac{\pi}{\theta_U - \theta_L} I_{[\theta_L, \theta_U]}(\theta) + (1 - \pi)g(\theta) . \qquad (4.15)$$

Sargent and Carlin (1996) derive expressions similar to (4.11) and (4.12) under the percentile restrictions of the previous subsection. However, these rather weak restrictions lead to prior classes that, while plausible, are often too broad for practical use. As such, we might consider a sequence of increasingly tight restrictions on the shape and smoothness of permissible priors, which in turn enable increasingly informative results. For example, we might retain the mixture form (4.15), but now restrict $g(\theta)$ to some particular parametric family. Carlin and Sargent (1996) refer to such a prior as "semiparametric" because the parametric form for g does not cover the indifference zone $[\theta_L, \theta_U]$, although since we have adopted another parametric form over this range (the uniform) one might argue that "biparametric" or simply "mixture" would be better names.

We leave it as an exercise to show that requiring $G \in \mathcal{H}_c$ is equivalent to requiring $BF \leq \left(\frac{p}{1-p}\right)\left(\frac{1-\pi}{\pi}\right)$, where

$$BF = \frac{\frac{1}{\theta_U - \theta_L} \int_{\theta_L}^{\theta_U} f(x|\theta)d\theta}{\int f(x|\theta)g(\theta)d\theta} \ , \tag{4.16}$$

the Bayes factor in favor of the null hypothesis. Equation (4.16) expresses the Bayes factor as the ratio of the marginal densities under the competing hypotheses; it is also expressible as the ratio of posterior to prior odds in favor of the null. As such, BF gives the extent to which the data have revised our prior beliefs concerning the two hypotheses. Note that if we take $\pi = 1/2$ (equal prior weighting of null and alternative), then a Bayes factor of 1 suggests equal posterior support for the two hypotheses. In this case, we require a Bayes factor of $1/19$ or smaller to insure that $P(H_0|x)$ does not exceed 0.05.

In practice, familiar models from the exponential family are often appropriate (either exactly or asymptotically) for the likelihood $f(x|\theta)$. This naturally leads to consideration of the restricted class of *conjugate* priors $g(\theta)$, to obtain a closed form for the integral in the denominator of (4.16). Since a normal approximation to the likelihood for θ is often suitable for even moderate sample sizes, we illustrate in the case of a conjugate normal prior. The fact that g is defined only on the complement of the indifference zone presents a slight complication, but, fortunately, the calculations remain tractable under a renormalized prior with the proper support. That is, we take

$$g(\theta) = \frac{N(\theta|\mu, \tau^2)}{1 - \left[\Phi\left(\frac{\theta_U - \mu}{\tau}\right) - \Phi\left(\frac{\theta_L - \mu}{\tau}\right)\right]} \ , \quad \theta \notin [\theta_L, \theta_U] \ ,$$

where the numerator denotes the density of a normal distribution with mean μ and variance τ^2, and Φ denotes the cdf of a standard normal distribution.

To obtain a computational form for equation (4.16), suppose we can approximate the likelihood satisfactorily with a $N(\theta|\hat{\theta}, \hat{\sigma}^2)$ density, where $\hat{\theta}$ is the maximum likelihood estimate (MLE) of θ and $\hat{\sigma}^2$ is a corresponding standard error estimate. Probability calculus then shows that

$$\int f(x|\theta)g(\theta)d\theta = \frac{1}{\sqrt{2\pi(\hat{\sigma}^2 + \tau^2)}} \exp\left[-\frac{(\mu - \hat{\theta})^2}{2(\hat{\sigma}^2 + \tau^2)}\right]$$
$$\times \left\{1 - \left[\Phi\left(\frac{\theta_U - \eta}{\nu}\right) - \Phi\left(\frac{\theta_L - \eta}{\nu}\right)\right]\right\} \ , \tag{4.17}$$

where $\eta = (\hat{\sigma}^2 \mu + \tau^2 \hat{\theta})/(\hat{\sigma}^2 + \tau^2)$ and $\nu^2 = \hat{\sigma}^2 \tau^2/(\hat{\sigma}^2 + \tau^2)$. (Note that η and ν^2 are respectively the posterior mean and variance under the fully parametric normal/normal model, described in the next subsection.) Since $\int_{\theta_L}^{\theta_U} f(x|\theta)d\theta = \Phi\left(\frac{\theta_U - \hat{\theta}}{\hat{\sigma}}\right) - \Phi\left(\frac{\theta_L - \hat{\theta}}{\hat{\sigma}}\right)$, we can now obtain the Bayes factor

(4.16) without numerical integration, provided that there are subroutines available to evaluate the normal density and cdf.

As a final approach, we might abandon the mixture prior form (4.15) in favor of a single parametric family $h(\theta)$, preferably chosen as conjugate with the likelihood $f(x|\theta)$. If such a choice is possible, we obtain simple closed form expressions for Bayes factors and tail probabilities whose sensitivity to changes in the prior parameters can be easily examined. For example, for our $N(\theta|\hat{\theta}, \hat{\sigma}^2)$ likelihood under a $N(\theta|\mu, \tau^2)$ prior, the posterior probability of H_0 is nothing but

$$P(\theta \in [\theta_L, \theta_U]|x) = \Phi\left(\frac{\theta_U - \eta}{\nu}\right) - \Phi\left(\frac{\theta_L - \eta}{\nu}\right),$$

where η and ν^2 are again as defined beneath equation (4.17). Posterior probabilities that correspond to stopping to reject the hypotheses $H_L : \theta < \theta_L$ and $H_U : \theta > \theta_U$ arise similarly.

4.3 Model assessment

We have already presented several tools for Bayesian model assessment in Subsection 2.5.1. It is not our intention to review all of these tools, but rather to point out how their computation is greatly facilitated by modern Monte Carlo computational methods. Because most Bayesian models encountered in practice require some form of sampling to evaluate the posterior distributions of interest, this means that common model checks will be available at little extra cost, both in terms of programming and runtime.

Consider for example the simple Bayesian residual,

$$r_i = y_i - E(y_i|\mathbf{z}), \ i = 1, \ldots, n,$$

where $\mathbf{z}$ is the sample of data used to fit the model and $\mathbf{y} = (y_1, \ldots, y_n)'$ is an independent validation sample. Clearly, calculation of r_i requires an expectation with respect to the posterior predictive distribution $p(y_i|\mathbf{z})$, which will rarely be available in closed form. However, notice that we can write

$$
\begin{aligned}
E(y_i|\mathbf{z}) &= \int y_i p(y_i|\mathbf{z}) dy_i \\
&= \int \int y_i f(y_i|\boldsymbol{\theta}) p(\boldsymbol{\theta}|\mathbf{z}) d\boldsymbol{\theta} dy_i \\
&= \int E(y_i|\boldsymbol{\theta}) p(\boldsymbol{\theta}|\mathbf{z}) d\boldsymbol{\theta} \\
&\approx \frac{1}{G} \sum_{g=1}^{G} E(y_i|\boldsymbol{\theta}^{(g)}),
\end{aligned}
$$

where the $\boldsymbol{\theta}^{(g)}$ are samples from the posterior distribution $p(\boldsymbol{\theta}|\mathbf{z})$ (for

MCMC algorithms, we would of course include only post-convergence samples). The equality in the second line holds due to the conditional independence of $\mathbf{y}$ and $\mathbf{z}$ given $\boldsymbol{\theta}$, while that in the third line arises from reversing the order of integration. But $E(y_i|\boldsymbol{\theta})$ typically *will* be available in closed form, since this is nothing but the mean structure of the likelihood. The fourth line thus arises as a Monte Carlo integration.

Even if $E(y_i|\boldsymbol{\theta})$ is not available in closed form, we can still estimate $E(y_i|\mathbf{z})$ provided we can draw samples $y_i^{(g)}$ from $f(y_i|\boldsymbol{\theta}^{(g)})$. Such sampling is naturally appended onto the algorithm generating the $\boldsymbol{\theta}^{(g)}$ samples themselves. In this case we have

$$E(y_i|\mathbf{z}) = \int\int y_i f(y_i|\boldsymbol{\theta})p(\boldsymbol{\theta}|\mathbf{z})d\boldsymbol{\theta}dy_i$$

$$\approx \frac{1}{G}\sum_{g=1}^{G} y_i^{(g)} ,$$

since $(y_i^{(g)}, \boldsymbol{\theta}^{(g)})$ constitute a sample from $p(y_i, \boldsymbol{\theta}|\mathbf{z})$. This estimator will not be as accurate as the first (because both integrals are now being done via Monte Carlo), but will still be simulation consistent (i.e., converge to the true value with probability 1 as $G \to \infty$).

Next, consider the cross-validation residual given in (2.27),

$$r_i = y_i - E(y_i|\mathbf{y}_{(i)}), \ i = 1,\ldots,n,$$

where we recall that $\mathbf{y}_{(i)}$ denotes the vector of all the data except the ith value. Now we have

$$E(y_i|\mathbf{y}_{(i)}) = \int E(y_i|\boldsymbol{\theta})p(\boldsymbol{\theta}|\mathbf{y}_{(i)})d\boldsymbol{\theta}$$

$$\approx \int E(y_i|\boldsymbol{\theta})p(\boldsymbol{\theta}|\mathbf{y})d\boldsymbol{\theta}$$

$$\approx \frac{1}{G}\sum_{g=1}^{G} E(y_i|\boldsymbol{\theta}^{(g)}) ,$$

where the $\boldsymbol{\theta}^{(g)}$ are samples from the complete data posterior $p(\boldsymbol{\theta}|\mathbf{y})$. The approximation in the second line should be adequate unless the dataset is small and y_i is an extreme outlier. It enables the use of the same $\boldsymbol{\theta}^{(g)}$ samples (already produced by the Monte Carlo algorithm) for estimating each conditional mean $E(y_i|\mathbf{y}_{(i)})$, hence each residual r_i, for $i = 1,\ldots,n$.

Finally, to obtain the cross-validation standardized residual in (2.28), we require not only $E(y_i|\mathbf{y}_{(i)})$, but also $Var(y_i|\mathbf{y}_{(i)})$. This quantity is estimable by first rewriting $Var(y_i|\mathbf{y}_{(i)}) = E(y_i^2|\mathbf{y}_{(i)}) - [E(y_i|\mathbf{y}_{(i)})]^2$, and then observing

$$
\begin{aligned}
E(y_i^2|\mathbf{y}_{(i)}) &= \int E(y_i^2|\boldsymbol{\theta})p(\boldsymbol{\theta}|\mathbf{y}_{(i)})d\boldsymbol{\theta} \\
&= \int \{Var(y_i|\boldsymbol{\theta}) + [E(y_i|\boldsymbol{\theta})]^2\}p(\boldsymbol{\theta}|\mathbf{y}_{(i)})d\boldsymbol{\theta} \\
&\approx \frac{1}{G}\sum_{g=1}^{G}\left\{Var(y_i|\boldsymbol{\theta}^{(g)}) + [E(y_i|\boldsymbol{\theta}^{(g)})]^2\right\} \ ,
\end{aligned}
$$

again approximating the reduced data posterior with the full data posterior and performing a Monte Carlo integration.

Other posterior quantities from Subsection 2.5.1, such as conditional predictive ordinates $f(y_i|\mathbf{y}_{(i)})$ and posterior predictive model checks (e.g., Bayesian p-values) can be derived similarly. The details of some of these calculations are left as exercises.

4.4 Bayes factors via marginal density estimation

As in the previous section, our goal in this section is not to review the methods for model choice and averaging that have already been presented in Subsections 2.3.3 and 2.5.2, respectively. Rather, we provide several computational approaches for obtaining these quantities, especially the Bayes factor. In Section 4.6, we will present some more advanced approaches that are not only implemented via MCMC methods, but also motivated by these methods, in the sense that they are applicable in very highly-parametrized model settings where the traditional model selection method offered by Bayes factors is unavailable or infeasible.

Recall the use of the Bayes factor as a method for choosing between two competing models M_1 and M_2, given in equation (2.19) as

$$
BF = \frac{p(\mathbf{y} \mid M_1)}{p(\mathbf{y} \mid M_2)} \ , \tag{4.18}
$$

the ratio of the observed marginal densities for the two models. For large sample sizes n, the necessary integrals over $\boldsymbol{\theta}_i$ may be available conveniently and accurately via asymptotic approximation, producing estimated Bayes factors accurate to order $O(1/n)$, where n is the number of factors contributing to the likelihood function. In any case, equation (4.10) reveals that asymptotic approximations are helpful in discovering the sensitivity of Bayes factors to prior specification, whether they played a role in the original calculation or not.

For moderate sample sizes n or reasonably challenging models, however, such approximations are not appropriate, and sampling-based methods must be used to obtain estimates of the marginal likelihoods needed to evaluate BF. This turns out to be a surprisingly difficult problem; unlike posterior and predictive distributions, marginal distributions are not easily estimated from the output of an MCMC algorithm. As a result, many different approaches have been suggested in the literature, all of which involve augmenting or otherwise "tricking" a sampler into producing the required marginal density estimates. As with our Subsection 3.4.6 discussion of convergence diagnostics, we do not attempt a technical review of every method, but we do comment on their strengths and weaknesses, so that the reader can judge which will likely be the most appropriate in a given problem setting. Kass and Raftery (1995) provide a comprehensive review of Bayes factors, including many of the computational methods described below.

4.4.1 Direct methods

In what follows, we suppress the dependence on the model indicator M in our notation because all our calculations must be repeated for both models appearing in expression (4.18). Observe that since

$$p(\mathbf{y}) = \int f(\mathbf{y} \mid \boldsymbol{\theta}) p(\boldsymbol{\theta}) d\boldsymbol{\theta} \, ,$$

we could generate observations $\{\boldsymbol{\theta}^{(g)}\}_{g=1}^{G}$ from the prior and compute the estimate

$$\hat{p}(\mathbf{y}) = \frac{1}{G} \sum_{g=1}^{G} f(\mathbf{y} \mid \boldsymbol{\theta}^{(g)}) \, , \tag{4.19}$$

a simple Monte Carlo integration. Unfortunately, the conditional likelihood $f(\mathbf{y} \mid \boldsymbol{\theta})$ will typically be very peaked compared to the prior $p(\boldsymbol{\theta})$, so that (4.19) will be a very inefficient estimator (most of the terms in the sum will be near 0). A better approach would be to use samples from the posterior distribution, as suggested by Newton and Raftery (1994). These authors develop the estimator

$$\hat{p}(\mathbf{y}) = \left[\frac{1}{G} \sum_{g=1}^{G} \frac{1}{f(\mathbf{y} \mid \boldsymbol{\theta}^{(g)})} \right]^{-1} \, , \tag{4.20}$$

the harmonic mean of the posterior sample conditional likelihoods. This approach, while efficient, can be quite unstable since a few of the conditional likelihood terms in the sum will still be near 0. Theoretically, this difficulty corresponds to this estimator's failure to obey a Gaussian central limit theorem as $G \rightarrow \infty$. To correct this, Newton and Raftery suggest a compromise between methods (4.19) and (4.20) wherein we use a mixture of the prior and posterior densities for each model, $\tilde{p}(\boldsymbol{\theta}) = \delta p(\boldsymbol{\theta}) + (1 - \delta) p(\boldsymbol{\theta}|\mathbf{y})$, as an importance sampling density. Defining $w(\boldsymbol{\theta}) = p(\boldsymbol{\theta})/\tilde{p}(\boldsymbol{\theta})$, we then have

$$\hat{p}(\mathbf{y}) = \frac{\sum_{g=1}^{G} f(\mathbf{y} \mid \boldsymbol{\theta}^{(g)}) \, w(\boldsymbol{\theta}^{(g)})}{\sum_{g=1}^{G} w(\boldsymbol{\theta}^{(g)})} \, .$$

This estimator is more stable and does satisfy a central limit theorem, although in this form it does require sampling from both the posterior and the prior.

A useful generalization of (4.20) was provided by Gelfand and Dey (1994), who began with the identity

$$[p(\mathbf{y})]^{-1} = \int \frac{h(\boldsymbol{\theta})}{f(\mathbf{y}|\boldsymbol{\theta})\,p(\boldsymbol{\theta})}\,p(\boldsymbol{\theta}|\mathbf{y})d\boldsymbol{\theta}\;,$$

which holds for any proper density h. Given samples $\boldsymbol{\theta}^{(g)}$ from the posterior distribution, this suggests the estimator

$$\hat{p}(\mathbf{y}) = \left[\frac{1}{G}\sum_{g=1}^{G}\frac{h(\boldsymbol{\theta}^{(g)})}{f(\mathbf{y}|\boldsymbol{\theta}^{(g)})\,p(\boldsymbol{\theta}^{(g)})}\right]^{-1}. \tag{4.21}$$

Notice that taking $h(\boldsymbol{\theta}) = p(\boldsymbol{\theta})$, the prior density, produces (4.20). However, this is a poor choice from the standpoint of importance sampling theory, which instead suggests choosing h to roughly match the posterior density. Gelfand and Dey suggest a multivariate normal or t density with mean and covariance matrix estimated from the $\boldsymbol{\theta}^{(g)}$ samples. Kass and Raftery (1995) observe that this estimator does satisfy a central limit theorem provided that

$$\int \frac{h^2(\boldsymbol{\theta})}{f(\mathbf{y}|\boldsymbol{\theta})\,p(\boldsymbol{\theta})}\,d\boldsymbol{\theta} < \infty\;,$$

i.e., the tails of h are relatively thin. The estimator does seem to perform well in practice unless $\boldsymbol{\theta}$ is of too high a dimension, complicating the selection of a good h density.

The material in this section is closely related to what Tan et al. (2008) refer to as *inverse Bayes formulae*, which use a reexpression of Bayes' Rule to express a particular marginal distribution in terms of available conditional distributions. Such formulae can allow one to avoid MCMC computation (for which we have seen convergence assessment is often problematic) for both posterior estimation and model comparison in a fairly broad class of missing data problems. In particular, these authors stress the use of noniterative Monte Carlo and EM computational methods. Still, even with data augmentation there is a limit to the generality of the hierarchical models one can analyze with this technology.

4.4.2 Using Gibbs sampler output

In the case of models fit via Gibbs sampling with closed-form full conditional distributions, Chib (1995) offers an alternative that avoids the specification of the h function above. The method begins by simply rewriting

Bayes' rule as

$$p(\mathbf{y}) = \frac{f(\mathbf{y}|\boldsymbol{\theta})\, p(\boldsymbol{\theta})}{p(\boldsymbol{\theta}|\mathbf{y})} \ .$$

Only the denominator on the right-hand side is unknown, so an estimate of the posterior would produce an estimate of the marginal density, as we desire. But because this identity holds for *any* $\boldsymbol{\theta}$ value, we require only a posterior density estimate at a single point – say, $\boldsymbol{\theta}'$. Therefore, we have

$$\log \hat{p}(\mathbf{y}) = \log f(\mathbf{y}|\boldsymbol{\theta}') + \log p(\boldsymbol{\theta}') - \log \hat{p}(\boldsymbol{\theta}'|\mathbf{y}) \ , \qquad (4.22)$$

where we have switched to the log scale to improve computational accuracy. While in theory $\boldsymbol{\theta}'$ is arbitrary, Chib (1995) suggests choosing it as a point of high posterior density, again to maximize accuracy in (4.22).

It remains to show how to obtain the estimate $\hat{p}(\boldsymbol{\theta}'|\mathbf{y})$. We first describe the technique in the case where the parameter vector can be decomposed into two blocks (similar to the data augmentation scenario of Tanner and Wong, 1987). That is, we suppose that $\boldsymbol{\theta} = (\boldsymbol{\theta}_1, \boldsymbol{\theta}_2)$, where $p(\boldsymbol{\theta}_1|\mathbf{y}, \boldsymbol{\theta}_2)$ and $p(\boldsymbol{\theta}_2|\mathbf{y}, \boldsymbol{\theta}_1)$ are both available in closed form. Writing

$$p(\boldsymbol{\theta}_1', \boldsymbol{\theta}_2'|\mathbf{y}) = p(\boldsymbol{\theta}_2'|\mathbf{y}, \boldsymbol{\theta}_1')p(\boldsymbol{\theta}_1'|\mathbf{y}) \ , \qquad (4.23)$$

we observe that the first term on the right-hand side is available explicitly at $\boldsymbol{\theta}'$, while the second can be estimated via the "Rao-Blackwellized" mixture estimate (3.9), namely,

$$\hat{p}(\boldsymbol{\theta}_1'|\mathbf{y}) = \frac{1}{G} \sum_{g=1}^{G} p(\boldsymbol{\theta}_1'|\mathbf{y}, \boldsymbol{\theta}_2^{(g)}) \ , \qquad (4.24)$$

since $\boldsymbol{\theta}_2^{(g)} \sim p(\boldsymbol{\theta}_2|\mathbf{y})$, $g = 1, \ldots, G$. Thus our marginal density estimate in (4.22) becomes

$$\log \hat{p}(\mathbf{y}) = \log f(\mathbf{y}|\boldsymbol{\theta}_1', \boldsymbol{\theta}_2') + \log p(\boldsymbol{\theta}_1', \boldsymbol{\theta}_2') - \log p(\boldsymbol{\theta}_2'|\mathbf{y}, \boldsymbol{\theta}_1') - \log \hat{p}(\boldsymbol{\theta}_1'|\mathbf{y}) \ .$$

Exponentiating produces the final marginal density estimate.

Next, suppose there are three parameter blocks, $\boldsymbol{\theta} = (\boldsymbol{\theta}_1, \boldsymbol{\theta}_2, \boldsymbol{\theta}_3)$. The decomposition of the joint posterior density (4.23) now becomes

$$p(\boldsymbol{\theta}_1', \boldsymbol{\theta}_2', \boldsymbol{\theta}_3'|\mathbf{y}) = p(\boldsymbol{\theta}_3'|\mathbf{y}, \boldsymbol{\theta}_1', \boldsymbol{\theta}_2')p(\boldsymbol{\theta}_2'|\mathbf{y}, \boldsymbol{\theta}_1')p(\boldsymbol{\theta}_1'|\mathbf{y}) \ .$$

Again the first term is available explicitly, while the third term may be estimated as a mixture of $p(\boldsymbol{\theta}_1'|\mathbf{y}, \boldsymbol{\theta}_2^{(g)}, \boldsymbol{\theta}_3^{(g)})$, $g = 1, \ldots, G$, similar to (4.24) above. For the second term, we may write

$$p(\boldsymbol{\theta}_2'|\mathbf{y}, \boldsymbol{\theta}_1') = \int p(\boldsymbol{\theta}_2'|\mathbf{y}, \boldsymbol{\theta}_1', \boldsymbol{\theta}_3)p(\boldsymbol{\theta}_3|\mathbf{y}, \boldsymbol{\theta}_1')d\boldsymbol{\theta}_3 \ ,$$

suggesting the estimator

$$\hat{p}(\boldsymbol{\theta}_2'|\mathbf{y}, \boldsymbol{\theta}_1') = \frac{1}{G} \sum_{g=1}^{G} p(\boldsymbol{\theta}_2'|\mathbf{y}, \boldsymbol{\theta}_1', \boldsymbol{\theta}_3^{*(g)}) \ ,$$

where $\boldsymbol{\theta}_3^{*(g)} \sim p(\boldsymbol{\theta}_3|\mathbf{y},\boldsymbol{\theta}_1')$. Such draws are *not* available from the original posterior sample, which instead contains $\boldsymbol{\theta}_3^{(g)} \sim p(\boldsymbol{\theta}_3|\mathbf{y})$. However, we may produce them simply by continuing the Gibbs sampler for an additional G iterations with only two full conditional distributions, namely

$$p(\boldsymbol{\theta}_2|\mathbf{y},\boldsymbol{\theta}_1',\boldsymbol{\theta}_3) \text{ and } p(\boldsymbol{\theta}_3|\mathbf{y},\boldsymbol{\theta}_1',\boldsymbol{\theta}_2) \ .$$

Thus, while additional sampling is required, new computer code is not; we need only continue with a portion of the old code. The final marginal density estimate then arises from

$$\begin{aligned}
\log \hat{p}(\mathbf{y}) = {} & \log f(\mathbf{y}|\boldsymbol{\theta}_1',\boldsymbol{\theta}_2',\boldsymbol{\theta}_3') + \log p(\boldsymbol{\theta}_1',\boldsymbol{\theta}_2',\boldsymbol{\theta}_3') \\
& - \log p(\boldsymbol{\theta}_3'|\mathbf{y},\boldsymbol{\theta}_1',\boldsymbol{\theta}_2') - \log \hat{p}(\boldsymbol{\theta}_2'|\mathbf{y},\boldsymbol{\theta}_1') - \log \hat{p}(\boldsymbol{\theta}_1'|\mathbf{y}) \ .
\end{aligned}$$

The extension to B parameter blocks requires a similar factoring of the joint posterior into B components, with $(B-1)$ Gibbs sampling runs of G samples each to estimate the various factors. We note that clever partitioning of the parameter vector into only a few blocks (each still having a closed form full conditional) can increase computational accuracy and reduce programming and sampling time as well.

Extension of this basic approach to more difficult model settings is possible. For instance, Basu and Chib (2003) apply it to the Dirichlet process (DP) mixture model setting of Subsection 2.6. This paper resolves the issue of calculation of the likelihood ordinate using a collapsed sequential importance sampling (SIS) algorithm.

4.4.3 Using Metropolis-Hastings output

Equations like (4.24) require us to know the normalizing constant for the full conditional distribution $p(\boldsymbol{\theta}_1|\mathbf{y},\boldsymbol{\theta}_2)$, thus precluding their use with full conditionals updated using Metropolis-Hastings (rather than Gibbs) steps. To remedy this, Chib and Jeliazkov (2001) extend the approach, which takes a particularly simple form in the case where the parameter vector $\boldsymbol{\theta}$ can be updated in a single block. Let

$$\alpha(\boldsymbol{\theta},\boldsymbol{\theta}^*|\mathbf{y}) = \min\left\{1, \frac{p(\boldsymbol{\theta}^*|\mathbf{y})q(\boldsymbol{\theta}^*,\boldsymbol{\theta}|\mathbf{y})}{p(\boldsymbol{\theta}|\mathbf{y})q(\boldsymbol{\theta},\boldsymbol{\theta}^*|\mathbf{y})}\right\} \ ,$$

the probability of accepting a Metropolis-Hastings candidate $\boldsymbol{\theta}^*$ generated from a candidate density $q(\boldsymbol{\theta},\boldsymbol{\theta}^*|\mathbf{y})$ (note that this density is allowed to depend on the data $\mathbf{y}$). Chib and Jeliazkov (2001) then show

$$p(\boldsymbol{\theta}'|\mathbf{y}) = \frac{E_1\left\{\alpha(\boldsymbol{\theta},\boldsymbol{\theta}'|\mathbf{y})q(\boldsymbol{\theta},\boldsymbol{\theta}'|\mathbf{y})\right\}}{E_2\left\{\alpha(\boldsymbol{\theta}',\boldsymbol{\theta}|\mathbf{y})\right\}} \ , \tag{4.25}$$

where E_1 is the expectation with respect to the posterior $p(\boldsymbol{\theta}|\mathbf{y})$ and E_2 is the expectation with respect to the candidate density $q(\boldsymbol{\theta}',\boldsymbol{\theta}|\mathbf{y})$. The numerator is then estimated by averaging the product in braces with respect to draws from the posterior, while the denominator is estimated by averaging the acceptance probability with respect to draws from $q(\boldsymbol{\theta}', \boldsymbol{\theta}|\mathbf{y})$, given the fixed value $\boldsymbol{\theta}'$. Note that this calculation does not require knowledge of the normalizing constant for $p(\boldsymbol{\theta}|\mathbf{y})$. Plugging this estimate of (4.25) into (4.22) completes the estimation of the marginal likelihood. When there are two more more blocks, Chib and Jeliazkov (2001) illustrate an extended version of this algorithm using multiple MCMC runs, similar to the Chib (1995) approach for the Gibbs sampler outlined in the previous subsection.

4.5 Bayes factors via sampling over the model space

The methods of the previous section all seek to estimate the marginal density $p(\mathbf{y})$ for each model, and subsequently calculate the Bayes factor via equation (4.18). They also operate on a posterior sample that has already been produced by some noniterative or MCMC method, though the methods of Chib (1995) and Chib and Jeliazkov (2001) will often require multiple runs of slightly different versions of the MCMC algorithm to produce the necessary output. But for some complicated or high-dimensional model settings, such as spatial models using Markov random field priors (which involve large numbers of random effect parameters that cannot be analytically integrated out of the likelihood nor readily updated in blocks), these methods may be infeasible.

An alternative approach favored by many authors is to include the model indicator M as a parameter in the sampling algorithm itself. This of course complicates the initial sampling process, but has the important benefit of producing a stream of samples $\{M^{(g)}\}_{g=1}^{G}$ from $p(M|\mathbf{y})$, the marginal posterior distribution of the model indicator. Hence, the ratio

$$\hat{p}(M = j|\mathbf{y}) = \frac{\text{number of } M^{(g)} = j}{\text{total number of } M^{(g)}}, \ \ j = 1, \ldots, K, \qquad (4.26)$$

provides a simple estimate of each posterior model probability, which may then be used to compute the Bayes factor between any two of the models, say j and j', via

$$\widehat{BF}_{jj'} = \frac{\hat{p}(M = j|\mathbf{y})/\hat{p}(M = j'|\mathbf{y})}{p(M = j)/p(M = j')}, \qquad (4.27)$$

the original formula used to define the Bayes factor in (2.18). Estimated variances of the estimates in (4.26) are easy to obtain even if the $M^{(g)}$ output stream exhibits autocorrelation through the batching formula (3.30) or other methods mentioned in Subsection 3.4.6.

Sampling over the model space alone

Most of the methods we will discuss involve algorithmic searches over both the model and the parameter spaces simultaneously. This is a natural way to think about the problem, but like any other augmented sampler such an approach risks a less well-identified parameter space, increased correlations, and hence slower convergence than a sampler operating on any one of the models alone. Moreover, if our interest truly lies only in computing posterior model probabilities $p(M = j|\mathbf{y})$ or a Bayes factor, the parameter samples are not needed, relegating the $\boldsymbol{\theta}_j$ to nuisance parameter status.

These thoughts motivate the creation of samplers that operate on the model space alone. This in turn requires us to integrate the $\boldsymbol{\theta}_j$ out of the model before sampling begins. To obtain such marginalized expressions in closed form requires fairly specialized likelihood and prior settings, but several authors have made headway in this area for surprisingly broad classes of models. For example, Madigan and York (1995) offer an algorithm for searching over a space of graphical models for discrete data, an approach they refer to as Markov chain Monte Carlo model composition, or $(MC)^3$. Raftery, Madigan, and Hoeting (1997) work instead in the multiple regression setting with conjugate priors, again enabling a model-space-only search. They compare the $(MC)^3$ approach with the "Occam's Window" method of Madigan and Raftery (1994). Finally, Clyde, DeSimone, and Parmigiani (1996) use importance sampling (not MCMC) to search for the most promising models in a hierarchical normal linear model setting. They employ an orthogonalization of the design matrix, which enables impressive gains in efficiency over the Metropolis-based $(MC)^3$ method.

Sampling over model and parameter space

Unfortunately, most model settings are too complicated to allow the entire parameter vector $\boldsymbol{\theta}$ to be integrated out of the joint posterior in closed form, and thus require that any MCMC model search be over the model and parameter space jointly. Such searches date at least to the work of Carlin and Polson (1991), who included M as a parameter in the Gibbs sampler in concert with the scale mixture of normals idea mentioned in Subsection 4.2.1. In this way, they computed Bayes factors and compared marginal posterior densities for a parameter of interest under changing specification of the model error densities and related prior densities. Their algorithm required that the models share the same parametrization, however, and so would not be appropriate for comparing two different mean structures (say, a linear and a quadratic). George and McCulloch (1993) circumvented this problem for multiple regression models by introducing latent indicator variables that determine whether or not a particular regression coefficient may be safely estimated by 0, a process they referred to as *stochastic search variable selection* (SSVS). Unfortunately, in order to satisfy the Markov convergence requirement of the Gibbs sampler, a regressor can never completely "disappear" from the model, so the Bayes factors obtained necessarily depend on values of user-chosen tuning constants.

4.5.1 Product space search

Carlin and Chib (1995), whom we abbreviate as "CC", present a Gibbs sampling method that avoids these theoretical convergence difficulties and accommodates completely general model settings. Suppose there are K candidate models, and corresponding to each, there is a distinct parameter vector $\boldsymbol{\theta}_j$ having dimension n_j, $j = 1, \ldots, K$. Our interest lies in $p(M = j|\mathbf{y})$, $j = 1, \ldots, K$, the posterior probabilities of each of the K models, and possibly the model-specific posterior distributions $p(\boldsymbol{\theta}_j|M = j, \mathbf{y})$ as well.

Supposing that $\boldsymbol{\theta}_j \in \Re^{n_j}$, the approach is essentially to sample over the model indicator and the *product* space $\prod_{j=1}^{K} \Re^{n_j}$. This entails viewing the prior distributions as model specific and part of the Bayesian model specification. That is, corresponding to model j we write the likelihood as $f(\mathbf{y}|\boldsymbol{\theta}_j, M = j)$ and the prior as $p(\boldsymbol{\theta}_j|M = j)$. Since we are assuming that M merely provides an indicator as to which particular $\boldsymbol{\theta}_j$ is relevant to $\mathbf{y}$, we have that $\mathbf{y}$ is independent of $\{\boldsymbol{\theta}_{i \neq j}\}$ given that $M = j$. In addition, since our primary goal is the computation of Bayes factors, we assume that each prior $p(\boldsymbol{\theta}_j|M = j)$ is proper (though possibly quite vague). For simplicity we assume complete independence among the various $\boldsymbol{\theta}_j$ given the model indicator M, and thus may complete the Bayesian model specification by choosing proper "pseudopriors," $p(\boldsymbol{\theta}_j|M \neq j)$. Writing $\boldsymbol{\theta} = \{\boldsymbol{\theta}_1, \ldots, \boldsymbol{\theta}_K\}$, our conditional independence assumptions imply that

$$
\begin{aligned}
p(\mathbf{y}|M = j) &= \int f(\mathbf{y}|\boldsymbol{\theta}, M = j)p(\boldsymbol{\theta}|M = j)d\boldsymbol{\theta} \\
&= \int f(\mathbf{y}|\boldsymbol{\theta}_j, M = j)p(\boldsymbol{\theta}_j|M = j)d\boldsymbol{\theta}_j ,
\end{aligned} \qquad (4.28)
$$

and so the form given to $p(\boldsymbol{\theta}_j|M \neq j)$ is irrelevant. Thus, as the name suggests, a pseudoprior is not really a prior at all, but only a conveniently chosen linking density, used to completely define the joint model specification. Hence the joint distribution of $\mathbf{y}$ and $\boldsymbol{\theta}$ given $M = j$ is

$$
p(\mathbf{y}, \boldsymbol{\theta}, M = j) = f(\mathbf{y}|\boldsymbol{\theta}_j, M = j)\left\{ \prod_{i=1}^{K} p(\boldsymbol{\theta}_i|M = j) \right\} \pi_j ,
$$

where $\pi_j \equiv P(M = j)$ is the prior probability assigned to model j, satisfying $\sum_{j=1}^{K} \pi_j = 1$.

In order to implement the Gibbs sampler, we need the full conditional distributions of each $\boldsymbol{\theta}_j$ and M. The former is given by

$$
p(\boldsymbol{\theta}_j|\boldsymbol{\theta}_{i \neq j}, M, \mathbf{y}) \propto \begin{cases} f(\mathbf{y}|\boldsymbol{\theta}_j, M = j)p(\boldsymbol{\theta}_j|M = j) , & M = j \\ p(\boldsymbol{\theta}_j|M \neq j) , & M \neq j \end{cases} .
$$

That is, when $M = j$ we generate from the usual model j full conditional; when $M \neq j$ we generate from the pseudoprior. Both of these generations are straightforward provided $p(\boldsymbol{\theta}_j|M = j)$ is taken to be conjugate with its likelihood. The full conditional for M is

$$p(M = j|\boldsymbol{\theta}, \mathbf{y}) = \frac{f(\mathbf{y}|\boldsymbol{\theta}_j, M = j) \left\{\prod_{i=1}^{K} p(\boldsymbol{\theta}_i|M = j)\right\} \pi_j}{\sum_{k=1}^{K} f(\mathbf{y}|\boldsymbol{\theta}_k, M = k) \left\{\prod_{i=1}^{K} p(\boldsymbol{\theta}_i|M = k)\right\} \pi_k}. \quad (4.29)$$

Since M is a discrete finite parameter, its generation is routine as well. Therefore, all the required full conditional distributions are well defined and, under the usual regularity conditions (Roberts and Smith, 1993), the algorithm will produce samples from the correct joint posterior distribution. In particular, equations (4.26) and (4.27) can now be used to estimate the posterior model probabilities and the Bayes factor between any two models, respectively.

Notes on implementation

Notice that, in contrast to equation (4.26), summarization of the totality of $\boldsymbol{\theta}_j^{(g)}$ samples is not appropriate. This is because what is of interest in our setting is not the marginal posterior densities $p(\boldsymbol{\theta}_j|\mathbf{y})$, but rather the *conditional* posterior densities $p(\boldsymbol{\theta}_j|M = j, \mathbf{y})$. However, suppose that in addition to $\boldsymbol{\theta}$ we have a vector of nuisance parameters η, common to all models. Then the fully marginal posterior density of η, $p(\eta|\mathbf{y})$, may be of some interest. Because the data are informative about η regardless of the value of M, a pseudoprior for η is not required. Still, care must be taken to ensure that η has the same interpretation in both models. For example, suppose we wish to choose between the two nested regression models

$$M = 1: \quad y_i = \alpha + \epsilon_i, \qquad \epsilon_i \overset{iid}{\sim} \text{Normal}(0, \sigma^2),$$

and

$$M = 2: \quad y_i = \alpha + \beta x_i + \epsilon_i, \quad \epsilon_i \overset{iid}{\sim} \text{Normal}(0, \tau^2),$$

both for $i = 1, \ldots, n$. In the above notation, $\boldsymbol{\theta}_1 = \sigma$, $\boldsymbol{\theta}_2 = (\beta, \tau)$, and $\eta = \alpha$. Notice that α is playing the role of a grand mean in model 1, but is merely an intercept in model 2. The corresponding posteriors could be quite different if, for example, the observed y_i values were centered near 0, while those for x_i were centered far from 0. Besides being unmeaningful from a practical point of view, the resulting $p(\alpha|\mathbf{y})$ would likely be bimodal, a shape that could wreak great havoc with convergence of the MCMC algorithm, since jumps between $M = 1$ and 2 would be extremely unlikely.

Poor choices of the pseudopriors $p(\boldsymbol{\theta}_j|M \neq j)$ can have a similar deleterious effect on convergence. Good choices will produce $\boldsymbol{\theta}_j^{(g)}$ values that are consistent with the data, so that $p(M = j|\boldsymbol{\theta}, \mathbf{y})$ will still be reasonably large at the next M update step. Failure to generate competitive pseudoprior values will again result in intolerably high autocorrelations in the $M^{(g)}$ chain, hence slow convergence. As such, matching the pseudopriors as nearly as possible to the true model-specific posteriors is recommended.

This can be done using first-order (normal) approximations or other parametric forms designed to mimic the output from K individual preliminary MCMC runs. Note that we are *not* using the data to help select the prior, but only the pseudoprior.

If for a particular dataset one of the $p(M = j|\mathbf{y})$ is extremely large, the realized chain will exhibit slow convergence due to the resulting near-absorbing state in the algorithm. In this case, the π_j may be adjusted to correct the imbalance; the final Bayes factor computed from (4.18) will still reflect the true odds in favor of $M = j$ suggested by the data.

Finally, one might plausibly consider the modified version of the above algorithm that skips the generation of actual pseudoprior values, and simply keeps $\boldsymbol{\theta}_j^{(g)}$ at its current value when $M^{(g)} \neq j$. While this MCMC algorithm is no longer a Gibbs sampler, Green and O'Hagan (1998) show that it does still converge to the correct stationary distribution (i.e., its transition kernel does satisfy detailed balance). Under this alternative we would be free to choose $p(\boldsymbol{\theta}_j|M=i) = p(\boldsymbol{\theta}_j|M=j)$ for all i, meaning that all of these terms cancel from the M update step in equation (4.29). This pseudoprior-free version of the algorithm thus avoids the need for preliminary runs, and greatly reduces the generation and storage burdens in the final model choice run. Unfortunately, Green and O'Hagan also find it to perform quite poorly, since without the variability in candidate $\boldsymbol{\theta}_j^{(g)}$ values created by the pseudopriors, switches between models are rare. For more recent developments and refinements along these same lines, see Congdon (2007b).

Scott (2002) and Congdon (2006) suggest skipping the pseudoprior generation and simply estimating the marginal model probabilities (4.28) using independent, model-specific MCMC runs. However, Robert and Marin (2008) show that such algorithms are biased, because they do not sample over the correct *joint* posterior $p(\boldsymbol{\theta}, M|\mathbf{y})$, required for validity of the product space approach.

4.5.2 "Metropolized" product space search

Dellaportas, Forster, and Ntzoufras (2002) propose a hybrid Gibbs-Metropolis strategy. In their strategy, the model selection step is based on a proposal for a move from model j to j', followed by acceptance or rejection of this proposal. That is, the method is a "Metropolized Carlin and Chib" (MCC) approach, which proceeds as follows:

1. Let the current state be $(j, \boldsymbol{\theta}_j)$, where $\boldsymbol{\theta}_j$ is of dimension n_j.

2. Propose a new model j' with probability $h(j, j')$.

3. Generate $\boldsymbol{\theta}_{j'}$ from a pseudoprior $p(\boldsymbol{\theta}_{j'}|M \neq j')$ as in Carlin and Chib's product space search method.

4. Accept the proposed move (from j to j') with probability

$$\alpha_{j \to j'} = \min \left\{ 1, \frac{f(\mathbf{y}|\boldsymbol{\theta}_{j'}, M = j')p(\boldsymbol{\theta}_{j'}|M = j')p(\boldsymbol{\theta}_j|M = j')\pi_{j'}h(j',j)}{f(\mathbf{y}|\boldsymbol{\theta}_j, M = j)p(\boldsymbol{\theta}_j|M = j)p(\boldsymbol{\theta}_{j'}|M = j)\pi_j h(j,j')} \right\}.$$

Thus by "Metropolizing" the model selection step, the MCC method needs to sample only from the pseudoprior for the proposed model j'. In this method, the move is a Gibbs step or a sequence of Gibbs steps when $j' = j$. Posterior model probabilities and Bayes factors can be estimated as before.

4.5.3 Reversible jump MCMC

This method, originally due to Green (1995), is another strategy that samples over the model and parameter space, but which avoids the full product space search of the Carlin and Chib (1995) method (and the associated pseudoprior specification and sampling), at the cost of a less straightforward algorithm operating on the *union* space, $\mathcal{M} \times \bigcup_{j \in \mathcal{M}} \Theta_j$. It generates a Markov chain that can "jump" between models with parameter spaces of different dimensions, while retaining the aperiodicity, irreducibility, and detailed balance conditions necessary for MCMC convergence.

A typical reversible jump algorithm proceeds as follows.

1. Let the current state of the Markov chain be $(j, \boldsymbol{\theta}_j)$, where $\boldsymbol{\theta}_j$ is of dimension n_j.

2. Propose a new model j' with probability $h(j, j')$.

3. Generate $\mathbf{u}$ from a proposal density $q(\mathbf{u}|\boldsymbol{\theta}_j, j, j')$.

4. Set $(\boldsymbol{\theta}'_{j'}, \mathbf{u}') = \mathbf{g}_{j,j'}(\boldsymbol{\theta}_j, \mathbf{u})$, where $\mathbf{g}_{j,j'}$ is a deterministic function that is 1-1 and onto. This is a "dimension matching" function, specified so that $n_j + dim(\mathbf{u}) = n_{j'} + dim(\mathbf{u}')$.

5. Accept the proposed move (from j to j') with probability $\alpha_{j \to j'}$, which is the minimum of 1 and

$$\frac{f(\mathbf{y}|\boldsymbol{\theta}'_{j'}, M = j')p(\boldsymbol{\theta}'_{j'}|M = j')\pi_{j'}h(j',j)q(\mathbf{u}'|\boldsymbol{\theta}_{j'}, j', j)}{f(\mathbf{y}|\boldsymbol{\theta}_j, M = j)p(\boldsymbol{\theta}_j|M = j)\pi_j h(j,j')q(\mathbf{u}|\boldsymbol{\theta}_j, j, j')} \left| \frac{\partial \mathbf{g}(\boldsymbol{\theta}_j, \mathbf{u})}{\partial(\boldsymbol{\theta}_j, \mathbf{u})} \right|.$$

When $j' = j$, the move can be either a standard Metropolis-Hastings or Gibbs step. Posterior model probabilities and Bayes factors may be estimated from the output of this algorithm as in the previous subsection.

The "dimension matching" aspect of this algorithm (step 4 above) is a bit obscure and merits further discussion. Suppose we are comparing two models, for which $\theta_1 \in \Re$ and $\theta_2 \in \Re^2$. If θ_1 is a subvector of θ_2, then when moving from $j = 1$ to $j' = 2$, we might simply draw $u \sim q(u)$ and set

$$\theta'_2 = (\theta_1, u).$$

That is, the dimension matching g is the identity function, and so the Jacobian in step 5 is equal to 1. Thus, if we set $h(1,1) = h(1,2) = h(2,1) = h(2,2) = 1/2$, we have

$$\alpha_{1\to 2} = \min\left\{1, \frac{f(\mathbf{y}|\boldsymbol{\theta}_2', M = 2)p(\boldsymbol{\theta}_2'|M = 2)\pi_2}{f(\mathbf{y}|\boldsymbol{\theta}_1, M = 1)p(\boldsymbol{\theta}_1|M = 1)\pi_1 q(u)}\right\} ,$$

with a corresponding expression for $\alpha_{2\to 1}$.

In many cases, however, θ_1 will not naturally be thought of as a subvector of θ_2. Green (1995) considers the case of a *changepoint model* (see e.g. Table 3.5 and the associated exercises in Chapter 3), in which the choice is between a time series model having a single, constant mean level θ_1, and one having two levels – say, $\theta_{2,1}$ before the changepoint and $\theta_{2,2}$ afterward. In this setting, when moving from Model 2 to 1, we would not likely want to use *either* $\theta_{2,1}$ or $\theta_{2,2}$ as the proposal value θ_1'. A more plausible choice might be

$$\theta_1' = \frac{\theta_{2,1} + \theta_{2,2}}{2} , \tag{4.30}$$

since the average of the pre- and post-changepoint levels should provide a competitive value for the single level in Model 1. To ensure reversibility of this move, when going from Model 1 to 2 we might sample $u \sim q(u)$ and set

$$\theta_{2,1}' = \theta_1 - u \quad \text{and} \quad \theta_{2,2}' = \theta_1 + u ,$$

since this is a 1-1 and onto function corresponding to the deterministic down move (4.30).

Several variations or simplifications of reversible jump MCMC have been proposed for various model classes; see e.g. Richardson and Green (1997) in the context of mixture modeling, and Knorr-Held and Rasser (2000) for a spatial disease mapping application. Also, the "jump diffusion" approach of Phillips and Smith (1996) can be thought of as a variant on the reversible jump idea.

As with other Metropolis-Hastings algorithms, transformations to various parameters are often helpful in specifying proposal densities in reversible jump algorithms (say, taking the log of a variance parameter). It may also be helpful to apply reversible jump to a somewhat reduced model, where we analytically integrate certain parameters out of the model and use (lower-dimensional) proposal densities for the parameters that remain. We will generally not have closed forms for the full conditional distributions of these "leftover" parameters, but this is not an issue because, unlike the Gibbs-based CC method, reversible jump does not require them. Here an example might be hierarchical normal random effects models with conjugate priors: the random effects (and perhaps even the fixed effects) may be integrated out, permitting the algorithm to sample only the model indicator and the few remaining (variance) parameters.

4.5.4 Using partial analytic structure

Godsill (2001) proposes use of a "composite model space," which is essentially the setting of Carlin and Chib (1995) except that parameters are allowed to be "shared" between different models. A standard Gibbs sampler applied to this composite model produces the CC method, while a more sophisticated Metropolis-Hastings approach produces a version of the reversible jump algorithm that avoids the "dimension matching" step present in its original formulation (see step 4 in Subsection 4.5.3 above). This step is often helpful for challenging problems (e.g., when moving to a model containing many parameters whose values would not plausibly equal any of those in the current model), but may be unnecessary for simpler problems.

Along these lines, Godsill (2001) outlines a reversible jump method that takes advantage of *partial analytic structure* (PAS) in the Bayesian model. This procedure is applicable when there exists a subvector $(\boldsymbol{\theta}_{j'})_{\mathcal{U}}$ of the parameter vector $\boldsymbol{\theta}_{j'}$ for model j' such that $p((\boldsymbol{\theta}_{j'})_{\mathcal{U}}|(\boldsymbol{\theta}_{j'})_{-\mathcal{U}}, M = j', \mathbf{y})$ is available in closed form, and in the current model j, there exists an equivalent subvector $(\boldsymbol{\theta}_j)_{-\mathcal{U}}$ (the elements of $\boldsymbol{\theta}_j$ *not* in subvector $\mathcal{U}$) of the same dimension as $(\boldsymbol{\theta}_{j'})_{-\mathcal{U}}$. Operationally,

1. Let the current state be $(j, \boldsymbol{\theta}_j)$, where $\boldsymbol{\theta}_j$ is of dimension n_j.

2. Propose a new model j' with probability $h(j, j')$.

3. Set $(\boldsymbol{\theta}_{j'})_{-\mathcal{U}} = (\boldsymbol{\theta}_j)_{-\mathcal{U}}$.

4. Accept the proposed move with probability

$$\alpha_{j \to j'} = \min \left\{ 1, \ \frac{p(j'|(\boldsymbol{\theta}_{j'})_{-\mathcal{U}}, \mathbf{y})h(j', j)}{p(j|(\boldsymbol{\theta}_j)_{-\mathcal{U}}, \mathbf{y})h(j, j')} \right\}, \qquad (4.31)$$

   where $p(j|(\boldsymbol{\theta}_j)_{-\mathcal{U}}, \mathbf{y}) = \int p(j, (\boldsymbol{\theta}_j)_{\mathcal{U}}|(\boldsymbol{\theta}_j)_{-\mathcal{U}}, \mathbf{y}) \, d(\boldsymbol{\theta}_j)_{\mathcal{U}}$.

5. If the model move is accepted, update the parameters of the new model $(\boldsymbol{\theta}_{j'})_{\mathcal{U}}$ and $(\boldsymbol{\theta}_{j'})_{-\mathcal{U}}$ using standard Gibbs or Metropolis-Hastings steps; otherwise, update the parameters of the old model $(\boldsymbol{\theta}_j)_{\mathcal{U}}$ and $(\boldsymbol{\theta}_j)_{-\mathcal{U}}$ using standard Gibbs or Metropolis-Hastings steps.

Note that model move proposals of the form $j \to j$ always have acceptance probability 1, and therefore when the current model is proposed, this algorithm simplifies to standard Gibbs or Metropolis-Hastings steps. Note that multiple proposal densities may be needed for $(\boldsymbol{\theta}_j)_{\mathcal{U}}$ across models since, while this parameter is common to all of them, its interpretation and posterior support may differ. Troughton and Godsill (1998) give an example of this algorithm, in which the update step of $(\boldsymbol{\theta}_j)_{\mathcal{U}}$ is skipped when a proposed model move is rejected.

Summary and recommendations

Han and Carlin (2001) review and compare many of the methods described in this and the previous subsections in the context of two examples, the first a simple regression example, and the second a more challenging hierarchical longitudinal model (see Section 7.3). The methods described in this section that sample jointly over model and parameter space (such as product space search and reversible jump) often converge very slowly, due to the difficulty in finding suitable pseudoprior (or proposal) densities. Marginalizing random effects out of the model can be helpful in this regard, but this tends to create new problems: the marginalization often leads to a very complicated form for the model switch acceptance ratio (step 5 in Subsection 4.5.3), thus increasing the chance of an algebraic or computational error. Even with the marginalization, such methods remain difficult to tune. The user will often need a rough idea of the posterior model probabilities $p(M = j|\mathbf{y})$ in order to set the prior model probabilities π_j in such a way that the sampler spends roughly equal time visiting each of the candidate models. Preliminary model-specific runs are also typically required to specify proposal (or pseudoprior) densities for each model.

By contrast, the marginal likelihood methods of Section 4.4 appear relatively easy to program and tune. These methods do not require preliminary runs (only a point of high posterior density, $\boldsymbol{\theta}'$), and in the case of the Gibbs sampler, only a rearrangement of existing computer code. Estimating standard errors is more problematic (the authors' suggested approach involves a spectral density estimate and the delta method), but simply replicating the entire procedure a few times with different random number seeds generally provides an acceptable idea of the procedure's order of accuracy.

In their numerical illustrations, Han and Carlin (2001) found that the RJ and PAS methods ran more quickly than the other model space search methods, but the marginal likelihood methods seemed to produce the highest degree of accuracy for roughly comparable runtimes. This is in keeping with the intuition that some gain in precision should accrue to MCMC methods that avoid a model space search.

As such, we recommend the marginal likelihood methods as relatively easy and safe approaches when choosing among a collection of standard (e.g., hierarchical linear) models. We hasten to add, however, that the blocking required by these methods may preclude their use in some settings, such as spatial models using Markov random field priors (which involve large numbers of random effect parameters that cannot be analytically integrated out of the likelihood nor readily updated in blocks; see Subsection 7.7.2). In such cases, reversible jump may offer the only feasible alternative for estimating a Bayes factor. The marginal likelihood methods would also seem impractical if the number of candidate models were very large (e.g., in variable selection problems having 2^p possible models, corresponding to each of p predictors being either included or excluded). But, as alluded to earlier, we caution that the ability of joint model and parameter space search methods to sample effectively over such large spaces is very much in doubt; see for example Clyde et al. (1996).

4.6 Other model selection methods

The Bayes factor estimation methods discussed in the previous two sections require substantial time and effort (both human and computer) for a rather modest payoff, namely, a collection of posterior model probability estimates (possibly augmented with associated standard error estimates). Besides being mere single number summaries of relative model worth, Bayes factors are not interpretable with improper priors on any components of the parameter vector, because if the prior distribution is improper, then the marginal distribution of the data necessarily is as well. Even for proper priors, the Bayes factor has been criticized on theoretical grounds; see for example Gelfand and Dey (1994) and Draper (1995). One might conclude that none of the methods considered thus far is appropriate for everyday, "rough and ready" model comparison, and instead search for more computationally realistic alternatives.

One such alternative might be more informal, perhaps graphical methods for model selection. These could be based on the marginal distributions $m(y_i)$, $i = 1, \ldots, n$ (as in Berger 1985, p. 199), provided they exist. Alternatively, we could use the conditional predictive distributions $f(y_i|\mathbf{y}_{(i)})$, $i = 1, \ldots, n$ (as in Gelfand et al., 1992), since they will be proper when $p(\boldsymbol{\theta}|\mathbf{y}_{(i)})$ is, and they indicate the likelihood of each datapoint in the presence of all the rest. For example, the product (or the sum of the logs) of the observed conditional predictive ordinate (CPO) values $f(y_i^{obs}|\mathbf{y}_{(i)})$ given in (2.29) could be compared across models, with the larger result indicating the preferred model. Alternatively, sums of the squares or absolute values of the standardized residuals d_i given in (2.28) could be compared across models, with the model having the smaller value now being preferred. We could improve this procedure by accounting for differing model size – say, by plotting the numerator of (2.28) versus its denominator for a variety of candidate models on the same set of axes. In this way we could judge the accuracy of each model relative to its precision. For example, an overfitted model (i.e., one with redundant predictor variables) would tend to have residuals of roughly the same size as an adequate one, but with higher variances (due to the "collinearity" of the predictors). We exemplify this conditional predictive approach in our Section 8.1 case study.

4.6.1 Penalized likelihood criteria: AIC, BIC, and DIC

For some advanced models, even cross-validatory predictive selection methods may be unavailable. For example, in the spatial models of Subsection 7.7.2, the presence of certain model parameters identified only by the prior leads to an information deficit that causes the conditional predictive distributions (2.29) to be improper. In many cases, informal likelihood or penalized likelihood criteria may offer a feasible alternative. Log-likelihood summaries are easy to estimate using posterior samples $\{\boldsymbol{\theta}^{(g)}, g = 1, \ldots, G\}$, since we may think of $\ell \equiv \log L(\boldsymbol{\theta})$ as a parametric function of interest, and subsequently compute

$$\hat{\ell} \equiv E[\log L(\boldsymbol{\theta})|\mathbf{y}] \approx \frac{1}{G} \sum_{g=1}^{G} \log L(\boldsymbol{\theta}^{(g)}) \qquad (4.32)$$

as an overall measure of model fit to be compared across models. To account for differing model size, we could penalize $\hat{\ell}$ using the same sort of penalties as the Bayesian information (Schwarz) criterion (2.20) or the Akaike information criterion (2.21). In the case of the former, for example, we would have

$$\widehat{BIC} = 2\hat{\ell} - p \log n ,$$

where as usual p is the number of parameters in the model, and n is the number of datapoints.

Unfortunately, a problem arises here for the case of hierarchical models: what exactly are p and n? For example, in a longitudinal setting (see Section 7.3) in which we have s_i observations on patient i, $i = 1, \ldots, m$, shall we set $n = \sum_i s_i$ (the total number of observations), or $n = m$ (the number of patients)? If the observations on every patient were independent, the former choice would seem most appropriate, while if they were perfectly correlated within each patient, we might instead choose the latter. But of course the true state of nature is likely somewhere in between. Similarly, if we had a collection of m random effects, one for each patient, what does this contribute to p? If the random effects had nothing in common (i.e., they were essentially like fixed effects), they would contribute the full m parameters to p, but if the data (or prior) indicated they were all essentially identical, they would contribute little more than one "effective parameter" to the total model size p. Pauler (1998) obtained results in the case of hierarchical normal linear models, but results for general models remain elusive. In particular, Volinsky and Raftery (2000) showed that *either* definition above (m or $\sum_i s_i$) can be justified asymptotically in the case of survival models with censored data.

In part to help address this problem, Spiegelhalter et al. (2002) suggested the *Deviance Information Criterion* (DIC), a generalization of the Akaike information criterion (AIC) that is based on the posterior distribution of the deviance statistic (2.24). As introduced in Subsection 2.4.2, the DIC approach captures the *fit* of a model by the posterior expectation of the deviance, $\overline{D} = E_{\theta|y}[D]$, and the *complexity* of a model by the effective number of parameters p_D, defined in equation (2.25) as

$$p_D = \overline{D} - D(\overline{\boldsymbol{\theta}}) .$$

The DIC is then defined as in equation (2.26) analogous to AIC as

$$DIC = \overline{D} + p_D .$$

Because we desire models that exhibit good fit but also a reasonable degree of parsimony, smaller values of DIC indicate preferred models. As with other penalized likelihood criteria, DIC is not intended for identification of the "correct" model, but rather merely as a method of comparing a collection of alternative formulations (all of which may be incorrect).

An asymptotic justification of DIC is straightforward in cases where the number of observations n grows with respect to the number of parameters p, and where the prior $p(\theta)$ is non-hierarchical and completely specified (i.e., having no unknown parameters). Here we may expand $D(\theta)$ around $\bar{\theta}$ to give, to second order,

$$D(\theta) \approx D(\bar{\theta}) - 2(\theta - \bar{\theta})^T L' - (\theta - \bar{\theta})^T L''(\theta - \bar{\theta}) \qquad (4.33)$$

where $L = \log p(\mathbf{y}|\theta) = -D(\theta)/2$, and L' and L'' are the first derivative vector and second derivative matrix with respect to θ. However, from the Bayesian Central Limit Theorem (Theorem 3.1) we have that $\theta \mid \mathbf{y}$ is approximately distributed as $N\left(\hat{\theta}, [-L'']^{-1}\right)$, where $\bar{\theta} = \hat{\theta}$ are the maximum likelihood estimates such that $L' = 0$. This in turn implies that $(\theta - \hat{\theta})^T(-L'')(\theta - \hat{\theta})$ has an approximate chi-squared distribution with p degrees of freedom. Thus, writing $D_{\text{non}}(\theta)$ to represent the deviance for a non-hierarchical model, from (4.33) we have that

$$D_{\text{non}}(\theta) \approx D(\hat{\theta}) - (\theta - \hat{\theta})^T L''(\theta - \hat{\theta}) = D(\hat{\theta}) + \chi_p^2 .$$

Rearranging this expression and taking expectations with respect to the posterior distribution of θ, we have

$$p \approx E_{\theta|y}[D_{\text{non}}(\theta)] - D(\hat{\theta}) , \qquad (4.34)$$

so that the number of parameters is approximately the expected deviance $\overline{D} = E_{\theta|y}[D_{\text{non}}(\theta)]$ minus the fitted deviance. But since $AIC = D(\hat{\theta}) + 2p$, from (4.34) we obtain $AIC \approx \overline{D} + p$, the expected deviance plus the number of parameters. The DIC approach for hierarchical models thus follows this equation and equation (4.34), but substituting the posterior mean $\bar{\theta}$ for the maximum likelihood estimate $\hat{\theta}$. It is a generalization of Akaike's criterion, because for non-hierarchical models, $\bar{\theta} \approx \hat{\theta}$, $p_D \approx p$, and DIC $\approx$ AIC. DIC can also be shown to have much in common with the hierarchical model selection tools previously suggested by Ye (1998) and Hodges and Sargent (2001), though the DIC idea applies much more generally.

As with all penalized likelihood criteria, DIC consists of two terms, one representing "goodness of fit" and the other a penalty for increasing model complexity. As mentioned in Subsection 2.4.2, DIC is scale-free, so as with AIC and BIC, only *differences* in DIC across models are meaningful. (Of course, p_D *does* have a scale, namely, the size of the effective parameter space.) DIC is also a very general tool, and can be readily calculated for each model being considered without analytic adaptation, complicated loss functions, additional MCMC sampling (say, of predictive values), or any matrix inversion.

The general applicability, attractive interpretations, and, perhaps most importantly, ready availability of p_D and DIC within the WinBUGS package led to their widespread use by data analysts even before the publication of the original paper by Spiegelhalter et al. (2002); see e.g. Erkanli et al. (1999). Still, many practical issues have led some to question the appropriateness of DIC for arbitrarily general Bayesian models. For example, DIC is not invariant to parametrization, so (as with prior elicitation) the most plausible parametrization must be carefully chosen beforehand. Unknown scale parameters and other innocuous restructuring of the model can also lead to small changes in the computed DIC value. Determining an appropriate variance estimate for DIC also remains a vexing practical problem. Zhu and Carlin (2000) experiment with various delta method approaches to this problem in the context of spatio-temporal models of the sort considered in Section 7.7.2, but conclude that a "brute force" replication approach may be the only suitably accurate method in such complicated settings. That is, we would independently replicate the calculation of DIC a large number of times N, obtaining a sequence of DIC estimates $\{DIC_l, \, l = 1, \ldots, N\}$, and estimate $Var(DIC)$ by the sample variance,

$$\widehat{Var}(DIC) = \frac{1}{N-1} \sum_{l=1}^{N} (DIC_l - \overline{DIC})^2 \, .$$

Finally, and perhaps most embarrassingly, p_D can occasionally emerge as *negative*, even though clearly an "effective model size" should be between 0 and the sheer number of parameters in the model. This situation is rare in standard model classes, but does arise in certain situations where the joint posterior departs markedly from normality, thus violating the terms of the asymptotic argument presented above. For example, negative p_Ds have been observed with non-log-concave likelihoods in the presence of substantial conflict between prior and data, when the posterior distribution for a parameter is extremely asymmetric (or symmetric but bimodal), and in situations where the posterior mean is a very poor summary statistic and thus leads to an unreasonably large deviance estimate $D(\bar{\theta})$. The following example provides an illustration of this problem.

Example 4.8 Consider a DIC analysis of the three error models for Fisher's sleep data compared in Example 4.7, namely the normal, the t_2, and the DE. Recall in that example we fit the latter two distributions as normal scale mixtures, but WinBUGS will also allow them to be fit directly by making two more copies of the data vector (say, Y4 and Y5) and augmenting our previous code with

```
BUGS code      Y4[i] ~ dt(theta[4], tau0[4], 2)
               Y5[i] ~ ddexp(theta[5], tau0[5])
```

Model	$\overline{D}$	$\hat{D}$	p_D	DIC
1 (normal)	33.99	32.24	1.75	35.73
2 (t_2, normal scale mixture)	24.54	20.65	3.89	28.44
3 (DE, normal scale mixture)	25.15	27.10	−1.95	23.20
4 (t_2, direct fit)	30.19	28.50	1.69	31.88
5 (DE, direct fit)	30.05	27.85	2.20	32.25

Table 4.3 *Comparison of DIC and related statistics for three error distributions, Fisher's sleep data.*

and then adding priors for the requisite new parameters simply by increasing the upper bound in the code's last `for` loop, i.e.,

BUGS code
```
for (k in 1:5) {
    theta[k] ~ dnorm(0,0.00001)   # Vague normal on mean
    sigma[k] ~ dunif(0,100)       # Uniform on sigma
    tau0[k] <- 1/(sigma[k]*sigma[k])
}
```

Table 4.3 shows the resulting fit ($\overline{D}$), complexity (p_D), and overall model choice (DIC) scores for the five models. The t_2 and DE models emerge as DIC-better than the normal, regardless of whether they are fit directly or as a normal scale mixture. In the t_2 case, the results for p_D suggest that the scale mixing parameters contribute only an extra 2 to 2.5 effective parameters (Model 2 versus Model 4). The smaller DIC value obtained in the scale mixing case implies that the mixing parameters are worthwhile in an overall model selection sense. The DIC scores for these two models need not be equal since, although they lead to the same marginal likelihood for the Y_i, the mixing prior across the λ_i creates a different *joint* distribution with a different effective dimension.

Turning to the DE errors, the p_D estimate is actually *negative* in the scale mixing case, the nonsensical result alluded to above. The non-log-concavity of the DE model seems to be causing some problems here. Specifically, it appears that the posterior mean is a poor summary statistic for these very asymmetric posteriors, and the resulting very large deviance has led to the nonsensical negative p_D estimate. As such, the very low DIC value obtained for this model (Model 3) is probably not to be trusted, since it benefits from the unrealistically negative p_D score. ∎

As of the present writing, the use of p_D and DIC remains common, though with a general caveat that they should be avoided for models lying far outside the exponential family, where it is not at all clear that a normal approximation to the posterior is sensible. Moreover, Celeux et al. (2006) observe that in missing data settings, multiple DICs can be defined, depending on whether one integrates out the missing data or treats it as something to be estimated along with the parameters $\boldsymbol{\theta}$. In advanced mixture modeling settings, these authors find none of the resulting DICs to emerge as obviously superior in all cases. See van der Linde (2004, 2005) for further discussion on the general applicability of DIC and related theoretical support. See also Hodges and Sargent (2001) and Lu et al. (2007) for an alternate method of counting parameters using the constraint case formulation in equation (3.23).

4.6.2 Predictive model selection

Besides being rather ad hoc, a problem with penalized likelihood approaches is that the usual choices of penalty function are motivated by asymptotic arguments that are sometimes hard to justify in practice. While p_D does provide one definition of the "effective size" of hierarchical models, it is difficult to justify in many nonparametric or other highly parametrized settings (e.g., Cox-type survival models).

Suppose we return to the cross-validatory approach initially presented in Subsection 2.5.1, and mimic the basic log-likelihood calculation in (4.32). That is, if we think of the product of all n of the CPO values $f(y_i|\mathbf{y}_{(i)})$ given in (2.29) as a "psuedo marginal likelihood," this gives a cross-validatory summary measure of fit. The *log pseudo marginal likelihood* (LPML), originally suggested by Geisser and Eddy (1979), is simply the log of this measure,

$$\text{LPML} = \log \left\{ \prod_{i=1}^{n} f(y_i|\mathbf{y}_{(i)}) \right\} = \sum_{i=1}^{n} \log f(y_i|\mathbf{y}_{(i)}) , \qquad (4.35)$$

the log being added primarily for computational convenience. LPML is sometimes used in place of $\overline{D}$ or DIC; for example, Draper and Krnjajic (2007, Sec. 4.1) have shown that DIC approximates the LPML for approximately Gaussian posteriors. Unlike Bayes factors, the LPML remains well defined under improper priors (provided the posterior does), and is quite stable computationally. Ibrahim, Chen, and Sinha (2001) offer a detailed discussion of the use of CPO and LPML with survival data; see also Gelfand and Mallick (1995), Sinha and Dey (1997), and Zhao et al. (2006).

As an alternative, we may work with the full posterior predictive distribution (i.e., conditional on all the observed data $\mathbf{y}$), as is done in Bayesian p-value calculations like (2.32). Following Laud and Ibrahim (1995), intuitively appealing model complexity penalty terms can emerge without resorting to complex asymptotics. As in the corresponding part of Subsection 2.5.1, the basic distribution we work with is

$$f(\mathbf{y}_{new}|\mathbf{y}_{obs}) = \int f(\mathbf{y}_{new}|\boldsymbol{\theta}) \, p(\boldsymbol{\theta}|\mathbf{y}_{obs}) \, d\boldsymbol{\theta} , \qquad (4.36)$$

where $\boldsymbol{\theta}$ denotes the collection of model parameters, and $\mathbf{y}_{new}$ is viewed as a replicate of the observed data vector $\mathbf{y}_{obs}$. The model selection criterion first selects a discrepancy function $d(\mathbf{y}_{new}, \mathbf{y}_{obs})$, then computes

$$E[d(\mathbf{y}_{new}, \mathbf{y}_{obs}) | \mathbf{y}_{obs}, M_i] , \qquad (4.37)$$

and selects the model which minimizes (4.37). For Gaussian likelihoods, Laud and Ibrahim (1995, p. 250) suggested

$$d(\mathbf{y}_{new}, \mathbf{y}_{obs}) = (\mathbf{y}_{new} - \mathbf{y}_{obs})^T (\mathbf{y}_{new} - \mathbf{y}_{obs}) . \qquad (4.38)$$

For a non-Gaussian generalized linear mixed model, we may prefer to replace (4.38) by the corresponding deviance criterion. For example, with a Poisson likelihood, we would set

$$d(\mathbf{y}_{new}, \mathbf{y}_{obs}) = 2 \sum_l \{y_{l,obs} \log(y_{l,obs}/y_{l,new}) - (y_{l,obs} - y_{l,new})\} , \quad (4.39)$$

where l indexes the components of $\mathbf{y}$. Routine calculation shows that the lth term in the summation in (4.39) is strictly convex in $y_{l,new}$ if $y_{l,obs} > 0$. To avoid problems with extreme values in the sample space, we replace (4.39) by

$$\tilde{d}(\mathbf{y}_{new}, \mathbf{y}_{obs})$$
$$= 2 \sum_l \left\{ \left(y_{l,obs} + \tfrac{1}{2}\right) \log \left(\frac{y_{l,obs} + \frac{1}{2}}{y_{l,new} + \frac{1}{2}}\right) - (y_{l,obs} - y_{l,new}) \right\} .$$

Suppose we write

$$\begin{aligned} E[\tilde{d}(\mathbf{y}_{new}, \mathbf{y}_{obs}) \,|\, \mathbf{y}_{obs}, M_i] &= \tilde{d}(E[\mathbf{y}_{new}|\mathbf{y}_{obs}, M_i], \mathbf{y}_{obs}) \\ &+ E[\tilde{d}(\mathbf{y}_{new}, \mathbf{y}_{obs}) \,|\, \mathbf{y}_{obs}, M_i] - \tilde{d}(E[\mathbf{y}_{new}|\mathbf{y}_{obs}, M_i], \mathbf{y}_{obs}) . \end{aligned} \qquad (4.40)$$

Intuitive interpretations may be given to the terms in (4.40). The left-hand side is the expected predictive deviance (EPD) for model M_i. The first term on the right-hand side is essentially the likelihood ratio statistic with the MLE for $E(\mathbf{y}_{new}|\boldsymbol{\theta}_i, M_i)$, which need not exist, replaced by $E(\mathbf{y}_{new}|\mathbf{y}_{obs}, M_i)$. Jensen's inequality shows that the second term minus the third is strictly positive, and is the penalty associated with M_i. This difference becomes

$$\begin{aligned} 2 \sum_l \left(y_{l,obs} + \tfrac{1}{2}\right) \\ \times \left\{ \log E \left[y_{l,new} + \tfrac{1}{2} | \mathbf{y}_{obs}\right] - E \left[\log \left(y_{l,new} + \tfrac{1}{2}\right) | \mathbf{y}_{obs}\right] \right\} . \end{aligned} \qquad (4.41)$$

Again, each term in the summation is positive. A second order Taylor series expansion shows that (4.41) is approximately

$$\sum_l \frac{y_{l,obs} + \frac{1}{2}}{[E(y_{l,new} + \frac{1}{2} | \mathbf{y}_{obs})]^2} \cdot Var(y_{l,new} | \mathbf{y}_{obs}) . \qquad (4.42)$$

Hence (4.41) can be viewed as a weighted predictive variability penalty, a natural choice in that if M_i is too large (i.e., contains too many explanatory terms resulting in substantial multicollinearity), predictive variances will increase. Lastly, (4.42) is approximately

$$E\left\{\sum_{l}\frac{(y_{l,new}-E(y_{l,new}|\mathbf{y}_{obs}))^2}{E(y_{l,new}+\frac{1}{2}\mid\mathbf{y}_{obs})}\,\Bigg|\,\mathbf{y}_{obs}\right\}$$

and thus may be viewed as a predictive *corrected* goodness-of-fit statistic.

Computation of (4.40) requires calculation of $E(y_{l,new}|\mathbf{y}_{obs},M_i)$ as well as $E[\log(y_{l,new}+\frac{1}{2})|\mathbf{y}_{obs},M_i]$. Such predictive expectations are routinely obtained as Monte Carlo integrations.

Finally, we remark that Gelfand and Ghosh (1998) extend the above approach to a completely formal decision-theoretic setting. Their method seeks to choose the model minimizing a particular posterior predictive loss. Like DIC and the less formal method above, the resulting criterion consists of a (possibly weighted) sum of a goodness-of-fit term and a model complexity penalty term.

4.7 Exercises

1. Steensma et al. (2005) presented the data in Table 4.4, from a randomized controlled trial comparing two dosing schedules for the drug erythropoiten. Serum hemoglobin (HGB, in g/dL) was recorded for $N = 365$ cancer patients with anemia over $T = 22$ weeks. The full data file is available at www.biostat.umn.edu/~brad/data/HGB_data.txt. We wish to fit a hierarchical simple linear regression model of the form

$$Y_{ij} = \beta_{0i} + \beta_{1i}(X_j - \mu_X) + \epsilon_{ij}, \ i = 1,\dots,365, \ j = 1,\dots,22, \quad (4.43)$$

where $X_j = j$, the week index, $\epsilon_{ij} \overset{iid}{\sim} N(0,\tau)$, $\beta_{0i} \overset{iid}{\sim} N(\mu_0,\tau_0)$, and $\beta_{1i} \overset{iid}{\sim} N(\mu_1,\tau_1)$. That is, as in Example 2.13 (and Example 7.2 in Chapter 7), we allow each subject's HGB trajectory to have its own slope and intercept, but borrow strength from the ensemble by treating these as normal random effects. Note that we also center the week index around its own mean, $\mu_X = 11.5$.

(a) Use WinBUGS to fit the above model, assuming vague priors for the hyperparameters τ, μ_0, τ_0, μ_1, and τ_1. Run multiple chains to assess convergence, and interpret the resulting posterior distributions for the grand slope μ_1 and the individual-specific slopes β_{1i}. How do the interpretations of these parameters differ? Use the compare function in WinBUGS to examine these for all participants. Are all participants' HGB measurements improving over time?

(b) Note that many of the Y_{ij} are missing; under the assumption that they are missing at random, WinBUGS can impute them according to the fitted model. Monitor the hemoglobin values estimated for participant

| Patient | HGB Measurements by Week j, $j = 1, \ldots, 22$ | | | | | |
i	$Y_{i,1}$	$Y_{i,2}$	$Y_{i,3}$	$\ldots$	$Y_{i,21}$	$Y_{i,22}$
1	10.2	9.8	10.3	$\cdots$	NA	11.4
2	10.1	9.4	9.1	$\cdots$	NA	11.6
3	9.9	10	10.2	$\cdots$	NA	12.4
4	10.7	9.7	11.4	$\cdots$	NA	NA
$\vdots$	$\vdots$	$\vdots$	$\vdots$	$\vdots$	$\vdots$	$\vdots$
365	10.7	10.5	10.4	$\cdots$	14.2	NA

Table 4.4 $T = 22$ *weekly hemoglobin (HGB) measurements.*

10, who is missing data from weeks 5 and 16 through 22. Compare the standard deviation of the estimate at week 5 to those at the end of the study, and explain any differences.

(c) As you've already noticed, the WinBUGS data file actually contains information on one more variable, newarm, a binary variable indicating which dosing schedule (or *treatment arm*, 1 or 2) was used for each patient. Modify model (4.43) above to allow different grand intercepts and/or slopes for the two treatment arms. Discuss any resulting problems with MCMC convergence, how they might arise, and how they can be handled. Are the two groups significantly different in any respect? Draw a quick plot in R comparing the fitted grand means in the two groups. Would a DIC comparison of this "full" model and the "reduced" model in part (a) be sensible here?

2. Refer to the data in Table 2.3, also available on the web in WinBUGS format at http://www.biostat.umn.edu/~brad/data.html. These pharmacokinetic (PK) data of Wakefield et al. (1994) were initially presented in Example 2.13, and analyzed there using an approximate linear hierarchical model. Recall these are the plasma concentrations Y_{ij} of the drug cadralazine at up to $T = 8$ different time lags x_{ij} following the administration of a single dose of 30 mg in $N = 10$ cardiac failure patients. Here, $i = 1, \ldots, 10$ indexes the patient, while $j = 1, \ldots, n_i$ indexes the observations, where $5 \leq n_i \leq 8$. Wakefield et al. suggest a "one-compartment" nonlinear pharmacokinetic model wherein the mean plasma concentration $\eta_{ij}(x_{ij})$ at time x_{ij} is given by

$$\eta_{ij}(x_{ij}) = 30\alpha_i^{-1} \exp(-\beta_i x_{ij}/\alpha_i) .$$

Subsequent unpublished work by these same authors suggests this model is best fit on the log scale. That is, we suppose

$$Z_{ij} \equiv \log Y_{ij} = \log \eta_{ij}(x_{ij}) + \epsilon_{ij} ,$$

where $\epsilon_{ij} \stackrel{ind}{\sim} N(0, \tau_i)$. The mean structure for the Z_{ij}'s thus emerges as

$$
\begin{aligned}
\log \eta_{ij}(x_{ij}) &= \log\left[30\alpha_i^{-1}\exp(-\beta_i x_{ij}/\alpha_i)\right] \\
&= \log 30 - \log \alpha_i - \beta_i x_{ij}/\alpha_i \\
&= \log 30 - a_i - \exp(b_i - a_i)x_{ij} ,
\end{aligned}
$$

where $a_i = \log \alpha_i$ and $b_i = \log \beta_i$.

(a) Assuming the two sets of random effects are independently distributed as $a_i \stackrel{iid}{\sim} N(\mu_a, \tau_a)$ and $b_i \stackrel{iid}{\sim} N(\mu_b, \tau_b)$, use WinBUGS or BRugs to analyze these data. You may adopt a vague prior structure, though the original authors suggest moderately informative independent $G(1, 0.04)$ priors for τ_a and τ_b.

Note that the full conditional distributions of the random effects are not simple conjugate forms nor guaranteed to be log-concave, so BUGS' Metropolis capability is required. Besides the usual posterior summaries and convergence checks, investigate the acceptance rate of the Metropolis algorithm, and the predictive distribution of $Y_{2,7}$ and $Y_{2,8}$, the missing observations on outlying patient 2. How do your results compare to those in Figure 2.22? You may also want to compare model fit, complexity, and overall quality via $\overline{D}, p_D$, and DIC.

(b) The assumption that the random effects a_i and b_i are independent within individuals was probably unrealistic. Instead, follow the original analysis of Wakefield et al. (1994), as well as that in Example 2.13, and assume the $\boldsymbol{\theta}_i \equiv (a_i, b_i)'$ are i.i.d. from a $N_2(\boldsymbol{\mu}, \Omega)$ distribution, where $\boldsymbol{\mu} = (\mu_a, \mu_b)$. Again adopt a corresponding vague conjugate prior specification, namely $\boldsymbol{\mu} \sim N_2(\mathbf{0}, C)$ and $\Omega \sim \text{Wishart}((\rho R)^{-1}, \rho)$ with $C \approx \mathbf{0}$, $\rho = 2$, and $R = Diag(0.01, 0.01)$.

Describe the changes (if any) in your answers from those in part (a).

3. Consider again the binary dugong modeling of Example 4.4. Suppose we wished to obtain side-by-side boxplots of the posteriors of the effect of log-age, β_1, across all three link functions (logit, probit, and cloglog). Modify the WinBUGS code given in that example so that the comparison tool on the Inference menu can be used to obtain this display.

(*Hint:* Use three copies of the dataset, one for each link function. This will also enable computation of three separate DIC p_D scores, since the DIC tool in WinBUGS will then decompose DIC by the three observation variable names.)

4. Spiegelhalter et al. (1995b) analyze the flour beetle mortality data in Table 3.3 using WinBUGS. These authors use only the usual, two-parameter parametrization for $p_i \equiv P(death|w_i)$, but compare the logit, probit, and complementary log-log link functions using the centered covariate $z_i = w_i - \bar{w}$.

Location	Y	X
1	0	1.9823
2	0	2.8549
3	0	2.7966
...	...	...
601	0	2.8189
602	1	2.0212
603	1	1.3979

Table 4.5 *Forestry co-presence data.*

(a) The full conditional distributions for α and β have no closed form, but WinBUGS does recognize them as being log-concave, and thus capable of being sampled using the Gilks and Wild (1992) adaptive rejection algorithm. Prove this log-concavity under the logit model.

(b) Following the model of Example 4.4, actually carry out the data analysis in WinBUGS (the program above, along with properly formatted data and initial value lists, are included in the "examples" section of the help materials). Do the estimated posteriors for the dosage effect β substantially differ for different link functions?

5. Consider www.biostat.umn.edu/~brad/data/copresence_data.txt, a data set for which a few records are shown in Table 4.5. Here, Y is a binary variable indicating co-presence of two species in a particular forest at $n = 603$ sampled locations. The lone predictor variable, X, is the log of the distance of each location to the forest edge. Suppose we use a logistic model for p, the probability of co-presence, namely

$$logit(p_i) = \beta_0 + \beta_1 X_i, \ i = 1, \ldots, n.$$

(a) Again following the model of Example 4.4, fit this model in WinBUGS, using vague priors. Is proximity to the forest edge a significant predictor of species co-presence?

(b) Fit the same model in R using a Metropolis-Hastings algorithm. Report parameter estimates and mean acceptance ratios for each chain. How many samples do you need (both here and in part (a)) to obtain results for β_1 reliable to two digits? Three digits?

(c) Replace the logit link above with the complementary log-log link, $\log[-\log(1 - p_i)]$, and compare the two posteriors for β_1. Also plot the two fitted curves as in Figure 4.6, and compare the models more formally using DIC or some other Bayesian model choice statistic. Does the choice of link function matter much for these data?

6. Consider the data displayed in Table 4.6, originally collected by Treloar (1974) and reproduced in Bates and Watts (1988). These data record the "velocity" y_i of an enzymatic reaction (in counts/min/min) as a function of substrate concentration x_i (in ppm), where the enzyme has been treated with puromycin.

Case (i)	x_i	y_i	Case (i)	x_i	y_i
1	0.02	76	7	0.22	159
2	0.02	47	8	0.22	152
3	0.06	97	9	0.56	191
4	0.06	107	10	0.56	201
5	0.11	123	11	1.10	207
6	0.11	139	12	1.10	200

Table 4.6 *Puromycin experiment data.*

A common model for analyzing biochemical kinetics data of this type is the *Michaelis-Menten* model, wherein we adopt the mean structure

$$\mu_i = \gamma + \alpha x_i/(\theta + x_i) \, ,$$

where $\alpha, \gamma \in \Re$ and $\theta \in \Re^+$. In the nomenclature of Example 4.6, we have design matrix

$$F_\theta^T = \left(\begin{array}{ccc} 1 & \cdots & 1 \\ \frac{x_1}{\theta + x_1} & \cdots & \frac{x_n}{\theta + x_n} \end{array} \right)$$

and $\beta^T = (\gamma, \alpha)$. For this model, obtain estimates of the marginal posterior density of the parameter α, and also the marginal posterior density of the mean velocity at $X = 0.5$, a concentration not represented in the original data, assuming that the error density of the data is

(a) normal
(b) Student's t with 2 degrees of freedom
(c) double exponential.

(*Hint:* The "duplicate the data" trick in Example 4.7 will not work in WinBUGS anymore if you handle the missing data prediction aspect in the usual way, i.e., by increasing N from 12 to 13 and adding 0.5 to the X vector and NA to the Y vector; you'll have to think of something else!)

7. For the point null prior partitioning setting described in Subsection 4.2.2, show that requiring $G \in \mathcal{H}_c$ as given in (4.12) is equivalent to requiring $BF \le \left(\frac{p}{1-p} \right) \left(\frac{1-\pi}{\pi} \right)$, where BF is the Bayes factor in favor of the null hypothesis.

8. For the interval null prior partitioning setting of Subsection 4.2.2, derive an expression for the set of priors $\mathcal{H}_c$ that correspond to rejecting H_0, similar to expression (4.12) for the point null case.

9. Suppose we are estimating a set of residuals r_i using a Monte Carlo approach, as described in Section 4.3. Due to a small sample size n, we are concerned that the approximation

$$E(y_i|\mathbf{y}_{(i)}) = \int E(y_i|\boldsymbol{\theta})p(\boldsymbol{\theta}|\mathbf{y}_{(i)})d\boldsymbol{\theta} \approx \int E(y_i|\boldsymbol{\theta})p(\boldsymbol{\theta}|\mathbf{y})d\boldsymbol{\theta}$$

will not be accurate. A faculty member suggests the importance sampling estimate

$$
\begin{aligned}
E(y_i|\mathbf{y}_{(i)}) &= \int E(y_i|\boldsymbol{\theta})\frac{p(\boldsymbol{\theta}|\mathbf{y}_{(i)})}{p(\boldsymbol{\theta}|\mathbf{y})}p(\boldsymbol{\theta}|\mathbf{y})d\boldsymbol{\theta} \\
&\approx \frac{1}{G}\sum_{g=1}^{G}E(y_i|\boldsymbol{\theta}^{(g)})\frac{p(\boldsymbol{\theta}^{(g)}|\mathbf{y}_{(i)})}{p(\boldsymbol{\theta}^{(g)}|\mathbf{y})}
\end{aligned}
$$

as an alternative. Is this a practical solution? What other approaches might we try? (*Hint:* See equation (3.4).)

10. Refer again to the cross-protocol data and model in Example 2.12.

    (a) Create WinBUGS code that will estimate cross validation ("leave one out") residuals and CPO values via importance sampling, as in equation (3.4). Compare your findings to those obtained via the "exact" and "approximate" methods in Example 2.16. Does your importance sampling approximation appear to be sufficiently accurate for the purpose of outlier identification? (*Hint:* Your solution to the previous question may be helpful.)

    (b) Redo the "exact" calculation using the BRugs package, following the approach suggested for the stack loss data in Chapter 2, Exercise 22.

11. Suppose we have a convergent MCMC algorithm for drawing samples from $p(\boldsymbol{\theta}|\mathbf{y}) \propto f(\mathbf{y}|\boldsymbol{\theta})\pi(\boldsymbol{\theta})$. We wish to locate potential outliers by computing the conditional predictive ordinate $f(y_i|\mathbf{y}_{(i)})$ given in equation (2.29) for each $i = 1, \ldots, n$. Give a computational formula we could use to obtain a simulation-consistent estimate of $f(y_i|\mathbf{y}_{(i)})$. What criterion might we use to classify y_i as a suspected outlier?

12. In the previous problem, suppose we wish to evaluate the model using the model check p'_D given in equation (2.33) with an independent validation data sample $\mathbf{z}$. Give a computational formula we could use to obtain a simulation-consistent estimate of p'_D.

13. Consider again the data in Table 3.3. Define model 1 to be the three variable model given in Example 3.7, and model 2 to be the reduced model having $m_1 = 1$ (i.e., the standard logistic regression model). Compute the Bayes factor in favor of model 1 using

Case (i)	y_i	x_i	z_i	Case (i)	y_i	x_i	z_i
1	3040	29.2	25.4	22	3840	30.7	30.7
2	2470	24.7	22.2	23	3800	32.7	32.6
3	3610	32.3	32.2	24	4600	32.6	32.5
4	3480	31.3	31.0	25	1900	22.1	20.8
5	3810	31.5	30.9	26	2530	25.3	23.1
6	2330	24.5	23.9	27	2920	30.8	29.8
7	1800	19.9	19.2	28	4990	38.9	38.1
8	3110	27.3	27.2	29	1670	22.1	21.3
9	3160	27.1	26.3	30	3310	29.2	28.5
10	2310	24.0	23.9	31	3450	30.1	29.2
11	4360	33.8	33.2	32	3600	31.4	31.4
12	1880	21.5	21.0	33	2850	26.7	25.9
13	3670	32.2	29.0	34	1590	22.1	21.4
14	1740	22.5	22.0	35	3770	30.3	29.8
15	2250	27.5	23.8	36	3850	32.0	30.6
16	2650	25.6	25.3	37	2480	23.2	22.6
17	4970	34.5	34.2	38	3570	30.3	30.3
18	2620	26.2	25.7	39	2620	29.9	23.8
19	2900	26.7	26.4	40	1890	20.8	18.4
20	1670	21.1	20.0	41	3030	33.2	29.4
21	2540	24.1	23.9	42	3030	28.2	28.2

Table 4.7 *Radiata pine compressive strength data.*

(a) the harmonic mean estimator, (4.20)

(b) the importance sampling estimator, (4.21).

Does the second perform better, as expected?

14. Consider the dataset of Williams (1959), displayed in Table 4.7. For $n = 42$ specimens of radiata pine, the maximum compressive strength parallel to the grain y_i was measured, along with the specimen's density, x_i, and its density adjusted for resin content, z_i (resin contributes much to density but little to strength of the wood). For $i = 1, \ldots, n$, we wish to compare the two models $M = 1$ and $M = 2$ where

$$M = 1: \quad y_i = \alpha + \beta(x_i - \bar{x}) + \epsilon_i, \quad \epsilon_i \overset{iid}{\sim} \text{Normal}(0, \sigma^2),$$

and

$$M = 2: \quad y_i = \gamma + \delta(z_i - \bar{z}) + \epsilon_i, \quad \epsilon_i \overset{iid}{\sim} \text{Normal}(0, \tau^2).$$

We desire prior distributions that are roughly centered around the appropriate least squares parameter estimate, but are extremely vague (though still proper). As such, we place

$$N\left((3000, 185)^T, \ Diag(10^6, 10^4)\right)$$

priors on $(\alpha, \beta)^T$ and $(\gamma, \delta)^T$, and

$$IG\left(3, \ (2 \cdot 300^2)^{-1}\right)$$

priors on σ^2 and τ^2, having both mean and standard deviation equal to 300^2.

Compute the Bayes factor in favor of model 2 (the adjusted density model) using

(a) the marginal density estimation approach of Subsection 4.4.2
(b) the product space search approach of Subsection 4.5.1
(c) the reversible jump approach of Subsection 4.5.3.

How do the methods compare in terms of accuracy? Ease of use? Which would you attempt first in a future problem?

15. In the previous problem, implement both models in WinBUGS, and use DIC instead of Bayes factors to choose between them. Is your preference between the two models materially altered by this change? What are the advantages and disadvantages of using DIC instead of a Bayes factor here?

16. Consider again the stack loss data originally presented in Example 2.16. Suppose we wish to compare the assumptions of normal, t_4, and DE errors for these data.

(a) Repeat the outlier analysis in Example 4.7 for these data. Do the λ_i posteriors for the nonnormal models effectively identify the outliers?

(b) Use the Comparison tool to obtain a side-by-side comparison of the boxplots of the posterior median θ across the three error models, similar to Figure 4.9. Then do a similar comparison of the width of the central 95% credible interval for the errors, $q_{.975}(\sigma) - q_{.025}(\sigma)$, and explain any differences. (*Hint:* Dump the $\sigma^{(g)}$ Gibbs samples into R using the Coda function in WinBUGS, convert them to the quantile difference scale, and then do the boxplot comparison. Use the qnorm and qt functions in R; the DE quantiles can be determined analytically.)

(c) Repeat the DIC analysis in Example 4.8 for these data. What model is DIC-best? Are negative p_D values again a problem here?

## CHAPTER 5

The empirical Bayes approach

5.1 Introduction

As detailed in Chapter 2, in addition to the likelihood, Bayesian analysis depends on a prior distribution for the model parameters. This prior can be nonparametric or parametric, depending on unknown parameters that may in turn be drawn from some second-stage prior. This sequence of parameters and priors constitutes a hierarchical model. The hierarchy must stop at some point, with all remaining prior parameters assumed known. Rather than make this assumption, the empirical Bayes (EB) approach uses the observed data to estimate these final stage parameters (or to directly estimate the Bayes decision rule) and then proceeds as though the prior were known.

Though in principle EB can be implemented for hierarchical models having any number of levels, for simplicity we present the approach for the standard two-stage model. That is, we assume a likelihood $f(\mathbf{y}|\boldsymbol{\theta})$ for the observed data $\mathbf{y}$ given a vector of unknown parameters $\boldsymbol{\theta}$, and a prior for $\boldsymbol{\theta}$ with cdf $G(\boldsymbol{\theta})$ and corresponding density or mass function $g(\boldsymbol{\theta}|\boldsymbol{\eta})$, where $\boldsymbol{\eta}$ is a vector of hyperparameters. With $\boldsymbol{\eta}$ known, the Bayesian uses Bayes' Theorem (2.1) to compute the posterior distribution,

$$p(\boldsymbol{\theta}|\mathbf{y}, \boldsymbol{\eta}) = \frac{f(\mathbf{y}|\boldsymbol{\theta})g(\boldsymbol{\theta}|\boldsymbol{\eta})}{m(\mathbf{y}|\boldsymbol{\eta})} \ , \tag{5.1}$$

where $m(\mathbf{y}|\boldsymbol{\eta})$ denotes the *marginal* distribution of $\mathbf{y}$,

$$m(\mathbf{y}|\boldsymbol{\eta}) = \int f(\mathbf{y}|\boldsymbol{\theta})g(\boldsymbol{\theta}|\boldsymbol{\eta})d\boldsymbol{\theta} \ , \tag{5.2}$$

or, in more general notation, $m_G(\mathbf{y}) = \int f(\mathbf{y}|\boldsymbol{\theta})dG(\boldsymbol{\theta})$.

If $\boldsymbol{\eta}$ is unknown, the fully Bayesian (Chapter 2) approach would adopt a hyperprior distribution, $h(\boldsymbol{\eta})$, and compute the posterior distribution as

$$p(\boldsymbol{\theta}|\mathbf{y}) = \frac{\int f(\mathbf{y}|\boldsymbol{\theta})g(\boldsymbol{\theta}|\boldsymbol{\eta})h(\boldsymbol{\eta})d\boldsymbol{\eta}}{\int \int f(\mathbf{y}|\mathbf{u})g(\mathbf{u}|\boldsymbol{\eta})h(\boldsymbol{\eta})d\mathbf{u}d\boldsymbol{\eta}} = \int p(\boldsymbol{\theta}|\mathbf{y}, \boldsymbol{\eta})h(\boldsymbol{\eta}|\mathbf{y})d\boldsymbol{\eta}. \tag{5.3}$$

The second representation shows that the posterior is a mixture of conditional posteriors (5.1) given a fixed $\boldsymbol{\eta}$, with mixing via the marginal posterior distribution of $\boldsymbol{\eta}$.

In empirical Bayes analysis, we instead use the marginal distribution (5.2) to estimate $\boldsymbol{\eta}$ by $\hat{\boldsymbol{\eta}} \equiv \hat{\boldsymbol{\eta}}(\mathbf{y})$ (e.g., the marginal maximum likelihood estimator). Inference is then based on the *estimated posterior* distribution $p(\boldsymbol{\theta}|\mathbf{y}, \hat{\boldsymbol{\eta}})$. The EB approach thus essentially replaces the integration in the rightmost part of (5.3) by a maximization, a substantial computational simplification. The name "empirical Bayes" arises from the fact that we are using the data to estimate the hyperparameter $\boldsymbol{\eta}$. Adjustments to account for the uncertainty induced by estimating $\boldsymbol{\eta}$ may be required, especially to produce valid interval estimates (see Section 5.4).

The development in the preceding paragraph actually describes what Morris (1983a) refers to as the *parametric* EB (PEB) approach. That is, we assume the distribution $g(\boldsymbol{\theta}|\boldsymbol{\eta})$ takes a parametric form, so that choosing a (data-based) value for $\boldsymbol{\eta}$ is all that is required to completely specify the estimated posterior distribution. For more flexibility, one can implement a *nonparametric* EB (NPEB) approach, in which $G(\boldsymbol{\theta})$ has an unknown form. Pioneered and championed by Robbins (1955, 1983), and further generalized and modernized by Maritz and Lwin (1989, Section 3.4), van Houwelingen (1977), and others, this method first attempts to represent the posterior mean in terms of the marginal distribution, and then uses the data to estimate the Bayes rule directly. Recent advances substitute a nonparametric estimate of $G(\boldsymbol{\theta})$, generalizing the approach to a broader class of models and inferential goals. The parametric setup is discussed in the next section, followed by the nonparametric formulation.

5.2 Parametric EB (PEB) point estimation

Parametric empirical Bayes (PEB) models use a family of prior distributions $G(\theta|\eta)$ indexed by a low-dimensional parameter η. In the context of our compound sampling framework (5.27), if η were known, this would imply that the posterior for θ_i depends on the data only through y_i, namely,

$$p(\theta_i|y_i, \eta) = \frac{f_i(y_i|\theta_i)g(\theta_i|\eta)}{m_i(y_i|\eta)} \ . \tag{5.4}$$

But because η is unknown, we use the marginal distribution of *all* the data $m(\mathbf{y}|\eta)$ to compute an estimate $\hat{\eta}$, usually obtained as an MLE or method of moments estimate. Plugging this value into (5.4), we get the *estimated posterior* $p(\theta_i|y_i, \hat{\eta})$. This estimated posterior drives all Bayesian inferences; for example, a point estimate $\hat{\theta}_i$ is the estimated posterior mean (or mode, or median). Note that $\hat{\theta}_i$ depends on *all* the data through $\hat{\eta} = \hat{\eta}(\mathbf{y})$. Morris (1983a) and Casella (1985) provide excellent introductions to PEB analysis.

5.2.1 Gaussian/Gaussian models

We now consider the two-stage Gaussian/Gaussian model,

$$Y_i \mid \theta_i \overset{iid}{\sim} N(\theta_i, \sigma^2), \quad i = 1, \ldots, k, \quad \text{and} \tag{5.5}$$

$$\theta_i \mid \mu \overset{iid}{\sim} N(\mu, \tau^2), \quad i = 1, \ldots, k.$$

First, we assume both τ^2 and σ^2 are known, so that $\eta = \mu$. Calculations similar to those for the posterior density in Example 2.2 show that the Y_i are marginally independent and identically distributed as $N(\mu, \sigma^2 + \tau^2)$ random variables. This implies that the joint marginal density of $\mathbf{y} = (y_1, \ldots, y_k)'$ is the product of the individual $m(y_i \mid \mu)$,

$$m(\mathbf{y} \mid \mu) = \frac{1}{[2\pi(\sigma^2 + \tau^2)]^{k/2}} \exp\left[-\frac{1}{2(\sigma^2 + \tau^2)} \sum_{i=1}^{k} (y_i - \mu)^2 \right],$$

so that $\hat{\mu} = \bar{y} = \frac{1}{k}\sum_{i=1}^{k} y_i$ is the marginal MLE of μ. Thus the estimated posterior distribution is

$$p(\theta_i \mid y_i, \hat{\mu}) = N(B\hat{\mu} + (1 - B)y_i, \ (1 - B)\sigma^2),$$

where $B = \sigma^2/(\sigma^2 + \tau^2)$. This produces the PEB point estimate of θ_i,

$$\hat{\theta}_i^{\mu} = B\bar{y} + (1 - B)y_i \tag{5.6}$$

$$= \bar{y} + (1 - B)(y_i - \bar{y}). \tag{5.7}$$

Notice that this formula is identical to that derived in Example 2.2, except that a known prior mean (μ) is replaced by the sample mean using all of the data. Inference on a single component (a single θ_i) depends on data from all the components (borrowing information). Formula (5.6) shows the estimate to be a weighted average of the estimated prior mean ($\bar{y}$) and the standard estimate (y_i), while (5.7) shows that the standard estimate is shrunk back toward the common mean a fraction of the distance between the two. The MLE is equivalent to setting $B = 0$, while complete pooling results from setting $B = 1$.

The PEB estimates borrow information from all the coordinates in making inference about a single coordinate. In a slightly more general setting below, we shall show that these PEB estimates outperform the usual MLE when the two-stage model holds. This result is not surprising because we are taking advantage of the connection between the k coordinate-specific θs provided by G. What is surprising is that estimates such as those above can outperform the MLE even when G is misspecified, or even when the θs are not sampled from a prior.

We now generalize the above by retaining (5.5), but assuming that τ is also unknown. The marginal likelihood must now provide information on $\eta = (\mu, \tau)$. We are immediately confronted with a decision on what estimate to use for τ (or τ^2 or B). If we use the marginal MLE (MMLE) for τ^2, then the invariance property of MLEs implies that the MMLE for B results from plugging $\hat{\tau}^2$ into its formula. For model (5.5), it is easy to show that the MMLE for μ is again $\bar{y}$, while the MMLE for τ^2 is

$$\hat{\tau}^2 = (s^2 - \sigma^2)^+ = \max\{0, s^2 - \sigma^2\}, \qquad (5.8)$$

where $s^2 = \frac{1}{k}\sum_{i=1}^k (y_i - \bar{y})^2$. This "positive part" estimator (5.8) reports the variation in the data over and above that expected if all the θs are equal (i.e., σ^2). Transforming (5.8), we find that the MMLE for B is

$$\hat{B} = \frac{\sigma^2}{\sigma^2 + \hat{\tau}^2} = \frac{\sigma^2}{\sigma^2 + (s^2 - \sigma^2)^+}. \qquad (5.9)$$

Clearly $0 \leq \hat{B} \leq 1$; this would not have been the case had we used the unbiased estimate of τ^2, which divides $\sum_{i=1}^k (y_i - \bar{y})^2$ by $k - 1$ instead of k and does not take the positive part. Because we know the target quantity B must lie in $[0, 1]$, this provides another illustration of the problems with unbiased estimates.

Substituting (5.9) for B in (5.7) produces the EB estimates

$$\hat{\theta}_i^{\mu\tau} = \bar{y} + (1 - \hat{B})(y_i - \bar{y}). \qquad (5.10)$$

As in the case where τ is known, the target for shrinkage is $\bar{y}$, but now the amount of shrinkage is controlled by the estimated heterogeneity in the data. The more the heterogeneity exceeds σ^2, the smaller $\hat{B}$ becomes, and the less the shrinkage. On the other hand, if the estimated heterogeneity is less than σ^2, $\hat{B} = 1$ and all estimates equal $\bar{y}$ (i.e., the data are pooled). As heterogeneity increases, the weight on the estimated coordinate-specific estimates increases to 1.

As discussed by Laird and Louis (1987), representation (5.7) shows a direct analogy with regression to the mean. In the model, Y_i and θ_i have correlation (the *intra-class* correlation) equal to $(1 - B)$, and an estimate for θ_i based on Y_i regresses back toward the common mean. In the usual regression model, we estimate the means and correlations from a data set where both the Ys and θs (the response variables) are observed. In Bayes and empirical Bayes analysis, a distribution on the θs provides sufficient information for inference even when the θs are not observed.

5.2.2 Computation via the EM algorithm

For the two-stage model, EB analysis requires maximization of the marginal likelihood given in (5.2). For many models, standard iterative maximum likelihood methods can be used directly to produce the MMLE and the observed information matrix. In more complicated settings, the *EM algorithm* (where EM stands for "expectation-maximization") offers an alternative approach, as we now describe.

Versions of the EM algorithm have been used for decades. Dempster,

Laird, and Rubin (1977) consolidated the theory and provided instructive examples. The EM algorithm is attractive when the function to be maximized can be represented as a missing data likelihood, as is the case for (5.2). It converts such a maximum likelihood situation into one with a "pseudo-complete" log-likelihood or score function, produces the MLE for this case, and continues recursively until convergence. The approach is most effective when finding the complete data MLE is relatively straightforward. Several authors have shown how to compute the observed information matrix within the EM structure, or provided various extensions intended to accelerate its convergence or ease its implementation.

Since the marginal likelihood (5.2) can be represented as a missing data likelihood, the EM algorithm is effective. It also provides a conceptual introduction to the more advanced Monte Carlo methods for producing the full posterior distribution presented in Section 3.4. Here we sketch the approach and provide basic examples.

EM for PEB

Consider a model where, if we observed $\boldsymbol{\theta}$, the MLE for η would be relatively straightforward. For example, in the compound sampling model (5.27), given $\boldsymbol{\theta}$ the MLE for η can be computed using only the prior g. Suppressing the dependency on $\mathbf{y}$ on the left-hand side of equations, let

$$S(\boldsymbol{\theta}|\eta) \quad = \quad \frac{\partial}{\partial \eta} \log(g(\boldsymbol{\theta}|\eta)) \tag{5.11}$$

be the score function. For the "E-step," let $\eta^{(j)}$ denote the current estimate of the hyperparameter at iteration j, and compute

$$\bar{S}(\eta|\eta^{(j)}) \quad = \quad E(S(\boldsymbol{\theta}|\eta) \mid \mathbf{y}, \eta^{(j)}) . \tag{5.12}$$

This step involves standard Bayesian calculations to arrive at the conditional expectation of the sufficient statistics for η based on $\boldsymbol{\theta}$.

The "M-step" then uses $\bar{S}$ in (5.12) to compute a new estimate of the hyperparameter, and the recursion proceeds until convergence. That is,

$$\eta^{(j+1)} = \{\eta : \bar{S}(\eta|\eta^{(j)}) = 0\} , \quad j = 1, 2, \ldots \tag{5.13}$$

If both the E and M steps are relatively straightforward (the most attractive case being when the conditional distributions in (5.12) and (5.13) are exponential families), the EM approach works very well.

Dempster et al. (1977), Meilijson (1989), Tanner (1993, p.43), and many other researchers have shown that, at every iteration, the marginal likelihood either increases or stays constant, i.e.,

$$m(\mathbf{y} \mid \eta^{(j+1)}) \geq m(\mathbf{y} \mid \eta^{(j)}) \text{ for all } j ,$$

and, thus, the algorithm converges monotonically. This convergence could be to a local (rather than global) maximum, however, and so, as with many optimization algorithms, multiple starting points are recommended.

Example 5.1 Consider the Gaussian/Gaussian model (5.5) with unequal sampling variances, and let $\eta = (\mu, T)$, where for notational convenience we use T for τ^2. For model component i, -2 times the log-likelihood for η as a function of θ_i is $\log(T) + (\theta_i - \mu)^2/T$, producing the score vector

$$S(\theta_i|\eta) = -\frac{1}{2} \left(\begin{array}{c} -2\frac{(\theta_i-\mu)}{T} \\ \frac{1}{T} - \frac{(\theta_i-\mu)^2}{T^2} \end{array} \right). \qquad (5.14)$$

Expanding the second component gives

$$\frac{1}{T} - \frac{\theta_i^2 - 2\theta_i\mu + \mu^2}{T^2}.$$

The MLE for η depends on the sufficient statistics $\sum_i \theta_i$ and $\sum_i \theta_i^2$. The EM proceeds by computing $\bar{S}$ as in (5.12) using the conditional posterior distribution $p(\theta_i|y_i, \mu^{(j)}, T^{(j)})$, substituting the posterior mean for θ_i and the posterior variance plus the square of the mean for θ_i^2, solving the "pseudo-complete"-data MLE equations, and continuing the recursion. Note that the algorithm requires expected sufficient statistics, so θ_i^2 must be replaced by its expected value, *not* by the square of the expectation of θ_i (as though the imputed θs were observed). ■

For this model and generalizations, the EM partitions the likelihood maximization problem into components that are relatively easy to handle. For example, unequal sampling variances produce no additional difficulty. The EM is most attractive when both the E and the M steps are relatively more straightforward than dealing directly with the marginal likelihood. However, even when the E and M steps are non-trivial, they may be handled numerically, resulting in potentially simplified or stabilized computations.

Computing the observed information

Though the EM algorithm can be very effective for finding the MMLE, it never operates directly on the marginal likelihood or score function. Therefore, it does not provide a direct estimate of the observed information needed to produce the asymptotic variance of the MMLE. Several approaches to estimating observed information have been proposed; all are based on the standard decomposition of a variance into an expected conditional variance plus the variance of a conditional expectation.

Using the notation for hierarchical models and suppressing dependence on η, the decomposition of the Fisher information for η is

$$I_{\boldsymbol{\theta}} = E(I_{\boldsymbol{\theta}|\mathbf{Y}} + I_{\mathbf{Y}}), \qquad (5.15)$$

where I is the information that would result from direct observation of $\boldsymbol{\theta}$.

The first term on the right is the amount of information on $\boldsymbol{\theta}$ provided by the conditional distribution of the complete data given $\mathbf{y}$, while the second term is the information on η in the marginal likelihood (the required information for the MMLE).

Louis (1982) estimates observed information versions of the left-hand side and the first right-hand term in (5.15), obtaining the second term by subtraction. Meng and Rubin (1991) generalize the approach, producing the *Supplemented EM* (SEM) algorithm. It requires only the EM computer code and a subroutine to compute the left-hand side. Meilijson (1989) shows how to take advantage of the special case where $\mathbf{y}$ is comprised of independent components to produce an especially straightforward computation. With i indexing components, this approach uses (5.12) for each component, producing $\bar{S}_i$, and then

$$\hat{I}_{\mathbf{Y}} = \sum_i \bar{S}_i \bar{S}_i^T, \tag{5.16}$$

requiring no additional computational burden. This gradient approach can be applied to (5.14) in the Gaussian example. With unequal sampling variances (σ_i^2), validity of the approach requires either that they are sampled from a distribution (a "supermodel"), producing an i.i.d. marginal model, or weaker conditions ensuring large sample convergence. Otherwise, the more complicated approaches of Louis (1982) or Meng and Rubin (1991) must be used.

Speeding convergence and generalizations

Several authors (see, e.g., Tanner, 1993, pp. 55–57) provide methods for speeding convergence of the EM algorithm using the information decomposition (5.15) to compute an acceleration matrix. Though the approach is successful in many applications, its use moves away from the basic attraction of the EM algorithm: its ability to decompose a complicated likelihood maximization into relatively straightforward components.

Several other extensions to the EM algorithm further enhance its effectiveness. Meng and Rubin (1993) describe an algorithm designed to avoid the nested EM iterations required when the M-step is not available in closed form. Called the *Expectation/Conditional Maximization* (ECM) algorithm, it replaces the M-step with a set of conditional maximization steps. Meng and Rubin (1992) review this extension and others, including the Supplemented ECM (SECM) algorithm, which supplements the ECM algorithm to estimate the asymptotic variance matrix of the MMLE in the same way that the SEM algorithm supplements EM.

Finally, the EM algorithm also provides a starting point for Monte Carlo methods. For example, if the E-step cannot be computed analytically but $\boldsymbol{\theta}$ values can be sampled from the appropriate conditional distribution, then $\bar{S}$ can be estimated by Monte Carlo integration (see Section 3.3 for details).

Wei and Tanner (1990) refer to this approach as the Monte Carlo EM (MCEM) algorithm. In fact, the entire conditional distribution of S could be estimated in this way, providing input for computing the full posterior distribution using data augmentation approach (Tanner and Wong, 1987). Meng and Rubin (1992) observe that their *Partitioned ECM* (PECM) algorithm, which partitions the parameter vector into k subcomponents in order to reduce the dimensionality of the maximizations, is essentially a deterministic version of the Gibbs sampler, as already discussed in Section 3.4.

Example 5.2 We apply our Gaussian/Gaussian EB model (5.5) to the data in Table 5.1. Originally analyzed in Louis (1984), these data come from the Hypertension Detection and Follow-up Program (HDFP). This study randomized between two intervention strategies: referred care (RC), in which participants were told they had hypertension (elevated blood pressure) and sent back to their physician, and stepped care (SC), in which participants were entered into an intensive and hierarchical series of lifestyle and treatment interventions. The data values Y are MLE estimates of 1000 times the log-odds ratios in comparing five-year death rates between the SC and RC regimens in twelve strata defined by initial diastolic blood pressure (I = 90–104, II = 105–114, III = 115+), race (B/W), and gender (F/M). In a stratum with five-year death rates r_{sc} and r_{rc}, we have

$$Y = 1000 \times \log \left(\frac{r_{sc} \cdot (1 - r_{rc})}{(1 - r_{sc}) \cdot r_{rc}} \right).$$

These estimated log-relative risks have estimated sampling standard deviations which we denote by $\hat{\sigma}$. These vary from stratum to stratum, and control the amount of shrinkage in a generalization of the empirical Bayes rule (5.10) wherein $\hat{B}$ is replaced by $\hat{B}_i = \hat{\sigma}_i^2 / (\hat{\sigma}_i^2 + \hat{\tau}^2)$. In this unequal sampling variance situation, MMLEs for μ and τ^2 require EM iteration, as described above.

Table 5.1 displays both the EB posterior means, $\hat{\theta}^{\mu\tau}$, and a set of *constrained* EB estimates, $\hat{\theta}^{con}$. Like the usual EB estimate, the constrained estimate is a weighted average of the estimated prior mean and the data, but with an adjusted prior mean estimate (here, –229 instead of –188) and a different weight on the data ($A/100$). Constrained estimates are designed to minimize squared error loss (SEL) subject to the constraint that their sample mean and variance are good estimates of their underlying empirical distribution function of the θs. We defer the specifics of constrained EB estimation and histogram estimates more generally to Section 7.1.

Notice from Table 5.1 and Figure 5.1 that the subgroup estimates with high variance for the MLE tend to produce the most extreme observed estimates and are shrunken most radically toward the estimated population mean (i.e., the value of $1 - \hat{B}$ is relatively small). The PSD column reports "naive" posterior standard deviations that have not been adjusted

Group		Y	$\hat{\theta}^{\mu\tau}$	D	$\hat{\sigma}$	PSD	$\hat{\theta}^{con}$	A
I	BM	−129	−157	54	170	125	−149	81
	BF	−304	−240	44	206	137	−285	74
	WM	−242	−220	59	153	117	−240	84
	WF	−355	−253	39	231	144	−316	69
II	BM	−274	−213	29	290	155	−255	59
	BF	−529	−266	23	337	161	−383	51
	WM	−41	−156	22	349	162	−136	50
	WF	809	−61	13	479	171	127	34
III	BM	−558	−273	23	337	161	−398	51
	BF	−235	−197	18	389	166	−231	44
	WM	336	−122	13	483	171	−38	34
	WF	1251	−103	6	730	178	43	18

Table 5.1 *Analysis of 1000 × log(odds ratios) for five-year survival in the HDFP study:* $\hat{\mu} = -188$, $\hat{\tau}^2 = (183)^2$, $D = 100(1 - B)$, *and* $PSD = \hat{\sigma}D^{\frac{1}{2}}/10$, *the posterior standard deviation.* $\hat{\theta}^{\mu\tau} = -188 + \frac{D}{100}(Y + 188)$, *the usual PEB point estimate, while* $\hat{\theta}^{con} = -229 + \frac{A}{100}(Y + 229)$, *the Louis (1984) constrained EB estimate.*

for having estimated parameters of the prior. Notice that these are far more homogeneous than the $\hat{\sigma}$ values; highly variable estimates have been stabilized. However, as we shall show in Section 5.4, these values need to be increased to account for having estimated the prior parameters. This increase will be greater for subgroups with a smaller value of $1 - B$ and for those farther from $\hat{\mu}$.

Typical relations between observed data and empirical Bayes estimates are shown in Figure 5.1, which plots the raw Y values on the center axis, connected by solid lines to the corresponding EB estimates $\hat{\theta}^{\mu\tau}$ on the upper axis. The most extreme estimates are shrunk the most severely, and rankings of the estimates can change (i.e., the lines can cross) due to the unequal sampling variances of the Ys. Efron and Morris (1977) give an example of this phenomenon related to geographic distribution of disease. The constrained EB estimates $\hat{\theta}^{con}$, plotted on the figure's lower axis, shrink observed estimates less than the posterior means, and produce ranks different from either the raw data or the standard EB estimates. ∎

The Gaussian/Gaussian model exhibits the more general phenomenon that EB estimates provide a compromise between pooling all data (and thus decreasing variance while increasing the bias of each coordinate-specific estimate) and using only coordinate-specific data for the coordinate-specific estimates (and increasing variance while decreasing bias). Complete pooling occurs when $\hat{B} = 1$, i.e., when the chi-square statistic for testing equality

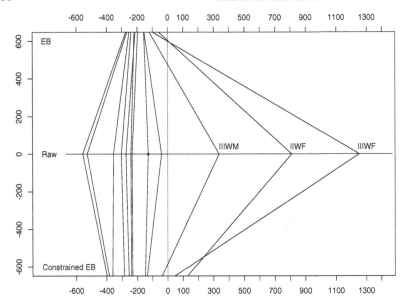

Figure 5.1 *HDFP raw data and EB estimates.*

of the k means is less than or equal to $k/(k-1)$. For large k and $\tau = 0$, the probability of complete pooling (equivalent to accepting the hypothesis that the coordinate-specific means are equal) is approximately .50. Thus, the test has type I error of .50, but this apparent liberality is compensated for by the smooth departure from complete pooling.

Proper selection of hyperparameter estimates is one inherent difficulty with the PEB approach. We have already proposed that using the positive part estimator for B in (5.9) is more sensible than using an unbiased estimate, but many other options for $\widehat{B}$ are possible, and it may be difficult to guess which will be best in a given situation. By contrast, the fully Bayesian model that adds another level to the distributional hierarchy (i.e., including a hyperprior $h(\mu, \tau^2)$) replaces estimation by integration, thus avoiding the problem of selecting an estimation criterion. This solution also deals effectively with the nonidentically distributed case (for example, when the sampling variances are different), and automatically injects uncertainty induced by estimating the hyperparameters into the posterior distribution. However, this does require selection of an effective hyperprior, an issue we shall revisit in Subsection 5.4.2 and the subsequent Example 5.4.

5.2.3 EB performance of the PEB

To see how well the empirical Bayes rule works when model (5.5) holds, we can compute the preposterior (Bayes) risk (B.6) for a single component, or for the average over all components. Since the Y_i are i.i.d., the two computations will be equal and we take advantage of this equality.

Calculating the preposterior risk requires integration over both the data and parameters, and we may integrate in either order. As shown in Appendix B, integrating only over the distribution of the parameters conditional on the data gives the posterior risk, while integrating only over the sampling distribution of the data conditional on the parameters gives the frequentist risk. Each is revealing. First, we consider the posterior risk for the ith component of the EB estimator (5.10). Under squared error loss, the usual posterior "variance plus squared bias" formula gives

$$(1 - B)\sigma^2 + [B\mu + (1 - B)y_i - \widehat{B}\bar{y} - (1 - \widehat{B})y_i]^2 \ .$$

Regrouping terms gives

$$(1-B)\sigma^2 + B^2(\bar{y}-\mu)^2 + (\widehat{B}-B)^2(y_i-\bar{y})^2 - 2B(\widehat{B}-B)(y_i-\bar{y})(\bar{y}-\mu). \quad (5.17)$$

The first term is the usual Bayes posterior risk. The second term accounts for estimating μ, the third for estimating B, and the fourth for correlation in these estimates. Notice that the last two depend on how far y_i is from $\bar{y}$, and that the second and third are very similar to those used to compute confidence intervals for regression lines (see, e.g., Snedecor and Cochran, 1980, Section 9.9).

To get the preposterior risk, we take the expectation of (5.17) with respect to the marginal distribution of $\mathbf{Y}$. The computation is far easier if we first average over all coordinates, to obtain

$$\frac{1}{k}\sum_{i=1}^{k} E((\theta_i - \hat{\theta}_i^{\mu\tau})^2 \mid \mathbf{y}) = (1 - B)\sigma^2 + B^2(\bar{y} - \mu)^2 + (\widehat{B} - B)^2 SS_y, \quad (5.18)$$

where $SS_y = \sum_{i=1}^{k}(y_i - \bar{y})^2$ because the last term in (5.17) sums to 0.

The marginal distribution of each Y_i is Gaussian with mean μ and variance $\sigma^2 + \tau^2 = \sigma^2/B$. For ease of calculation, we take the estimate of B to be of the form

$$\widehat{B} = \frac{c\sigma^2}{SS_y} \quad (5.19)$$

(i.e., we do not use a "positive part," as in (5.9)). We then expand the third term in (5.18) and take expectations using properties of the gamma distribution to obtain the preposterior risk,

$$r(G, \hat{\theta}^{\mu\tau}) = (1 - B)\sigma^2 + \frac{B\sigma^2}{k} + \frac{B\sigma^2}{k}\left(\frac{c^2}{k-3} - 2c + k - 1\right), \quad (5.20)$$

provided that $k \geq 4$. Again, the first term is the usual Bayes risk, the second accounts for estimating μ, and the third accounts for estimating B. The unbiased estimate of B uses $c = k - 3$, giving $2B\sigma^2/k$ for the last term. This is also its minimum value for this restricted class of estimators, but overall performance can be improved by improving the estimate of B (for example, by truncating it at 1).

Formula (5.20) shows the Bayes/EB advantage, when the compound sampling model holds. The usual MLE ($\hat{\theta} = y_i$) has preposterior risk σ^2, so for $c = k - 3$, the savings in risk is

$$B\sigma^2 \left(1 - \frac{3}{k}\right) ,$$

which can be substantial. Recall that this savings applies to each coordinate θ_i, though we have averaged over coordinates in our calculations.

5.2.4 Stein estimation

The above results show that there is great potential to improve on the performance of the standard maximum likelihood estimate when the two-stage structure holds and the assumed form of the prior distribution is correct. The prior provides a connection among the coordinate-specific parameters.

Now consider the other extreme where there is no prior distribution; we assume only that $Y_i|\theta_i \overset{ind}{\sim} N(\theta_i, \sigma^2)$, $i = 1, \ldots, k$, where σ^2 is assumed known. We investigate the purely frequentist properties of the EB procedure for the fixed vector $\boldsymbol{\theta} = (\theta_1, \ldots, \theta_k)'$. For there to be any hope of improvement over the maximum likelihood estimate $\hat{\boldsymbol{\theta}}^{MLE}(\mathbf{Y}) = \mathbf{Y}$, we must provide a connection among the coordinates, and, without a prior, we do this on the "other end" by assessing *summed squared error loss* (SSEL),

$$SSEL(\mathbf{Y}, \boldsymbol{\theta}) = \frac{1}{k} \sum_{i=1}^{k} (\hat{\theta}_i(\mathbf{Y}) - \theta_i)^2 . \qquad (5.21)$$

Hence, while the parameters are unconnected in the model, our inferences regarding them *are* connected through the loss function. Note that for convenience, we are actually computing *averaged* squared error loss, but will still refer to it as SSEL.

This loss structure is "societal" in that it combines performance over all coordinates and does not pay specific attention to a single coordinate as would, say, the maximum of the coordinate-specific losses. The results that follow are dependent on a loss function that averages over coordinates; the data analyst must decide if such a loss function is relevant.

Consider the frequentist risk (equation (B.4) in Appendix B) for SSEL,

$$R(\boldsymbol{\theta}, \hat{\boldsymbol{\theta}}(\mathbf{Y})) = E(SSEL(\mathbf{Y}, \boldsymbol{\theta}) \mid \boldsymbol{\theta}) , \qquad (5.22)$$

and our Gaussian sampling distribution. It is easy to see that

$$R\left(\boldsymbol{\theta}, \hat{\boldsymbol{\theta}}^{MLE}(\mathbf{Y})\right) = \frac{1}{k} \sum_{i=1}^{k} E((Y_i - \theta_i)^2 | \boldsymbol{\theta})$$

$$= \frac{1}{k}\sum_{i=1}^{k}\sigma^2 = \sigma^2,$$

which, of course, is constant regardless of the true value of $\boldsymbol{\theta}$. Stein (1955) proved the remarkable result that for $k \geq 3$, $\hat{\boldsymbol{\theta}}^{MLE}(\mathbf{Y})$ is *inadmissible* as an estimator of $\boldsymbol{\theta}$. That is, there exists another estimator with frequentist risk no larger than σ^2 for every possible $\boldsymbol{\theta}$ value. One such dominating estimator was derived by James and Stein (1961) as

$$\hat{\theta}_i^{JS}(\mathbf{Y}) = \left[1 - \frac{(k-2)\sigma^2}{||\mathbf{Y}||^2}\right]Y_i, \qquad (5.23)$$

where $||\mathbf{Y}||^2 \equiv \sum Y_i^2$. These authors showed that

$$
\begin{aligned}
R\left(\boldsymbol{\theta}, \hat{\boldsymbol{\theta}}^{JS}(\mathbf{Y})\right) &= \frac{1}{k}\sum_{i=1}^{k}E((\hat{\theta}_i^{JS} - \theta_i)^2|\boldsymbol{\theta}) \\
&= \sigma^2\left[1 - \frac{k-2}{k}E_{\mathbf{Y}|\boldsymbol{\theta}}\left(\frac{(k-2)\sigma^2}{||\mathbf{Y}||^2}\right)\right], \qquad (5.24)
\end{aligned}
$$

which is less than σ^2 provided $k \geq 3$. The subscript on the expectation in (5.24) indicates that it is with respect to the full sampling distribution $f(\mathbf{Y}|\boldsymbol{\theta})$, but it actually only requires knowledge of the simpler, univariate distribution of $||\mathbf{Y}||$ given $||\boldsymbol{\theta}||$.

The risk improvement in (5.24) is often referred to as "the Stein effect." While it is indeed surprising given the mutual independence of the k components of $\mathbf{Y}$, it does depend strongly on a loss function that combines over these components. The reader may also wonder about its connection to empirical Bayes analysis. This connection was elucidated in a series of papers by Efron and Morris (1971; 1972a,b; 1973a,b; 1975; 1977). Among many other things, these authors showed that $\hat{\boldsymbol{\theta}}^{JS}$ is exactly the PEB point estimator in the Gaussian/Gaussian model (5.5) that assumes μ to be known and equal to 0, and estimates B by

$$\widehat{B} = \frac{k-2}{||\mathbf{Y}||^2}, \qquad (5.25)$$

an unbiased estimator of B (the proof of this fact is left as an exercise). As suggested by the previous subsection, EB performance of the James-Stein estimator can be improved by using a positive part estimator of B, leading to

$$\hat{\theta}_i^{JS^+}(\mathbf{Y}) = \left[1 - \min\left(\frac{(k-2)\sigma^2}{||\mathbf{Y}||^2}, 1\right)\right]Y_i,$$

so that estimates of θ_i are never shrunk *past* the prior mean, 0. (Notice that this improvement implies that the original James-Stein estimator is itself inadmissible!) Another way to generalize $\hat{\theta}_i^{JS}(\mathbf{Y})$ is to shrink toward

prior means other than $\mu = 0$, or toward the *sample* mean in the absence
of a reliable prior estimate for μ. This latter estimator is given by

$$\hat{\theta}_i^{JS'}(\mathbf{Y}) = \bar{Y} + \left[1 - \frac{(k-3)\sigma^2}{||\mathbf{Y} - \bar{Y}||^2}\right](Y_i - \bar{Y}) .$$

Note that this estimator is a particular form of our Gaussian/Gaussian EB
point estimator (5.10). For SSEL, one can show that

$$R\left(\boldsymbol{\theta}, \hat{\boldsymbol{\theta}}^{JS'}(\mathbf{Y})\right) = \sigma^2 \left[1 - \frac{k-3}{k} E_{\mathbf{Y}|\boldsymbol{\theta}}\left(\frac{k-3}{||\mathbf{Y} - \bar{Y}||^2}\right)\right] ,$$

where the expectation depends only on $||\boldsymbol{\theta} - \bar{\theta}||^2$. Hence $\hat{\boldsymbol{\theta}}^{JS'}(\mathbf{Y})$ dominates
the MLE $\mathbf{Y}$ provided $k \geq 4$, a slightly larger dataset being required to "pay"
for our estimation of μ.

The basic result (5.24) can be proven directly (see, e.g., Cox and Hinkley,
1974, pp.446–8), or indirectly by showing that the expectation of (5.24)
when the θ_i are i.i.d. from a $N(0, \tau^2)$ prior matches the preposterior risk
of $\hat{\boldsymbol{\theta}}^{JS}$ under this prior for all τ. But, these expectations depend only on
$||\boldsymbol{\theta}||^2$, which is distributed as a multiple of a central chi-square distribution.
Completeness of the chi-square distribution dictates that, almost surely,
there can be only one solution to this relation, and the Stein result follows.
We leave the details as an exercise.

Figure 5.2 plots the frequentist risk of the James-Stein estimator when
$\sigma^2 = 1$ and $k = 5$. As $||\boldsymbol{\theta}||^2$ increases, the risk improvement relative to the
MLE decreases and eventually disappears, but performance is never worse
than the MLE $\mathbf{Y}$. Because $\mathbf{Y}$ is also the minimax rule, the James-Stein
estimator is *subminimax* for $k \geq 3$ (see Subsection B.1.4 in Appendix B).

Coordinate-specific versus coordinate average loss

The PEB results in Subsection 5.2.3 show how effective the empirical Bayes
approach can be in estimating individual parameters when the assumed
compound structure holds. The Stein results in Subsection 5.2.4 show that
the EB estimators perform well for all $\boldsymbol{\theta}$ vectors when evaluated under
SSEL. However, the coordinate-specific risk of these estimators can be high
if a relatively small number of the θs are outliers (an unlikely event if the
θ_i are sampled from a Gaussian prior). In fact, one can show that

$$\sup_{\boldsymbol{\theta}} \max_{1 \leq i \leq k} R(\theta_i, \hat{\theta}_i^{JS}) = \frac{k\sigma^2}{4} , \tag{5.26}$$

where $R(\theta_i, \hat{\theta}_i^{JS})$ is defined as a single component of the sum in (5.24). The
supremum is attained by setting $\theta_1 = \theta_2 = \cdots = \theta_{k-1} = 0$ and letting
$\theta_k \to \infty$ at rate $\sqrt{k}$ (see the exercises). In this limiting case, the *average*
expected squared error loss is $\frac{\sigma^2}{4}$, a substantial improvement over the MLE.

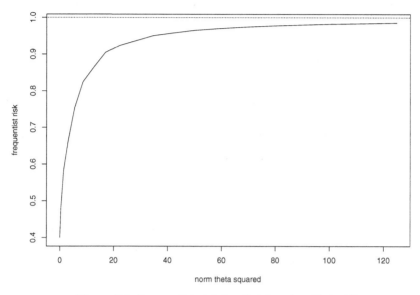

Figure 5.2 *Frequentist risk for Stein's rule with $k = 5$.*

But, for the MLE, the supremum is σ^2, and Stein's rule can produce a substantial coordinate-specific penalty in this extreme situation. This poor performance for outlying parameters has motivated *limited translation rules* (Efron and Morris, 1972a), which are compromises between MLEs and Stein rules that limit component-wise risk. Selection of priors and likelihoods that are more robust against outliers is another possible remedy, and one previously considered in Section 4.2.

In most applications, even if the assumed prior does not hold, the coordinate parameters are not so completely unrelated as in the situation producing (5.26), and so, when carefully applied, the empirical Bayes approach is effective even for coordinate-specific evaluations. Many empirical case studies support this point; for a sampler of applications, see Rubin (1980) on law school admissions, Tomberlin (1988) on insurance rating, Efron and Morris (1973b, 1977) on batting averages and toxoplasmosis, Devine et al. (1994a,b; 1996) on disease mapping and accident mapping, van Houwelingen and Thorogood (1995) on kidney graft survival modeling, and Zaslavsky (1993) on the census undercount.

In other applications, SSEL may be the most reasonable performance benchmark. For example, Hui and Berger (1983) use the approach to stabilize residual variance estimates based on a small number of degrees of freedom. The estimated variances are then used to produce weights in a subsequent analysis. Overall performance is important here, with little interest in individual variance estimates.

5.3 Nonparametric EB (NPEB) point estimation

5.3.1 Compound sampling models

Because, for EB analysis, $G(\boldsymbol{\theta})$ is not completely known and at least some of its features must be estimated, repeated draws from G are required to obtain sufficient information. Consider then a more general formulation of the compound sampling model (5.5),

$$Y_i \mid \theta_i \overset{ind}{\sim} f_i(y_i \mid \theta_i), \quad \text{and} \tag{5.27}$$

$$\theta_i \overset{iid}{\sim} G(\cdot),$$

where $i = 1, \ldots, k$. As we have already seen, compound sampling arises in a wide variety of hierarchical modeling applications. Model (5.27) can be extended by allowing regression and correlation structures in G, or allowing the Y_i to be correlated given the θ_i.

Suppose we seek point estimates for the θ_i, and that the prior cdf G has corresponding density function g. Writing $\mathbf{Y} = (Y_1, \ldots, Y_k)$ and $\boldsymbol{\theta} = (\theta_1, \ldots, \theta_k)$, it is left as an exercise to show that the marginal density of the data $m_G(\mathbf{y})$ satisfies the equation

$$m_G(\mathbf{y}) = \prod_{i=1}^{k} m_G(y_i), \tag{5.28}$$

where $m_G(y_i) = \int f_i(y_i \mid \theta_i) dG(\theta_i)$. That is, the Y_i are marginally independent (and identically distributed as well if $f_i = f$ for all i).

5.3.2 Simple NPEB (Robbins' method)

Consider the basic compound model (5.27) where G is completely unspecified, and $Y_i|\theta_i$ is Poisson(θ_i). That is,

$$f(y_i \mid \theta_i) = \frac{e^{-\theta_i} \theta_i^{y_i}}{y_i!}, \quad y_i = 0, 1, 2, \ldots.$$

As shown in Appendix B, the Bayes estimate of θ_i under squared error loss is the posterior mean,

$$\begin{aligned}
\theta_i^B &= E(\theta_i \mid \mathbf{y}) = E(\theta_i \mid y_i) \\
&= \frac{\int (u^{y_i+1} e^{-u}/y_i!) dG(u)}{\int (u^{y_i} e^{-u}/y_i!) dG(u)} \\
&= \frac{(y_i + 1) m_G(y_i + 1)}{m_G(y_i)}, \tag{5.29}
\end{aligned}$$

where $m_G(y_i)$ is the marginal distribution of y_i. Thus, thanks to the special structure of this model, we have written the Bayes rule in terms of the data and the marginal density, which is directly estimable from the observed data.

Taking advantage of this structure, Robbins (1955) proposed the completely nonparametric estimate computed by estimating the marginal probabilities by their empirical frequencies, namely,

$$\hat{\theta}_i^B = (y_i + 1)\frac{\#(ys \text{ equal to } y_i + 1)}{\#(ys \text{ equal to } y_i)} . \tag{5.30}$$

This formula exemplifies *borrowing information*. Because the estimate of the marginal density depends on all of the y_i, the estimate for each component θ_i is influenced by data from other components.

The foregoing is one of several models for which the analyst can make empirical Bayes inferences with no knowledge of or assumptions concerning G. Maritz and Lwin (1989) refer to such estimators as *simple EB estimators*. The added flexibility and robustness of such nonparametric procedures is very attractive, provided they can be obtained with little loss of efficiency.

The Robbins estimate was considered a breakthrough, in part because he showed that (5.30) is *asymptotically optimal* in that, as $k \to \infty$, its Bayes risk (see (B.6) in Appendix B) converges to the Bayes risk for the true Bayes rule for known G. However, in Example 5.3 we show that this estimator actually performs poorly, even when k is large. This poor performance can be explained in part by the estimator's failure to incorporate constraints imposed by the hierarchical structure. For example, though the Bayes estimator (5.29) is monotone in y, the Robbins estimate (5.30) need not be. Equation (5.29) also imposes certain convexity conditions not accounted for by (5.30). Several authors (van Houwelingen, 1977; Maritz and Lwin, 1989, Subsection 3.4.5) have developed modifications of the basic estimator that attempt to smooth it. One can also generalize the Robbins procedure to include models where the sampling distribution f is continuous.

Large-sample success of the Robbins approach and its refinements depends strongly on the model form and use of the posterior mean (the Bayes rule under squared error loss). A far more general and effective approach adopts the parametric EB template of first estimating G by $\widehat{G}$ and then using (5.29) with G replaced by $\widehat{G}$. That is, we use $m_{\widehat{G}}(\cdot)$ in place of $\widehat{m}_G(\cdot)$. This approach accommodates general models and loss functions, imposes all necessary constraints (such as monotonicity and convexity), and provides a unified approach for discrete and continuous f. To be fully nonparametric with respect to the prior, the approach requires a fully nonparametric estimate of G. Laird (1978) proved the important result that the G which maximizes the likelihood (5.28) is a discrete distribution with at most k mass points; $\widehat{G}$ is discrete even if G is continuous. We now introduce this

nonparametric maximimum likelihood (NPML) estimate, and consider it
in more detail in Subsection 7.1.1.

EM for NPEB

The EM algorithm is ideally suited for computing the NPML estimate.
To initiate the EM algorithm, recall that the prior $G(\boldsymbol{\theta})$ that maximizes
the likelihood (5.28) in the nonparametric setting must be a discrete distribution with at most k mass points. Thus, assume the prior has mass
points $\boldsymbol{\mu}^{(0)} = (\mu_1^{(0)}, \dots, \mu_J^{(0)})$ and probabilities $\boldsymbol{\pi}^{(0)} = (\pi_1^{(0)}, \dots, \pi_J^{(0)})$,
where $J = k$. Then, for the $(\nu + 1)^{st}$ iteration, let

$$
\begin{aligned}
w_{ij}^{(\nu+1)} &= P(\theta_i = \mu_j^{(\nu)} \mid \boldsymbol{\mu}^{(\nu)}, \boldsymbol{\pi}^{(\nu)}, y_i), \\
\pi_j^{(\nu+1)} &= w_{+j}^{(\nu+1)},
\end{aligned}
$$

and let $\mu_j^{(\nu+1)}$ maximize $\sum_i w_{ij}^{(\nu+1)} \log(f_i(y_i|\mu_j))$. Note that the updated πs
and μs maximize weighted likelihoods, and that the updated μs are not
necessarily equal to the previous values. The ws are particularly straightforward to compute using Bayes' theorem:

$$
w_{ij}^{(\nu+1)} \propto \pi_j^{(\nu)} f_i(y_i|\mu_j^{(\nu)}) .
$$

Normalization requires dividing by the sum over index j.

For the Poisson sampling distribution, we have

$$
\mu_j^{(\nu+1)} = \frac{\sum_i w_{ij}^{(\nu)} y_i}{\sum_i w_{ij}^{(\nu)}},
$$

a weighted average of the y_i.

The $\mu_j^{(0)}$ should be spread out to extend somewhat beyond the range
of the data. The number of mass points in the NPML is generally much
smaller than k. Therefore, though the number of mass points will remain
constant at J, some combination of a subset of the $\pi^{(\nu)}$ approaching 0 and
the $\mu_j^{(\nu)}$ approaching each other will occur as the EM iterations proceed.
The EM algorithm has no problem dealing with this convergence to a ridge
in the parameter space, though convergence will be very slow.

Example 5.3 We consider counts of accident insurance policies reporting
y_i claims during a particular year (Table 5.2, taken from Simar, 1976). The
data are discrete, and the Poisson likelihood with individual accident rates
drawn from a prior distribution (producing data exhibiting extra-Poisson
variation) is a good candidate model.

We compute the empirical Bayes, posterior mean estimates of the rate parameters θ_i using several priors estimated from the data and the "Robbins
rule" that directly produces the estimates. The table shows the Robbins
simple EB estimate, as well as the NPEB estimate obtained by plugging

y_i	count	$\hat{m}_G(y_i)$	$E_G(\theta_i\|y_i)$		
			Robbins	Gamma	NPML
0	7840	.82867	.168	.159	.168
1	1317	.13920	.363	.417	.372
2	239	.02526	.527	.675	.610
3	42	.00444	1.333	.933	1.001
4	14	.00148	1.429	1.191	1.952
5	4	.00042	6.000	1.449	2.836
6	4	.00042	1.750	1.707	3.123
7	1	.00011	0.000	1.965	3.142

Table 5.2 *Accident Data: Observed counts and empirical Bayes posterior means for each number of claims per year for $k = 9461$ policies issued by La Royal Belge Insurance Company (adapted from Simar, 1976). The y_i are the observed frequencies, $\hat{P}_G$ is the observed relative frequency, "Robbins" is the Robbins NPEB rule, "Gamma" is the PEB posterior mean estimate based on the Poisson/gamma model, and "NPML" is the posterior mean estimate based on the EB rule for the nonparametric prior.*

the NPML for G into the Bayes rule. As mentioned in Section 5.1, another option is to use a parametric form for G, say $G(\theta_i|\eta)$, and estimate only the parameter η. Given our Poisson likelihood, the conjugate $G(\alpha, \beta)$ prior is the most natural choice. Proceeding in the manner of Example 2.7, we can then estimate $\eta = (\alpha, \beta)$ from the marginal distribution $m(\mathbf{y}|\alpha, \beta)$, plug them into the formula for the Bayes rule, and produce a parametric EB estimate. The specifics of this approach are left as Exercise 9.

Table 5.2 reports the empirical relative frequencies for the various observed y_i values, as well as the results of modeling the data using the Robbins rule, the PEB Poisson/gamma model, and the NPML approach. Despite the size of our dataset ($k = 9461$), the Robbins rule performs erratically for all but the smallest observed values of y_i. It imposes no restrictions on the estimates, and fails to exploit the fact that the marginal distribution of $\mathbf{y}$ is generated by a two-stage process with a Poisson likelihood. By first estimating the prior, either parametrically or nonparametrically, constraints imposed by the two-stage process (monotonicity and convexity of the posterior mean for the Poisson model) are automatically imposed. Furthermore, the analyst is not restricted to use of the posterior mean.

The marginal method of moments estimated mean and variance for the gamma prior are 0.2144 and 0.0160, respectively. Because the gamma distribution is conjugate, the EB estimate is a weighted average of the data y and the prior mean, with a weight of 0.7421 on the prior. For small values of y_i, all three estimates of $E(\theta_i|y_i)$ in Table 5.2 are in close agreement. The predicted accident rate for those with no accidents ($y_i = 0$) is ap-

mass point, θ	0.089	0.580	3.176	3.669
mass, $\hat{g}(\theta)$	.7600	.2362	.0037	.0002

Table 5.3 *NPML prior estimate for the Simar Accident Data.*

proximately 0.16, whereas the estimate based on the MLE is 0 because $\hat{\theta}_i^{MLE} = y_i$. The Poisson/gamma and NPML methods part company for y values greater than 3. Some insight into the plateau in the NPML estimates for larger y is provided by the estimated prior $\hat{g}$, displayed in Table 5.3. Virtually all of the mass is on "safe" drivers (the first two mass points). As y increases, the posterior loads mass on the highest mass point (3.669), but the posterior mean can never go beyond this value. ∎

Of course, Table 5.2 provides no information on the inferential performance of these rules. Maritz and Lwin (1989, p. 86) provide simulation comparisons for the Robbins, Poisson/gamma and several "improved Robbins" rules (they do not consider the NPML), showing that the EB approach is very effective. Tomberlin (1988) uses empirical Bayes to predict accident rates cross-classified by gender, age, and rating territory. He shows that the approach is superior to maximum likelihood in predicting the next year's rates, especially for data cells with a small number of accidents.

5.4 Interval estimation

Obtaining an empirical Bayes confidence interval (EBCI) is, in principle, very simple. Given an estimated posterior $p(\theta_i|y_i, \hat{\eta})$, we could use it as we would any other posterior distribution to obtain an HPD or equal tail credible set for θ_i. That is, if $q_\alpha(y_i, \eta)$ is such that

$$P\left(\theta_i \leq q_\alpha(y_i, \eta) \mid \theta_i \sim p(\theta_i|y_i, \eta)\right) = \alpha \,,$$

then a $100(1 - \alpha)\%$ equal tail *naive EBCI* for θ_i is

$$\left(q_{\alpha/2}(y_i, \hat{\eta}), \quad q_{1-(\alpha/2)}(y_i, \hat{\eta})\right) \,.$$

Why is this interval "naive"? From introductory mathematical statistics we have that

$$\mathrm{Var}(\theta_i|\mathbf{y}) = E_{\eta|\mathbf{y}}\left[\mathrm{Var}(\theta_i|y_i, \eta)\right] + \mathrm{Var}_{\eta|y}\left[E(\theta_i|y_i, \eta)\right] \,. \tag{5.31}$$

In the Gaussian/Gaussian case, a 95% naive EBCI would be

$$E(\theta_i \mid y_i, \hat{\eta}) \pm 1.96\sqrt{\mathrm{Var}(\theta_i \mid y_i, \hat{\eta})} \,.$$

The term under the square root approximates $E_{\eta|y}\left[\mathrm{Var}(\theta_i \mid y_i, \eta)\right]$, since the EB approach simply replaces integration with maximization. However, this corresponds to only the *first* term in (5.31); the naive EBCI is ignoring the posterior uncertainty about η (i.e., the second term in (5.31)).

Hence, the naive interval may be too short, and have lower than advertised coverage probability.

To remedy this, we must first define the notion of "EB coverage."

Definition 5.1 $t_\alpha(\mathbf{y})$ is a $(1 - \alpha)100\%$ *unconditional EB confidence set* for $g(\theta)$ if and only if for each η,

$$P_{\mathbf{y}, \theta | \eta}(g(\theta) \in t_\alpha(\mathbf{y})) \approx 1 - \alpha .$$

So we are evaluating the performance of the EBCI over the variability inherent in *both* θ and the data. But is this too weak a requirement? Many authors have suggested instead conditioning on some data summary, $b(\mathbf{y})$. For example, taking $b(\mathbf{y}) = \mathbf{y}$ produces fully conditional (Bayesian) coverage. Many likelihood theorists would take $b(\mathbf{y})$ equal to an appropriate ancillary statistic instead (see Hill, 1990). Or we might simply take $b(\mathbf{y}) = y_i$, on the grounds that y_i is sufficient for θ_i when η is known.

Definition 5.2 $t_\alpha(\mathbf{y})$ is a $100(1 - \alpha)\%$ *conditional EB confidence set* for $g(\theta)$ given $b(\mathbf{y})$ if, for each $b(\mathbf{y}) = b$ and η,

$$P_{\mathbf{y}, \theta | b(y) = b, \eta}(g(\theta) \in t_\alpha(\mathbf{y})) \approx 1 - \alpha .$$

Naive EB intervals typically fail to attain their nominal coverage probability, in either the conditional or unconditional EB sense (or in the frequentist sense). One cannot usually show this analytically, but it is often easy to check via simulation. We devote the remainder of this section to outlining methods that have been proposed for "correcting" the naive EBCI.

5.4.1 Morris' approach

For the Gaussian/Gaussian model (5.5), Morris (1983a) suggests basing the EBCI on the *modified* estimated posterior that uses the "naive" mean, but inflates the variance to capture the second term in equation (5.31). This distribution is $p'(\theta_i | y_i, \hat{\eta}) = N(\hat{\theta}_i^{EB}, V^*)$, where

$$V^* = \sigma^2 \left(1 - \frac{k-1}{k} \widehat{B} \right) + \frac{2}{k-3} \widehat{B}^2 (Y_i - \overline{Y})^2 . \tag{5.32}$$

The first term in (5.32) approximates the first term in (5.31), and is essentially $Var(\theta_i | y_i, \hat{\eta})$, the naive EB variance estimate. The second term in (5.32) approximates $Var_{\eta | \mathbf{y}} [E(\theta_i | y_i, \eta)]$, the second term in (5.31), and thus serves to "correct" the naive EB interval by widening it somewhat. Notice the amount of correction decreases to zero as k increases, as the estimated shrinkage factor $\widehat{B}$ decreases, or as the data-value approaches the estimated prior mean.

Morris actually proposes estimating shrinkage by

$$\widehat{B} = \frac{k-3}{k-1} \left(\frac{\sigma^2}{\sigma^2 + \hat{\tau}^2} \right), \text{ where } \hat{\tau}^2 = \left[\frac{1}{k-1} \sum_i (Y_i - \overline{Y})^2 - \sigma^2 \right] ,$$

which is the result of several "ingenious adhockeries" (see Lindley, 1983) designed to better approximate (5.31). It is important to remember that this entire derivation assumes σ^2 is known; if it is unknown, a data-based estimate may be substituted. Morris offers evidence (though not a formal proof) that his intervals do attain the desired nominal coverage. Extension of these ideas to non-Gaussian and higher dimensional settings is possible, but awkward; see Morris (1988) and Christiansen and Morris (1997a).

5.4.2 Marginal posterior approach

This approach is similar to Morris' in that we mimic a fully Bayesian calculation, but a bit more generally. Since η is unknown, we place a hyperprior $\psi(\eta)$ on η, and base inference about θ_i on the *marginal posterior*,

$$l_h(\theta_i|\mathbf{y}) = \int p(\theta_i|\mathbf{y}, \eta)h(\eta|\mathbf{y})d\eta \ ,$$

where $h(\eta|\mathbf{y}) \propto m(\mathbf{y}|\eta)\psi(\eta)$.

Several simplifications to this procedure are available for many models. First, if $\hat{\eta} = \hat{\eta}(\mathbf{Y})$ is sufficient for η in the marginal family $m(\mathbf{y}|\eta)$ and has density $\rho(\hat{\eta}|\eta)$, then we can replace $h(\eta|\mathbf{y})$ by $h(\eta|\hat{\eta}) \propto \rho(\hat{\eta}|\eta)\psi(\eta)$, which conditions only on a univariate statistic. Second, note that

$$
\begin{aligned}
p(\theta_i|\mathbf{y}, \eta) &\propto & p(\theta_i, y_i \mid y_{j\neq i}, \eta) \\
&=& p(y_i \mid \theta_i, y_{j\neq i}, \eta) \, p(\theta_i \mid y_{j\neq i}, \eta) \\
&=& f(y_i \mid \theta_i, \eta) \, g(\theta_i \mid \eta) \ , \quad\quad (5.33)
\end{aligned}
$$

the product of the sampling model and the prior. Hence $p(\theta_i \mid \mathbf{y}, \eta)$ depends only on y_i and η, and the marginal posterior can be written as

$$l_h(\theta_i \mid \mathbf{y}) = l_h(\theta_i \mid y_i, \hat{\eta}) = \int p(\theta_i \mid y_i, \eta)h(\eta \mid \hat{\eta})d\eta \ . \quad\quad (5.34)$$

Appropriate percentiles of the marginal posterior l_h (instead of the estimated posterior) determine the EBCI. This is an intuitively reasonable approach because l_h accounts explicitly for the uncertainty in η. In other words, mixing $p(\theta_i \mid y, \eta)$ with respect to h should produce wider intervals.

As indicated by expression (5.33), the first term in the integral in equation (5.34) will typically be known due to the conjugate structure of the hierarchy. However, two issues remain. First, will h be available? And second, even if so, can the integral be computed analytically?

Deely and Lindley (1981) were the first to answer both of these questions affirmatively, obtaining closed-form results in the Poisson/gamma and "signal detection" (i.e., where θ_i is a 0–1 random variable) cases. They referred to the approach as "Bayes empirical Bayes," since placing a hyperprior on η is essentially a fully Bayesian solution to the EB problem. However, the first general method for implementing the marginal posterior approach is

$$\underbrace{\eta \to \boldsymbol{\theta} \to \mathbf{Y} \to \hat{\eta}}_{\text{data process}} \to \underbrace{\left\{ \begin{array}{c} \boldsymbol{\theta}_1^* \\ \vdots \\ \boldsymbol{\theta}_N^* \end{array} \right\} \to \left\{ \begin{array}{c} \mathbf{y}_1^* \\ \vdots \\ \mathbf{y}_N^* \end{array} \right\} \to \left\{ \begin{array}{c} \eta_1^* \\ \vdots \\ \eta_N^* \end{array} \right\}}_{\text{bootstrap process}}.$$

Figure 5.3 *Diagram of the Laird and Louis Type III Parametric Bootstrap.*

due to Laird and Louis (1987), who used the bootstrap. Their idea was to take $h(\eta \mid \hat{\eta}) = \rho(\eta \mid \hat{\eta})$, the sampling density of $\hat{\eta}$ given η *with the arguments interchanged.* We then approximate l_ρ by observing

$$l_\rho(\theta_i \mid y_i, \hat{\eta}) = \int p(\theta_i \mid y_i, \eta)\rho(\eta \mid \hat{\eta})d\eta$$

$$\approx \frac{1}{N} \sum_{j=1}^{N} p(\theta_i \mid y_i, \eta_j^*), \qquad (5.35)$$

where $\eta_j^* \overset{\text{iid}}{\sim} \rho(\cdot \mid \hat{\eta})$, $j = 1, \ldots N$. Notice that this estimator converges to l_ρ as $N \to \infty$ by the Law of Large Numbers. The η_j^* values are easily generated via what the authors refer to as the *Type III Parametric Bootstrap*: given $\hat{\eta}$, we draw $\theta_i^* \sim g(\theta \mid \hat{\eta})$, and then draw $y_i^* \sim f(y \mid \theta_i^*)$, for $i = 1, \ldots k$. From the bootstrapped data sample $\mathbf{y}^* = (y_1^*, \ldots, y_i^*)$ we may compute $\eta^* = \hat{\eta}(\mathbf{y}^*)$. Repeating this process N times, we obtain $\eta_j^* \overset{\text{iid}}{\sim} \rho(\eta \mid \hat{\eta})$, $j = 1, \ldots, N$ for use in equation (5.35), the quantiles of which, in turn, provide our corrected EBCI.

Figure 5.3 provides a graphical illustration of the parametric bootstrap process: it simply mimics the process generating the data, replacing the unknown η by its estimate $\hat{\eta}$. The "parametric" name arises because the process is entirely generated from a single parameter estimate $\hat{\eta}$ rather than by resampling from the data vector $\mathbf{y}$ itself (a "nonparametric bootstrap"). Hence, we can easily draw from $\rho(\cdot \mid \hat{\eta})$ even when this function is not available analytically (e.g., when $\hat{\eta}$ itself must be computed numerically, as in many marginal MLE settings).

Despite this computational convenience, the reader might well question the wisdom of taking $h = \rho$. An ostensibly more sensible approach would be to choose a reasonable $\psi(\eta)$, compute $h(\eta \mid \hat{\eta}) \propto \rho(\hat{\eta} \mid \eta)\psi(\eta)$, and obtain an estimator of the corresponding l_h. This marginal posterior would match a prespecified hyperprior Bayes solution, and, unlike l_ρ, would not be sensitive to the precise choice of $\hat{\eta}$. However, if *EB coverage* is the objective, then l_ρ may be as good as l_h. After all, taking quantiles of l_h will lengthen the naive EBCI, but need not "correct" the interval to level $1 - \alpha$, as we desire. Still, Carlin and Gelfand (1990) show how to use the Type

III parametric bootstrap to match any l_h, *provided* the sampling density $\rho(\cdot|\hat{\eta})$ *is* available in closed form.

5.4.3 Bias correction approach

A problem with the marginal posterior approach is that it is difficult to say how to pick a *good* hyperprior $\psi(\eta)$ (i.e., one that will result in l_h quantiles that actually achieve the nominal EB coverage rate). In fact, if $\hat{\eta}(\mathbf{y})$ is badly biased, the naive EBCI may not be too short, but too *long*. Thus, in general, what is needed is not a method that will widen the naive interval, but one that will *correct* it.

Suppose we attempt to tackle the problem of EB coverage directly. Recall that $q_\alpha(y_i, \eta)$ is the αth quantile of $p(\theta_i \mid y_i, \eta)$. Define

$$r(\hat{\eta}, \eta, y_i, \alpha) = P(\theta_i \leq q_\alpha(y_i, \hat{\eta}) \mid \theta_i \sim p(\theta_i|y_i, \eta)) \,,$$

and

$$R(\eta, y_i, \alpha) = E_{\hat{\eta}|y_i, \eta} \left\{ r(\hat{\eta}, \eta, y_i, \alpha) \right\} \,.$$

Hence R is the true EB coverage, conditional on $b(\mathbf{y}) = y_i$, of the naive EB tail area. Usually the naive EBCI is too short, i.e.,

$$R(\eta, y_i, \alpha) \begin{cases} > \alpha, & \alpha \text{ small (say .025)} \\ < \alpha, & \alpha \text{ large (say .975)} \end{cases} \,.$$

If we solved $R(\eta, y_i, \alpha') = \alpha$ for α', then using this α' with $\hat{\eta}$ would conditionally "correct the bias" in our naive procedure and give us intervals with the desired conditional EB coverage. But of course η is unknown, so instead we might solve

$$R(\hat{\eta}, y_i, \alpha') = \alpha \tag{5.36}$$

for $\alpha' = \alpha'(\hat{\eta}, y_i, \alpha)$, and take the naive interval with α replaced by α' as our corrected confidence interval. We refer to this interval as a *conditionally bias corrected* EBCI. For *unconditional* EB correction, we can replace R by

$$R^*(\eta, \alpha) = E_{\hat{\eta}, y_i|\eta} \left\{ r(\hat{\eta}, \eta, y_i, \alpha) \right\} \,,$$

and solve $R^*(\hat{\eta}, \alpha') = \alpha$ for α'. The naive interval with α replaced by this α' is called the *unconditionally bias corrected* EBCI.

Implementation

If $p(\hat{\eta} \mid y_i, \eta)$ is available in closed form (e.g., the Gaussian/Gaussian and exponential/inverse gamma models), then solving (5.36) requires only traditional numerical integration and a rootfinding algorithm. If $p(\hat{\eta} \mid y_i, \eta)$ is *not* available (e.g., if $\hat{\eta}$ itself is not available analytically, as in the Poisson/gamma and beta/binomial models), then solving (5.36) must be done via Monte Carlo methods. In particular, Carlin and Gelfand (1991a) show how the Type III parametric bootstrap may be used in this regard. Notice that in either case, because $g(\theta_i \mid \eta)$ is typically chosen to be conjugate with $f(y_i \mid \theta_i)$,

$$r(\hat{\eta}, \eta, y_i, \alpha) = F_{y_i, \eta}\left[F_{y_i, \hat{\eta}}^{-1}(\alpha)\right] ,$$

where $F_{y_i, \eta}$ is the cdf corresponding to $p(\theta_i \mid y_i, \eta)$. So even when Monte Carlo methods must be employed, mathematical subroutines can be employed at this innermost step.

To evaluate whether bias correction is truly effective, we must check whether

$$\alpha^* \equiv E_{\hat{\eta} \mid y_i, \eta}\left[r\left(\hat{\eta}, \eta, y_i, \alpha'(\hat{\eta}, y_i, \alpha)\right)\right] = \alpha . \tag{5.37}$$

If $\hat{\eta} \to \eta$ as $k \to \infty$ (i.e., $\hat{\eta}$ is consistent for η), then (5.37) would certainly hold for large k, but, in this event, the naive EB interval would do fine as well! For fixed k, Carlin and Gelfand (1990) provide conditions (basically, stochastic ordering of $p(\theta_i \mid y_i, \eta)$ and $p(\hat{\eta} \mid y_i, \eta)$ in η) under which

$$\alpha + \min(I_1, I_2) \leq \alpha^* \leq \alpha + \max(I_1, I_2) ,$$

where

$$I_1 = \int_{\hat{\eta} > \eta} \left[\alpha'(\hat{\eta}, y_i, \alpha) - r(\hat{\eta}, \eta, y_i, \alpha'(\eta, y_i, \alpha))\right] p(\hat{\eta} \mid y_i, \alpha) d\hat{\eta}$$

and

$$I_2 = \int_{\hat{\eta} < \eta} \left[\alpha'(\hat{\eta}, y_i, \alpha) - r(\hat{\eta}, \eta, y_i, \alpha'(\eta, y_i, \alpha))\right] p(\hat{\eta} \mid y_i, \alpha) d\hat{\eta} .$$

That is, since $I_1 \times I_2 < 0$, the true coverage of the bias corrected EB tail area, α^*, lies in an interval containing the nominal coverage level α.

Example 5.4 We now consider the exponential/inverse gamma model, where $Y_1, \ldots, Y_k \overset{iid}{\sim} \text{Expo}(\theta_i)$ and $\theta_1, \ldots, \theta_k \overset{iid}{\sim} \text{IG}(\eta, 1)$. The marginal distribution for y_i is then $m(y_i \mid \eta) = \eta/(y_i + 1)^{\eta+1}$, $y_i > 0$, so that the MLE of η is $\hat{\eta} = k/\sum_{i=1}^{k} \log(y_i + 1)$.

Consider bias correcting the lower tail of the EBCI. Solving $R^*(\eta, \alpha') = \alpha$ for α' (unconditional bias correction), and plotting the values obtained as a function of η, we see that $\alpha'(\eta, \alpha)$ is close to α for η near 1, but decreases steadily as η increases, and that this decrease is more pronounced for larger values of α. The behavior for the upper tail is similar, except that $\alpha'(\eta, \alpha)$ is now an increasing function of η (recall α is now near 0.95).

To evaluate the ability of the various methods considered in this section to achieve nominal EB coverage in this example, we simulated their unconditional EB coverage probabilities and average interval lengths for two nominal coverage levels (90% and 95%), where we have set the true value of $\eta = 2$ and $k = 5$. Our simulation used 3000 replications, and set $N = 400$ in those methods that required a parametric bootstrap. In addition to the methods presented in this section, we included the classical (frequentist) interval and two intervals that arise from matching a specific

Interval Method	Average Lower Endpoint	Average Upper Endpoint	Average Interval Length	Average Uncond'l Cov. Prob.
$\gamma = .90$				
Classical	.335	19.5	19.2	.901
Naive EB	.355	3.87	3.51	.839
Bias correction	.331	4.74	4.41	.897
Laird and Louis	.339	5.15	4.81	.904
ψ_1-matching	.287	3.23	2.95	.868
ψ_2-matching	.311	4.00	3.69	.894
$\gamma = .95$				
Classical	.268	39.1	38.8	.952
Naive EB	.306	5.53	5.22	.900
Bias correction	.285	7.84	7.55	.952
Laird and Louis	.283	7.79	7.50	.954
ψ_1-matching	.246	4.46	4.51	.930
ψ_2-matching	.265	5.93	5.66	.951

Table 5.4 *Comparison of simulated unconditional EB coverage probabilities, exponential/inverse gamma model.*

hyperprior Bayes solution as in equation (5.34). The two hyperpriors we consider are both noninformative and improper, namely, $\psi_1(\eta) = 1$, $\eta > 0$ (so that $h_1(\eta \mid \hat{\eta}) = G(k+1, \hat{\eta}/k)$), and $\psi_2(\eta) = 1/\eta$, $\eta > 0$ (so that $h_2(\eta \mid \hat{\eta}) = G(k, \hat{\eta}/k)$).

Looking at the results in Table 5.4, we see that the classical method faithfully produces intervals with the proper coverage level, but which are extremely long relative to all the EB methods. The naive EB intervals also perform as expected, having coverage probabilities significantly below the nominal level. The intervals based on bias correction and the Laird and Louis bootstrap approach both perform well in obtaining nominal coverage, with the former being somewhat shorter for $\gamma = .90$ and the latter a bit shorter for $\gamma = .95$. Finally, of the two hyperprior matching intervals, only the one based on ψ_2 performs well, with the ψ_1 intervals being too short. This highlights the difficulty in choosing hyperpriors that will have the desired result with respect to *empirical* Bayes coverage.

On a related note, the proper choice of marginal hyperparameter estimate $\hat{\eta}$ can also have an impact on coverage, and this impact is difficult to predict. Additional simulations (not shown) suggest that the best choice for $\hat{\eta}$ is not the marginal MLE, but the marginal uniformly minimum variance unbiased estimate (UMVUE), $\hat{\eta} = (k-1)/\sum_{i=1}^{k} \log(y_i + 1)$. Ironically, the MLE-based intervals were, in general, a bit too long for the bias correction method, but too short for the Laird and Louis method. ∎

5.5 Bayesian processing and performance

To this point we have developed the Bayes and EB approaches to data analysis, and evaluated their effectiveness on their own terms (e.g., by checking the Bayes or EB coverage of the resulting confidence sets). In the remainder of this chapter, we consider a more formidable challenge: can they also emerge as effective in traditional, frequentist terms? One might expect the news to be bad here, but, for the most part, we will show that when carefully applied (usually with low-information priors), the Bayes and EB formalisms remain effective using the more traditional criteria.

In what follows, we employ the criteria outlined in Appendix B, our review of decision theory. There, using squared error loss (SEL) and its frequentist expectation, mean squared error (MSE), we develop basic examples of how shrinkage and other features of Bayes and EB procedures can offer improved performance. Here, we elaborate on this idea and extend it beyond MSE performance to many other aspects of modern statistical modeling, such as the appropriate tracking and incorporation of relevant uncertainties into a final analysis, convenient structuring of sensitivity analyses, and documentation of assumptions and models. Note that none of these features are unique to, nor place any requirements on, one's philosophical outlook.

We start with two basic, but possibly nonintuitive, examples of how Bayes rules process data. These examples serve to alert the analyst to the complex nature of multilevel inference, and drive home the point that once a model is specified, it is important to let the probability calculus take over and produce inferences. We then proceed to a frequentist and empirical Bayes evaluation of the Bayesian approach to point and interval estimation.

Most of the common examples of Bayesian analysis are like Example 2.2 in that the posterior mean exhibits *shrinkage* from the data value toward the prior mean. However, univariate models with nonconjugate priors and multivariate models with conjugate priors can possibly exhibit nonintuitive behavior. In this section, we give two examples based on the Gaussian sampling distribution. Schmittlein (1989) provides additional examples.

5.5.1 Univariate stretching with a two-point prior

Consider a Gaussian sampling distribution with unit variance and a two-point prior with mass at $\pm\delta$. Based on one observation y, the posterior

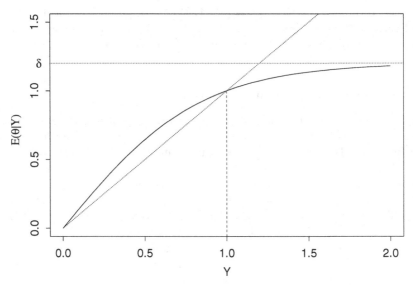

Figure 5.4 *Example of stretching: Gaussian likelihood with two-point prior.*

distribution is

$$P(\theta = \delta \mid y) = \frac{e^{2\delta y}}{1 + e^{2\delta y}},$$

with one minus this probability on $-\delta$. The posterior expectation of θ is thus

$$\delta \cdot \frac{e^{2\delta y} - 1}{1 + e^{2\delta y}}.$$

If $\delta > 1$, for an interval of y-values around 0, the posterior mean is *farther* from 0 than is y, a situation we term *stretching*. Figure 5.4 illustrates the phenomenon when $\delta = 1.2$ for positive y values (the plot for negative y is a mirror image of this one). Notice that $E(\theta|y)$ stretches (exceeds y) for y less than about 1.0. For larger y, the posterior mean does exhibit the usual shrinkage back toward zero, and approaches an asymptote of $E(\theta|y) = \delta$ as $y \to \infty$.

Stretching does violate intuition about what should happen with a unimodal sampling distribution. The explanation lies in use of the posterior mean (resulting from squared-error loss) in a situation where the prior is two-point. In many applications, the posterior probability itself is a sufficient summary and squared error loss is inappropriate.

5.5.2 Multivariate Gaussian model

We now consider another example where intuition may not provide the correct answer, and the Bayesian formalism is needed to sort things out. Consider again the model of Subsection 4.1.1, wherein the coordinate-specific parameters and the data are both p-variate Gaussian. That is,

$$\mathbf{Y} \mid \boldsymbol{\theta} \sim N_p(\boldsymbol{\theta}, \boldsymbol{\Sigma}) \quad \text{and} \quad \boldsymbol{\theta} \mid \mathbf{A} \sim N_p(\mathbf{0}, \mathbf{A}) .$$

Writing the posterior mean and variance in their "shrinkage" form, we have

$$
\begin{aligned}
E(\boldsymbol{\theta} \mid \mathbf{y}) &= (\mathbf{I} - \mathbf{B})\mathbf{y} \\
Var(\boldsymbol{\theta} \mid \mathbf{y}) &= (\mathbf{I} - \mathbf{B})\boldsymbol{\Sigma}
\end{aligned}
\tag{5.38}
$$

where $\mathbf{B} = \boldsymbol{\Sigma}(\boldsymbol{\Sigma} + \mathbf{A})^{-1}$. If both $\boldsymbol{\Sigma}$ and $\mathbf{A}$ are diagonal, there is no linkage among coordinates, and coordinate-specific posteriors can be computed without reference to other coordinates. In general, however, the full multivariate structure must be maintained. Preserving the full structure increases efficiency (DeSouza, 1991), and can produce the nonintuitive result that coordinate-specific estimates can be stretched away from the prior mean (as in the previous example), or shrunk toward it but *beyond* it, a situation we term *crossing*. There is always shrinking toward the prior mean for the full multivariate structure, however, in that the posterior mean in (5.38) has a Euclidean norm smaller than does the data $\mathbf{y}$.

Example 5.5 Consider the bivariate case ($p = 2$) with $\boldsymbol{\Sigma} = \mathbf{I}$ and the diagonal elements of $\mathbf{A}$ equal (i.e., the two coordinates of $\boldsymbol{\theta}$ have equal prior variance). Then $\mathbf{I} - \mathbf{B}$ has the form

$$
\mathbf{I} - \mathbf{B} = \begin{pmatrix} a & b \\ b & a \end{pmatrix},
\tag{5.39}
$$

where $0 \leq a \leq 1$ and $a > |b|$. The posterior mean shows bivariate shrinkage of the observed data toward the prior mean. The point determined by the posterior mean is inside the circle with center at the prior mean, and radius the distance of the observed data from the prior mean (since $\mathbf{I} - \mathbf{B}$ has eigenvalues at $\pm b$). But the posterior mean of the first coordinate is

$$\hat{\theta}_1 = aY_1 + bY_2 .$$

Thus, stretching for this coordinate will occur if $Y_2 > \frac{1-a}{b}Y_1$. Crossing will occur if $Y_2 < -\frac{a}{b}Y_1$. ■

Though we have presented a two-stage Bayesian example, the same properties hold for empirical Bayes (see, e.g., Waternaux et al., 1989) and multilevel hierarchical Bayesian analyses.

5.6 Frequentist performance

Besides having good Bayesian properties, estimators derived using Bayesian methods can have excellent frequentist and EB properties, and produce improvements over estimates generated by frequentist or likelihood-based approaches. We start with two basic examples and then a generalization, still in a basic framework. These show that the Bayesian approach can be very effective in producing attractive frequentist properties, and suggest that this performance advantage holds for more complex models.

5.6.1 Gaussian/Gaussian model

The familiar Gaussian/Gaussian model provides a simple illustration of how the Bayesian formalism can produce procedures with good frequentist properties, in many situations better than those based on maximum likelihood or unbiasedness theory. We consider the basic model wherein $Y|\theta \sim N(\theta, 1)$ and $\theta \sim N(0, \tau^2)$ (that is, the setting of Example 2.2 where $\sigma^2 = 1$ and $\mu = 0$). We compare the two decision rules $\delta_1(y) = y$ and $\delta_2(y) = (1 - B)y$, where $B = 1/(1 + \tau^2)$. Here, δ_1 is the MLE and UMVUE of θ, while δ_2 is the posterior mode, median, and mean (the latter implying it is the Bayes rule under squared error loss). We then have

$$
\begin{aligned}
MSE(\delta_1) &= R(\theta, \delta_1) = E_{Y|\theta}(\theta - \delta_1(Y))^2 = E_{Y|\theta}(\theta - Y)^2 \\
&= Var_{Y|\theta}(Y) = 1 \,,
\end{aligned}
$$

and also

$$
\begin{aligned}
MSE(\delta_2) &= R(\theta, \delta_2) = E_{Y|\theta}(\theta - \delta_2(Y))^2 \\
&= E_{Y|\theta}[\theta - (1 - B)Y]^2 \\
&= E_{Y|\theta}[(1 - B)(\theta - Y) + B\theta]^2 \\
&= (1 - B)^2 E_{Y|\theta}(\theta - Y)^2 \\
&\quad + 2(1 - B)B\theta E_{Y|\theta}(\theta - Y) + B^2\theta^2 \\
&= (1 - B)^2 + B^2\theta^2 \,,
\end{aligned}
$$

which is exactly the variance of δ_2 plus the square of its bias. It is easy to show that $MSE(\delta_2) \leq MSE(\delta_1)$ if and only if $|\theta| \leq \sqrt{(2 - B)/B}$. That is, the Bayes rule has smaller *frequentist* risk provided the true mean, θ, is close to the prior mean, 0.

This situation is illustrated for the case where $B = 0.5$ in Figure 5.5, where the dummy argument t is used in place of θ. $MSE(\delta_1)$ is shown by the solid horizontal line at 1, while $MSE(\delta_2)$ is the dotted parabola centered at 0. For comparison, a dashed line corresponding to the MSE for a third rule, $\delta_3(y) = 2$, is also given. This rather silly rule always estimates θ to be 2, ignoring the data completely. Clearly its risk is given by

$$
MSE(\delta_3) = R(\theta, \delta_3) = E_{Y|\theta}(\theta - 2)^2 = (\theta - 2)^2 \,,
$$

which is 0 if θ happens to actually be 2, but increases much more steeply than $MSE(\delta_2)$ as θ moves away from 2. This rule is admissible, because no other rule could possibly have lower MSE for all θ (thanks to the 0 value at $\delta = 2$), but the large penalty paid for other θ values make it

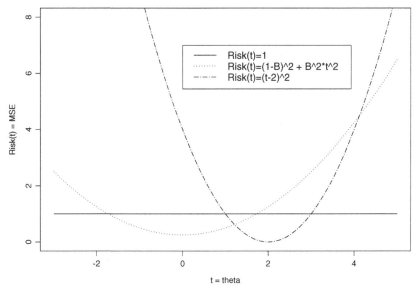

Figure 5.5 *MSE (risk under squared error loss) for three estimators in the Gaussian/Gaussian model.*

unattractive for general use. The Bayes rule δ_2 can also be shown to be admissible, but notice that it would be inadmissible if $B < 0$, since then $MSE(\delta_2) > 1 = MSE(\delta_1)$ for all θ. Fortunately, B cannot be negative since τ^2 cannot be; the point here is only to show that crossing (shrinking past the prior mean) is a poor idea in this example.

For the more general setting of Example 2.3, the Bayes rule will have smaller MSE than the MLE when

$$\theta \in \mu \pm \sqrt{\frac{\sigma^2 + 2n\tau^2}{n}} \ .$$

For the special case where $\tau^2 = \sigma^2$ (i.e., the prior is "worth" one observation), the limits simplify to $\mu \pm \sigma\sqrt{2 + \frac{1}{n}}$, a broad region of superiority.

5.6.2 Beta/binomial model

As a second example, consider the estimation of the event probability in a binomial distribution. For the Bayesian analysis, we use the conjugate $Beta(a, b)$ prior distribution, and reparametrize from (a, b) to (μ, M) where $\mu = a/(a + b)$, the prior mean, and $M = a + b$, a measure of prior precision (i.e., increasing M implies decreasing prior variance). Based on squared-

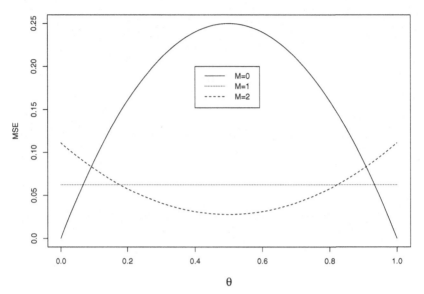

Figure 5.6 *MSE for three estimators in the beta/binomial model with $n = 1$ and $\mu = 0.5$.*

error loss, the Bayes estimate is the posterior mean, namely,

$$\hat{\theta}^{\mu,M} \equiv \frac{M\mu + X}{M + n} = \frac{M}{M + n}\mu + \frac{n}{M + n}\frac{X}{n} . \tag{5.40}$$

Therefore, $\hat{\theta}^{\mu,M}$ is a weighted average of the prior mean and the maximum likelihood estimate, X/n. Irrespective of the value of μ, the MLE of θ is $\hat{\theta}^{\mu,0}$. The MSE of the Bayes estimate, $E_{X|\theta}(\hat{\theta}^{\mu,M} - \theta)^2$, is then given by

$$\left(\frac{n}{M + n}\right)^2 \frac{\theta(1 - \theta)}{n} + \left(\frac{M}{M + n}\right)^2 (\mu - \theta)^2 . \tag{5.41}$$

This equation shows the usual variance plus squared bias decomposition of mean squared error.

Figure 5.6 shows the risk curve for $n = 1$, $\mu = 0.5$, and $M = 0, 1, 2$. If one uses the MLE with $n = 1$, the MLE must be either 0 or 1; no experienced data analyst would use such an estimator. Not surprisingly, the MSE is 0 for $\theta = 0$ or 1, but it rises to .25 at $\theta = .5$. The Bayes rule with $M = 1$ (dotted line) has lower MSE than the MLE for θ in the interval $.5 \pm \frac{\sqrt{3}}{4} = (.067, .933)$. When $M = 2$ (dashed line), the region where the Bayes rule improves on the MLE shrinks toward 0.5, but the amount of improvement is greater. This suggests that adding a little bias to a rule in order to reduce variance can pay dividends.

Next, look at Figure 5.7, where μ is again 0.5, but now $n = 20$ and $M =$

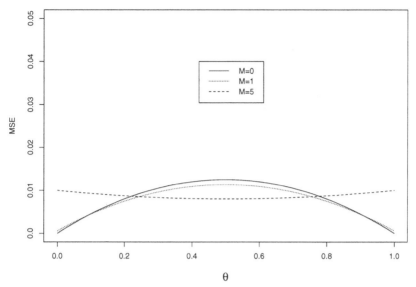

Figure 5.7 *MSE for three estimators in the beta/binomial model with $n = 20$ and $\mu = 0.5$.*

$0, 1, 5$. Due to the increased sample size, all three estimators have smaller MSE than for $n = 1$, and the MLE performs quite well. The Bayes rule with $M = 1$ produces modest benefits for θ near 0.5, with little penalty for θ near 0 or 1, but it takes a larger M (i.e., more weight on the prior mean) for the Bayes rule to be very different from the MLE. This is demonstrated by the curve for $M = 5$, which shows a benefit near 0.5, purchased by a substantial penalty for θ near 0 or 1. The data analyst (and others who need to be convinced by the analysis) would need to be quite confident that θ is near 0.5 for this estimator to be attractive.

Using the Bayes rule with "fair" prior mean $\mu = 0.5$ and small precision $M = 1$ pays big dividends when $n = 1$ and essentially reproduces the MLE for $n = 20$. Most would agree that the MLE needs an adjustment if $n = 1$, a smaller adjustment if $n = 2$, and so on. Bayes estimates with diffuse priors produce big benefits for small n (where variance reduction is important).

Finally, Figure 5.8 shows the costs and benefits of using a Bayes rule with an asymmetric prior having $\mu = 0.1$ when $n = 20$. As in the symmetric prior case ($\mu = 0.5$), setting $M = 1$ essentially reproduces the MLE. With $M = 5$, modest additional benefits accrue for θ below about 0.44 (and a little above 0), but performance is disastrous for θ greater than 0.6. Such an estimator might be attractive in some application settings, but would require near certainty that $\theta < 0.5$. We remark that the Bayes preposterior risk is the integral of these curves with respect to the prior distribution.

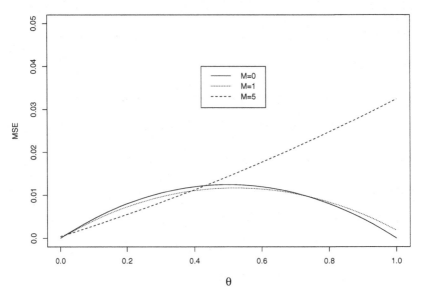

Figure 5.8 *MSE for three estimators in the beta/binomial model with $n = 20$ and $\mu = 0.1$.*

These integrals produce the preposterior performance of various estimates, a subject to which we return in Section 5.7.

In summary, Bayesian point estimators based on informative priors can be risky from a frequentist standpoint, showing that there is no statistical "free lunch" when incorporating subjective information. However, even for univariate analyses without compound sampling, the Bayesian formalism with weak prior information produces benefits for the frequentist.

A note on robustness

In both the binomial and Gaussian examples, either an empirical Bayes or hierarchical Bayes approach produces a degree of robustness to prior misspecification, at least within the parametric family, by tuning the hyperparameters to the data. In addition, for exponential sampling distributions with quadratic variance functions, Morris (1983b) has shown broad robustness outside of the conjugate prior family.

5.7 Empirical Bayes performance

In Section 5.4, we showed how empirical Bayes (EB) confidence intervals may be tuned to have a desired level of *empirical Bayes* coverage, given an assumed "true" prior distribution for θ. In this section, we investigate the

EB coverage of Bayesian confidence sets more generally, evaluating

$$P_{Y,\theta}[\theta \in C(Y)] = E_\theta E_{Y|\theta}[I_{\{\theta \in C(Y)\}}]$$

for various modeling scenarios. We again study the Gaussian/Gaussian model, leaving others as exercises. Since EB coverage is the subject of interest, we investigate the compound sampling framework, which we recall for this problem as

$$Y_i|\theta_i \overset{\text{ind}}{\sim} N(\theta_i, 1), \ i = 1, \ldots, k, \quad \text{and} \tag{5.42}$$

$$\theta_i \overset{\text{iid}}{\sim} N(\mu, \tau^2), \ i = 1, \ldots, k, \tag{5.43}$$

where for the purpose of illustration we set $k = 5$. Parallel to our development earlier in this chapter, we begin by evaluating the performance of Bayes and empirical Bayes point estimators of the θ_i in this setting, and subsequently investigate the corresponding interval estimates.

5.7.1 Point estimation

As discussed above, the traditional approach to estimation does not combine the data in the compound sampling setting. More specifically, the natural frequentist point estimate based on the likelihood (5.43) would be

$$\hat{\theta}_i^F = y_i,$$

the MLE and UMVUE of θ_i. By contrast, the Bayes estimate would be the posterior mean $\hat{\theta}_i^B = B\mu + (1 - B)y_i$, where $B \equiv 1/(1 + \tau^2)$. Because we must specify specific values for μ and τ^2 in order to compute $\hat{\theta}_i^B$, we select the two most obvious choices $\mu = 0$ and $\tau^2 = 1$. Thus we have

$$\hat{\theta}_i^B = y_i/2,$$

an estimator that shrinks each observation halfway to zero. Note that this corresponds to a fairly strong prior belief, since the prior mean is receiving the same posterior weight as the data.

To avoid choosing values for μ and τ^2, we know we must either estimate them from the data or place hyperprior distributions on them. The former approach results in the empirical Bayes estimator

$$\hat{\theta}_i^{EB} = \hat{B}\bar{y} + (1 - \hat{B})y_i,$$

where $\hat{B} \equiv 1/(1 + \hat{\tau}^2)$ and of course $\bar{y} = \frac{1}{k}\sum_{i=1}^k y_i$. Here we adopt the usual "positive part" estimator for τ^2, namely, $\hat{\tau}^2 = (s^2 - 1)^+ = \max\{0, s^2 - 1\}$, where $s^2 = \frac{1}{k-1}\sum_{i=1}^k (y_i - \bar{y})^2$. Alternatively, the modified EB estimate suggested by Morris (1983a) takes the form

$$\hat{\theta}_i^M = \hat{B}_M \bar{y} + (1 - \hat{B}_M)y_i,$$

where $\hat{B}_M = [(k - 3)/(k - 1)]\hat{B}$.

As described earlier in this chapter, the Morris estimator is an attempt to approximate the hierarchical Bayes rule. But in our normal/normal setting with $k = 5$ and using the flat hyperprior $\pi(\mu, \tau^2) = 1$, Berger (1985, p. 187) shows that an exact closed form expression can be obtained for the hierarchical Bayes point estimator, namely,

$$\hat{\theta}_i^{HB} = B^* \bar{y} + (1 - B^*) y_i \; ,$$

where

$$B^* = \frac{k-3}{(k-1)s^2} [1 - g(s^2)]$$

and

$$g(s^2) = \frac{(k-1)s^2}{2} \left\{ \exp \left[\frac{(k-1)s^2}{2} \right] - 1 \right\}^{-1} .$$

As in Example 5.4, we used simulation to evaluate the EB properties of our five estimators. That is, after choosing "true" values for μ and τ^2, we sampled $\boldsymbol{\theta}_j = (\theta_{1j}, \ldots, \theta_{kj})$ from the prior (5.42) followed by $\mathbf{Y}_j = (Y_{1j}, \ldots, Y_{kj})$ from the likelihood (5.43), for $j = 1, \ldots, N$. This resulted in N simulated values for each of our five estimators, which we then used to estimate the EB risk under squared error loss associated with each. For example, for our final estimator we estimated the average risk

$$r = E_{\boldsymbol{\theta}} E_{\mathbf{Y}|\boldsymbol{\theta}} \left[\frac{1}{k} \sum_{i=1}^{k} (\hat{\theta}_i^{HB} - \theta_i)^2 \right]$$

by the simulated value

$$\hat{r} = \frac{1}{Nk} \sum_{j=1}^{N} \sum_{i=1}^{K} (\hat{\theta}_{ij}^{HB} - \theta_{ij})^2$$

with associated Monte Carlo standard error estimate

$$\hat{se}(\hat{r}) = \sqrt{ \frac{1}{(Nk)(Nk-1)} \sum_{j=1}^{N} \sum_{i=1}^{K} \left[(\hat{\theta}_{ij}^{HB} - \theta_{ij})^2 - \hat{r} \right]^2 } .$$

Using $N = 10,000$ replications, we simulated EB risks for four possible (μ, τ^2) combinations: (0,1), (0,10), (5,1), and (5,10). The results of our simulation study are shown in Table 5.5. The fact that the frequentist estimator always has simulated risk near 1 serves as a check on our calculations, since

$$
\begin{aligned}
r(\hat{\boldsymbol{\theta}}^F) &= E_{\boldsymbol{\theta}} E_{\mathbf{Y}|\boldsymbol{\theta}} \left[\frac{1}{k} \sum_{i=1}^{k} (Y_i - \theta_i)^2 \right] \\
&= E_{\boldsymbol{\theta}} \left[\frac{1}{k} \sum_{i=1}^{k} Var_{Y_i|\theta_i}(Y_i) \right]
\end{aligned}
$$

	$\hat{r}$	$\hat{se}(\hat{r})$
$(\mu, \tau^2) = (0, 1):$		
Frequentist	1.005	0.006
Bayes $(B = 0.5)$	0.500	0.003
Empirical Bayes	0.705	0.005
Morris EB	0.750	0.005
Hierarchical Bayes	0.762	0.005
$(\mu, \tau^2) = (0, 10):$		
Frequentist	0.994	0.006
Bayes $(B = 0.5)$	2.736	0.017
Empirical Bayes	0.976	0.006
Morris EB	0.953	0.006
Hierarchical Bayes	0.951	0.006
$(\mu, \tau^2) = (5, 1):$		
Frequentist	1.000	0.006
Bayes $(B = 0.5)$	6.774	0.016
Empirical Bayes	0.704	0.005
Morris EB	0.746	0.005
Hierarchical Bayes	0.758	0.005
$(\mu, \tau^2) = (5, 10):$		
Frequentist	0.987	0.006
Bayes $(B = 0.5)$	9.053	0.041
Empirical Bayes	0.975	0.006
Morris EB	0.947	0.006
Hierarchical Bayes	0.945	0.006

Table 5.5 *Simulated EB risks, Gaussian/Gaussian model with $k = 5$.*

$$= E_{\boldsymbol{\theta}} \left[\frac{1}{k}(k) \right] = 1 .$$

The simple Bayes estimator $\hat{\boldsymbol{\theta}}^B$ performs extremely well in the first scenario (when the true μ and τ^2 are exactly those used by the estimator), but poorly when the true prior variance is inflated, and very poorly when the mean is shifted as well. As in our Section 5.6 study of frequentist risk (MSE), shrinking toward the wrong prior mean (or even overenthusiastic shrinking toward the right one) can have disastrous consequences.

By contrast, the EB performance of all three of the remaining estimators is quite impressive, being superior to that of the frequentist estimator when there is a high degree of similarity among the θ_is (i.e., τ^2 is small), and no worse otherwise. There is some indication that the hierarchical Bayes estimator does better when τ^2 is large, presumably because these values are strongly favored by the flat hyperprior assumed for this estimator. Conversely, the uncorrected EB estimator performs better when τ^2 is small. This is likely due to its tendency toward overshrinkage; in the two cases where $\tau^2 = 1$, $\widehat{B}$ was observed to equal 1 (implying $\hat{\theta}^{EB} = \bar{y}$) in roughly 27% of the replications.

5.7.2 Interval estimation

In this subsection, we continue our investigation of the Gaussian/ Gaussian compound sampling model, but in the case of evaluating the EB coverage of confidence intervals for θ_i. Once again, the frequentist procedure does no pooling of the data, resulting in the nominal $100(1 - \alpha)\%$ interval

$$y_i \pm z_{\alpha/2} \,,$$

where, as usual, $z_{\alpha/2}$ is the upper $(\alpha/2)$-point of the standard normal distribution.

For the Bayes interval, the symmetry of the normal posterior distribution implies that the HPD interval and equal-tail interval are equivalent in this case. Since the posterior has variance $(1 - B) = 0.5$, we obtain the interval

$$y_i/2 \pm z_{\alpha/2}/\sqrt{2} \,,$$

which will always be only 71% as long as the frequentist interval, but will carry the risk of being inappropriately centered.

For the uncorrected ("naive") EB interval, from Section 5.4 we take the upper and lower $(\alpha/2)$-points of the *estimated* posterior, obtaining

$$\hat{\theta}_i^{EB} \pm z_{\alpha/2}\sqrt{1 - \widehat{B}} \,,$$

where $\hat{\theta}_i^{EB}$ is as given in the previous subsection. Similarly, the Morris EB interval is based on the corrected estimated posterior having mean $\hat{\theta}_i^M$ and variance

$$\widehat{V}_i^M = \left(1 - \frac{k-1}{k}\widehat{B}_M\right) + \frac{2}{k-3}\widehat{B}_M^2(y_i - \bar{y})^2 \,,$$

as given in Subsection 5.4.1. Thus, the Morris EB interval is

$$\hat{\theta}_i^M \pm z_{\alpha/2}\sqrt{\widehat{V}_i^M} \,.$$

Finally, a flat hyperprior for (μ, τ^2) again allows the marginal posterior for θ_i (and, hence, the hyperprior Bayes interval) to emerge in closed form, namely,

$$\hat{\theta}_i^{HB} \pm z_{\alpha/2}\sqrt{\widehat{V}_i^{HB}} \,,$$

where $\hat{\theta}_i^{HB}$ is the same as before, and

$$\widehat{V}_i^{HB} = \left(1 - \frac{k-1}{k} B^*\right) + G(y_i - \bar{y})^2 .$$

In this equation, B^* is the same as in the previous subsection and

$$G = \frac{2(k-3)}{[(k-1)s^2]^2} \left\{ 1 + \left[\frac{(k-1)s^2}{2}(B^* - 1) - 1 \right] g(s^2) \right\} ,$$

where $g(s^2)$ is also the same as before.

	Simulated Probability			Average
	Lower	Central	Upper	Length
$(\mu, \tau^2) = (0, 1)$:				
Frequentist	0.027	0.948	0.025	3.920
Bayes $(B = 0.5)$	0.026	0.950	0.025	2.772
Naive EB	0.186	0.629	0.185	1.964
Morris EB	0.019	0.964	0.018	3.605
Hierarchical Bayes	0.011	0.978	0.011	4.024
$(\mu, \tau^2) = (0, 10)$:				
Frequentist	0.024	0.950	0.025	3.920
Bayes $(B = 0.5)$	0.198	0.598	0.203	2.772
Naive EB	0.045	0.909	0.046	3.535
Morris EB	0.025	0.950	0.025	3.850
Hierarchical Bayes	0.021	0.957	0.022	3.972
$(\mu, \tau^2) = (5, 1)$:				
Frequentist	0.023	0.951	0.025	3.920
Bayes $(B = 0.5)$	0.000	0.057	0.943	2.772
Naive EB	0.184	0.632	0.184	1.966
Morris EB	0.017	0.965	0.018	3.605
Hierarchical Bayes	0.010	0.980	0.010	4.024
$(\mu, \tau^2) = (5, 10)$:				
Frequentist	0.024	0.952	0.025	3.920
Bayes $(B = 0.5)$	0.009	0.240	0.751	2.772
Naive EB	0.043	0.912	0.045	3.533
Morris EB	0.023	0.952	0.025	3.850
Hierarchical Bayes	0.020	0.959	0.021	3.973

Table 5.6 *EB coverages and interval lengths of nominal 95% intervals, Gaussian/Gaussian model with $k = 5$.*

Table 5.6 gives the results of using $N = 10,000$ replications to simulate EB coverages and interval lengths for the same four (μ, τ^2) combinations as presented in Table 5.5. The lower, central, and upper probabilities in the table correspond to the proportion of times a simulated θ_{ij} value fell below, in, or above the corresponding confidence interval. Each interval uses a nominal coverage level of 95%. That the frequentist intervals are achieving this coverage to within the accuracy of the simulation serves as a check because the intervals are guaranteed to have the proper level of frequentist coverage and thus must also have this coverage level when averaged over any proper prior (here, a $N(\mu, \tau^2)$ distribution). Next, and as we might have guessed from the previous simulation, the Bayes interval performs very well in the first case, providing the proper coverage level and the narrowest interval of the five. However, in the other three cases (where its prior is not supported by the data), its coverage performance ranges from very poor to abysmal.

The naive EB interval lives up to its name in Table 5.6, missing the nominal coverage level by a small amount when τ^2 is large, and by much more when τ^2 is small. As discussed in Section 5.4, this interval is failing to capture the variability introduced by estimating μ and τ^2. However, Morris' attempt at correcting this interval fares much better, providing a slightly conservative level of coverage and shorter intervals than the frequentist method – noticeably so in cases where τ^2 is small. Finally, the hierarchical Bayes method performs well when τ^2 is large, but appears overly conservative for small τ^2. Apparently if EB coverage with respect to a normal prior is the goal, a hyperprior that places too much mass on large τ^2 values will result in intervals that are a bit too long. A lighter-tailed hyperprior for τ, such as an inverse gamma, or the bias correction method for EB confidence intervals described in Section 5.4, would likely produce better results.

In summary, our intent in the last two sections has been to show that procedures derived using Bayesian machinery and vague priors typically enjoy good frequentist and EB (preposterior) risk properties. The Bayesian approach also allows the proper propagation of uncertainty throughout the various stages of the model, and provides a convenient roadmap for obtaining estimates in complicated models. By contrast, frequentist approaches tend to rely on clever "tricks," such as sufficient statistics and pivotal quantities, that are often precluded in complex or high-dimensional models. All of this suggests the use of Bayesian methodology as an engine for producing procedures that will work well in a variety of situations. Indeed, this seems to have been the intent of the developers of empirical Bayes methodology, who recognized the benefits of the Bayesian approach but rejected the unyielding subjectivism of the era's leading Bayesian thinkers.

5.8 Exercises

1. (Berger, 1985, p. 298) Suppose $Y_i \overset{ind}{\sim} N(\theta_i, 1)$, $i = 1, \ldots, k$, and that the θ_i are i.i.d. from a common prior G. Define the marginal density $m(y_i)$ in the usual way as

$$m(y_i) = \int N(y_i \mid \theta_i, 1) \, dG(\theta_i) \, .$$

   (a) Show that, given G, the posterior mean for θ_i can be written as

$$E(\theta_i | y_i) = y_i + \frac{m'(y_i)}{m(y_i)} \, ,$$

   where m' is the derivative of m.

   (b) Suggest a related nonparametric EB estimator for θ_i.

2. Under the compound sampling model (5.27) with $f_i = f$ for all i, show that (5.28) holds (i.e., that the Y_is are marginally i.i.d.). What is the computational significance of this result for parametric EB data analysis?

3. Consider the gamma/inverse gamma model, i.e., $Y_1, \ldots, Y_k \overset{ind}{\sim} G(\alpha, \theta_i)$, α a known tuning constant, and $\theta_1, \ldots, \theta_k \overset{iid}{\sim} IG(\eta, 1)$.

   (a) Find the marginal density of y_i, $m(y_i | \eta)$.

   (b) Suppose $\alpha = 2$. Find the marginal MLE of η, $\hat{\eta}$.

4. In the Gaussian/Gaussian model (5.5), if $\sigma^2 = 1$ and $\mu = 0$, the Y_is are marginally independent with distribution

$$Y_i | \tau \overset{iid}{\sim} N(0, 1 + \tau^2) \equiv N(0, 1/B) \, .$$

   (a) Find the marginal MLE of B, $\widehat{B}_{MLE}$, and the resulting PEB point estimates $\hat{\theta}_i^{EB}$, $i = 1, \ldots, k$.

   (b) Show that while $\widehat{B}$ in (5.25) is not equal to $\widehat{B}_{MLE}$, it *is* unbiased for B with respect to $m(\mathbf{y}|B)$. Show further that using this $\widehat{B}$ in the usual PEB point estimation fashion produces the James-Stein estimator (5.23).

5. Prove result (5.24) using the completeness of the noncentral chi-square distribution.

6. Show how to evaluate (5.24) for a general θ. (*Hint:* A non-central chi-square can be represented as a Poisson mixture of central chi-squares with mixing on the degrees of freedom.)

7. Prove result (5.26) on the maximum coordinate-specific loss for the James-Stein estimate.

8. Consider again Fisher's sleep data:

$$1.2, 2.4, 1.3, 1.3, 0.0, 1.0, 1.8, 0.8, 4.6, 1.4 .$$

Suppose these $k = 10$ observations arose from the Gaussian/Gaussian PEB model,

$$Y_i | \theta_i \overset{ind}{\sim} N(\theta_i, \sigma^2), \quad i = 1, \ldots, k ,$$

$$\theta_i \overset{iid}{\sim} N(\mu, \tau^2), \quad i = 1, \ldots, k .$$

Assuming $\sigma^2 = 1$ for these data, compute

(a) $\hat{\theta}^{JS}$

(b) $\hat{\theta}^{JS'}$

(c) A naive 95% EBCI for θ_9.

(d) A Morris 95% EBCI for θ_9.

Compare the two point and interval estimates you obtain.

9. Consider the PEB model

$$Y_i | \theta_i \overset{ind}{\sim} Poisson(\theta_i t_i), \quad \theta_i \overset{iid}{\sim} G(a, b), \quad i = 1, \ldots, k.$$

(a) Find the marginal distribution of $\mathbf{Y} = (Y_1, \ldots, Y_k)^T$.

(b) Use the method of moments to obtain closed form expressions for the hyperparameter estimates $\hat{a}$ and $\hat{b}$. (*Hint:* Define the rates $r_i \equiv Y_i/t_i$, and equate their first two moments, $\bar{r}$ and s_r^2, to the corresponding moments in the marginal family.)

(c) Let $k = 5$, $a = b = 3$, and $t_i = 1$ for all i, and perform a simulation study to determine the actual unconditional EB coverage of the 90% equal-tail naive EBCI for θ_1.

10. Consider again the data in Table 3.1, also reproduced here for convenience (Table 5.7). These are the numbers of pump failures, Y_i, observed in t_i thousands of hours for $k = 10$ different systems of a certain nuclear power plant. The observations are listed in increasing order of raw failure rate r_i, the classical point estimate of the true failure rate θ_i for the ith system.

(a) Using the statistical model and results of the previous question, compute PEB point estimates and 90% equal-tail naive interval estimates for the true failure rates of systems 1, 5, 6, and 10.

(b) Using the approach of Laird and Louis (1987) and their Type III parametric bootstrap, obtain corrected 90% EBCIs for θ_5 and θ_6. Why does the θ_5 interval require more drastic correction?

11. In the Gaussian/Gaussian EM example,

i	Y_i	t_i	r_i
1	5	94.320	.053
2	1	15.720	.064
3	5	62.880	.080
4	14	125.760	.111
5	3	5.240	.573
6	19	31.440	.604
7	1	1.048	.954
8	1	1.048	.954
9	4	2.096	1.910
10	22	10.480	2.099

Table 5.7 *Pump Failure Data (from Gaver and O'Muircheartaigh, 1987).*

(a) Complete the derivation of the method for finding MMLEs of the prior mean and variance.

(b) Use the Meilijson approach to find the observed information.

(c) For this model, a direct (but still iterative) approach to finding the MMLE is straightforward. Find the gradient for the marginal distribution and propose an iterative algorithm for finding the MMLEs.

(d) Find the observed information based on the marginal distribution.

(e) Compare the EM and direct approaches.

12. Do a simulation comparison of the MLE and three EB approaches: Robbins, the Poisson/gamma, and the Gaussian/Gaussian model for estimating the rate parameter in a Poisson distribution. Note that the Gaussian/Gaussian model is incorrect. Investigate two values of k (10, 20) and several true prior distributions. Study gamma distributions with large and small means and large and small coefficients of variation. Study two distributions that are mixtures of these gammas. Compare approaches relative to SSEL and the maximum coordinate-specific loss and discuss results. You need to design the simulations.

Extra credit: Include in your evaluation a Maritz (forced monotonicity) improvement of the Robbins rule, and the rule based on the NPML (see Section 7.1 and Subsection 5.3.2 above).

13. In the bivariate Gaussian example of Subsection 5.5.2, show that if $\Sigma = I$, then $\mathbf{B}$ has the form (5.39).

14. In the beta/binomial point estimation setting of Subsection 5.6.2,

(a) For $\mu = .5$ and general n and M, find the region where the Bayes rule has smaller risk than the MLE.

(b) Find the Bayes rule when the loss function is

$$L(\theta, a) = \frac{(a - \theta)^2}{\theta(1 - \theta)} .$$

15. In the interval estimation setting involving a beta prior and a binomial likelihood,

   (a) Under what condition on the prior will the HPD interval for θ be one-sided when $X = n$? Find this interval.

   (b) Outline a specific interval-halving algorithm to find the HPD interval (θ_L, θ_U). Use your algorithm to find the 95% HPD interval when $a = b = 1$, $n = 5$, and $x = 1$. What percentage reduction in width does the HPD interval offer over the equal tail interval in this case?

16. Study the frequentist and Bayesian coverage probabilities for Bayesian intervals based on the Gaussian/Gaussian model with a $N(\mu, \tau^2)$ prior and a sampling distribution, conditional on θ, that is $N(\theta, \sigma^2)$.

   (a) Show you can assume $\mu = 0$ and $\sigma^2 = 1$ without loss of generality .

   (b) Show that the highest posterior probability interval is also equal-tailed.

   (c) Evaluate the frequentist coverage of the Bayesian interval for combinations of the prior variance, τ^2, and the (fixed) parameter of interest, θ. For example, plot coverage versus θ for several values of τ^2.

   (d) Evaluate the Bayesian posterior coverage and preposterior coverage of the interval based on $\mu = 0$, when the true prior mean is not 0.

   (*Hint:* All probabilities can be represented as Gaussian integrals, and the interval endpoints come directly from the Gaussian cumulative distribution function.)

17. Show that if the sampling variance is unknown and has an inverse gamma prior distribution, then in the limit as information in this inverse gamma goes to 0 and as $\tau^2 \to \infty$, Bayesian intervals are Student's t-intervals.

## CHAPTER 6

Bayesian design

6.1 Principles of design

When discussing experimental design, many of the controversies surrounding Bayesian analysis abate. Most statisticians agree that although different approaches to statistical inference engender different design goals, "everyone is a Bayesian in the design phase," even if only informally. Because no data have been collected, all evaluations are preposterior and require integration over both the data (a frequentist act) and the parameters (a Bayesian act). In the terminology of Rubin (1984), this double integration is a "Bayesianly justifiable frequentist calculation." As such, experimental design shares a kinship with the other material in this chapter, and so we include it here.

Whether the analytic goal is frequentist or Bayes, design can be structured by a likelihood, a loss function, a class of decision rules, and either formal or informal use of a prior distribution. For example, in point estimation under SEL, the frequentist may want to find a sample size sufficiently large to control the frequentist risk over a range of parameter values. The Bayesian will want a sample size sufficient to control the preposterior risk of the Bayes or other rule. Both may want procedures that are robust.

Though design is not a principal focus of this book, to show the linkage between Bayesian and frequentist approaches, we outline basic examples of Bayesian design for both frequentist and Bayesian goals. Verdinelli (1992) provides a review of developments in Bayesian design.

6.1.1 Bayesian design for frequentist analysis

Though everyone may be a Bayesian in the design stage, the same is most assuredly not true regarding analysis. Therefore, designs that posit a prior distribution for parameters, but design for a frequentist analysis, have an important role. We consider a basic example of finding the sample size to achieve desired type I error and power in a classical hypothesis testing model. More complicated models are addressed by Carlin and Louis (1985), Louis and Bailey (1990), and Shih (1995).

Consider one-sided hypothesis testing for the mean of n normal variables, $Y_1, \ldots, Y_n \overset{iid}{\sim} N(\theta, \sigma^2)$. That is, $H_0 : \theta = 0$ versus $H_a : \theta > 0$. Standard computations show that for a fixed θ, the n that achieves Type I error α and power $1 - \beta$ satisfies

$$\Phi\left(\frac{\sqrt{n}\theta}{\sigma} - z_\alpha\right) = 1 - \beta, \tag{6.1}$$

where Φ is the standard normal cdf and, as usual, z_α is the upper α-point of this distribution. The standard frequentist design produces

$$n_\theta = \sigma^2 \frac{(z_\beta + z_\alpha)^2}{\theta^2}. \tag{6.2}$$

Because the required n depends on θ, a frequentist design will typically be based on either a typical or a conservative value of θ. However, the former can produce too small a sample size by underrepresenting uncertainty in θ, while the latter can be inefficient.

Now consider the Bayesian design for this frequentist analysis. Given a prior $g(\theta)$ for the treatment effect, Spiegelhalter (2000) suggests a simple Monte Carlo approach wherein we draw $\theta_j^* \overset{iid}{\sim} g(\theta_i), j = 1, \ldots, N$, and subsequently apply formula (6.1) to each outcome. This produces a collection of sample sizes $n_{\theta,j}^*$, the distribution of which reflects the uncertainty in the optimal sample size resulting from prior uncertainty. Adding a prior for σ^2 as well creates no further complication.

A slightly more formal approach would be to set the *expected* power under the chosen prior g equal to the desired level, and solve for $\tilde{n}$, i.e.,

$$\int \Phi\left(\frac{\sqrt{n}\theta}{\sigma} - z_\alpha\right) g(\theta) d\theta = 1 - \beta. \tag{6.3}$$

Shih (1995) applies a generalization of this approach to clinical trial design when the endpoint is the time to an event. Prior distributions can be placed on the treatment effect, event rates, and other inputs.

Solving for $\tilde{n}$ in (6.3) requires interval-halving or some other recursive approach. However, if G is $N(\mu, \tau^2)$, then for $\sigma = 1$, $\tilde{n}$ satisfies

$$\sqrt{\tilde{n}}\mu - z_\alpha = z_\beta\sqrt{1 + \tilde{n}\tau^2}. \tag{6.4}$$

For all α and β, $\tilde{n} \geq n_\theta$, and if $\tau = 0$, then $\tilde{n} = n_\theta$.

Formulas (6.1) and (6.3) produce a sample size that ensures expected power is sufficiently high. Alternatively, one could find a sample size that, for example, makes $\text{P[power} > \beta \mid n] \geq \gamma$.

Example 6.1 Consider comparison of two Poisson distributions, one with rate λ and the other with rate μ. Assume interest lies in the parameter $\delta = \lambda/\mu$ or a monotone transformation thereof. With data $S_n = X_1 + \cdots + X_n$ and $T_n = Y_1 + \cdots + Y_n$, respectively, $W_n = S_n + T_n$ is ancillary for δ (and sufficient for $\eta = \lambda + \mu$).

Irrespective of whether one is a Bayesian or frequentist, inference can be based on the conditional distribution of S_n given W_n,

$$S_n \mid W_n \quad \sim \quad Bin(W_n, \, \theta),$$

where $\theta = \delta/(\delta + 1)$. For this conditional model, W_n is the sample size and there is no fixed value of n that guarantees W_n will be sufficiently large. However, Bayesian design for Bayesian analysis puts priors on both δ and η and finds a sample size that controls preposterior performance. Bayesian design for frequentist analysis puts a prior on η and finds a sample size n which produces acceptable frequentist performance in the conditional binomial distribution with parameters $(W_n, \, \theta)$ averaged over the distribution of W_n or otherwise using the distribution of W_n to pick a sample size. Carlin and Louis (1985) and Louis and Bailey (1990) apply these ideas to reducing problems of multiple comparisons in the analysis of rodent tumorigenicity experiments.

If sequential stopping is a design option, one can stop at $\mathcal{N}$, the first n such that W_n gives acceptable frequentist or Bayesian performance. The prior on η allows computation of the distribution of $\mathcal{N}$. ■

6.1.2 Bayesian design for Bayesian analysis

Having decided that Bayesian methods are appropriate for use at the design stage, it becomes natural to consider them for the analysis stage as well. But in this case, we need to keep this new analysis plan in mind as we design the experiment. Consider, for instance, estimation under SEL for the model

$$Y_1, \ldots, Y_n \mid \theta \quad \overset{iid}{\sim} \quad f(y \mid \theta),$$

where θ has prior distribution G, and f is an exponential family with sufficient statistic $S_n = Y_1 + \cdots + Y_n$. The Bayes estimate is then $\hat{\theta}(S_n) = E_G(\theta \mid S_n)$, with posterior and preposterior risk given by

$$\rho(G, S_n) = V_G(\theta \mid S_n) \quad \text{and} \quad r(G, S_n) = E_{\mathbf{y}}[V_G(\theta \mid S_n)] \, ,$$

where this last expectation is taken with respect to the marginal distribution of $\mathbf{y}$. We begin with the following simple design goal: find the smallest sample size n such that

$$r(G, S_n) \leq C^2 \, . \tag{6.5}$$

For many models, the calculations can often be accomplished in closed form, as the following two examples illustrate.

Example 6.2 Consider first the normal/normal model of Example 2.3, where both G and f are normal with variances τ^2 and σ^2, respectively. Then $\rho(G, S_n) = (1 - B_n)\sigma^2/n$, where $B_n = \sigma^2/[\sigma^2 + n\tau^2]$. This risk depends neither on the prior mean μ nor the observed data S_n. Thus,

$\rho(G, S_n) = r(G, S_n)$, and (6.5) produces

$$n \approx \left(\frac{\sigma}{C}\right)^2 \left[1 - \left(\frac{C}{\tau}\right)^2\right]^+.$$

As expected, increasing C decreases n, with the opposite relation for τ. As $\tau \to \infty$, $n \to (\frac{\sigma}{C})^2$, the usual frequentist design. ∎

Example 6.3 Now consider the Poisson/gamma model of Example 2.7, where we assume that G is gamma with mean μ and variance μ^2/α, and $f(\cdot \mid \theta)$ is $Po(\theta)$. Then

$$\rho(G, S_n) = \frac{\alpha + S_n}{(n + \alpha/\mu)^2},$$

$$\text{and} \quad r(G, S_n) = \frac{\mu^2}{\alpha + n\mu}.$$

Plugging this into (6.5), we obtain

$$n \approx \frac{1}{\mu}\left[\left(\frac{\mu}{C}\right)^2 - \alpha\right]^+.$$

For this model, the optimal Bayesian design depends on both the prior mean and variance. As $\alpha \to 0$, the prior variance increases to infinity and $n \to \mu/C^2$, which is the frequentist design for $\theta = \mu$. ∎

Using preposterior Bayes risk as the design criterion produces a sufficient sample size for controlling the double integral over the sample and parameter spaces. Integrating first over the sample space represents the preposterior risk as the frequentist risk $R(\theta, S_n)$ averaged over the parameter space. The Bayesian formalism brings in uncertainty, so that generally the sample size will be larger than that obtained by a frequentist design based on a "typical" (i.e., expected) parameter value, but perhaps smaller than that based on a "conservative" value.

A fully Bayesian decision-theoretic approach to the design problem invokes the notions of expected utility and the cost of sampling. While many of the associated theoretical considerations are beyond the scope of our book, the basic ideas are not difficult to explain or understand; our discussion here follows that in Müller and Parmigiani (1995). Suppose $U(n, \mathbf{y}, \theta)$ is the payoff from an experiment having sample size n, data $\mathbf{y} = (y_1, \ldots, y_n)'$ and parameter θ, while $C(n, \mathbf{y}, \theta)$ is the cost associated with this experiment. A reasonable choice for $U(n, \mathbf{y}, \theta)$ might be the negative of the loss function; i.e., under SEL we would have $U(n, \mathbf{y}, \theta) = -(\hat{\theta}_n(\mathbf{y}) - \theta)^2$. Cost might be measured simply on a per sample basis, i.e., $C(n, \mathbf{y}, \theta) = cn$ for some $c > 0$. Then, a plausible way of choosing the best sample size n would be to maximize the *expected utility* associated with this experiment,

$$\mathcal{U}(n) = \int \int [U(n, \mathbf{y}, \theta) - C(n, \mathbf{y}, \theta)] f(\mathbf{y}|\theta) g(\theta) d\mathbf{y} d\theta \ , \qquad (6.6)$$

where, as usual, $f(\mathbf{y}|\theta)$ is the likelihood and $g(\theta)$ the chosen prior. That is, we are choosing the design that maximizes the net payoff we expect using this sample size n, where as before we take the preposterior expectation over both the parameter θ and the (as yet unobserved) data $\mathbf{y}$. Provided the per sample cost $c > 0$, $\mathcal{U}(n)$ will typically be concave down, with expected utility at first increasing in n, but eventually declining as the cost of continued sampling becomes prohibitive. Of course, since it may be difficult to assess c in practice, one might instead simply set $c = 0$, meaning that $\mathcal{U}(n)$ would increase without bound. We could then select n as the first sample size that delivers a given utility, equivalent to the constrained minimization in condition (6.5) above when $U(n, \mathbf{y}, \theta) = -l(\theta, \hat{\theta}_n(\mathbf{y}))$.

Another difficulty is that one or both of the integrals in (6.6) may be intractable and hence require numerical integration methods. Fortunately, Monte Carlo draws are readily available by simple composition. Similar to the process outlined in Figure 5.3, we would repeatedly draw θ_j^* from $g(\theta)$ followed by $\mathbf{y}_j^*$ from $f(\mathbf{y}|\theta_j^*)$, producing a joint preposterior sample $\{(\theta_j^*, \mathbf{y}_j^*), j = 1, \ldots N\}$. The associated Monte Carlo estimate of $\mathcal{U}(n)$ is

$$\widehat{\mathcal{U}}(n) = \frac{1}{N} \sum_{j=1}^{N} [U(n, \mathbf{y}_j^*, \theta_j^*) - C(n, \mathbf{y}_j^*, \theta_j^*)] \ . \qquad (6.7)$$

Repeating this calculation over a grid of n values $\{n_i, i = 1, \ldots, I\}$, we obtain a pointwise Monte Carlo estimate of the true curve. The optimal sample size $\tilde{n}$ is the one for which $\widehat{\mathcal{U}}(n_i)$ is a max.

When exact pointwise evaluation of $\widehat{\mathcal{U}}(n_i)$ is prohibitively time consuming due to the size of I or the dimension of $(\mathbf{y}, \theta)$, Müller and Parmigiani (1995) recommend simply drawing a *single* (y_j^*, θ_j^*) from the joint distribution for a randomly selected design n_i, and plotting the integrands $u_i \equiv U(n, \mathbf{y}_j^*, \theta_j^*) - C(n, \mathbf{y}_j^*, \theta_j^*)$ versus n_i. Using either traditional parametric (e.g., nonlinear regression) or nonparametric (e.g., loess smoothing) methods, we may then fit a smooth curve $\tilde{\mathcal{U}}(n)$ to the resulting point cloud, and finally take $\tilde{n} = \operatorname{argmax} \tilde{\mathcal{U}}(n)$. Notice that by ignoring the integrals in (6.6) we are essentially using a Monte Carlo sample size of just $N = 1$. However, results are obtained much faster, due to the dramatic reduction in function evaluations, and may even be more accurate in higher dimensions, where the "brute force" Monte Carlo approach (6.7) may not produce a curve smooth enough for a deterministic optimization subroutine.

Example 6.4 Müller and Parmigiani (1995) consider optimal selection of the sample size for a binomial experiment. Let $f(y|\theta)$ by the usual binomial likelihood $\binom{n}{y} \theta^y (1 - \theta)^{n-y}$ and $g(\theta)$ be the chosen prior – say, a $Unif(0, 1)$

(a noninformative choice). If we adopt absolute error loss (AEL), an appropriate choice of payoff function is

$$U(n, \mathbf{y}, \theta) = -|m_y - \theta| \, ,$$

where m_y is the posterior median, the Bayes rule under AEL; see Appendix B, problem 3. Adopting a sampling cost of .0008 per observation, (6.6) then becomes

$$\mathcal{U}(n) = \int_0^1 \sum_{y=0}^n [-|m_y - \theta| - .0008n] \binom{n}{y} \theta^y (1-\theta)^{n-y} d\theta \, . \qquad (6.8)$$

Here the y sum is over a finite range, so Monte Carlo integration is needed only for θ. We thus set up a grid of n_i values from 0 to 120, and for each, draw $\theta_j^* \stackrel{iid}{\sim} Unif(0,1), j = 1, \ldots N$, and obtain

$$\widehat{\mathcal{U}}(n_i) = \frac{1}{N} \sum_{j=1}^N \sum_{y=0}^{n_i} [-|m_y - \theta_j^*| - .0008n_i] \binom{n_i}{y} (\theta_j^*)^y (1 - \theta_j^*)^{n_i - y} \, , \quad (6.9)$$

where m_y can be derived numerically from the $Beta(y + 1, n_i - y + 1)$ posterior. Again, $\tilde{n} = \text{argmax}\, \widehat{\mathcal{U}}(n_i)$.

Müller and Parmigiani (1995) actually note that in this case the θ integral can be done in closed form after interchanging the order of the integral and the sum in (6.8), so Monte Carlo methods are not actually needed at all. Alternatively, the quick estimate using $N = 1$ and smoothing the resulting (n_i, u_i) point cloud performs quite well in this example. ∎

In medical decision-making, utility maximization is often done in a *sequential* framework, wherein we must alternately evaluate the evidence provided by the data we currently have, and then decide whether to take additional samples, a process whose cost may or may not be outweighed by the utility gain provided by these samples. Bayesian evaluation of such sequential problems is both conceptually and computationally challenging, and is the subject of Subsection 7.6 below.

6.2 Bayesian clinical trial design

Perhaps the most valuable contribution of Bayesian methods to healthcare evaluation involves study design. Drug and medical device clinical trialists are increasingly confronted with data that feature complex correlation structures, and are costly and difficult to obtain. In such settings, Bayesian trial designs are attractive because they can incorporate historical data or information from published literature, thus saving time and expense and minimizing the number of subjects exposed to an inferior treatment. Bayesian designs can also adapt to unexpected changes in the protocol, and allow the investigator to explore the plausibility of various outcome scenarios before any patients are enrolled in the trial. Recently, the U.S. Food and Drug Administration (FDA) Center for Devices has encouraged hierarchical Bayesian statistical approaches that allow for the incorporation of such valuable historical data into the design and analysis of new device trials.

While the Bayesian advantages of flexibility and borrowing of strength (both from previous data and across subgroups) have been well-known to clinical trialists for some time, they have proven elusive to obtain in practice due to the difficulty in converting historical information into prior distributions, and in computing the necessary posterior summaries. Still, ever-increasing pressure to minimize the financial and ethical cost of clinical trials encourages greater development and use of Bayesian thinking in their design and analysis. In the case of medical device trials, where data are often scanty and expensive to obtain, the recently released FDA guidance document (http://www.fda.gov/cdrh/osb/guidance/1601.html) is a significant push for Bayesian methods, which already make up roughly 10% of new device approvals (Berry, 2006). While the area of drug trials has been slower to embrace the methods, even here they are gaining traction. Bayes is an especially natural approach for incorporating historical controls into the analysis, an area for which the classical (frequentist) literature is very limited (though see Pocock, 1976; Prentice et al., 2006; and Neaton et al., 2007, for notable exceptions).

In this section, we illustrate Bayesian analysis and sample size calculations using BRugs. This function's ability to iteratively call BUGS allows us to repeatedly estimate the posterior given various artificial data samples, and hence simulate the operating characteristics (power and Type I error rates) of our Bayesian designs. We then provide an application of the approach using a logistic model in an AIDS drug trial setting where incorporation of available historical information is crucial. A second example, incorporating interim analysis and a more difficult statistical model (Cox regression), is deferred until Section 6.3.2.

6.2.1 Classical versus Bayesian trial design

Classical analysis of outcomes observed in a randomized controlled trial is based upon the frequentist theory of hypothesis testing and its perspective of probability as a long-run frequency behavior. Frequentists use data collected in a particular trial to test a null hypothesis of no difference between the effects of treatment and control. Such approaches typically enjoy low bias, but combining information across trials is awkward; each trial must essentially stand on its own. In order to detect moderate distinctions in effectiveness, classical trials will often require prohibitively large sample sizes and follow-up periods. Moreover, classical sample size calculations de-

pend upon the accurate assessment of uncertain quantities, but potential
inaccuracies in these assessments are not formally acknowledged.

Bayesian designs enable incorporation of existing trial results or other
knowledge through the specification of prior distributions, potentially reducing time, expense, and unethical exposure of patients to inferior treatments. The Bayesian approach also better emulates the way clinical conclusions are reached in the scientific community, with evidence supporting
or condemning the effectiveness of a particular intervention accumulating
gradually over time. As Spiegelhalter et al. (1994) put it, "data from a
single trial report add to available evidence rather than form the basis for
decision-making in themselves." Furthermore, a Bayesian approach is better able to adapt to circumstances where randomization is unethical, or
a trial needs to be stopped early. Sample size calculations may be based
on prior distributions that are uninformative, optimistic, or skeptical with
respect to the effectiveness of the intervention of interest, allowing investigators to quantify the likely gain from incorporating prior knowledge into
the inference, and assess the need for a new trial. Finally, acceptance of
Bayesian trial designs by the scientific community may spur more thorough reviews of historical data, and thus help increase overall scientific
openness and honesty (Spiegelhalter, Abrams, and Myles, 2004, p.181).

Classical Approach

Sample size formulation in the frequentist paradigm is based upon the convergence of various test statistics to standard (typically normal) probability
distributions for large sample sizes. For example, for continuous, approximately normally distributed responses, recall, the frequentist per group
sample size is given by (6.2),

$$n = \sigma^2 \frac{(z_\beta + z_\alpha)^2}{\Delta^2} \, ,$$

where we seek a procedure having power $1 - \beta$ and type I error rate α,
and we now write θ as Δ to emphasize its nature as a treatment difference.
Since we do not know σ^2, it must be assumed or estimated from prior data.
Similarly, because Δ is also unknown, one typically chooses a value that is
pragmatically attainable, yet small enough to distinguish between groups
with truly disparate conditions. Often the value chosen is the *minimally
clinically significant difference*, i.e., the smallest improvement the clinician
would value enough to justify the treatment's expense and toxicities. See
the fine new book edited by Cook and DeMets (2008) for a comprehensive
introduction to the classical approach to clinical trial design, monitoring,
and analysis.

Frequentists use prior information (say, from previous trials or knowledge
of the underlying biological mechanism) as well as characteristics of the
population under study when choosing Δ. Yet uncertainty about Δ is not

accounted for in the trial's stated operating characteristics (power and Type I error) or their uncertainty; a prior distribution on Δ could help quantify this uncertainty. The information from expert opinion or prior data used to specify Δ is also not incorporated into the analysis of the trial's results.

6.2.2 Bayesian assurance

O'Hagan and Stevens (2001) lay out a Bayesian formulation of the sample size determination problem that generalizes a traditional frequentist sample size calculation based on hypothesis testing. These authors do this in the context of assessing the *cost effectiveness* of a particular treatment relative to control. Specifically, they let e_{ij} be the observed efficacy and c_{ij} be the cost of treatment i for patient j, $j = 1, \ldots, n_i$ and $i = 1, 2$, where $i = 1$ denotes control and $i = 2$ denotes treatment. These authors then assume the bivariate normal model

$$\begin{pmatrix} e_{ij} \\ c_{ij} \end{pmatrix} \sim N_2 \left(\begin{pmatrix} \mu_i \\ \gamma_i \end{pmatrix}, \Sigma \right) ,$$

where $\Sigma_{11} = \sigma_i^2, \Sigma_{22} = \tau_i^2$, and $\Sigma_{12} = \Sigma_{21} = \rho_i \sigma_i \tau_i$. Given the scale in which the costs c_{ij} are expressed (dollars, patient lives, etc.), suppose K is the maximum amount we are prepared to pay to obtain one unit of increase in efficacy e_{ij}. Then our cost effectiveness assessment must be based on the net benefit

$$\beta = K(\mu_2 - \mu_1) - (\gamma_2 - \gamma_1) .$$

The treatment $(i = 2)$ is cost effective if $\beta > 0$. Denoting all the data $\{e_{ij}, c_{ij}\}$ as $\mathbf{y}$, suppose we require $P(\beta > 0 | \mathbf{y}) > \omega$. Then this Bayesian analysis objective is analogous to rejecting $H_0 : \beta = 0$ in favor of the one-sided alternative $H_a : \beta > 0$ at a p-value of $\alpha = 1 - \omega$. O'Hagan and Stevens (2001) refer to this as the *analysis objective*, and the prior used to calculate the posterior probability as the *analysis prior*.

Now, the sample sizes n_i in each group must be such that, averaging over all datasets we might see, the probability of a positive result is at least δ. That is, using subscripts to more clearly indicate the random variable with respect to which an expectation is taken, we require

$$P_{\mathbf{Y}} \left[P_{\boldsymbol{\xi}}(\mathbf{a}' \boldsymbol{\xi} > 0 | \mathbf{y}) > \omega \right] > \delta , \tag{6.10}$$

where $\boldsymbol{\xi} = (\mu_1, \gamma_1, \mu_2, \gamma_2)'$ and $\mathbf{a} = (-K, 1, K, -1)'$, so that $\beta = \mathbf{a}' \boldsymbol{\xi}$.

The authors refer to the left hand side of (6.10) as the *Bayesian assurance*; note it is the Bayesian analogue of power, albeit averaged over the prior distribution used to calculate the marginal distribution of the data $\mathbf{Y}$. O'Hagan and Stevens (2001) observe that this prior distribution need *not* be the same as the one used to calculate the inner, posterior probability in (6.10). For instance, in order to satisfy a skeptical regulatory agency, we might need to use a very vague prior at the analysis stage (to compute the inner probability). However, at the design stage (to compute the outer probability), the best Bayesian assurance calculation would likely arise from a much more informative prior that reflects our knowledge of what we honestly think the data $\mathbf{Y}$ are going to look like. O'Hagan and Stevens (2001) formalize this notion by specifying *two* priors,

$$\text{design prior:} \quad \boldsymbol{\xi} \sim N(\mathbf{m}_d, V_d) \,,$$
$$\text{analysis prior:} \quad \boldsymbol{\xi} \sim N(\mathbf{m}_a, V_a) \,,$$

where $\mathbf{m}_d, V_d, \mathbf{m}_a$, and V_a are all assumed known. Note that $V_a^{-1} = 0$ produces a vague (zero precision) analysis prior, while $V_d = 0$ produces a point fitting prior. Under these two conditions, Bayesian assurance is equal to frequentist power at the proposed true $\mathbf{m}_a$ value. These fully specified, conjugate forms enable closed form posterior and marginal distributions for $\boldsymbol{\xi}$ and $\mathbf{Y}$, respectively, that in turn facilitate calculation of the assurance. For instance, writing $\bar{\mathbf{y}} = (\bar{e}_1, \bar{c}_1, \bar{e}_2, \bar{c}_2)'$, one can show that the analysis objective $P(\beta > 0 | \mathbf{y}) > \omega$ is met if and only if

$$\mathbf{a}'V^*(V_a^{-1}\mathbf{m}_a + S^{-1}\bar{\mathbf{y}}) \geq z_{1-\omega}\sqrt{\mathbf{a}'V^*\mathbf{a}} \,, \tag{6.11}$$

where $z_{1-\omega}$ is the upper $(1 - \omega)$-point of a standard normal distribution, $V^* = (V_a^{-1} + S^{-1})^{-1}$, the posterior variance of $\boldsymbol{\xi}$, and S is the sampling variance matrix of $\bar{\mathbf{y}}$,

$$S = \begin{pmatrix} \sigma_1^2/n_1 & \rho_1\sigma_1\tau_1/n_1 & 0 & 0 \\ \rho_1\sigma_1\tau_1/n_1 & \tau_1^2/n_1 & 0 & 0 \\ 0 & 0 & \sigma_2^2/n_2 & \rho_2\sigma_2\tau_2/n_2 \\ 0 & 0 & \rho_2\sigma_2\tau_2/n_2 & \tau_2^2/n_2 \end{pmatrix} .$$

Note that (6.11) depends on the observed data $\bar{\mathbf{y}}$; averaging over the marginal distribution of such datasets arising from the design prior, one can show that the Bayesian assurance will be at least δ if and only if

$$\mathbf{a}'V^*(V_a^{-1}\mathbf{m}_a + S^{-1}\mathbf{m}_d) \geq z_{1-\omega}\sqrt{\mathbf{a}'V^*\mathbf{a}} + z_{1-\delta}\sqrt{\mathbf{a}'V^*S^{-1}(V_d + S)S^{-1}V^*\mathbf{a}} \,. \tag{6.12}$$

This expression is complex, but easily calculated without need for MCMC sampling.

To simplify things a bit, suppose we consider the frequentist setting of a flat analysis prior ($V_a^{-1} = 0$ in the limit) and a point design prior ($V_d = 0$). The former condition implies $V^* = S$, while the latter implies a precise specification of $\boldsymbol{\xi} = \mathbf{m}_d$. Expression (6.12) then readily simplifies to

$$\mathbf{a}'\mathbf{m}_d \geq (z_{1-\omega} + z_{1-\delta})\sqrt{\mathbf{a}'S\mathbf{a}} \,. \tag{6.13}$$

Note that the determination of suitable values of n_1 and n_2 will typically require an iterative search, and even then, neither (6.12) nor (6.13) will typically lead to a unique solution unless we place additional constraints on the problem. However, if we require $n = n_1 = n_2$, equal sample sizes in the treatment and control groups, then the common sample size in the frequentist case of (6.13) must satisfy

$$n \geq \frac{(z_{1-\omega} + z_{1-\delta})^2 \mathbf{a}' S_1 \mathbf{a}}{(\mathbf{a}' \mathbf{m}_d)^2} , \qquad (6.14)$$

where S_1 is the single-observation variance matrix obtained by setting $n_1 = n_2 = 1$ in the previous expression for S. Details of these calculations are left as Exercise 3.

Obviously the results of this subsection depend heavily on the specific model used, which is somewhat artificial and specialized (normal distributions with known variance matrices). However, the principles involved, namely choosing design and analysis priors and using them to determine sample sizes through fixed definitions of trial success (analysis objective) and Bayesian power (design objective), are quite general. In the next subsection, we consider using these principles in a more complex setting where Monte Carlo methods must be used to solve for the necessary sample size(s).

6.2.3 Bayesian indifference zone methods

As just seen, traditional design concepts (significance level, power, true treatment difference, etc.) have analogues in the Bayesian approach, but are defined relative to the posterior distribution. In this subsection we adapt the framework of Freedman et al. (1984), who begin by replacing the unrealistic point null $H_0 : \Delta = 0$ with a *range* of null Δ's, $\Delta \in [\delta_L, \delta_U]$, over which we are indifferent between the intervention and the control. The upper bound, δ_U, represents the amount of improvement required by the intervention to suggest clinical superiority over control, while δ_L denotes the threshold below which the intervention would be considered clinically inferior. Trial results can be based upon the location of the 95% posterior credible interval for Δ with respect to this *indifference zone* $[\delta_L, \delta_U]$, as demonstrated in Figure 6.1. Exactly six cases are possible, with stronger evidence required to "accept" one hypothesis than merely reject the other. Note that to conclude "equivalence" the 95% interval must lie entirely *within* the indifference zone; if the interval straddles *both* ends of the zone, posterior evidence is too weak to make a decision of any kind.

Consider a trial where increased Δ implies increased benefit associated with intervention. We sometimes take $\delta_L = 0$ and $\delta_U > 0$, but we might also center the indifference zone around 0, i.e., $\delta_L = 0 - \xi$ and $\delta_U = 0 + \xi$. In the latter case, for a fixed n and under a proper prior on Δ, expanding the indifference zone by increasing ξ corresponds to a decrease in type I error (since rejection of the control becomes more difficult) but also a decrease in power. On the other hand, decreasing ξ powers the trial for a more desirable effect difference, yet corresponds to an increase in type

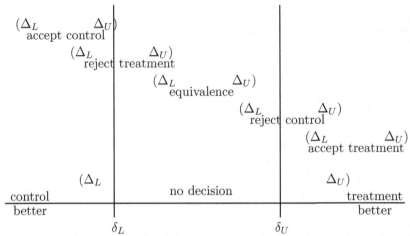

Figure 6.1 *Indifference zone* (δ_L, δ_U) *and corresponding conclusions for a clinical trial based on the location of the 95% posterior credible interval for* Δ.

I error. One often strives for an indifference zone with various appealing "symmetry" properties; we illustrate in Sections 6.3.1 and 6.3.2 below. However, note that symmetry within the indifference zone for regression coefficients in a logistic or Cox proportional hazards model does not yield corresponding symmetries on the (nonlinearly transformed) odds or hazard ratio scales.

Next, the likelihood for the observed trial outcomes must be formulated. Consider first the case of a binary trial endpoint; say, whether the patient experiences progression of disease or not during the trial, for $i = 1, \ldots, N$. Then $Y_i \sim Bernoulli(\pi_i)$, where π_i is the probability of disease progression for the ith patient. Now let x_i be an indicator variable for the intervention group. One possible model for π_i assumes

$$logit(\pi_i) = \log\left(\frac{\pi_i}{1 - \pi_i}\right) = \lambda_0 + \lambda_1 x_i \, , \qquad (6.15)$$

where λ_0 and λ_1 are random hyperparameters. If $x_i = 0$ for control and 1 for treatment, then λ_1 captures the intervention effect. Inference is typically based on e^{λ_1}, the ratio of odds for disease progression between the two groups. Therefore, e^{λ_1} plays the role of Δ above.

Evaluating the posterior distribution of e^{λ_1} requires us to specify priors for the regression parameters. Because $\lambda_k \in \Re$ for $k = 0, 1$, normal priors could be used, with informative content added as indicated below. Note also that any important prognostic factor z_i may be added to $logit(\pi_i)$ as a $\lambda_2 z_i$ term, although e^{λ_1} would now need to be interpreted as the odds ratio of disease progression for individuals with identical z_i.

If instead of binary outcomes, we have continuous measurements (e.g., blood pressure, weight, etc.), then a normal likelihood may be more appropriate. Now Δ would be expressed as the difference in group means, prior knowledge on the likelihood mean would likely be incorporated using a normal prior, and the likelihood variance might use the standard inverse gamma prior. Time-to-event outcomes may also be of interest, and employ Weibull, gamma, or Cox partial likelihoods; Section 6.3.2 offers an illustration.

When determining the priors for λ_0 and λ_1, information may be plentiful for the former (since it is determined by the rate in the control group), but not the latter (since this parameter captures the improvement of the new therapy over control). Spiegelhalter et al. (1994) conducted a thorough examination of the utility of evidence external to the trial, and our overall strategy follows theirs. That is, we first review available historical evidence on both the treatment and control groups. This review helps us determine our fitting priors, the prior distributions that will be used when the data are ultimately collected and the posterior distribution computed. Even when not directed to by a regulatory agency, we will often wish to compare the results obtained under an informative fitting prior (i.e., one that incorporates available historical data) with those from a noninformative one, to see the impact these historical data have on the posterior.

Now let us take a step back to the design stage. Suppose we wish to power a study to deliver any of the six outcomes in (or combinations thereof) with a given probability. For any fixed, "true" values of λ_0 and λ_1 and proposed treatment allocation $\{x_i\}_{i=1}^N$, we can simulate the frequentist power of our Bayesian procedure by computing the π_i from equation (6.15), and then generating fake data values Y_{ij}^* repeatedly from the binomial likelihood for $j = 1, \ldots, N_{rep}$. Each fake data vector $\mathbf{Y}_j^* = (Y_{1j}^*, \ldots, Y_{Nj}^*)'$ leads to a 95% posterior interval for λ_1, and hence one of the six decisions in Figure 6.1. Repeating this for each of the N_{rep} datasets, we can compute the empirical probability of each of the six outcomes, and thus estimate any power we desire (a type I error calculation arises by setting $\lambda_1 = 0$, the null value) in conjunction with the appropriate superiority hypothesis. Thus, our Bayesian sample size problem comes down to choosing a design (i.e., a sample size N and indifference zone) that delivers some prespecified acceptable frequentist properties. The use of an informative fitting prior is likely to pay dividends (greater power) in cases where the "truth" is congruent with this prior.

As described above, a fully Bayesian version of this procedure would replace the fixed, true values (λ_0, λ_1) by draws $\{(\lambda_{0j}^*, \lambda_{1j}^*), j = 1, \ldots, N_{rep}\}$ from their prior distributions. This would acknowledge the uncertainty in these parameters; we will never know "the truth" at the design stage. However, like O'Hagan and Stevens (2001), we recognize that the design prior (i.e., the prior used to generate the fake λ_{0j}^* and λ_{1j}^*) need *not* be the same as the fitting prior, with the latter typically being the vaguer of the two.

Having selected design and fitting priors, all that remains is to summarize the (frequentist or Bayesian) power and type I error, and select a sample size N that delivers satisfactory levels of each. Note that the posterior calculation for each fake data vector $\mathbf{Y}_j^*$ may require MCMC methods, meaning that the WinBUGS software would need to be called N_{rep} times. A feasible solution here (and the one we adopt) is to write the outer, fake data-generating loop in R, and call BUGS repeatedly using commands from the BRugs package.

All of the preceding discussion refers to trial designs having only a single monitoring point, after all N data values have been collected. Adding multiple looks at the data is problematic for classical, frequentist trials, due to the impact this has on the procedure's Type I error rate, which in turn determines the trial's outcome. Early solutions to this problem by Pocock (1977) and O'Brien and Fleming (1979) involve prespecifying the number of interim analyses and computing a stopping boundary for each that restricts overall Type I error. Lan and DeMets (1983) avoid prespecifying the number of interim looks by utilizing a Type I error "spending function." By contrast, because the posterior credible interval in Figure 6.1 can be evaluated and the corresponding decision reached at any time during the trial, strictly speaking, our Bayesian method is unaffected by multiple looks at the accumulating data. However, if the *frequentist* properties of this Bayesian procedure are of interest, any planned sequence of looks must be incorporated into the design simulation exercise described above, where we check the stopping criterion after each look. Section 6.3.2 offers an illustration.

6.2.4 Other Bayesian approaches

Using predictive probabilities

Berry (2006) is a strong advocate for the use of predictive probabilities in making decisions based on accumulating clinical trial data. Such an outlook is helpful in cases where, perhaps due to especially acute ethical concerns, we are under pressure to terminate trials of ineffective treatments early (say, because the treatment is especially toxic or expensive). The basic idea is to compute the probability that a treatment will *ever* emerge as superior given the patient recruitment outlook and the data accumulated so far; if this probability is too small, the trial is stopped. In the past, frequentist have sometimes referred to this as *stochastic curtailment*; applied Bayesians have instead tended to use the phrase *stopping for futility*.

To fix ideas, consider again the simple binomial case where each patient i is either a success on the study treatment ($Y_i = 1$) or a failure ($Y_i = 0$).

Assuming the patients are independent with common success probability π, we obtain the familiar binomial likelihood for $X = \sum_i Y_i$. Now suppose we have observed n_1 patients to date, of which X_1 have been successes, so that $X_1 \sim Bin(n_1, \pi)$. Under a conjugate $Beta(a, b)$ prior for π, we of course obtain a $Beta(X_1 + a, n_1 - X_1 + b)$ posterior for π. Posterior inference and decision making would now arise in the usual way (say, after adopting an indifference zone for π and proceeding as indicated by Figure 6.1.

Now suppose that the trial has yet to reach a definitive conclusion, and we wish to decide whether or not to randomize an additional n_2 statistically independent patients into the protocol. Because we know Bayes' Rule may be used sequentially in this case, the current $Beta(X_1 + a, n_1 - X_1 + b)$ posterior now serves as the prior for π, to be combined with a $Bin(n_2, \pi)$ likelihood for X_2. Posterior inference would now focus on the resulting $Beta(X_1 + X_2 + a, n_1 + n_2 - X_1 - X_2 + b)$ updated posterior. The predictive point of view argues that the appropriate calculation at this point is to sample values π_j^* from the "prior" (actually, the interim posterior) $Beta(X_1 + a, n_1 - X_1 + b)$, followed by fake data values X_{2j}^* repeatedly from the $Bin(n_2, \pi_j^*)$ likelihood. Repeating this process for $j = 1, \ldots, N_{rep}$ produces the collection of *posterior predictive* distributions

$$p(\theta^* | X_1, X_{2j}^*) = Beta(X_1 + X_{2j}^* + a, n_1 + n_2 - X_1 - X_{2j}^* + b) .$$

Inference is now based on the appropriate summary of these distributions.

Example 6.5 A major Minneapolis medical device company, MedGuidant Scientific, wishes to run a safety study on one of its new cardiac pacemakers. Specifically, the company wishes to show that men receiving its new product will be very likely to be free from adverse events (AEs) during the three months immediately following implantation of the device. (Here, adverse events are limited to those for which the device is directly responsible, and which require additional action by the implanting physician; if a patient dies after being hit by a bus, this would be truly unfortunate, but would not count as an "adverse event.") Letting π be the probability a patient does *not* experience an AE in the first three months, we seek a 95% equal-tail Bayesian confidence interval for π, $(\pi_{.025}, \pi_{.975})$. Suppose our trial protocol uses the following decision rule:

$$\text{Device is safe from AEs at 3 months} \Longleftrightarrow \pi_{.025} > 0.85 .$$

That is, if the lower confidence bound for the chance of freedom from AEs is at least 85%, the trial succeeds; otherwise it fails.

Now suppose we already have a preliminary study, Study A, whose results are given in Table 6.1. In our above notation, we have $X_1 = 110$ and $n_1 = 117$. Our task is now to evaluate whether it is worth running a second study, Study B, which would enroll an additional n_2 patients. If we begin with a $Uniform(0, 1) = Beta(1, 1)$ prior for π, the interim posterior is then $Beta(X_1 + a, n_1 - X_1 + b) = Beta(111, 8)$. Sampling π_j^* values from this

	No AE	AE	Total
Count	110	7	117
(%)	(94)	(6)	

Table 6.1 *Historical AE data, Safety Study A.*

| | | Safety | |
		$\overline{AE}$	AE
Long-Term Efficacy	$\overline{R}$	θ_1	θ_2
	R	θ_3	θ_4

Table 6.2 *Probabilities θ_j of possible patient responses at six months, two-endpoint study. R denotes atrial fibrillation recurrence within six months, while $\overline{R}$ denotes no such R recurrence; AE denotes occurrence of an adverse event within six months, while $\overline{AE}$ denotes no such AE.*

prior followed by potential Study B values X_{2j}^* from the $Bin(n_2, \pi_j^*)$ likelihood produces the necessary $Beta(111 + X_{2j}^*, 8 + n_2 - X_{2j}^*)$ posteriors and, hence, simulated lower confidence limits $\pi_{.025,j}^*$ from the posterior predictive distribution for $j = 1, \ldots, N_{rep}$. The empirical predictive probability of trial success is then

$$\widehat{P}(\pi_{.025} > 0.85) = \frac{\text{number of } \pi_{.025,j}^* > 0.85}{N_{rep}} . \qquad (6.16)$$

If this number is less than some prespecified cutoff (say, 0.70), the trial would be declared *futile* at this point, and it would be abandoned without randomizing the additional n_2 patients. ∎

Handling multiple endpoints

The basic approach described above is readily extended to the situation where a trial must be powered to evaluate *two* endpoints – say, safety and long-term efficacy. If the the responses on both endpoints can be reasonably treated as discrete, a sensible and relatively simple Bayesian approach might use a *Dirichlet-multinomial* model, a straightforward extension of the beta-binomial model for binary responses. Specifically, once the joint distribution of the multiple endpoints is specified via cross-classification, the multinomial becomes the natural likelihood model. The Dirichlet distribution (see Appendix A, Section A.2) then offers a convenient conjugate prior whose specification is similar to that of the beta distribution.

As a specific illustration, consider the joint probabilities given in Table 6.2. This 2×2 table reflects the four possible outcomes arising from a device trial having two endpoints: recurrence (R) or non-recurrence $(\bar{R})$ of atrial fibrillation within six months of device implantation, and presence (AE) or absence $(\overline{AE})$ of an adverse event at six months. Suppose we again let $i = 1, \ldots, N$ index patients, and $j = 1, \ldots, 4$ index the probabilities θ_j in the table. The prior, which we notate as $\boldsymbol{\theta} \sim Dirichlet(\boldsymbol{\alpha})$, has mean components $E(\theta_j) = \alpha_j / \alpha.$, where $\alpha. = \sum_{j=1}^{4} \alpha_j$. The likelihood is written as $\mathbf{Y}|\boldsymbol{\theta} \sim Mult(n, \boldsymbol{\theta})$ where $\mathbf{Y} = (Y_1, \ldots, Y_4)$. Thanks to the conjugacy of the Dirichlet prior with the multinomial, the posterior then emerges as a $Dirichlet(\mathbf{Y} + \boldsymbol{\alpha})$. Because the marginal distributions of a Dirichlet are themselves beta-distributed, all marginal posteriors $p(\theta_j|\mathbf{Y})$ are available in closed form, again without resort to MCMC or any other complicated computational method.

Next, suppose we choose fixed constants $\delta(AE)$, the largest permissible increase in the adverse event rate, $\delta(\bar{R})$, the desired increase in long-term efficacy. These values are set in consultation with clinical experts and study managers familiar with the trial's goals. Now, given fixed probabilities p_L and p_U (say, $p_L = 0.05$ and $p_U = 0.95$), following Thall et al. (1995) we monitor both safety and efficacy using *multiple* stopping criteria as follows:

A. If $Pr[\eta_S(AE) + \delta(AE) \leq \eta_E(AE) \,|\, \mathbf{Y}] \geq p_{U,1}(AE)$, we stop and declare the treatment *unsafe*.

B. If $Pr[\eta_S(\bar{R}) + \delta(\bar{R}) \leq \eta_E(\bar{R}) \,|\, \mathbf{Y}] \leq p_L(\bar{R})$, we stop and declare the trial *futile*.

C. If $Pr[\eta_S(\bar{R}) \leq \eta_E(\bar{R}) \,|\, \mathbf{Y}] \geq p_U(\bar{R})$, we stop and declare the treatment *efficacious*.

D. If $Pr[\eta_S(AE) \geq \eta_E(AE) \,|\, \mathbf{Y}] \geq p_{U,2}(AE)$, we stop and declare the treatment *safe*.

In these expressions, the ηs are sums of the θ_j corresponding to the safety (AE) or efficacy $(\bar{R})$ outcome indicated, while the subscripts indicate experimental (E) or standard (S) treatment. The overall stopping rule is then

Stop the trial if A or B or (C and D).

That is, we stop if the treatment emerges as unsafe, or futile, or *both* efficacious and safe. Otherwise, we continue to enroll and follow patients until a maximum sample size N_{max} is reached. Clearly this procedure depends crucially on the control constants $\delta(AE)$, $\delta(\bar{R})$, p_L, and p_U, and simulation of the trial's Bayesian and frequentist operating characteristics (power, Type I error, etc.) will be necessary to investigate their precise impact. Fortunately, these operating characteristics can be simulated in much the same way as described above for the case of a single stopping criterion. That is, we first sample fake $\boldsymbol{\theta}^*$ values from the prior, followed by fake $\mathbf{Y}^*$ values from the multinomial likelihood. We then compute the relevant posterior probabilities, see if they lead to stopping, and, if so, for what reason. Repeating this for N_{rep} fake datasets, we can obtain empirical probabilities of stopping for a lack of safety, for futility, or for safety and efficacy for the given sample number of patients N. Also as before, we can repeat this calculation for various choices of N, creating a Bayesian sample size table, or for trial designs having multiple looks at the accumulating data.

6.2.5 Extensions

Naturally, many extensions of the basic approach described in this section are possible. Perhaps the best way to illustrate many of them is in the context of specific examples, and this is the approach we take in Section 6.3 below. For now, we simply list a few issues of immediate importance and provide some associated references.

An interesting feature of the Bayesian approach to clinical trial design is that, strictly speaking, there is no need to randomize patients to treatment assignment. This has ethical implications, since frequentist trials' need to randomize is what forces researchers to obtain "informed consent" from all participants, essentially reminding them that by entering the trial, they may not end up getting the new treatment, and this decision is based entirely on random chance. Through the use of the prior distribution, Bayesian methods can at least in principle overcome this problem, with expert opinion helping to guide treatment assignments based on patient's characteristics. The problem of course is that an expert may place all the sick patients on treatment and all the healthy patients on placebo, biasing our estimate of the treatment effect.

Kadane (1986; 1996) offers something of a compromise here. In a nutshell, he proposes basing treatment assignment on a *panel* of experts, obtaining five different priors for the new treatment's likely effect on each subject. If at least four of the five experts agree that either the new or the standard treatment is likely to be best for a particular subject, then this subject receives this treatment with probability 1; if not, the subject is randomized to treatment. This design retains much of the traditional frequentist advantage of randomization, but treats the most sensitive patients more ethically. While further experience is needed, increasing pressure to treat clinical trial enrollees more quickly and ethically suggests an expanding role for Bayesian methods here.

Practical developments in Bayesian trial design, interim monitoring, and analysis continue; see Berry et al. (2002) for an early case study in an industry setting. A particularly "hot" area of current development is *adaptive* designs, which permit changes to the protocol in the middle of the trial as ethical and medical conditions warrant. Again, this is awkward or impossible in a frequentist setting, since as mentioned several times already, changing the design changes the p-value, and so all possible changes must be anticipated by the trial's protocol. By contrast, the Bayesian paradigm permits major changes, such as dropping an unsuccessful arm or an ineffective dose, in the middle of the trial. Such flexibilities are especially useful in Phase II dose-finding trials. Trials that move seamlessly from Phase I to Phase II, or even from Phase II to Phase III, are also possible in a hierarchical Bayesian framework. More recent work suggests trials that are adaptive at the *patient* level, actually permitting patient-specific dose adaptation during the trial. Again, the ethical and medical benefits of such flexibility are immediately apparent; see Berry (2006) for a recent overview.

In summary, the Bayesian approach offers several advantages, most especially the potential for smaller and/or shorter trials, and thus earlier submissions to regulatory agencies. The approach also forces more serious pretrial planning, addressing the questions of why the trial is being run at all, and what the next steps will be after the trial is completed. In adaptive settings, drugs or doses that are ineffective may be eliminated earlier. Drawbacks to the Bayesian method include their higher conceptual and computational burdens, and possibly more set-up and review time. Nevertheless, their increasing use in both industry- and government-sponsored trials suggests their growing importance in practice.

6.3 Applications in drug and medical device trials

6.3.1 Binary endpoint drug trial

Design and analysis of the FIRST trial

We close this section with an illustration based on Community Programs for Clinical Research on AIDS (CPCRA) protocol 058, the Flexible Initial Retrovirus Suppressive Therapies (FIRST) trial. As explained in MacArthur et al. (2001), this was a large, long-term, randomized, prospective comparison of three different antiretroviral strategies (protease inhibitor (PI) only, non-nucleoside reverse transcriptase inhibitor (NNRTI), and PI plus NNRTI-containing regimens) in highly active, antiretroviral therapy-naive, HIV-1-infected persons. Patients within all three strategies were also assigned one or two nucleoside reverse transcriptase inhibitors (NRTIs), most commonly abacavir plus lamivudine (ABC+3TC) and didanosine plus stavudine (ddI+d4T). The three strategies (and specific antiretroviral regimens) for initial treatment were compared for long-term virological and immunological durability and safety, for the development of drug resistance, and for clinical disease progression. A novel aspect of the trial was that, before randomization, patients and their clinicians within the two strategy arms involving NNRTIs were given the option of preselecting the NNRTI drug, either nevirapine (NVP) or efavirenz (EFV), or allowing an additional randomization to NVP or EFV. Patients in the latter group were randomized to study-specified drugs when they were randomized to a strategy arm.

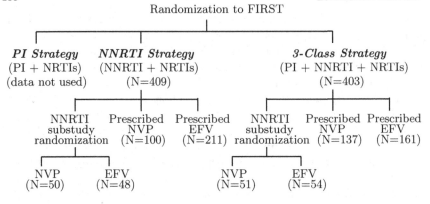

Figure 6.2 *Outline of FIRST design and randomization for eligible subjects (from Berg-Wolf et al., 2006).*

We illustrate Bayesian drug trial design using as historical data patients randomized within the FIRST NNRTI substudy to regimens including EFV ($n = 102$) or NVP ($n = 101$), as well as patients who chose a regimen including EFV ($n = 372$) or NVP ($n = 237$). Our goal is to compare the probabilities of virological suppression (HIV RNA < 50 copies/mL) under EFV and NVP at 32 weeks. Note that 54 of the patients randomized to EFV, 51 of the patients randomized to NVP, 161 of the patients whose physician chose EFV, and 137 of the patients whose physician chose NVP had a regimen that also included one PI. Therefore, 49.6% of patients were randomized to the PI plus NNRTI arm, 58% of patients were assigned a regimen containing EFV, and 75% of patients had their regimen preselected by a physician. Viral RNA load varied substantially at baseline. The data set excludes patients missing an eight-month plasma HIV RNA measurement. Figure 6.2 offers an pictorial representation of the study design. Note that the PI strategy did not include NNRTI regimens and therefore is not used in our illustration.

Let $Y_i = 1$ if patient i experiences virological suppression (VS) at the eight-month visit. We assume $Y_i \sim Bernoulli(\pi_i)$, where π_i denotes the probability of VS for $i = 1, \ldots, 812$. Following (6.15), we fit a general linear model with logistic link function,

$$\log \left(\frac{\pi_i}{1 - \pi_i} \right) = (\mathbf{Z}\boldsymbol{\lambda})_i , \qquad (6.17)$$

where $\boldsymbol{\lambda} = (\lambda_1, \lambda_2, \lambda_3, \lambda_4, \lambda_5)'$ and $\mathbf{Z}$ is a 812×5 design matrix. The covariates (in order) are EFV indicator, NVP indicator, an indicator for whether the patient's physician chose the regimen (EFV or NVP), an indicator for whether baseline HIV-1 RNA < 100,000 copies/mL, and an indicator for whether the patient was randomized to the PI + NNRTI strategy. A total

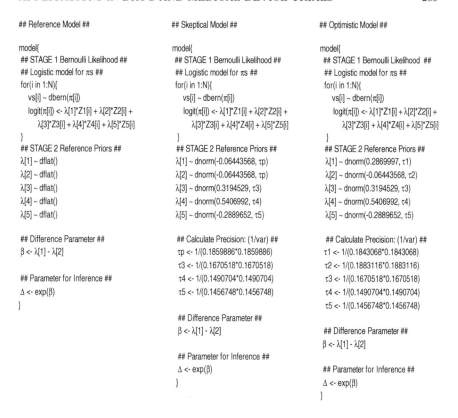

```
## Reference Model ##

model{
  ## STAGE 1 Bernoulli Likelihood ##
  ## Logistic model for πs ##
  for(i in 1:N){
    vs[i] ~ dbern(π[i])
    logit(π[i]) <- λ[1]*Z1[i] + λ[2]*Z2[i] +
      λ[3]*Z3[i] + λ[4]*Z4[i] + λ[5]*Z5[i]
  }
  ## STAGE 2 Reference Priors ##
  λ[1] ~ dflat()
  λ[2] ~ dflat()
  λ[3] ~ dflat()
  λ[4] ~ dflat()
  λ[5] ~ dflat()


  ## Difference Parameter ##
  β <- λ[1] - λ[2]

  ## Parameter for Inference ##
  Δ <- exp(β)
}
```

```
## Skeptical Model ##

model{
  ## STAGE 1 Bernoulli Likelihood ##
  ## Logistic model for πs ##
  for(i in 1:N){
    vs[i] ~ dbern(π[i])
    logit(π[i]) <- λ[1]*Z1[i] + λ[2]*Z2[i] +
      λ[3]*Z3[i] + λ[4]*Z4[i] + λ[5]*Z5[i]
  }
  ## STAGE 2 Reference Priors ##
  λ[1] ~ dnorm(-0.06443568, τp)
  λ[2] ~ dnorm(-0.06443568, τp)
  λ[3] ~ dnorm(0.3194529, τ3)
  λ[4] ~ dnorm(0.5406992, τ4)
  λ[5] ~ dnorm(-0.2889652, τ5)

  ## Calculate Precision: (1/var) ##
  τp <- 1/(0.1859886*0.1859886)
  τ3 <- 1/(0.1670518*0.1670518)
  τ4 <- 1/(0.1490704*0.1490704)
  τ5 <- 1/(0.1456748*0.1456748)

  ## Difference Parameter ##
  β <- λ[1] - λ[2]

  ## Parameter for Inference ##
  Δ <- exp(β)
}
```

```
## Optimistic Model ##

model{
  ## STAGE 1 Bernoulli Likelihood ##
  ## Logistic model for πs ##
  for(i in 1:N){
    vs[i] ~ dbern(π[i])
    logit(π[i]) <- λ[1]*Z1[i] + λ[2]*Z2[i] +
      λ[3]*Z3[i] + λ[4]*Z4[i] + λ[5]*Z5[i]
  }
  ## STAGE 2 Reference Priors ##
  λ[1] ~ dnorm(0.2869997, τ1)
  λ[2] ~ dnorm(-0.06443568, τ2)
  λ[3] ~ dnorm(0.3194529, τ3)
  λ[4] ~ dnorm(0.5406992, τ4)
  λ[5] ~ dnorm(-0.2889652, τ5)

  ## Calculate Precision: (1/var) ##
  τ1 <- 1/(0.1843068*0.1843068)
  τ2 <- 1/(0.1883116*0.1883116)
  τ3 <- 1/(0.1670518*0.1670518)
  τ4 <- 1/(0.1490704*0.1490704)
  τ5 <- 1/(0.1456748*0.1456748)

  ## Difference Parameter ##
  β <- λ[1] - λ[2]

  ## Parameter for Inference ##
  Δ <- exp(β)
}
```

Figure 6.3 BUGS *code for "reference"* **(left)**, *"skeptical"* **(center)**, *and "optimistic"* **(right)** *analysis models.*

of 471 patients were assigned a regimen containing EFV and 341 a regimen containing NVP. Improper uniform (noninformative) priors were then assigned to the logistic regression coefficients λ_j. Let $\beta = \lambda_1 - \lambda_2$, the difference in the regression coefficients for EFV and NVP. Our parameter of interest is the ratio of odds of VS for patients assigned a regimen containing EFV versus NVP. Let p_1 and p_2 be the marginal probabilities of VS for patients on EFV and NVP. Then the odds ratio with choice of regimen, baseline RNA, and strategy held constant follows as

$$\Delta = \frac{\frac{p_1}{1-p_1}}{\frac{p_2}{1-p_2}} = \frac{e^{\log(\frac{p_1}{1-p_1})}}{e^{\log(\frac{p_2}{1-p_2})}} = \frac{e^{\lambda_1+\lambda_3+\lambda_4+\lambda_5}}{e^{\lambda_2+\lambda_3+\lambda_4+\lambda_5}} = e^{\lambda_1-\lambda_2} = e^\beta. \qquad (6.18)$$

WinBUGS code for our model is shown in Figure 6.3 (first column); for full details, including the BRugs "wrapper" program, see the book's soft-

	Mean	Standard Deviation	MC Error of the Mean	2.5	Percentiles 50	97.5
λ_1	0.2870	0.1843	0.0022	-0.0763	0.2884	0.6454
λ_2	-0.0644	0.1883	0.0024	-0.4344	-0.0637	0.3039
λ_3	0.3195	0.1670	0.0018	-0.0068	0.3204	0.6508
λ_4	0.5407	0.1491	0.0013	0.2465	0.5417	0.8312
λ_5	-0.2890	0.1457	0.0012	-0.5715	-0.2893	-0.0052
β	0.3514	0.1492	0.0010	0.0595	0.3508	0.6452
Δ	1.4370	0.2158	0.0014	1.0610	1.4200	1.9060

Table 6.3 **BUGS** *empirical summary statistics, posteriors for* λ, β, *and* Δ, *FIRST data.*

ware page, www.biostat.umn.edu/~brad/software.html. Posterior distributions for the λ_j, β, and Δ were obtained from three parallel sampling chains of 6000 iterations each after a burn-in period of 3000 iterations. The results in Table 6.3 show a 95% posterior credible interval for β of [0.0595, 0.6452], corresponding to an odds ratio (Δ) interval of [1.06, 1.91], or an increase of between 6 and 91 percent in the odds of VS for patients receiving EFV compared to those receiving NVP. The posterior confidence interval for λ_3 very nearly excludes 0, suggesting a marginally significant benefit to those whose regimen was chosen by their physician before randomization. The strong association between VS and RNA viral load at baseline is reflected in the posterior of λ_4: those with lower baseline HIV-1 RNA were more likely to experience VS at eight months. Finally, patients randomized to the composite PI+NNRTI strategy were significantly worse off in terms VS at eight months. As a side note, there were also several modest pairwise correlations within the λ vector, the largest being an estimated posterior correlation of 0.68 between λ_1 and λ_2.

Using FIRST to design a new trial

FIRST was the first prospective, randomized trial comparing EFV and NVP for virological suppression of HIV-1 RNA when enrollment began in February 1999. However, its design makes it hard to estimate potentially important interactions, such as between the treatment effect and either the strategy used (PI+NNRTI versus NNTRI alone) or whether the patient's physician chose the regimen. In addition, results from two other similar randomized studies comparing the effectiveness of EFV and NVP in virological suppression at 48 weeks in antiretroviral-naive adults have now appeared. The SENC (Spanish Efavirenz versus Nevirapine Comparison) trial was an open-label pilot study reported by Núñez et al. (2002), and the 2NN (2

Non-Nucleoside reverse transcriptase inhibitors) study was a multicenter, open-label trial reported by van Leth et al. (2004). While we lack access to these trials' actual data, the published summaries for both SENC and 2NN suggested no significant difference between EFV and NVP at $\alpha = 0.05$, contrary to FIRST (two-sided frequentist p-value 0.011 for $H_0 : \beta = 0$). However, point estimates in both studies suggest that EFV outperformed NVP. Resolving the question of which NNRTI performs better for virological suppression at 32 to 48 weeks in antiretroviral-naive adults may require another randomized trial similar in structure to FIRST. Yet ensuring that this trial is adequately powered for frequentist analysis may also require a very large sample size. As such, we now illustrate a Bayesian design using data from FIRST that produces a sizeable reduction in sample size and cost without sacrificing sensitivity or specificity.

Suppose that this follow-up trial, which we name "SECOND," is designed to compare two NNRTI drug regimens with either EFV or NVP, plus possibly one or two NRTIs (ABC, 3TC, ddI, d4T) in antiretroviral-naive adults for virological suppression of HIV-1 RNA (< 50 copies/mL) after 32 weeks of treatment. We will incorporate the data from FIRST into the design and fitting priors for the λ_j. To facilitate a more randomized comparison of drug regimens than FIRST, only 50% of subjects enrolled will be allowed to have their physicians choose their drug regimen. To achieve balance with respect all prognostic factors, balanced permuted block randomization could be utilized such that within each treatment group, subjects will be randomized such that roughly 50% will have baseline HIV-1 RNA $< 100,000$ copies/mL and be assigned a PI. Therefore, if the sample size per group were $n = 32$, then $\mathbf{Z}$ will be constructed such that within each treatment group we will have four subgroups of eight subjects each (Figure 6.4). In the first subgroup, all eight subjects will have baseline HIV-1 RNA $< 100,000$ copies/mL and be assigned a PI; for four of these a physician will choose EFV or NVP. The second subgroup will consist of eight subjects who have baseline HIV-1 RNA $< 100,000$ copies/mL, but who are not assigned a PI; again, a physician will choose EFV or NVP for four of them. The pattern continues for the other two subgroups of subjects with baseline HIV-1 RNA $> 100,000$ copies/mL. (Note that 100,000 may not be the median baseline HIV-1 RNA load in the population, so the probability of VS over all subjects may not be well-estimated in this design even if the effects of RNA and NNRTI treatment are additive.)

Powering SECOND for a trial outcome

In order to evaluate the power of SECOND for a given per-group sample size n, we need to sample the λ_j from design priors, compute π_i following equation (6.17), generate fake data values Y_i^* in R from the binomial likelihood, compute the posterior of Δ given the Y_i^* and the fitting priors for

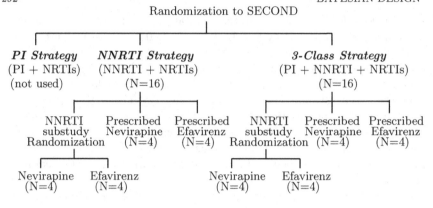

Figure 6.4 *Outline of randomization scheme to SECOND based on $n = 32$.*

the λ_j by using BRugs in R, and evaluate the 95% posterior credible interval for Δ. The conclusion of the trial for this fake dataset is determined by the position of this interval relative to the indifference zone, as shown in Figure 6.1. If we repeat this procedure for $j = 1, ..., N_{rep}$ iterations and collect the N_{rep} decisions in a vector in R, then for sufficiently large N_{rep} we can estimate the empirical probability of each of the six outcomes.

We may perform Bayesian sample size calculations for SECOND with two types of design priors. First, we might employ a classical interpretation and think of each λ_j as a fixed real number. This corresponds to a degenerate point prior, having some prespecified mean but zero variance. On the other hand, a fully Bayesian design prior can reflect uncertainty in the λ_j through a nondegenerate distribution. Given the rough comparability of the two trials, we will assume that the effect of choosing treatment, having baseline HIV-1 RNA $< 100,000$, and being assigned the composite strategy regimen (PI+NNRT) experienced in FIRST will be exactly duplicated in SECOND. Frequentist point priors for the nuisance parameters $(\lambda_3, \lambda_4, \lambda_5)$ then follow directly from Table 6.3 as normal distributions with zero variance: $\lambda_3 \sim N(0.3195, 0)$, $\lambda_4 \sim N(0.5407, 0)$, and $\lambda_5 \sim N(-0.2890, 0)$. Now we must model the true treatment effect differences of interest through the design prior. In our approach the prior mean for λ_2 is fixed as -0.0644, the estimate from FIRST in Table 6.3, for all "true" differences. Then the prior mean for λ_1 is incremented from -0.0644 by "true" β values of $log(1)$, $log(1.1)$, $log(1.2)$, $log(1.3)$, $log(1.4)$, $log(1.5)$, and $log(1.6)$. Thus, $\lambda_2 \sim N(-0.0644, 0)$, $\lambda_1 \sim N(\beta - 0.0644, 0)$, and sample size calculations will correspond to fixed "true" Δ values of 1, 1.1, 1.2, 1.3, 1.4, 1.5, and 1.6. Notice that when the true β equals $log(1.4)$, this model assigns the prior means $(-0.0644, 0.2721)$ to (λ_1, λ_2), roughly the same values seen in FIRST. An alternative approach would be to configure the design prior means relative to the FIRST regression coefficient for average treatment, roughly $(0.287 - 0.0644)/2$. Our approach illustrates how a true treatment difference can be incorporated into the model. Fully Bayesian versions of these design priors would also incorporate variance estimates from FIRST, using the squares of the "sd" values in Table 6.3 as the variances in the above normal distributions. A more sophisticated approach here would incorporate the aforementioned pairwise correlations within the λ vector, say, within a five-dimensional multivariate normal distribution.

Turning to the choice of fitting priors, we compare results under choices that are optimistic, skeptical, and noninformative (reference prior) with respect to the superiority of EFV for VS. Given that the results from FIRST suggest a significant preference of EFV over NVP, an "optimistic" model might assign independent normal fitting priors to the λs with means and variances obtained from Table 6.3. The model skeptical to the superiority of EFV would instead assume that the effects of EFV and NVP are equivalent by assigning the mean estimate for λ_2 from FIRST, -0.0644, to the fitting priors for *both* λ_1 and λ_2, as well as a pooled variance estimate. Fitting priors for the other three nuisance regression parameters for the skeptical model mirror those in the optimistic model. Figure 6.3 gives BUGS code for the skeptical (second column) and optimistic (third column) models. Readers new to BUGS should note well that the second parameter in any normal specification in this language is a *precision* (reciprocal variance), not a variance itself.

Next, we must set bounds (δ_L, δ_U) for the indifference zone, against which we compare the 95% posterior interval for fake dataset j, $(\Delta_{Lj}^*, \Delta_{Uj}^*)$, as illustrated in Figure 6.1. In this figure, NVP plays the role of "control" and EFV plays the role of "treatment." In order to achieve balance with respect to empirical probabilities of the six outcomes when the true Δ is 1, we need to evaluate each $(\Delta_{Lj}^*, \Delta_{Uj}^*)$ within an indifference zone that is symmetric about 0 with respect to β, yet asymmetrical about 1 with respect to $\Delta \equiv e^\beta$. As such, we define $\delta_U = 1 + \phi$ and $\delta_L = 1/(1 + \phi)$ for $\phi > 0$.

We can compute the empirical probability of each of the six outcomes by collecting each of the N_{rep} decisions in a vector in R. Defining $D_j = (d_{1j}, ..., d_{6j})$ as in Figure 6.5, $\overline{D} = \frac{1}{N_{rep}} \sum_{j=1}^{N_{rep}} D_j$ estimates the probabilities of each outcome for given per-group sample size n, design priors, analysis model, "true" β, and indifference zone parameter ϕ. We ran the Gibbs sampler for 1500 iterations following a burn-in of 500 for each of $N_{rep} = 300$ replicate fake datasets. When selecting the proper n and analysis model for SECOND, we need to consider both the empirical probability of satisfying equivalence when the true $\Delta = 1$, as well as the empirical probability of detecting a true difference relative to our indifference zone.

Tables 6.4 and 6.5 summarize the results for the analysis models using the

$d_{1j} = 1$ if $\Delta^*_{Uj} < \delta_L$ (accept control)

$d_{2j} = 1$ if $\Delta^*_{Uj} < \delta_U$ yet $\Delta^*_{Uj} > \delta_L$ and $\Delta^*_{Lj} < \delta_L$ (reject treatment)

$d_{3j} = 1$ if $\Delta^*_{Lj} > \delta_L$ and $\Delta^*_{Uj} < \delta_U$ (equivalence)

$d_{4j} = 1$ if $\Delta^*_{Lj} > \delta_L$ yet $\Delta^*_{Lj} < \delta_U$ and $\Delta^*_{Uj} > \delta_U$ (reject control)

$d_{5j} = 1$ if $\Delta^*_{Lj} > \delta_U$ (accept treatment)

$d_{6j} = 1$ if $\Delta^*_{Lj} < \delta_L$ and $\Delta^*_{Uj} > \delta_U$ (no decision)

Figure 6.5 *Definitions of components of decision rule D_j corresponding to the 95% posterior interval of Δ^*_j in Figure 6.1. In all cases, if the condition is not met, the component is set equal to 0.*

n	Accept NVP	Reject EFV	Equivalence	Reject NVP	Accept EFV	No Decision
320	0.00	0.44	0.00	0.50	0.00	0.06
480	0.00	0.42	0.21	0.37	0.00	0.00
640	0.00	0.31	0.41	0.28	0.00	0.01
800	0.00	0.18	0.63	0.19	0.00	0.00
960	0.00	0.12	0.74	0.14	0.00	0.00
1120	0.00	0.08	0.82	0.10	0.00	0.00

Table 6.4 *Empirical probabilities for the six outcomes in Figure 6.1 by n for the skeptical analysis model for true $\Delta = 1$, and $\phi = 0.30$.*

skeptical and optimistic fitting priors, respectively, for "true" $\Delta = 1$ and $\phi = 0.30$. Table 6.4 shows that using the skeptical fitting prior corresponds to a balanced empirical probability of accepting or rejecting one NNRTI over the other when the effect of both regimens is equivalent. By contrast, the effect of generating posteriors for Δ with the optimistic fitting prior is evident from Table 6.5. The 95% posterior intervals for Δ fall within the "reject NVP" zone more often than the "reject EFV" zone, even though the true odds of VS are the same in both groups. Notice that as n increases the effect of the optimistic prior diminishes, and this discrepancy is reduced. We emphasize that our analysis here (as elsewhere) is intended as illustrative, and the Type I error rates seen in the tables for the skeptical and optimistic models are likely higher than a regulatory agency would prefer in actual practice.

Both Tables 6.4 and 6.5 suggest that a per-group sample size of $n = 1120$ is not sufficient to deliver an equivalence probability much greater than 0.8 for $\phi = 0.30$. Therefore, Table 6.6 considers larger ϕ for all three fitting

n	Accept NVP	Reject EFV	Equivalence	Reject NVP	Accept EFV	No Decision
320	0.00	0.19	0.00	0.77	0.00	0.04
480	0.00	0.20	0.16	0.64	0.00	0.00
640	0.00	0.13	0.41	0.46	0.00	0.00
800	0.00	0.08	0.52	0.40	0.00	0.00
960	0.00	0.06	0.70	0.24	0.00	0.00
1120	0.00	0.05	0.78	0.17	0.00	0.00

Table 6.5 *Empirical probabilities for the six outcomes in Figure 6.1 by n for the optimistic analysis model for true $\Delta = 1$ and $\phi = 0.30$.*

n	Skeptical Model $\phi = 0.40$	Skeptical Model $\phi = 0.50$	Reference Model $\phi = 0.40$	Reference Model $\phi = 0.50$	Optimistic Model $\phi = 0.40$	Optimistic Model $\phi = 0.50$
320	0.43	0.73	0.05	0.36	0.30	0.62
480	0.71	0.90	0.42	0.70	0.55	0.81
640	0.80	0.93	0.72	0.93	0.75	0.93
800	0.93	0.98	0.83	0.95	0.84	0.96
960	0.95	1.00	0.88	0.98	0.93	0.99
1120	0.98	1.00	0.95	0.99	0.96	1.00

Table 6.6 *Empirical probabilities of equivalence in Figure 6.1 for the skeptical, reference, and optimistic analysis models by n and ϕ for true $\Delta = 1$.*

priors. The gain in specificity associated with the skeptical fitting prior is now apparent (though this may well be tempered by a relatively large Type I error rate). Estimated empirical probabilities for an equivalence conclusion when the true $\Delta = 1$ are uniformly greater under the skeptical model relative to the optimistic, and the gains are even larger relative to the reference model, though the effects have again dissipated by $n = 1120$. The gain associated with the skeptical model is more pronounced for smaller n. Interestingly, the optimistic model slightly outperforms the reference model, probably because the former gives narrower credible intervals. There is also apparently some disproportionate shifting of the 95% posterior intervals for Δ into and out of the indifference zone determined by $\phi = 0.4$ and 0.5 when using the optimistic prior as compared to the reference prior.

We now turn our attention from the "null" case of $\Delta = 1$ to the case of detecting an actual difference in the odds of VS. Figure 6.6 plots the empirical probabilities of accepting EFV by true Δ for $\phi = 0.20$. Each curve

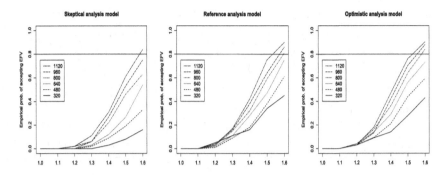

Figure 6.6 *Plots of empirical probability of accepting EFV by Δ for analysis models with skeptical, reference, and optimistic fitting priors, and using $\phi = 0.20$.*

corresponds to a particular sample size n. We see that the optimistic and reference models outperform the skeptical model for smaller n. For example, for the skeptical model, only one curve, $n = 1120$, intersects the horizontal reference line at probability 0.80 for $\Delta \leq 1.6$. Yet for the reference and optimistic models, curves representing $n = 960$ and $n = 800$ also reach probability 0.80. Curves produced from the optimistic prior appear only slightly more powerful than curves generated with the reference prior. This suggests that the optimistic fitting prior has marginal impact on this case (accept EFV) for large n.

Curves more analogous to traditional frequentist power curves can be constructed by summing across the acceptance and rejection outcomes (i.e., all outcomes except equivalence and no decision). The resulting "Bayesian power" curves from our simulations are displayed in the first row of Figure 6.7 for $\phi = 0.30$ and three per-group sample sizes, 160, 640, and 1120. Such plots allow us to assess the gain from incorporating our historical data from FIRST into the design of SECOND. The second row of this figure plots only the equivalence probabilities versus Δ.

Consider first the case where $\Delta = 1$. Type I error corresponds to a decision of nonequivalence. Recall that the indifference zone permits us to distinguish "absence of evidence" from "evidence of absence." As seen in the first plot of the second row, the sample size $n = 160$ is too small to deliver sufficiently narrow confidence intervals for an equivalence decision to ever be reached. As n increases, the confidence intervals narrow, resulting in a higher proportion of decisions, and equivalence is concluded more and more for the smaller Δ values. This is demonstrated in the symmetry that occurs when comparing the pairs of plots vertically for $n = 640$ and $n = 1120$. For $\Delta > 1$, we observe significant gain in power under the optimistic prior for the smallest n.

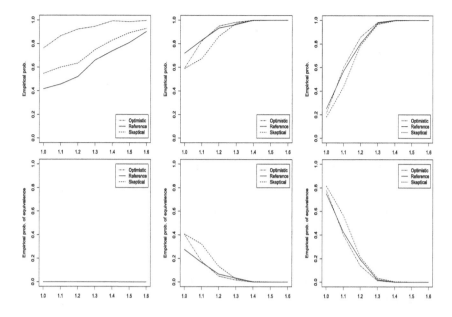

Figure 6.7 *Comparison of empirical probabilities of (top) nonequivalence decision and (bottom) equivalence decision by true* Δ *for analysis models with skeptical, reference, and optimistic fitting priors for* $\phi = 0.30$ *and per-group sample sizes (left)* $n = 160$, *(center)* $n = 640$, *and (right)* $n = 1120$.

6.3.2 Cox regression device trial with interim analysis

We now turn to the application of Bayesian methods in medical device trials. When reliable prior information on the treatment group is not available, the Bayesian approach may incorporate valuable information into trial analysis and design through the use of a historical control group. Such a group consists of the data observed from a previous investigation of the control therapy, or perhaps from data generated from a prior distribution trained by previous control therapy data. Typically some contemporaneous control data will also be collected during the trial. Pocock (1976) suggests, "The major problem with historical controls is that one is unable to ensure comparability between the groups of patients and methods of evaluation for the new and standard treatment." Yet, if the study populations and data collection methods are similar, and the trial periods are relatively near temporally, then pooling historical and current subjects may offer the best approach (Neaton et al., 2007).

In the device setting, where interventions are costly to produce and maintain, sample sizes are smaller, surgical risks exist, and randomization may not be ethical or feasible, the advantages of randomizing more patients to

Month	Surviving Fraction	Month	Surviving Fraction
0	1.00	18	0.38
3	0.73	21	0.38
6	0.62	24	0.23
9	0.56	27	0.06
12	0.53	30	0.06
15	0.44		

Table 6.7 *Surviving fraction by month, LVAD group, REMATCH trial.*

receive the "new" intervention are evident. The Bayesian design method discussed in this section easily handles sample size calculations for simultaneous use of both historical and contemporaneous controls.

Design and analysis of the REMATCH LVAD trial

We illustrate via a comparison of survival for patients randomized to two implantable ventricular assist devices, which support patients with end-stage congestive heart failure. Implantable circulatory-support devices used to support the left ventricle alone are known as LVADs (left ventricular assist devices). Patients awaiting cardiac transplantation have experienced short-term benefits from LVADs serving as a bridge to transplant. However, these devices have recently shown promise as long-term myocardial replacement therapies for patients ineligible for cardiac transplantation.

The Randomized Evaluation of Mechanical Assistance for the Treatment of Congestive Heart Failure study group (REMATCH) was the first to assess LVADs as permanent or "destination" therapies for the enhancement of survival and quality of life in a randomized, prospective trial (Rose et al., 2001). Patients with end-stage heart failure and contraindications to transplantation were randomized to receive LVAD ($n = 68$) or "optimal medical management" ($n = 61$) and monitored for 30 months. A committee developed medical therapy strategies for patients randomized to the optimal medical management group aimed at optimizing organ perfusion and minimizing symptoms of congestive heart failure.

The results suggest LVADs provide a significant improvement in survival. Kaplan-Meier analysis demonstrated a 48% reduction in the risk of death from any cause in the LVAD group compared to the medical management group (relative risk, 0.52; 95% CI [0.34, 0.78]; frequentist $p = 0.001$), an estimated 12-month survival of 52% versus 25%, and median survival of 408 versus 150 days. Of the fatalities in the medical management group, 93% were caused by left ventricular dysfunction, while 41% of the deaths in the LVAD group were due to sepsis, 17% were due to device failure, and only

2% were due to left ventricular dysfunction. Table 6.7 provides a summary of survival for patients randomized to the LVAD group by month.

Using the REMATCH LVAD trial to design a new trial

In March 2006, the Heart Failure Society of America (HFSA) and the FDA convened a workshop to discuss proper designs for mechanical circulatory support devices, including alternatives to randomized trials. Researchers have found it challenging to recruit patients to such studies, due to the paucity of suitable clinical sites and general reluctance to randomize patients to older, control devices that may be technologically inferior. In the summary report of one workshop panel, Neaton et al. (2007) discuss alternatives to fully randomized designs, including allowing the control device to be chosen prior to randomization, conducting smaller initial vanguard studies, and allowing crossovers in trials with optimal medical management controls. They also advocate trials that incorporate both randomized and nonrandomized components, such as comprehensive cohort designs (see, e.g., Olschewski et al., 1992). They acknowledge that prior opinion regarding certain variance parameters may be necessary, but stop short of recommending Bayesian analytic methods. We proceed with a partially randomized Bayesian design that utilizes historical (nonrandomized) controls, either observed or generated from a prior distribution trained by the observed historical data, as well as concurrent (randomized) controls.

Now suppose that shortly after the LVAD used in the REMATCH trial (LVAD1) was shown to be efficacious, a new enhanced device (LVAD2) is developed. The developers wish to compare LVAD1 (control) to LVAD2 (treatment) in a prospective randomized trial called COMPARE with design congruent to that of the REMATCH trial. Given the propinquity of the two trials and the equivalency of eligibility criteria, ignoring prior information from the 68 LVAD1 patients in the REMATCH study appears suboptimal. Also, given the nature of LVAD2 it may be unethical to subject patients to the risks of the implantation procedure to achieve the benefit of the older LVAD1. Therefore, COMPARE will terminate "early" if posterior inference implies that LVAD2 is superior, LVAD1 is superior, or the effect of LVAD2 is equivalent to LVAD1. We will obtain prior information for COMPARE from the article by Rose et al. (2001), not from our own analysis of the historical data (which is unavailable to us).

We use the MCMC-Bayes formulation of Cox regression explained in the BUGS manual in the context of the Leuk example (on the WinBUGS website, www.mrc-bsu.cam.ac.uk/bugs/documentation/exampVol1/node29.html) as the statistical model for our analysis of survival outcomes among patients in the LVAD1 and LVAD2 groups. The Leuk example is based on the model of Clayton (1994), who formulates the Cox model using the counting process notation introduced by Andersen and Gill (1982) in order to facilitate MCMC estimation of the baseline hazard and regression parameters. In this formulation, let z_i be the covariate, $z_i = 1$ for LVAD2 and 0 for LVAD1, and ω_i be the death indicator for the ith subject (1 if dead, 0 if right-censored), for $i = 1, \ldots, n$. Let t_i be the observed death time rounded to the nearest day for the ith subject such that $t_i \leq 900$ days if $\omega_i = 1$, and $t_i = 900$ days if $\omega_i = 0$. Data V_i for the ith subject consist of the triple $V_i = (t_i, z_i, \omega_i)$.

The statistical model supposes that death times follow a *counting process*. Let $N_i(t)$ be an indicator for whether subject i dies at or before time t, $N_i(t) = I[t_i \leq t, \omega_i = 1]$. The counting process $N(t) = \sum_{i=1}^{n} N_i(t)$ counts the number of deaths that have occurred up to time t. Then $N_1(t), N_2(t), \ldots,$ $N_n(t)$ are assumed independent and identically distributed, and $N(t)$ is right-continuous and piecewise constant with jumps of size $+1$. Furthermore, $N(0) = 0$ and $N(t) < \infty$. Let $I_i(t)$ be the corresponding intensity process at time t for the ith patient given by $I_i(t)dt = E(dN_i(t)|F_{t-})$, where $dN_i(t)$ is the increment of N_i over the small time interval $[t, t + dt)$, and F_{t-} represents the available data just before time t. We can parse t into small intervals such that $dN_i(t) = 1$ if subject i dies during this time interval and 0 otherwise. Thus, $E(dN_i(t)|F_{t-})$ corresponds to the probability of subject i failing in the interval $[t, t + dt)$. As $dt \to 0$, then this probability becomes the instantaneous hazard at time t for the ith subject.

Let the observed process $Y_i(t) = 1$ if subject i is observed at time t and 0 otherwise. We assume proportional hazards,

$$I_i(t) = Y_i(t)\lambda_0(t)\exp(\beta z_i) , \qquad (6.19)$$

where $\lambda_0(t)\exp(\beta z_i)$ is the semi-parametric regression model given by Cox and Oakes (1984), $\lambda_0(t)$ is the hazard rate for patients in the LVAD1 group (baseline hazard rate), β the regression parameter, and e^β represents the relative risk or hazard ratio of a patient with LVAD2 versus LVAD1. Let $\Delta = e^\beta$ be the parameter used for inference. Note that $\Delta = 1$ means no difference in hazard rate between LVAD2 and LVAD1; we expect $\Delta < 1$, reduced hazard for LVAD2. In what follows we use a flat fitting prior for β, and estimate the cumulative hazard $\Lambda_0(t) = \int_0^t \lambda_0(u)du$ nonparametrically.

Writing the entire dataset as $\mathbf{G} = \{(N_i(t), Y_i(t), z_i)\}_{i=1}^{n}$, the joint posterior for β and $\Lambda_0(\cdot)$ is $p(\beta, \Lambda_0(\cdot)|\mathbf{G}) \propto p(\mathbf{G}|\beta, \Lambda_0(\cdot))p(\beta)p(\Lambda_0(\cdot))$. We need only to specify the form of the likelihood $P(\mathbf{G}|\beta, \Lambda_0(\cdot))$ and prior distributions for β and $\Lambda_0(\cdot)$ in our BUGS model. Since noninformative censoring is assumed, the data likelihood is proportional to

$$\prod_{i=1}^{n} \left[\prod_{t \geq 0} I_i(t)^{dN_i(t)} \right] e^{-I_i(t)dt} .$$

This suggests that increments of the counting process $dN_i(t)$ in the time interval $[t, t+dt]$ are independent Poisson random variables with mean param-

```
model{
## Set up data ##
for(i in 1:N) {
for(j in 1:T) {
## risk set = 1 if obs.t >= t ##
Y[i,j] <- step(obs.t[i] - t[j] + 0.000001)
## counting process jump = 1 if obs.t in [ t[j], t[j+1] ) ##
##                i.e. if t[j] <= obs.t < t[j+1]    ##
dN[i, j] <- Y[i, j] * step(t[j + 1] - obs.t[i] - 0.000001) * obs[i]
}
}
## Model ##
for(j in 1:T) {
for(i in 1:N) {
dN[i, j] ~ dpois(Idt[i, j])
Idt[i, j] <- Y[i, j] * exp(β * Z[i]) * dΛ0[j]
}
dΛ0[j] ~ dgamma(dΛ0.star[j] * c, c)
}
for (j in 1 : T) {
dΛ0.star[j] <- r * (t[j + 1] - t[j])
}
β ~ dnorm(0.0,0.000001)
Δ <- exp(β)
}
```

Figure 6.8 WinBUGS *code for the Bayesian Cox regression model utilized in the design of COMPARE to estimate the posterior distribution of* Δ_j^*.

eter $I_i(t)dt$. Next, write $I_i(t)dt = Y_i(t)e^{\beta z_i}d\Lambda_0(t)$, where $d\Lambda_0(t) = \Lambda_0(t)dt$ is the increment or jump in the integrated baseline hazard function occurring during the time interval $[t, t + dt])$. Assuming a gamma distribution for the increments $d\Lambda_0(t)$ is convenient since this is the conjugate prior for the Poisson. In particular, we follow Kalbfleisch (1978) and adopt the conjugate independent increments prior $d\Lambda_0(t) \sim Gamma(cd\Lambda_0^*(t), c)$. We parameterize the $Gamma(u, k)$ distribution to have mean u/k and variance u/k^2, so that $d\Lambda_0^*(t)$ functions as a prior guess for $d\Lambda_0(t)$, and c signifies the degree of confidence in this guess (small values corresponding to low confidence). In our model, we take $d\Lambda_0^*(t) = rdt$ where r is a guess for the failure rate per unit time, and dt is the size of the time interval. BUGS code for the Cox model which will be used in BRugs for inference on Δ is given in Figure 6.8. Note that the statistical model utilizes a noninformative fitting prior for β; we also set $c = 0.001$, making r quite arbitrary.

Having established the statistical model, in order to power COMPARE for a decision we need to simulate realizations of **G**. Yet doing so from the counting process formulation by generating $d\Lambda_0(\cdot)$ from a gamma distribution, computing $I_i(t)dt$, and generating $dN_i(t)$ from $Poisson(I_i(t)dt)$ is rather awkward. Instead, we sample inter-event times from a lifetime distribution, and then formulate survival times by cumulatively summing over the generated inter-event times. Specifically, we let $x_i \sim Gamma(\upsilon = u, \kappa = k)$, for $i = 1, ..., n$ and constant u, k, and then generate LVAD1 and LVAD2 survival times as

$$\gamma_n(u, k) = \left(x_1, \sum_{i=1}^{2} x_i, ..., \sum_{i=1}^{n-1} x_i, \sum_{i=1}^{n} x_i \right).$$

This is referred to as a *gamma(u,k) process of size n*. The variability of $\gamma_n(u, k)$ depends heavily on the shape of inter-event time distribution. The gamma distribution becomes increasingly right-skewed as the shape parameter υ decreases toward zero. This produces a $\gamma_n(u, k)$ with less smoothness. Note that the median survival of $\gamma_n(u, k)$ depends upon the size of n.

Since prior information is available only through the summary information in the published literature, for this illustration we create a single historical control dataset that emulates the survival experience for the 68 subjects in Table 6.7. First we must identify the historical gamma process that is consistent with our data. After some experimentation, we found that a $\gamma_{68}(u = 0.6, k = 0.0499)$ process fit reasonably well. Survival summaries for 200 generations from this gamma process are displayed in the first plot in Figure 6.9. As a further check, we generated a sample of 100,000 median survival times from our $\gamma_{68}(u = 0.6, k = 0.0499)$. The $(0.025, 0.50, 0.975)$ quantiles for this sample were $(256, 408, 610)$ days, respectively, the median being in perfect agreement with that observed in REMATCH, also 408 days.

In the remainder of this section, we fix a single simulation from our $\gamma_{68}(u = 0.6, k = 0.0499)$ as the historical LVAD1 dataset, and refer to it as the *genuine* historical data. We take this dataset, shown in the center plot of Figure 6.9 along with the empirical REMATCH curve, as our assumed LVAD1 REMATCH survival data.

Conventionally, COMPARE would be designed assuming that the result from the REMATCH trial is "centered" with respect to the true survival effect of LVAD1. Yet, the survival curve from REMATCH may actually underestimate or exaggerate the true survival effect of LVAD1, leading to a design that is under- or overpowered. Therefore, we consider sample size calculations using two more historical control datasets, one "enthusiastic" and the other "pessimistic" relative to the survival experienced for patients on LVAD1 in REMATCH. Note the distinction with Section 6.3.1, where the "optimistic" and "skeptical" fitting priors corresponded to a treat-

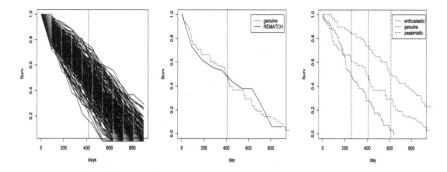

Figure 6.9 *(left) 200 empirical survival curves generated from a $\gamma_{68}(u = 0.6, k = 0.0499)$ process; (center) survival curve from Table 6.7 (solid) and survival data for the "genuine" historical dataset (dashed); (right) survival data by month for the "genuine", "enthusiastic", and "pessimistic" historical datasets. The vertical lines in all plots indicate median survival times.*

ment difference; here, prior information is incorporated only for the control group, LVAD1, since no prior data exists for LVAD2. We used the estimated $(0.025, 0.975)$ quantiles of median survival for $\gamma_{68}(u = 0.6, k = 0.0499)$ to determine the degree of pessimism or enthusiasm. Based on our earlier work, we simply chose two randomly generated $\gamma_{68}(u = 0.6, k = 0.0499)$ datasets having median survivals of 256 and 610 days, respectively, as our pessimistic and enthusiastic datasets. Survival curves for all three historical datasets are plotted together in the rightmost panel of Figure 6.9.

Powering COMPARE for a trial outcomes with interim evaluations

In order to compute empirical probabilities for the six trial outcomes in Figure 6.1, we first need to designate a "baseline" gamma process, $\gamma_{n_0}(\upsilon_0, \kappa_0)$. This design gamma process should be variable enough to simulate all conceivable results for subjects randomized to LVAD1 in the upcoming trial. Let n again denote sample size per group. The historical datasets described in the previous paragraph contain survival times for 68 subjects (the number randomized to LVAD1 in the REMATCH trial). Thus when computing empirical probabilities for the six trial outcomes, simulated events for $n-68$ contemporaneous controls need to be drawn from the baseline gamma process and merged with the historical control data to form the aggregated control group. Events will be simulated for all n subjects in the LVAD2 treatment group from a gamma process with the same shape parameter υ_0, but scale parameter κ_0 multiplied by e^{β}. This family of gamma processes serves as the design prior for survival. We also need to select the gamma process parameters, υ_0 and κ_0, and the LVAD2 effect, β. Similar

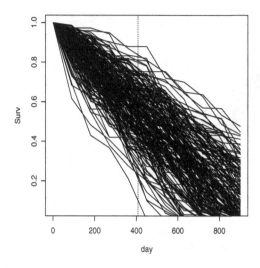

Figure 6.10 *200 empirical survival curves generated from the design gamma process,* $\gamma_{300}(0.08, 0.0255)$.

to our Section 6.3.1 approach, we treat these as fixed design constants. Our Bayesian procedure will estimate β from the fake data, but only using a flat fitting prior, because we lack any sort of prior information regarding LVAD2. Note that this means we are treating the genuine historical data and the COMPARE control data as if they are fully exchangeable; a more circumspect fitting prior might allow for the possibility of partial exchangeability. We also simplify matters by ignoring any differential lengths of follow-up caused by the trial's rate of enrollment, which could affect the trial's operating characteristics.

We select a $\gamma_{n_0=300}(v_0 = 0.08, \kappa_0 = 0.0255)$ design baseline gamma process for COMPARE, having inter-event time mean and standard deviation, 3.14 and 11.09 days, respectively. We presume that no more than 300 subjects per group could possibly be enrolled in COMPARE. From Figures 6.9 and 6.10 it appears that this baseline gamma process covers all reasonable survival curves (including the genuine, enthusiastic, and pessimistic curves) for subjects assigned to LVAD1 in a trial spanning nine months. To confirm this impression, we generated a sample of 100,000 survival medians from the $\gamma_{300}(0.08, 0.0255)$ process, and found the $(0.025, 0.975)$ quantile interval to be wider than that for the historical $\gamma_{68}(0.6, 0.0499)$ process, namely $(240, 776)$ versus $(256, 610)$.

We assume interim evaluations will take place after events for $n = 100, 150, 200, 250,$ and 300 patients (per group) occur. For a given historical LVAD1 dataset and fixed true β, fake datasets V_j^*, $j = 1, ..., N_{rep}$, each consisting of $i = 1, \ldots, 2n_{max} = 600$ survival events $V_{ij}^* = (t_{ij}^*, z_{ij}^*, \omega_{ij}^*)'$, are generated in R, where we write

$$
V_j^* = \begin{pmatrix} t_{1j}^* & z_{1j}^* & \omega_{1j}^* \\ \vdots & \vdots & \vdots \\ t_{600j}^* & z_{600j}^* & \omega_{600j}^* \end{pmatrix}.
$$

Within each V_j^*, 68 survival events are fixed historical events, $300 - 68 = 232$ are contemporaneous LVAD1 events generated from the $\gamma_{300}(0.08, 0.0255)$, and 300 are LVAD2 events generated from $\gamma_{300}(0.08, 0.0255e^\beta)$.

For each j, up to five posterior inferences on Δ_j^* are possible, corresponding to the five interim evaluations. Initially, 32 survival events are randomly sampled from the 232 concurrent controls ($z_{ij} = 0$) in V_j, and added to the 68 historical controls in the aggregate control group (LVAD1). Then $n = 100$ survival events are randomly sampled from the 300 LVAD2 events. Let $k = 1, ..., 5$ index the five interim sample sizes. From the $k = 1$ fake dataset, the posterior distribution of $\Delta_{jk}^* = e^{\beta_{jk}^*}$ is estimated using the Cox model BRugs code in Figure 6.8. For given δ_L and δ_U, if the bounds of the 95% posterior interval for $\Delta_{j,k=1}^*$ enable us to accept LVAD2, accept LVAD1, or conclude equivalence as indicated by Figure 6.5 (where LVAD2 plays the role of "control" and LVAD1 the role of "treatment"), then the jth fake "trial" is terminated at $n = 100$. Otherwise, 50 more concurrent LVAD1 and LVAD2 events are sampled, merged with the existing data, and the 95% posterior interval for $\Delta_{j,k=2}^*$ is evaluated. This process is repeated for $k = 3, 4$, and 5, where all 6 decisions (including "no decision" and the two "rejections") become possible for $k = 5$, since the trial *must* be stopped then.

Note that smaller values of Δ suggest improved survival for subjects assigned to LVAD2. Because we again want the indifference zone to be symmetric, we again let $\delta_U = 1 + \phi$ and $\delta_L = 1/(1 + \phi)$ for $\phi > 0$. Then the magnitude of ϕ signifies the percent improvement in hazard one LVAD must demonstrate in order for it to be considered superior to the other. Data monitoring allows us to compute empirical probabilities for all 18 possible outcomes by collecting each of the N_{rep} decisions in a matrix, only a slight generalization of our Section 6.3.1 approach.

We ran a single Gibbs chain for 2000 iterations after a burn-in of 1000 iterations for $N_{rep} = 500$ fake datasets. Tables 6.8 and 6.9 summarize the results, where again we emphasize their illustrative nature despite the occasionally high Type I error rates they imply. Beginning with the "genuine" data section of Table 6.8, it is evident that when true $\Delta = 1$, $\gamma_{300}(0.08, 0.0255)$ is somewhat "unbalanced," with LVAD1 being more (less) likely to be rejected (accepted) than LVAD2. This may be due to higher variability in the LVAD2 data; $n = 68$ of the LVAD1 controls are always taken from the "genuine" historical data, while all of the LVAD2 data arises from the design prior in Figure 6.10. More specifically, when

| | Accept | Reject | | Reject | Accept | No |
n	LVAD2	LVAD1	Equivalence	LVAD2	LVAD 1	Decision
With *enthusiastic* historical controls:						
100	0.02	—	0.14	—	0.23	—
150	0.04	—	0.20	—	0.00	—
200	0.06	—	0.14	—	0.00	—
250	0.02	—	0.04	—	0.00	—
300	0.03	0.02	0.02	0.04	0.00	0.00
With *genuine* historical controls:						
100	0.15	—	0.22	—	0.02	—
150	0.08	—	0.07	—	0.01	—
200	0.07	—	0.03	—	0.01	—
250	0.05	—	0.03	—	0.01	—
300	0.03	0.13	0.01	0.07	0.00	0.00
With *pessimistic* historical controls:						
100	0.53	—	0.09	—	0.00	—
150	0.03	—	0.07	—	0.01	—
200	0.02	—	0.04	—	0.02	—
250	0.01	—	0.02	—	0.01	—
300	0.00	0.11	0.02	0.01	0.01	0.00

Table 6.8 *Empirical probabilities of the* 18 *possible COMPARE outcomes using the baseline gamma process* $\gamma_{300}(0.08, 0.0255)$, *true* $\Delta = 1$, *and* $\phi = 0.50$.

$n = 100$, 68% of the controls are from REMATCH, but for $n = 300$ this percentage drops to 23%. Relatedly, the pessimistic historical control data appear more pessimistic (e.g., 53% LVAD2 acceptance for $n = 100$) than the enthusiastic data is enthusiastic (23% LVAD1 acceptance for this same n). Indeed, the total empirical probability of accepting LVAD2 when the true $\Delta = 1$ is roughly 0.6 for the pessimistic historical control data. In Table 6.9 where the true Δ is now 0.50 (clear improvement under LVAD2), most empirical probability lies within the accept LVAD2 region for all three historical control datasets. The pessimistic data delivers an empirical probability of 0.93 for LVAD2 acceptance after just one interim evaluation. By contrast, the enthusiastic data achieves an empirical LVAD2 acceptance probability of only 0.23 for $n = 100$, but this increases to 0.70 once we sum over all n. This shows the Bayesian method's ability to stop early when conditions warrant while maintaining full flexibility in the design.

We close this section with a brief mention of one potential downfall of our approach. Consider the case in which both randomized concurrent LVAD1

n	Accept LVAD2	Reject LVAD1	Equivalence	Reject LVAD2	Accept LVAD1	No Decision
With *enthusiastic* historical controls:						
100	0.23	—	0.10	—	0.01	—
150	0.26	—	0.08	—	0.00	—
200	0.11	—	0.03	—	0.00	—
250	0.07	—	0.01	—	0.00	—
300	0.03	0.06	0.00	0.01	0.00	0.00
With *genuine* historical controls:						
100	0.68	—	0.02	—	0.00	—
150	0.10	—	0.02	—	0.00	—
200	0.05	—	0.01	—	0.00	—
250	0.02	—	0.01	—	0.00	—
300	0.01	0.07	0.00	0.01	0.00	0.00
With *pessimistic* historical controls:						
100	0.93	—	0.00	—	0.00	—
150	0.01	—	0.00	—	0.00	—
200	0.00	—	0.01	—	0.00	—
250	0.00	—	0.00	—	0.00	—
300	0.00	0.04	0.01	0.00	0.00	0.00

Table 6.9 *Empirical probabilities of the* 18 *possible COMPARE outcomes using the baseline gamma process* $\gamma_{300}(0.08, 0.0255)$, *true* $\Delta = 0.67$, *and* $\phi = 0.50$.

and LVAD2 in patients in COMPARE perform worse for survival than the LVAD1 patients in REMATCH, yet the concurrent LVAD2 arm significantly outperforms the concurrent LVAD1 arm. COMPARE might then wrongly conclude that LVAD1 is equivalent or even superior to LVAD2 in the first few interim evaluations when n is small and the historical control group is more "informative." Such trials could benefit from future research on *temporally adaptive equivalence zones*, i.e., indifference zone boundaries that evolve over interim evaluations. For example, one could start with four boundaries that produce a narrow equivalence zone yet initially make it difficult to accept one treatment arm over the other. As more data accumulates, the equivalence and acceptance boundaries could converge to the indifference zone proposed by Freedman et al. (1984).

6.4 Exercises

1. Actually carry out the analysis in Example 6.4, where the prior $g(\theta)$ in (6.6) is

   (a) the $Unif(0,1)$ prior used in the example, $g(\theta) = 1$.

   (b) a mixture of two $Beta$ priors, $g(\theta) = .5 \cdot 3\theta^2 + .5 \cdot 30\theta^2(1 - \theta^2)$.

   What is the optimal sample size $\tilde{n}$ in each case? Is either prior amenable to a fully analytical solution (i.e, without resort to Monte Carlo sampling)?

2. Using equations (6.2) and (6.4), verify that $\tilde{n} \geq n_\theta$. Will this relation hold for all G? (*Hint:* see equation (6.3).)

3. Consider again the clinical trial design setting of O'Hagan and Stevens (2001), as described in Subsection 6.2.2.

   (a) Confirm that expression (6.11) is the result of the analysis objective $P(\beta > 0|\mathbf{y}) > \omega$ for this model. (*Hint:* Use expression (4.2), the Lindley and Smith (1972) result on the normality of the posterior in normal linear models with known variances, as well as standard formulae for linear transformations of multivariate normal distributions.)

   (b) Confirm that expression (6.12) is the result of the design objective (6.10) for this model. (*Hint:* This time use expression (4.1), the Lindley and Smith (1972) result on the normality of the *marginal* distribution of the data in normal linear models with known variances.)

   (c) Confirm that expression (6.14) is the explicit solution to the the frequentist version of this Bayesian sample size problem in the above setting when we insist $n = n_1 = n_2$.

4. Consider a clinical trial of a cholesterol-lowering drug, where we wish to use equal sample sizes in the experimental and control arms. Assume that LDL values (Y) in the population are $Y \sim Normal(\mu, \sigma^2)$, where the population variance is assumed known $(\sigma = 4.5)$. We consider the trial result to be positive if we obtain a posterior probability of at least ω that the mean LDL level of patients taking the new drug, μ_X, is less than patients taking the control drug, μ_C. Letting $\beta = \mu_C - \mu_X$, we set $H_0 : \beta = 0$ and $H_a : \beta > 0$. Suppose we set $\omega = 0.95$ and seek a procedure having Bayesian assurance of roughly $\delta = 0.80$.

   Following the approach of O'Hagan and Stevens (2001) as outlined in Subsection 6.2.2, and considering *only* clinical efficacy (not cost),

   (a) Calculate the necessary sample size from a frequentist point of view. Specifically, create a sample size table assuming fixed true differences in the means of 5, 10, 15, and 20 units.

(b) Next, take a fully Bayesian approach and reestimate the sample size after marginalizing over a $Normal(7.5, 2^2)$ design prior on β. (You may again use a vague analysis prior for β.)

(c) Finally, change the clinical endpoint to a binary variable, with success $(Y = 1)$ being reduction of LDL by at least 15%.

5. Consider again the simple beta-binomial setting of Example 6.5, where we have a binomial response and the preliminary data from Study A given in Table 6.1. We again seek to base posterior inference on the lower 0.025 quantile of the posterior for π, after utilizing the existing data from the n_1 Study A patients and the still-to-be-collected data on n_2 Study B patients.

(a) Write a short R program to estimate the posterior predictive probability of trial success (6.16) assuming $n_2 = 0$ (i.e., no further sampling of patients). What does this say about the information content of Study A, and the decision whether or not to enroll patients in Study B?

(b) Suppose we now wish to use $n_2 = 40$ new patients, but also limit the impact of the Study A data, in order to ensure that the new Study B data drive the trial's outcome. One way to downweight the effect of Study A would be to replace the existing $Beta(X_1 + a, n_1 - X_1 + b) = Beta(111, 8)$ Study B prior by a $Beta(wX_1 + a, w(n_1 - X_1) + b) = Beta(110w + 1, 7w + 1)$ prior for $0 \le w \le 1$. This reduces the Study A effective sample size to $117w$; each Study A patient is now only "worth" a fraction w of a Study B patient. Repeat the calculation of the previous part using $w = 0.5$, and $w = 0.1$. Compute the mean and variance of each of these priors, and investigate their impact on the posterior predictive likelihood of trial success.

(c) In the setting of part (b) above, create the "Bayesian sample size table" whose rows correspond to different downweighting levels w (say, $0.2, 0.4, \ldots, 1.0$), columns correspond to different sample sizes n_2 (say, $20, 40, \ldots, 100$), and whose entries are the simulated Bayesian power (posterior predictive probability of trial success) for the given row-column combination. Can one obtain reasonable power in this trial using a modest additional sample n_2 and also significant downweighting of the Study A data?

(d) Repeat the $n_2 = 40$ calculation in part (b), but now replacing the downweighting of the Study A information with a *shifting* of this information, specifically by using a $Beta(110 + s + 1, 7 - s + 1)$ prior where $0 \le s < 7$. This variant leaves the effective Study A sample size at 117, but shifts to more optimistic $(s > 0)$ or pessimistic $(s < 0)$ levels by increasing or decreasing the number of successes in the Study A data. For the cases $s = 3$ and $s = -3$, again compute the mean and variance of the resulting priors and investigate their impact on the posterior predictive likelihood of trial success.

(e) Repeat the approach of part (b) yet again, but this time replacing the $Beta(a,b)$ prior with the following nonconjugate specification:

$$\phi \sim N(\mu, \tau^2), \quad \text{where } \phi = logit(\pi) .$$

Choose μ and τ^2 to deliver a prior that is broadly consistent with the (non-downweighted, non-shifted) Study A data. Because the posterior will no longer be available in closed form, use BRugs to iteratively call BUGS from R, and thus obtain the necessary $\pi^*_{.025,j}$ values. Are your results roughly comparable to those from part (b)?

## CHAPTER 7

Special methods and models

Having learned the basics of specifying, fitting, evaluating, and comparing Bayesian models, we are now well equipped to tackle real-world data analysis. In this chapter, we outline the Bayesian treatment of several special methods and models useful in practice. We make no claim of comprehensive coverage; indeed, each of this chapter's sections could be expanded into a book of its own! Rather, our goal is to highlight the key differences between the Bayesian and classical approaches in these important areas, and provide links to more complete investigations. As general Bayesian modeling references, we list the recent books by O'Hagan and Forster (2004) and Gelman, Carlin, Stern, and Rubin (2004).

7.1 Estimating histograms and ranks

7.1.1 Bayesian ranking

Performance evaluation burgeons in many areas, including health services (Christiansen and Morris, 1997b; Daniels and Normand, 2006; Goldstein and Spiegelhalter, 1996), drug evaluation (DuMouchel, 1999), disease mapping (Conlon and Louis, 1999; Devine and Louis, 1994), and education (Draper and Gittoes, 2004; Lockwood et al., 2002). Ranking and selection is also fundamental to genomic analysis and many innovations have been generated by this application (Bureau et al., 2005; Carvalho et al., 2007; Efron and Tibshirani, 2002).

Goals of such investigations include valid and efficient estimation of population parameters, such as average performance (over clinics, physicians, health service regions, genes, or other "units of analysis"), estimation of between-unit variation (variance components), and unit-specific evaluations. The latter includes estimating unit specific performance, computing the probability that a unit's true, underlying performance is in a specific region, ranking units for use in profiling and league tables (Goldstein and Spiegelhalter, 1996), and identification of excellent and poor performers.

Bayesian models coupled with optimizing a loss function provide an effective framework for computing nonstandard inferences (such as ranks, percentiles, or histograms) and producing data-analytic performance assessments. (See Appendix B for a review of Bayesian decision theory.) Inferences depend on the posterior distribution, and how the posterior is used should depend on inferential goals. Gelman and Price (1999) showed that no single set of estimates can simultaneously optimize loss functions targeting the unit-specific parameters (e.g, unit-specific means, optimized by the posterior mean) *and* those targeting the ranks of these parameters. For example, as Shen and Louis (1998) and Liu et al. (2004) showed, ranking the unit-specific maximum likelihood estimates (MLEs) performs poorly, as does ranking Z-scores for testing whether a unit's mean equals the population mean.

In some situations, ranking the posterior means of unit-specific parameters can perform well, but in general a structured approach to estimate ranks is needed. Shen and Louis (1998) use squared-error loss (SEL) operating on the difference between the estimated and true ranks. But in many applications, interest focuses on identifying the relatively good (e.g., in the upper 10%) or relatively poor performers, a "down/up" classification. For example, quality improvement initiatives should be targeted at health care providers that have the highest likelihood of being the poorest performers; geography-specific, environmental assessments should be targeted at the most likely high incidence locations (Wright et al., 2003); genes with differential expression in the top 0.1% should be selected for further study.

In this section we develop ranks based on SEL, and also on the down/up classification loss function. We compare performance and show how a mathematical result substantially speeds up both computation and accuracy. We follow Lin et al. (2006) and Paddock et al. (2006), and present an example using data from the U.S. Renal Data System.

Here, we detail the general methodology that involves optimal ranking of units such as single nucleotide polymorphisms (SNPs), haplotypes, environmental factors, and interactions thereof based on estimated values. For example, the parameter θ_k in what follows might be the interaction effect size of SNP k $(k = 1, \ldots, K)$ with a particular environmental factor, such as smoking exposure. Our strategic approach involves ranking the SNPs, for example to identify the top 0.1% of them. If we observed the θs, we would know which SNPs to select for further study, however we must use statistical relations between the underlying θs and the observed data to guide the selection. Bayes and EB ranking/percentiling approaches are very effective for this goal (see Efron et al., 2001; Efron and Tibshirani, 2002; Lin et al., 2006).

To fix ideas, consider the basic, three-stage Bayesian hierarchical model for $\{\eta, \boldsymbol{\theta} = (\theta_1, \ldots, \theta_K), \mathbf{Y} = (Y_1, \ldots, Y_K)\}$:

$$\eta \sim h(\eta); \quad \boldsymbol{\theta} \sim G(\boldsymbol{\theta} \mid \eta); \quad [Y_k \mid \theta_k] \sim f_k(Y_k \mid \theta_k)$$

$$g(\boldsymbol{\theta} \mid \mathbf{Y}) = \frac{\int \{\prod_k f_k(Y_k \mid \theta_k)\} g(\boldsymbol{\theta} \mid \eta) dh(\eta \mid \mathbf{Y})}{\int \int \{\prod_k f_k(Y_k \mid \theta_k)\} g(\boldsymbol{\theta} \mid \eta) dh(\eta \mid \mathbf{Y}) d\boldsymbol{\theta}} \quad (7.1)$$

$$\text{and} \quad g_k(\theta_k \mid \mathbf{Y}) = \frac{\int f_k(Y_k \mid \theta_k) g(\theta_k \mid \eta) dh(\eta \mid \mathbf{Y})}{\int \int f_k(Y_k \mid u) g(u \mid \eta) dh(\eta \mid \mathbf{Y}) du}.$$

Here, η represents unknowns that specify the prior distribution, and the Y_k are independent conditional on $\boldsymbol{\theta}$. Generalizations include a regression model in the prior $\{G(\boldsymbol{\theta} \mid \mathbf{X}, \eta)\}$, correlations among SNPs, and non-exchangeable priors to identify "favored" or "candidate" SNPs. Lin et al. (2006) document our principal mathematical and simulation-based results on loss function based ranks for model (7.1) with η assumed known. Because evaluations are for moderate to large K, this restriction has essentially no impact.

Ranking: Represent the ranks by

$$R_k(\boldsymbol{\theta}) = \text{rank}(\theta_k) = \sum_{j=1}^{K} I_{\{\theta_k \geq \theta_j\}}; \ P_k(\boldsymbol{\theta}) = R_k(\boldsymbol{\theta})/(K+1). \quad (7.2)$$

The smallest θ has rank 1 and the largest has rank K (see Laird and Louis, 1989). The ranks (7.2) are monotone transform invariant (e.g., ranking the logs of parameters produces the original ranks), a desirable property for estimated ranks.

We don't get to see the underlying θs, and so must use statistical relations among them and the data $\mathbf{Y}$ to produce effective estimates. In general, different statistical summaries will produce different estimated ranks. For example, if all θ_k are equal, units with estimated θs that have relatively high variance will tend to be ranked at the extremes when ranks are based on these estimates; units with estimates that have relatively low variance will tend to be at the extremes when ranks are based on statistics testing the global null hypothesis that all θs are equal.

Appropriate ranks compromise between these two extremes, a compromise best structured by a formal loss function in the Bayesian context. Shen and Louis (1998) showed that when the posterior distributions of the θs are stochastically ordered, maximum likelihood estimate based ranks, posterior mean based ranks, SEL-optimal ranks and those based on most other rank-specific loss functions are identical.

Squared-error loss: Squared error loss (SEL) is optimized by the posterior mean of the target parameter. SEL-optimal ranks result from minimizing $K^{-1} \sum_k \{R_k^{est} - R_k(\boldsymbol{\theta})\}^2$, i.e., by setting R_k^{est} equal to

$$\bar{R}_k(\mathbf{Y}) = E_{\boldsymbol{\theta} \mid \mathbf{Y}}[R_k(\boldsymbol{\theta}) \mid \mathbf{Y}] = \sum_{j=1}^{K} \text{P}(\theta_k \geq \theta_j \mid \mathbf{Y}). \quad (7.3)$$

The $\bar{R}_k$ are shrunk towards the mid-rank, $(K+1)/2$, and generally are not integers (Shen and Louis, 1998) Optimal integer ranks are produced by ranking the $\bar{R}_k$, producing

$$\hat{R}_k(\mathbf{Y}) = \text{rank}(\bar{R}_k(\mathbf{Y})); \ \hat{P}_k = \hat{R}_k/(K+1). \tag{7.4}$$

Classification (above γ/below γ) loss: SEL evaluates general performance without specific attention to identifying the relatively low or high units. Ranks optimizing a 0/1 loss function targeting this goal are computed as follows. Let $0 < \gamma < 1$, $p_{k\ell} = pr(R_k = \ell \mid \mathbf{Y})$, and $\pi_k(\gamma) = \text{P}(P_k > [\gamma K]) = \sum_{\ell=[\gamma K]+1}^{K} p_{k\ell}$. Then

$$\tilde{P}_k(\gamma) = \text{rank}(\pi_k(\gamma))/(K+1). \tag{7.5}$$

The $\tilde{P}_k(\gamma)$ also optimize a loss function that, in addition to 0/1 scoring, computes squared error between the (above γ/below γ) boundary and the estimated percentile. We refer to this as "penalized 0/1 loss."

Ranking by "exceedance" probabilities

For large K, accurately computing $p_{k\ell}$ is very computer intensive and unstable. However, there is a substantially more efficient approach that has the added benefit of relating ranks to a substantive scale. Landrum et al. (2003) and Ribeiro and Diggle (2001) rank using $\text{P}(\theta_k > t \mid \mathbf{Y})$. Let $P_k^*(\gamma)$ be the percentiles induced by ranking the $\text{P}\{\theta_k \geq \bar{G}_K^{-1}(\gamma|\mathbf{Y})\}$, with $\bar{G}_\mathbf{Y}(t) = \frac{1}{K}\sum_{k=1}^{K} \text{P}(\theta_k \leq t|\mathbf{Y})$. Lin et al. (2006) show that $P_k^*(\gamma)$ is asymptotically (in K) equivalent to $\tilde{P}_k(\gamma)$. $\bar{G}$ is easily computed from MCMC output. After burn-in, we compute the empirical distribution function of the pooled θs.

Operating characteristic for (above γ/below γ) classification: The vector of P_k^{est} from any ranking method can be used to classify units into (above γ/below γ) groups and the posterior classification performance (operating characteristic) can be computed. Following Lin et al. (2006),

$$
\begin{aligned}
OC(\gamma \mid \mathbf{Y}) &= pr(P \leq \gamma \mid P^{est} > \gamma, \mathbf{Y}) + pr(P > \gamma \mid P^{est} \leq \gamma, \mathbf{Y}) \\
&= \gamma^{-1} pr(P \leq \gamma \mid P^{est} > \gamma, \mathbf{Y}). \tag{7.6}
\end{aligned}
$$

The first probability in the first line of this expression is similar (but not identical) to a false discovery rate (FDR) (Benjamini and Hochberg, 1995; Efron and Tibshirani, 2002; Efron et al., 2001), and the second to a "false non-discovery rate," but both are always well-defined. $OC(\gamma \mid \mathbf{Y})$ produces a standardized comparison across γ values. For example, when the data are completely uninformative it equals 1; for the model with a Gaussian prior and Gaussian sampling distribution $OC_{\tilde{P}}(\gamma \mid \mathbf{Y})$ is nearly constant in γ with value depending on the ratio of the prior to the sampling variance.

$OC(\gamma \mid \mathbf{Y})$ provides a data analytic performance evaluation. Direct computation sums the $\pi_k(\gamma \mid \mathbf{Y}) = pr(P_k > \gamma \mid \mathbf{Y})$ over a P^{est} identified set of indices. Plotting the $\pi_k(\gamma \mid \mathbf{Y})$ versus the P_k^{est} (Figure 7.1) displays percentile-specific, classification performance (confidence scores). $OC(\gamma)$ is

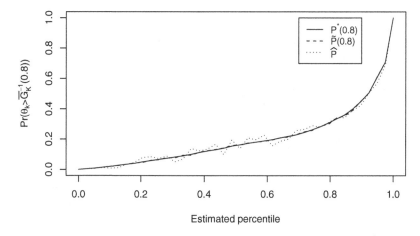

Figure 7.1 $\Pr(\theta_k > 0.18 \mid \mathbf{Y})$ *for three percentiling methods.*

the area between the $\pi_k(\gamma)$ and 1 for the k with $P^{est} \geq \gamma$ plus the area below the $\pi_k(\gamma)$ for $P^{est} < \gamma$. Using $\tilde{P}_k(\gamma)$ or $P_k^*(\gamma)$ produces a monotone plot and the minimum attainable OC. If the data $(\mathbf{Y})$ were completely uninformative, Figure 7.1 would be a horizontal line at the value 0.20; if the data were only weakly informative or the global null hypothesis holds, the plot would start slightly below 0.20 and rise to slightly above that value.

Performance evaluations and comparisons

In a Bayesian model, an estimator's performance can be evaluated by using the posterior distribution (data analytic evaluations) and by using the marginal distribution of the data (preposterior evaluations). For example, preposterior SEL performance is the sum of expected posterior variance plus expected squared posterior bias.

Lin et al. (2006) provide evaluations relative to the loss function used to produce the estimate and other potentially relevant loss functions. For example, performance with respect to SEL should be computed for the SEL optimizer and for other estimators. These comparisons help to determine the efficiency of an estimator that optimizes one loss function when evaluated for other loss functions. Procedures that are robust to the choice of loss functions will be attractive in applications. For loss functions that sum over unit-specific components, performance can also be evaluated for individual units and, in a frequentist evaluation, for individual $\boldsymbol{\theta}$ vectors.

Example 7.1 The standardized mortality ratio (SMR), the ratio of observed to expected deaths, is commonly used to compare health care providers geographic regions and other units. We illustrate the methods of this section using data from the U.S. Renal Data System (USRDS) for the years

1998 to 2001. We use information on the $K = 3173$ dialysis centers that contributed information for all four years. Annually, the USRDS estimates SMRs for several thousand dialysis centers and uses these as a quality screen (Lacson et al., 2001; United States Renal Data System, 2005). For a fair and valid comparison, unit-specific expected deaths must account for age, gender, diabetes, type of dialysis, severity of disease and other factors that are associated with death rates. Challenges include identifying a reference population relative to which we may compute the expected values (providers included, the time period covered), validation of the adjustment model, and attributing deaths to providers.

Our analyses here extend Liu et al. (2004) by combining evidence over multiple years via a Gaussian first-order autoregressive (AR1) model on log(SMR), by using a nonparametric prior, and by including estimates developed by Lin et al. (2006) that minimize errors in classifying providers above or below a percentile cutpoint. For single-year analyses we compare the results from the log-normal prior to those based on the nonparametric maximum likelihood (NPML) prior (Laird, 1978), whose computation via the EM algorithm we elucidate below.

Models

Let (Y_{kt}, m_{kt}) be the observed and expected deaths for provider k in year t, $k = 1, \ldots 3173$, $t = 0, 1, 2, 3$. The m_{kt} are computed from a case mix adjustment based on the hypothesis that all providers give the same quality of care for patients with identical covariates. Let ρ_{kt} be the underlying true SMR. Then

$$Y_{kt} \mid m_{kt}, \rho_{kt} \sim \text{Poisson}(\rho_{kt} m_{kt})$$
$$E(Y_{kt} \mid m_{kt}, \rho_{kt}) = m_{kt}\rho_{kt}.$$

When $\rho_{kt} = 1$ the provider performs as expected.

Let $\boldsymbol{\rho}_t = (\rho_{1t}, \ldots, \rho_{Kt})$ and $\boldsymbol{\rho} = (\boldsymbol{\rho}_0, \boldsymbol{\rho}_1, \boldsymbol{\rho}_2, \boldsymbol{\rho}_3)$. In the general Bayesian model $\boldsymbol{\rho} \sim G$; for the single-year analyses we assume $\boldsymbol{\rho}_t \overset{iid}{\sim} G_t$. To model correlation of ρ_{kt} between years we use a log-normal prior on $\theta_{kt} = \log(\rho_{kt})$. Specifically, we assume $\theta_{kt} \overset{iid}{\sim} N(\xi_t, \tau_t^2)$.

The AR1 Model

Let $\phi = \text{cor}(\theta_{k,t}, \theta_{k(t+1)})$, $-1 < \phi < 1$, and use the hierarchical model,

$$[\theta_{10}, \ldots, \theta_{K0} \mid \xi_0, \tau_0] \overset{iid}{\sim} N(\xi_0, \tau_0^2)$$
$$[\theta_{kt} \mid \theta_{k(t-1)}, \xi, \tau, \phi] \overset{ind}{\sim} N\left(\xi_t + \phi\tau_t\tau_{t-1}^{-1}\{\theta_{k(t-1)} - \xi_{t-1}\}, \{1 - \phi^2\}\tau_t^2\right)$$
$$[Y_{kt} \mid m_{kt}, \rho_{kt}] \overset{ind}{\sim} \text{Poisson}(m_{kt}\rho_{kt}) . \tag{7.7}$$

To complete the hierarchical model, we add $\xi_t \overset{iid}{\sim} N(0, V)$, $\phi \sim h_\phi(\cdot)$, and $\lambda_t = \tau_t^{-2} \overset{iid}{\sim} \text{Gamma}(\alpha, \mu/\alpha)$. Setting $\phi \equiv 0$ in (7.7) produces four year-specific analyses, each using the Liu et al. (2004) model. When $\phi > 0$ we have a standard AR1 model on the latent log(SMR)s and the posterior distribution combines evidence both across dialysis centers within year and within dialysis center across years.

EM algorithm for the NPML

Subsection 5.3.2 described the basic EM algorithm useful in computing the NPML. Here we expand slightly on those ideas and use the notation of this subsection.

Assume $\rho_k \sim G, k = 1, \ldots, K$ with G discrete having at most J mass points $u_1, \ldots, u_J$ with probabilities $p_1, \ldots, p_J$. To estimate the u's and p's, start with $u_1^{(0)}, \ldots, u_J^{(0)}$ and $p_1^{(0)}, \ldots, p_J^{(0)}$ and use the EM algorithm, for the recursion,

$$
\begin{aligned}
w_{kj}^{(v+1)} &= P(\rho_k = u_j^{(v)} | \text{data}) \\
w_{kj}^{(v+1)} &= \frac{\left(m_k u_j^{(v)}\right)^{y_k} e^{-m_k u_j^{(v)}} p_j^{(v)}}{\sum_l \left(m_k u_l^{(v)}\right)^{y_k} e^{-m_k u_l^{(v)}} p_j^{(v)}} \\
p_j^{(v+1)} &= \frac{w_{+j}^{(v+1)}}{w_{++}^{(v+1)}} \\
u_j^{(v+1)} &= \frac{\sum_k w_{kj}^{(v+1)} y_k}{\sum_k w_{kj}^{(v+1)} m_k}.
\end{aligned}
\tag{7.8}
$$

This recursion converges to a fixed point $\widehat{G}$ and, if unique, to the NPML. The recursion is stopped when the maximum relative change in each step for both the $u_j^{(v)}$ and the $p_j^{(v)}$, $j = 1, 2, \cdots, K$ is smaller than 0.001. At convergence, $\widehat{G}$ is both prior and the Shen and Louis (1998) histogram estimate $\widehat{G}_K$.

Care is needed in programming the recursion. The w-recursion is

$$
w_{kj}^{(v+1)} = \frac{\left(m_k u_j^{(v)}\right)^{y_k} e^{-m_k u_j^{(v)}} p_j^{(v)}}{\sum_l \left(m_k u_l^{(v)}\right)^{y_k} e^{-m_k u_l^{(v)}} p_j^{(v)}}.
$$

Because $e^{-m_k u_j^{(v)}}$ can be extremely small $(m_k u_j^{(v)}$ can be extremely large), to stabilize the computations we define,

$$
\bar{\rho}^{(v)} = \sum_j p_j(v) u_j^{(v)},
$$

and write
$$\left(m_k u_j^{(v)}\right)^{y_k} = \exp\left(y_k \log\left(m_k u_j^{(v)}\right)\right).$$

· The w-recursion becomes

$$\begin{aligned}
w_{kj}^{(v+1)} &= \frac{\left(u_j^{(v)}/\overline{\rho}^{(v)}\right)^{y_k} \exp\left(-m_k\left(u_j^{(v)} - \overline{\rho}^{(v)}\right)\right) p_j^{(v)}}{\sum_{l=1}^{J}\left(u_l^{(v)}/\overline{\rho}^{(v)}\right)^{y_k} \exp\left(-m_k\left(u_l^{(v)} - \overline{\rho}^{(v)}\right)\right) p_l^{(v)}} \\
&= \frac{p_j^{(v)} \exp\left(y_k \log\left(u_j^{(v)}/\overline{\rho}^{(v)}\right) - m_k\left(u_j^{(v)} - \overline{\rho}^{(v)}\right)\right)}{\sum_{l=1}^{J} p_l^{(v)} \exp\left(y_k \log\left(u_l^{(v)}/\overline{\rho}^{(v)}\right) - m_k\left(u_l^{(v)} - \overline{\rho}^{(v)}\right)\right)}.
\end{aligned}$$

Numerical implementation

We implemented a Gibbs sampler for model (7.7) in WinBUGS via the R package R2WinBUGS, using the coda package to diagnose convergence. We set $V = 10$, $\mu = 0.01$, and $\alpha = 0.05$, values that stabilize the computation while allowing sufficient adaptation to the data. With $V = 10$, the 95% *a priori* probability interval for ξ_t is $(-6.20, 6.20)$, or $(0.002, 492.75)$ on the SMR scale; the values for α and μ produce a distribution for τ^2 with center near 100, inducing large *a priori* variation for the θ_{kt}. For the AR1 model, reported results are based on the $N(0, 0.2)$ hyperprior for the Fisher's Z transform of ϕ, $0.5 \log\{(1 + \phi)/(1 - \phi)\}$. This produces a 95% *a priori* probability interval for ϕ of $(-0.70, 0.70)$. We also tried a $N(0, 2)$ hyperprior, which produced a wider 95% *a priori* interval $(-0.99, 0.99)$, but posterior results virtually identical to those based on the $N(0, .0.2)$.

Longitudinal variation

To measure variation in the percentile estimates within a dialysis center over the four years, we compute *longitudinal variation*,

$$LV_{P^{est}} = 1000\frac{1}{3K}\sum_{k=1}^{K}\sum_{t=0}^{3}(P_{kt}^{est} - P_{k\bullet}^{est})^2,$$

where P_{kt}^{est} is the estimated percentile for dialysis center k in year t and $P_{k\bullet}^{est}$ is the mean over the four years.

Single year and multiyear analyses

Using model (7.7) we estimated single-year based and AR1 model based percentiles. Table 7.1 reports that the ξ are near 0, as should be the case because we have used internal standardization (the typical $\log(SMR) = 0$). The within year, between provider variation in $100\log(SMR)$ is essentially constant at approximately $100\tau = 24$, producing a 95% *a priori* interval for the ρ_{kt} of $(0.79, 1.27)$. Additional case-mix adjustment might reduce

Parameter	Single Year: ($\phi \equiv 0$)				Multiyear: ($100\phi \sim {}_{88}90_{92}$)			
	1998	1999	2000	2001	1998	1999	2000	2001
100ξ	-2.8	-1.3	-2.3	-0.7	-3.1	-0.8	-1.7	-0.3
100τ	24.1	23.5	23.1	22.2	25.8	25.0	24.9	24.1
$100OC_{\tilde{P}}(0.8)$	62	61	60	62	49	47	46	50
$LV(\hat{P}_k)$		62				4		

Table 7.1 *Results for $\hat{P}_k$ and $\tilde{P}(0.8)$. In the multiyear results, $100OC_{\tilde{P}}(0.8)$ is for the indicated year, and $_{88}90_{92}$ denotes a distribution for 100ϕ having posterior median 90 and 95% posterior credible interval (88, 92).*

this unexplained, between-center variation. Use of the AR1 model to combine evidence over years (with the posterior distribution for ϕ concentrated around 0.90) reduces $OC_{\tilde{P}}(0.8)$ from 61 to 48, a 20 percent decrease. Classification performance using the $\hat{P}_k$ is very close to that for the optimal $\tilde{P}_k(0.8)$.

In Table 7.1, $100OC(0.8)$ is 62 and 49 for the single-year and AR1 models. Figure 7.2 displays the details behind this superior classification performance. In the upper range of $\tilde{P}_k(0.8)$, the curve for the AR1 model lies above that for the single year, in the lower range it lies below. For the AR1 model to dominate the single year at all values of $\tilde{P}_k(0.8)$, the curves would need to cross at $\tilde{P}_k(\gamma) = 0.8$, but the curves cross at about 0.7. This inconsistency appears in most simulations and data analyses and remains to be investigated. We conjecture that for the basic Gaussian/Gaussian model the expected curves do cross at γ.

We also computed $P^*(0.8)$ using the Gaussian prior for θ and 1998 data. The θ-threshold, $\bar{G}_K^{-1}(0.8) = 0.169$ (ρ-threshold = 1.184). In our analysis of the 1998 data the curve based on $P^*(0.8)$ and that based on $\tilde{P}(0.8)$ almost perfectly superimpose.

Longitudinal variation in ranks/percentiles (LV_{Pest}) is dramatically reduced for the AR1 model going from 62 for the year-by-year analysis to 4 for the multiyear. As a basis for comparison, if $\phi \to 1$, $LV_{\hat{P}} \to 0$ and if the data provide no information on the SMRs (the $\tau \to \infty$), then $LV_{\hat{P}} = 83$.

Comparisons using the 1998 data

We compared performance of percentiles based on ranked MLEs ($\rho_k^{mle} = \frac{Y_k}{m_k}$), to those based on ranked Z-scores testing the hypothesis $H_0 : \rho_{k0} \equiv 1$,

$$Z_{k0} = \sqrt{m_{k0}} \log \left(\frac{y_{k0}}{m_{k0}} + 0.25 \right), \tag{7.9}$$

those based on the posterior mean SMRs ($\rho_k^{pm} = E(\rho_k \mid \mathbf{Y})$), and the optimal percentiles, $\hat{P}_k$ and $\tilde{P}_k(\gamma)$. If we regard a dialysis center with ρ_k^{mle} greater than 1.5 as "flagged," then 379 (12%) of dialysis centers will be

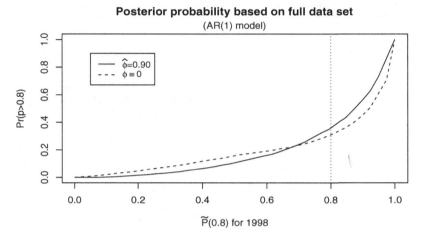

Figure 7.2 $\pi_k(0.8 \mid \mathbf{Y})$ versus $\tilde{P}_k(0.8)$ for 1998. Optimal percentiles and posterior probabilities computed with the single year model ($\phi \equiv 0$) and the AR1 model ($\hat{\phi} = 0.90$).

	$\hat{P}_k$	ρ^{pm}	ρ^{mle}	Z-score
kappa	0.90	0.94	0.78	0.83

Table 7.2 *Kappa statistics between (above γ/below γ) classifications with $\gamma = 0.80$ based on $\tilde{P}_k(\gamma)$ and other methods.*

identified; if we regard a dialysis center with Z-score greater than 1.645 as "flagged," then 647 (20.4%) of dialysis centers are identified.

Table 7.2 displays the kappa statistics for agreement in (above γ/below γ) classification between $\tilde{P}_k(\gamma)$ (optimal, with $\gamma = 0.80$) and the other methods. Classification using $\tilde{P}_k(\gamma)$, $\hat{P}_k$ and ρ^{pm} produce high agreement; use of the MLEs or Z-scores produce considerably less agreement.

Figure 7.3 compares posterior probabilities of correct classification $P(P_k > 0.8 \mid \mathbf{Y})$. The curve for $\tilde{P}_k(\gamma)$ is monotone and optimal because we construct $\tilde{P}_k(\gamma)$ by ranking these probabilities. The curves using ρ^{pm} and $\hat{P}_k$ to order the centers (not shown) are very close to the $\tilde{P}_k(\gamma)$ curve. The curves for MLE-based and Z-score-based percentiles are far from monotone and far from the optimal curve.

The MLE SMRs for centers with relatively small expected deaths have relatively large variances. To study the impact of large and small variances on estimated percentiles, Figure 7.3 also identifies them for the 147 dialy-

Comparison of $\tilde{P}(0.8)$, MLE and Z–score

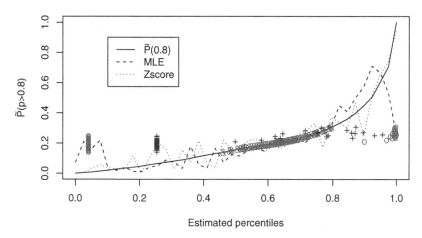

Figure 7.3 *$\pi_k(0.8)$ versus estimated percentiles for three ranking methods using the 1998 data: $\tilde{P}_k(\gamma)$, MLE-based, and Z-score-based. For small dialysis centers (fewer than five patients), the "o" denotes MLE-based percentiles, the "+" denotes Z-score-based percentiles, and the "&" denotes $\tilde{P}_k(\gamma)$.*

sis centers that treated fewer than five patients in 1998. These constitute 4.5% of all centers. Generally, MLE-based percentiles for these centers are at the extremes whereas Z-score based percentiles tend to be near 0.5. However, because the ρ_k posteriors for the high variance centers are concentrated near 1, the $\tilde{P}_k(\gamma)$ for these centers are near 0.5, and similarly for $\hat{P}_k$ and percentiles based on ρ^{pm}. For dialysis centers with a large number of patients and thereby a small variance, the the optimally estimated percentiles, $\hat{P}_k$ and $\tilde{P}_k(\gamma)$ spread out to cover full range from 0 to 1. There is better agreement between MLE-based, Z-score-based, and optimal percentiles when the small centers are removed from the dataset and estimates are recomputed.

Figure 7.4 displays estimates for the 40 providers at the $1/3174, 82/3174,$ $163/3174, \ldots, 3173/3174$ percentiles as determined by $\hat{P}_k$ (see Conlon and Louis, 1999, for a similar plot based on SMRs of disease rates in small areas). For each display, the Y-axis is $100\bar{P}_k$ with its 95% posterior interval. The X-axis for the upper left panel is $\hat{P}$, for the upper right is percentiles based on ρ^{pm}, for the lower left is percentiles based on ρ^{mle}, and for the lower right is percentiles based on Z-scores testing $\rho_k = 1$. Note that in the upper left display the $\bar{P}_k$ do not fill out the $(0, 1.0)$ percentile range; they are shrunk toward 0.50 by an amount that reflects estimation uncertainty. Also, the posterior probability intervals are very wide, indicating considerable uncertainty in estimating percentiles. The plotted points in the upper

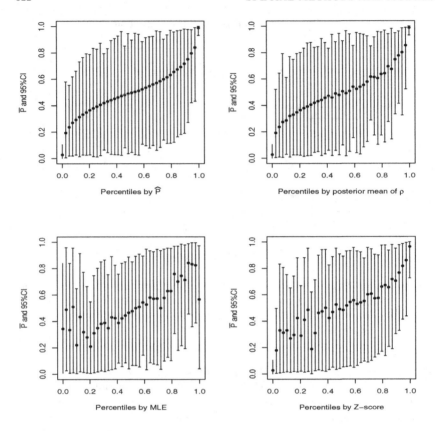

Figure 7.4 *SEL-based percentiles for 1998. For each display, the Y-axis is* $100\bar{P}_k$ *with its 95% probability interval. The X-axis for the upper left panel is* $\hat{P}$, *for the upper right is percentiles based on* ρ^{pm}, *for the lower left is percentiles based on the* ρ^{mle} *and for the lower right is percentiles based on Z-scores testing* $\rho_k = 1$.

left display are monotone because the x-axis is the ranked y-axis values. Plotted points in the upper right display are almost monotone; PM-based percentiles perform well. The lower left and lower right panels show considerable departure from monotonicity, indicating that MLE-based ranks and hypothesis test-based ranks are very far from optimal. Note also that the pattern of departures is quite different in the two panels, showing that these methods produce quite different ranks. Similar comparisons for SMRs estimated from the pooled 1998 to 2001 data would be qualitatively similar, but the departures from monotonicity would be less extreme.

Posteriors for θ, 1998

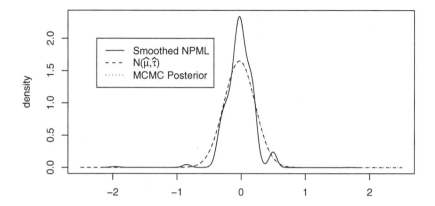

Figure 7.5 *Estimated priors for θ = log(ρ) using the 1998 data. The solid curve is a smoothed NPML using the "density" function in R with adjustment parameter = 10. The dashed curve is Gaussian using posterior medians for (μ, τ); the dotted curve (hidden by the dashed curve) is a mixture of Gaussians with (μ, τ) sampled from their MCMC computed joint posterior distribution.*

Parametric and nonparametric priors

We compare results using posterior distributions computed from parametric and nonparametric priors. Figure 7.5 displays estimated priors for θ based on the Gaussian distribution (one substituting posterior medians for (μ_0, τ_0), one produced by averaging over the posterior distribution of these parameters), and one based on a smoothed NPML. The posterior distribution of (μ_0, τ_0) has close to 0 variance, so the two parametric curves superimpose. The NPML is discrete and was smoothed using the `density` function in R with adjustment parameter 10. We use the discrete NPML to produce ranks/percentiles.

The smoothed NPML has at least two modes, with considerable mass at approximately $\theta = 0.5$ ($\rho = 1.65$), but this departure from the Gaussian distribution has little effect on classification performance. Using 1998 data, for the NPML $100 \times OC(0.8) \approx 67$ while for the Gaussian prior the value is 62. For performance evaluations of the NPML, see Paddock et al. (2006).

Finally, Figure 7.6 compares percentiles under the two priors. The dialysis centers at the top and bottom of the rankings (which have low variance MLEs) have percentiles that are generally same under the two priors. However, for the centers with middle-level rankings (higher MLE variance), the percentiles depend more on the prior. ∎

In summary, effective ranking or percentiling should be based on a loss function computed from the estimated and true ranks, or be asymptotically

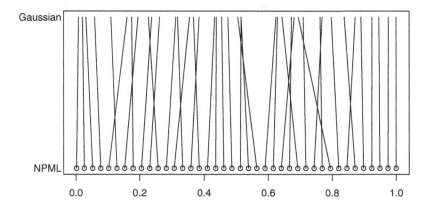

Figure 7.6 *Comparison of 1998 percentiles estimated with NPML and Gaussian prior. Circles represent 40 dialysis centers evenly spread across percentiles estimated with NPML prior. The percentiles of the same center are connected.*

equivalent to such loss function based estimates. Doing so produces optimal or near optimal performance and ensures desirable properties, such as monotone transform invariance. In general, percentiles based on MLEs or on posterior means of the target parameter can perform poorly. Similarly, hypothesis test-based percentiles perform poorly.

When posterior distributions are not stochastically ordered and the choice of ranking methods does matter, simulations have shown that though $\tilde{P}_k(\gamma)$ and $\hat{\tilde{P}}_k(\gamma)$ are optimal for their respective loss functions and outperform $\hat{P}_k$, $\hat{P}_k$ performs well for a broad range of γ values. And, $\tilde{P}_k(\gamma)$ can have poor SEL performance. However, for some scenarios the relative benefit of using an optimal procedure is considerable and so a choice of estimator should be guided by goals. Performance evaluations for three-level models with a hyperprior and robust analyses based on the nonparametric maximum likelihood prior or a fully Bayesian nonparametric prior (Paddock et al., 2006) showed that SEL-optimal ranks perform well over a wide range of prior specifications.

7.1.2 Histogram and triple goal estimates

Much of this book has concentrated on inferences for coordinate-specific parameters: an experimental unit, an individual, a clinic, a geographic region, etc. We now consider situations where the collection of parameters (the *ensemble*) is of primary or equal interest. As already seen, estimates of the histogram or *empirical distribution function* (edf) of the underlying rates (for example, to evaluate the number of rates above a threshold) or their rank ordering (for example, to prioritize environmental assessments) can be of equal importance. Other settings in which histograms or ranks might be of interest include estimating clinic-specific treatment effects, hospital or physician specific surgical death rates, school-specific teaching effectiveness, or county-specific numbers of children in poverty. For consistency, throughout this section we will always refer to the units as "regions."

Broadly put, our goal is to estimate the edf of the latent parameters that generated the observed data. One may also be interested in the individual parameters, but the initial goal is estimation of the edf (see Tukey, 1974; Louis, 1984; Laird and Louis, 1989; and Devine, Halloran, Louis, 1994). As shown by Shen and Louis (1998), Conlon and Louis (1999), and references therein, while posterior means are the most obvious and effective estimates for region-specific parameters, their edf is underdispersed and never valid for estimating the edf (or histogram) of the parameters, while the edf of the coordinate-specific MLEs is overdispersed. Ranking the posterior means can also produce suboptimal estimates of the parameter ranks; effective estimates of the parameter ranks should target them directly. Fortunately, structuring via Bayesian decision theory guides development of effective estimates of histograms and ranks, as we now describe.

Estimating the individual θs

Under squared error loss, the posterior means (PM),

$$\hat{\theta}_i^{PM} \equiv \eta_i = E[\theta_i \mid Y_i] \, ,$$

are optimal. These are the "traditional" parameter estimates. In our subsequent development, all estimates are posterior means of target features, but we reserve the "PM" label for estimates of the individual θs.

Estimating the EDF

Let G_k denote the edf of the θs that underlie the current data set. Then $G_k(t \mid \boldsymbol{\theta}) = \frac{1}{k} \sum_{i=1}^k I_{\{\theta_i \le t\}}, -\infty < t < \infty$. With $A(t)$ a candidate estimate of $G_k(t)$, under weighted integrated squared error loss, $\int w(t)[A(t) - G_k(t)]^2 dt$, the posterior mean is optimal:

$$\bar{G}_k(t) \quad = \quad E[G_k(t) \mid \mathbf{Y}] = \frac{1}{k} \sum P(\theta_i \le t \mid Y_i). \qquad (7.10)$$

With the added constraint that $A(t)$ is a discrete distribution with at most k mass points, Shen and Louis (1998) show that the optimal estimator $\hat{G}_k$ has mass $\frac{1}{k}$ at

$$\hat{U}_i = \bar{G}_k^{-1} \left(\frac{2j-1}{2k} \right), \, j = 1, \ldots, k. \qquad (7.11)$$

We now show that both the PMs and the Ys themselves are inappropriate as estimates of G_k. Except when θ is the conditional mean of Y_i, the cdf based on the Ys may have no direct relevance to the distribution of θ. Even when the Ys are relevant (when θ is the mean parameter), the histogram produced from the Y_i is overdispersed. It has the appropriate mean, but its variance is $E[V(Y \mid \theta)] + V[E(Y \mid \theta)] = E[V(Y \mid \theta)] + V[\theta]$; the first term produces overdispersion.

On the other hand, the histogram produced from the PMs is underdispersed. To see this, let $v_i = V(\theta_i \mid Y_i)$ and recall $\eta_i = E(\theta \mid Y_i)$. Then the moments produced by $\bar{G}_k$ (and approximately those by $\hat{G}_k$) are

$$
\begin{aligned}
\text{mean} \quad &= \quad \int t \, d\bar{G}_k(t) = \frac{1}{k} \sum \eta_i = \bar{\eta}, \\
\text{variance} \quad &= \quad \int (t - \bar{\eta})^2 \, d\bar{G}_k(t) \qquad\qquad (7.12) \\
&= \quad \frac{1}{k} \sum v_i + \frac{1}{k} \sum (\eta_i - \bar{\eta})^2.
\end{aligned}
$$

The sample variance of the PMs produces only the second term, so the histogram based on them is underdispersed. To correct for this underdispersion, Louis (1984) and Ghosh (1992) derived *constrained Bayes* (CB) estimates. These estimates minimize posterior expected (weighted) SEL for individual parameters subject to the constraint that their sample mean and variance match those in (7.12). Specifically, CB estimates $(a_1, \dots, a_k)$ minimize

$$
\sum w_i E[(a_i - \theta_i)^2 \mid \mathbf{Y}] = \sum w_i (a_i - \eta_i)^2 + \sum w_i v_i \ ,
$$

subject to the constraints

$$
\begin{aligned}
\frac{1}{k} \sum a_i &= \frac{1}{k} \sum \eta_i \\
\frac{1}{k} \sum (a_i - \bar{a})^2 &= \frac{1}{k} \sum v_i + \frac{1}{k} \sum (\eta_i - \bar{\eta})^2.
\end{aligned}
$$

The equally-weighted case ($w_i \equiv 1$) produces the closed-form solution

$$
\hat{\theta}_i^{CB} \quad = \quad \bar{\eta} + \left[1 + \frac{\sum v_i}{\sum (\eta_i - \bar{\eta})^2} \right]^{\frac{1}{2}} (\eta_i - \bar{\eta}).
$$

These CB estimates start with the PMs and adjust them to have the variance in (7.12). Because the term in square brackets is greater than 1, the estimates are more spread out around $\bar{\eta}$ than are the PMs. Table 5.1 and Figure 5.1 present an example of these estimates.

Building on the CB approach, Shen and Louis (1998) defined a "G then rank" (GR) approach by assigning the mass points of $\hat{G}_k$ (equation (7.11)) to coordinates using the ranks $\hat{R}$,

$$
\hat{\theta}_i^{GR} = \hat{U}_{\hat{R}_i}. \qquad\qquad (7.13)
$$

The GR estimates have an edf equal to $\hat{G}_k$ and ranks equal to the $\hat{R}_i$ and so are SEL-optimal for these features.* Though there is no apparent attention to reducing coordinate-specific SEL, assigning the $\hat{U}$s to coordinates by a permutation vector **z** to minimize $\sum(\hat{U}_{z_i} - \eta_i)^2$ produces the same assignments as in (7.13). This equivalence indicates that the GR approach pays some attention to estimating the individual θs and provides a structure for generalizing the approach to multivariate coordinate-specific parameters. As an added benefit of the GR approach, the $\hat{\theta}_i^{GR}$ are monotone transform equivariant, i.e., $\widehat{[h(\theta)]}^{GR} = h(\hat{\theta}_i^{GR})$ for monotone $h(\cdot)$.

Triple goal estimates

Because estimates of coordinate-specific parameters are inappropriate for producing histograms and ranks, no single set of values can simultaneously optimize all three goals. However, in many policy settings, communication and credibility is enhanced by reporting a single set of estimates with good performance for all three goals. To this end, Shen and Louis (1998) introduce "triple-goal" estimation criteria. Effective estimates are those that simultaneously produce a histogram that is a high-quality estimate of the parameter histogram, induced ranks that are high-quality estimates of the parameter ranks, and coordinate-specific parameter estimates that perform well. Shen and Louis (1998) evaluate and compare candidate estimates including the Ys, the PMs, CB, and GR estimates. Shen and Louis (1998) study properties of the various estimates, while Conlon and Louis (1999), Louis and Shen (1999), and Shen and Louis (2000) show how they work in practice. Here we summarize results.

As Shen and Louis (1998) show in the exchangeable case, the edf of the CB estimates is consistent for G if and only if the marginal distribution of η differs from G only by a location/scale change. The Gaussian/Gaussian model with θ the mean parameter qualifies, but in general CB estimates will not produce a consistent edf. However, $\bar{G}_k$ is consistent for G whenever G is identifiable from the marginal distributions (see Lindsay, 1983). Consistency is transform invariant.

PM dominates the other approaches for estimating individual θs under SEL, as it must. However, both GR and CB increase SEL by no more than 32% in all simulation cases, and both outperform use of the MLE (the Y_i). Comparison of CB and GR is more complicated. As predicted by the Shen and Louis (1998) results, for the Gaussian-Gaussian case, CB and GR are asymptotically equivalent, while for the gamma-gamma case, model CB has a slightly better large-sample performance than does GR. However, for finite k, simulations show that GR can perform better than CB for skewed or long-tailed prior distributions.

---

* Ranks other than the $\hat{R}$ can be used. For example, use of $\tilde{R}_k(\gamma)$ will produce optimal (above γ/below γ) classification.

As it must, GR dominates other approaches for estimating G_k and the advantage of GR increases with sample size. The Gaussian-Gaussian model provides the most favorable setup for CB relative to GR. Yet, the risk for CB exceeds GR by at least 39%. Performance for CB is worse in other situations, and in the Gaussian-lognormal case CB is even worse than PM. Importantly, the GR estimates are transform equivariant, and their performance in estimating G_k is transform invariant.

A start has been made on approaches to multiple-goal estimates for multivariate, coordinate-specific parameters. Ranks are not available in this situation, but coordinate-specific estimates based on the histogram are available. First, estimate G_k by a k-point discrete $\hat{G}_k$ that optimizes some histogram loss function. Then assign the mass points of $\hat{G}_k$ to coordinates so as to minimize SEL for the individual parameters. In the univariate parameter case, these estimates equal the GR estimates.

Finally, in addition to possible nonrobustness to misspecification of the sampling likelihood (a potential problem shared by all methods), inferences from hierarchical models may not be robust to *prior* misspecification, especially if coordinate-specific estimates have relatively high variance. This nonrobustness is most definitely a threat to the GR estimates, which depend on extreme percentiles of the posterior distribution. In the next subsection we discuss "robustifying" Bayes and EB procedures.

7.1.3 Robust prior distributions

Our ranking procedures and GR estimates, indeed all of Bayesian analysis, operate through the posterior distribution. Conditional on observed data, the posterior depends on the prior and the likelihood; validity and efficiency in turn depend on these. As such, we now focus on use of *robust* priors, i.e., priors that lead to results that are insensitive to moderate departures from model assumptions. A satisfactory trade-off between robustness and efficiency is especially important when estimating ranks or histograms because these inferences depend on the tails and other details of the posterior distribution that have less influence on posterior means and variances and so are more sensitive to prior misspecification.

A fully nonparametric approach is available, namely the aforementioned NPML; however, it is discrete with number of mass points no greater than the number of unique data points, and generally much smaller than this. More recently, a fully Bayesian approach using Dirichlet process (DP) priors and their mixtures has become popular (Subsection 2.6; see also Hanson, 2006). In this subsection, we investigate whether "robustified" prior distributions can produce both efficient and robust estimates under a variety of scenarios and goals associated with estimating unit-specific parameters (the θs), their EDF (G_k, equation (7.11)), and their SEL-minimizing ranks ($\hat{R}_k$, equation (7.4)).

Smoothing by roughening

Subsection 5.3.2 presented the EM algorithm approach to finding the NPML estimate of the prior distribution. Using it instead of a parametric prior (e.g., a conjugate) has been shown to be quite efficient and robust, especially for inferences based on only the first two moments of the posterior distribution (Butler and Louis, 1992). Because the NPML is always discrete with at most k mass points, smoothing is attractive for increased efficiency with small sample sizes. Approaches include directly smoothing the NPML, adding "pseudo-data" to the dataset and then computing the NPML, and model-based methods, such as use of a Dirichlet process prior for G.

Laird and Louis (1991) propose an alternative, which they call *smoothing-by-roughening* (SBR), based on the EM algorithm method of finding the NPML. Shen and Louis (1999) provide additional evaluations. Though initiating the EM by a discrete prior with k mass points is best for producing the NPML, a continuous prior can initiate the EM. In this situation, the initiating prior $(G^{(0)})$ and all subsequent iterates $(G^{(\nu)})$ are continuous. As $\nu \to \infty$, iterates converge very slowly to the NPML. Therefore, stopping the EM at iteration ν_k produces an estimate of the prior that is smoothed toward $G^{(0)}$ (equivalently, roughened away from $G^{(0)}$) with more smoothing for smaller values of ν_k. Note that the EM recursion uses the conditional expectation that produces $\bar{G}_k$. Specifically,

$$G^{(\nu+1)}(t) = E[G_k(t) \mid \mathbf{Y}, G^{(\nu)}], \qquad (7.14)$$

where $G^{(\nu)}$ plays the role of the prior.

As an example of smoothing-by-roughening and its relation to the NPML and ensemble estimates, Shen and Louis (1999) analyzed the baseball data from Efron and Morris (1973b). Batting averages are available for the first 45 at-bats of the 1970 season for 18 major league baseball players. The full-season batting average plays the role of the true parameter value. Batting averages are arc-sine transformed to stabilize their variance, and then these are centered around the mean of the transformed data. As a result, 0 is the center of the transformed observed batting average data.

The SBR starting distribution $G^{(0)}$ is $Unif(-2.5, 2.5)$, and is approximated by 200 equi-probable, equally-spaced mass points. Figure 7.7 shows the cdf for the marginal maximum likelihood estimate (MMLE) normal prior and the NPML. Notice that the NPML is concentrated at two points (i.e., it is very "rough"). The estimated normal prior is both smoothed and constrained to a narrow parametric form.

Figure 7.8 displays a sequence of roughened prior estimates $G^{(\nu)}$, along with the normal prior. Recall that the NPML is $G^{(\infty)}$. The sequence of estimates roughens from its $Unif(-2.5, 2.5)$ starting point toward the NPML. Convergence is extremely slow.

Finally, Figure 7.9 displays various choices for estimating G_k, including

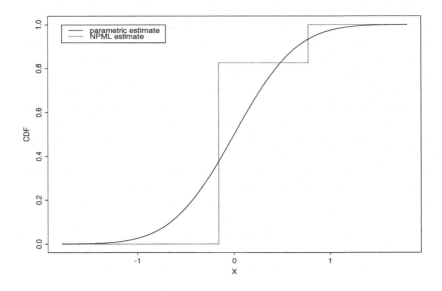

Figure 7.7 *Estimated parametric and nonparametric priors for the baseball data.*

the actual function (the edf of the transformed season batting averages). The remaining estimates (see the figure legend) are

- MLE: edf of the transformed, observed data
- PEB: The posterior means based on a MMLE normal prior
- NPEB: The posterior means based on the NPML prior
- CEB: The Louis and Ghosh constrained estimates
- SBR: $G^{(20)}$

Notice that the edf based on the MLE has a wider spread than that computed from the true parameter values. However, the posterior means from the PEB and the NPEB are shrunk too far toward the prior mean, and produce edfs with too small a spread. The CB approach gets the spread and shape about right, as does the SBR estimate from 20 iterations, though it produces a continuous distribution. Figure 5.1 presents another comparison of edfs based on the data (MLE), PEB and CB.

The SBR estimate is included in Figure 7.9 to show its alignment with the true G_k, even though it is not a candidate (since it is continuous and the target distribution is discrete). A discretized version similar to $\hat{G}_k$ (see equation 7.11) may be effective, but further evaluation is needed.

The primary potential of the SBR estimate is as an estimate of G, not G_k. Such a smoothed estimate has the potential to produce robust and efficient (empirical) Bayes procedures. If $G^{(0)}$ is chosen to be the prior mean of a

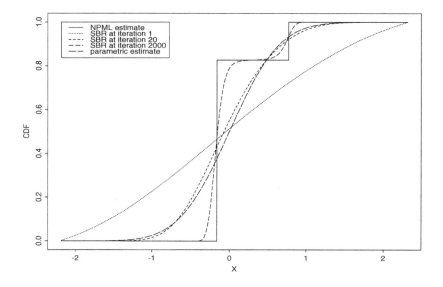

Figure 7.8 *Convergence of the smoothing-by-roughening (SBR) estimate from a uniform $G^{(0)}$ toward the NPML using the EM algorithm.*

hyperprior, the SBR approach is similar to use of a Dirichlet process prior. Indeed, employing this idea for estimation of the mean in a normal/normal two-stage model, setting $\nu_k = a + b\log(k)$ exactly matches the estimate based on a parametric hyperprior. The parameters a and b capture the relative precision of the hyperprior and the data, and the number of iterates is always of order $log(k)$.

Comparison of Histogram Estimates

We examined fully Bayesian and EB approaches to estimating histograms (see Paddock et al., 2006 for details). Figure 7.10 displays histogram estimates based on a Gaussian prior, a t-prior with 5 df, two Dirichlet process priors (DP-2 puts relatively high weight on the Gaussian shape), and SBR. The first row of the figure shows that all methods are quite competitive when the data are highly informative. DP-1 and SBR are more effective when the sampling variance increases. DP-2 strongly favors a Gaussian distribution at the expense of flexibility, rendering the empirical distribution estimates inaccurate.

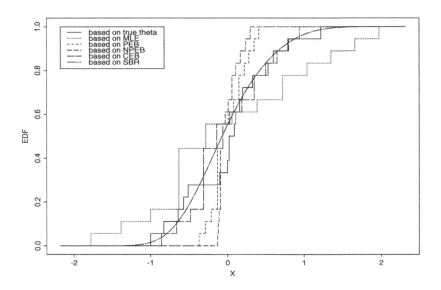

Figure 7.9 *Various estimates of G_k for the baseball data (see text for key to figure legend).*

Discussion

Evaluations show that the nonparametric priors using a fully Bayesian or EB approach are highly efficient relative to assuming the true parametric form (e.g., Gaussian) and are robust to departures from the parametric form. Overall, the choice of approach has relatively little effect on performance in estimating the θs and the ranks compared to estimating percentiles of G_k. When data are highly informative, all estimates are robust to model misspecification, However, when the data are less informative, inferences can be sensitive to such misspecification. In general, nonparametric approaches such as SBR and DP are highly efficient relative to the correct, parametric alternative; they are slightly less efficient for estimating G_K, but still perform well.

Nonparametric models often protect against errors caused by an incorrect parametric form for G. However, caution is required when applying DP or other Bayesian nonparametric models, because their performance depends on sensible specification of hyperparameters, especially when information on the prior is low (small sample size or high sampling variance). Similarly, SBR performance depends on reasonable specification of the initial distribution, $G^{(0)}$, and the number of iterations.

DP approaches are clearly the better choice when the true distribution is not Gaussian, and can also adapt well to the data even when the Gaussian

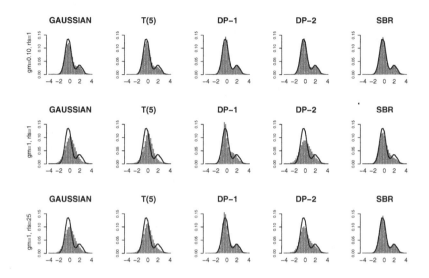

Figure 7.10 *Histogram estimates based on a Gaussian prior, a t prior with 5 df, two Dirichlet process priors, and the SBR approach (columns) for three sampling distribution scenarios. In the first and second rows, the sampling distributions all have the same variance (rls = 1); in the third row, the variances are different (rls = 25). The sampling distributions in the first row have a variance only 10% of that for the prior (gm = 0.10); in the second and third rows, the typical sampling variance equals the variance of the prior (gm = 1). The true distribution is indicated by the solid black line.*

assumption is reasonable. Hyperprior Bayesian approaches are, at least in principle, more attractive than SBR, because the posterior process incorporates residual uncertainty.

7.2 Order restricted inference

Another area for which Bayes and empirical Bayes methods are particularly well suited is that of order restricted statistical inference. By this we mean modeling scenarios that feature constraints on either the sample space (truncated data problems) or the parameter space (constrained parameter problems). Such problems arise in many biostatistical settings, such as life table smoothing, bioassay, longitudinal analysis, and response surface modeling. While a great deal of frequentist literature exists on such problems (see, e.g., the textbook by Robertson, Wright, and Dykstra, 1988), analysis is often difficult or even impossible because the order restrictions typically wreak havoc with the usual approaches for estimation and testing.

By contrast, the Bayesian approach to analyzing such problems is virtually unchanged: we simply add the constraints to our Bayesian model (likelihood times prior) in the form of indicator functions, and compute the posterior distribution of the parameter of interest as usual. But again, this conceptual simplicity could not be realized in practice before the advent of sampling based methodology.

As an example, consider the problem of smoothing a series of raw mortality rates y_i into a gradually increasing final sequence θ_i, suitable for publication as a reference table. This process is referred to as "graduation" in the actuarial science literature. A restricted parameter space arises because the unknown $\boldsymbol{\theta} = (\theta_1, \theta_2, \dots, \theta_k)^T$ must lie in the set $S = \{\boldsymbol{\theta} : \theta_1 < \theta_2 < \cdots < \theta_k\}$. Under the prior distribution $\pi(\boldsymbol{\theta}|\boldsymbol{\lambda})$, we have that $p(\mathbf{y}, \boldsymbol{\theta}|\boldsymbol{\lambda}) = f(\mathbf{y}|\boldsymbol{\theta})\pi(\boldsymbol{\theta}|\boldsymbol{\lambda})I_S(\boldsymbol{\theta})$ where $I_S(\boldsymbol{\theta})$ is the indicator function of the set S, so that $I_S(\boldsymbol{\theta})$ equals 1 when $\boldsymbol{\theta} \in S$, and equals 0 otherwise. Gelfand, Smith, and Lee (1992) point out that the complete conditional distribution for, say, θ_i satisfies

$$p(\theta_i|\mathbf{y}, \boldsymbol{\lambda}, \theta_{j \neq i}) \propto f(\mathbf{y}|\boldsymbol{\theta})\pi(\boldsymbol{\theta}|\boldsymbol{\lambda})I_{S_i}(\theta_i) , \qquad (7.15)$$

where S_i is the appropriate cross section of the constraint set S. In our increasing parameter case, for example, we have $S_i = \{\theta_i : \theta_{i-1} < \theta_i < \theta_{i+1}\}$. Hence, if π is chosen to be conjugate with the likelihood f so that the unconstrained complete conditional distribution emerges as a familiar standard form, then the constrained version given in (7.15) is simply the same standard distribution restricted to S_i. Carlin (1992) and Carlin and Klugman (1993) use this technology to obtain increasing and increasing convex graduations of raw mortality data for males over age 30. These papers employ vague prior distributions in order to maintain objectivity in the results and keep the prior elicitation burden to a minimum.

Finally, Geyer and Thompson (1992) show how Monte Carlo methods may be used to advantage in computing maximum likelihood estimates in exponential families where closed forms for the likelihood function are unavailable. Gelfand and Carlin (1993) apply these ideas to arbitrary constrained and missing data models, where integration and sampling is over the data space, instead of the parameter space. These ideas could be applied to the problem of empirical Bayes point estimation in nonconjugate scenarios, where again the likelihood contains one or more intractable integrals.

7.3 Longitudinal data models

A great many datasets arising in statistical and biostatistical practice consist of repeated measurements (usually ordered in time) on a collection of individuals. Data of this type are referred to as *longitudinal data*, and require special methods for handling the correlations that are typically present among the observations on a given individual. More precisely, suppose we let Y_{ij} denote the jth measurement on the ith individual in the study, $j = 1, \ldots, s_i$ and $i = 1, \ldots, n$. Arranging each individual's collection of observations in a vector $Y_i = (Y_{i1}, \ldots, Y_{is_i})^T$, we can fit the usual fixed effects model

$$Y_i = \mathbf{X}_i \alpha + \epsilon_i \, ,$$

where $\mathbf{X}_i$ is an $s_i \times p$ design matrix of covariates and α is a $p \times 1$ parameter vector. While the usual assumptions of normality, zero mean, and common variance might still be appropriate for the components of ϵ_i, generally they are correlated with covariance matrix $\mathbf{\Sigma}$. This matrix can be general or follow some simplifying pattern, such as compound symmetry or a first-order autoregression. Another difficulty is that data frequently exhibit more variability than can be adequately explained using a simple fixed effects model.

A popular alternative is provided by the *mixed* model

$$Y_i = \mathbf{X}_i \alpha + \mathbf{W}_i \beta_i + \epsilon_i \, , \tag{7.16}$$

where $\mathbf{W}_i$ is an $s_i \times q$ design matrix (q typically less than p), and β_i is a $q \times 1$ vector of subject-specific random effects, usually assumed to be normally distributed with mean vector $\mathbf{0}$ and covariance matrix $\mathbf{V}$. The β_is capture any subject-specific mean effects, and also enable the model to accurately reflect any extra variability in the data. Furthermore, it may now be realistic to assume that, given β_i, the components of ϵ_i *are* independent. This allows us to set $\mathbf{\Sigma} = \sigma^2 \mathbf{I}_{s_i}$ (as in the original paper by Laird and Ware, 1982), although an autoregressive or other structure for $\mathbf{\Sigma}$ may still be appropriate. Note that the marginal covariance matrix for Y_i includes terms induced both by the random effects (from $\mathbf{V}$) and the likelihood (from $\mathbf{\Sigma}$).

The model (7.16) can accommodate the case where not every individual has the same number of observations (unequal s_i), perhaps due to missed office visits, data transmission errors, or study withdrawal. In fact, features far broader than those suggested above are possible in principle. For example, heterogeneous error variances σ_i^2 may be employed, removing the assumption of a common variance shared by all subjects. The Gaussian errors assumption may be dropped in favor of other symmetric but heavier-tailed densities, such as the Student's t. Covariates whose values change and growth curves that are nonlinear over time are also possible within this framework.

An increasing number of commercially available computer packages (e.g., SAS Proc Mixed) enable likelihood analysis for mixed models. Though these packages employ sophisticated numerical methods to estimate the fixed effects and covariance matrices, in making inferences they are unable to include the variance inflation and other effects of having estimated the features of the prior distributions. By contrast, MCMC methods are ideally suited to this goal, since they are able to perform the necessary integrations in our high-dimensional (and typically analytically intractable) hierarchical model. Several papers (e.g., Zeger and Karim, 1991; Lange et al., 1992; Gilks, Wang et al., 1993; McNeil and Gore, 1996) have employed MCMC methods in longitudinal modeling settings. Chib and Carlin (1999) provide a method for generating the fixed effects α and all the random effects β_i in a single block, which greatly improves the convergence of the algorithm.

We require prior distributions for α, σ^2, and $\mathbf{V}$ to complete our model specification. Bayesians often think of the distribution on the random effects β_i as being part of the prior (the "first stage," or *structural* portion), with the distribution on $\mathbf{V}$ forming another part (the "second stage," or *subjective* portion). Together, these two pieces are sometimes referred to as a *hierarchical prior*.

While the case study in Section 8.1 provides a real-data illustration, here we offer a more "textbook" example implemented in WinBUGS in its Examples Volume I.

Example 7.2 Consider the longitudinal dataset on the weights of young laboratory rats given in Table 7.3, which was originally analyzed by Gelfand, Hills, Racine-Poon, and Smith (1990). While this dataset is already something of a "golden oldie" in the MCMC Bayesian literature, we include it because it is a problem of moderate difficulty that exploits the sampler in the context of the normal linear model, common in practice. In these data, Y_{ij} corresponds to the weight of the ith rat at measurement point j, while x_{ij} denotes its age in days at this point. We adopt the model

$$Y_{ij} \stackrel{ind}{\sim} N\left(\alpha_i + \beta_i x_{ij}, \sigma^2\right), \quad i = 1, \ldots, k, \; j = 1, \ldots, n_i, \qquad (7.17)$$

where $k = 30$ and $n_i = 5$ for all i (unlike humans, laboratory rats can be counted upon to show up for *all* of their office visits!). Thus, the model supposes a straight-line growth curve with homogeneous errors, but allows for the possibility of a *different* slope and intercept for each individual rat. Still, we believe all the growth curves to be similar, which we express statistically by assuming that the slope-intercept vectors are drawn from a common normal population, i.e.,

$$\boldsymbol{\theta}_i \equiv \begin{pmatrix} \alpha_i \\ \beta_i \end{pmatrix} \stackrel{iid}{\sim} N\left(\boldsymbol{\theta}_0 \equiv \begin{pmatrix} \alpha_0 \\ \beta_0 \end{pmatrix}, \Sigma\right), \quad i = 1, \ldots, k. \qquad (7.18)$$

Therefore, we have what a classical statistician might refer to as a *random effects model* (or a *random coefficient model*).

To effect a Bayesian analysis, we must complete the hierarchical structure at the second stage with a prior distribution for σ^2, and at the third stage with prior distributions for $\boldsymbol{\theta}_0$ and Σ. We do this by choosing conjugate

Rat	Weight Measurements				
i	Y_{i1}	Y_{i2}	Y_{i3}	Y_{i4}	Y_{i5}
1	151	199	246	283	320
2	145	199	249	293	354
3	147	214	263	312	328
4	155	200	237	272	297
5	135	188	230	280	323
6	159	210	252	298	331
7	141	189	231	275	305
8	159	201	248	297	338
9	177	236	285	340	376
10	134	182	220	260	296
11	160	208	261	313	352
12	143	188	220	273	314
13	154	200	244	289	325
14	171	221	270	326	358
15	163	216	242	281	312
16	160	207	248	288	324
17	142	187	234	280	316
18	156	203	243	283	317
19	157	212	259	307	336
20	152	203	246	286	321
21	154	205	253	298	334
22	139	190	225	267	302
23	146	191	229	272	302
24	157	211	250	285	323
25	132	185	237	286	331
26	160	207	257	303	345
27	169	216	261	295	333
28	157	205	248	289	316
29	137	180	219	258	291
30	153	200	244	286	324

Table 7.3 *Rat population growth data (from Gelfand et al., 1990, Table 3).*

forms for each of these distributions, namely,

$$\begin{aligned}
\sigma^2 &\sim IG(a, b) , \\
\boldsymbol{\theta}_0 &\sim N(\boldsymbol{\eta}, C) , \text{ and} \\
\Sigma^{-1} &\sim W\left((\rho R)^{-1}, \rho\right) ,
\end{aligned} \qquad (7.19)$$

where W denotes the *Wishart* distribution, a multivariate generalization of the gamma distribution. Here, R is a 2×2 matrix and $\rho \geq 2$ is a scalar "degrees of freedom" parameter. In more generic notation, a $p \times p$ symmetric and positive definite matrix V distributed as $W(D, n)$ has density function proportional to

$$\frac{|V|^{(n-p-1)/2}}{|D|^{n/2}} \exp\left[-\frac{1}{2} tr(D^{-1}V)\right] \tag{7.20}$$

provided that $n \geq p$. For this distribution, one can show $E(V_{ij}) = nD_{ij}$, $Var(V_{ij}) = n(D_{ij}^2 + D_{ii}D_{jj})$, and $Cov(V_{ij}, V_{kl}) = n(D_{ik}D_{jl} + D_{il}D_{jk})$. Thus, under the parametrization chosen in (7.19) we have $E(\Sigma^{-1}) = R^{-1}$, so R^{-1} is the expected prior precision of the θ_is (i.e., R is approximately the expected prior variance). Also, $Var(\Sigma_{ij})$ is decreasing in ρ, so small ρ values correspond to vaguer prior distributions. For more on the Wishart distribution, the reader is referred to Press (1982, Chapter 5).

The hyperparameters in our model are a, b, η, C, ρ, and R, all of which are assumed known. We seek the marginal posterior for θ_0 given only the observed data. The total number of unknown parameters in our model is 66: 30 α_is, 30 β_is, α_0, β_0, σ^2, and the 3 unique components of Σ. This number is far too high for any approximation or numerical quadrature method, and the noniterative Monte Carlo methods in Section 3.3 would be very cumbersome. By contrast, the Gibbs sampler is relatively straightforward to implement here thanks to the conjugacy of the prior distributions at each stage in the hierarchy. For example, in finding the full conditional for θ_i we are allowed to think of the remaining parameters as fixed. But as proved by Lindley and Smith (1972) and described in Subsection 4.1.1, with variance parameters known the posterior distribution for θ_i emerges as

$$\theta_i | \, \mathbf{y}, \, \theta_0, \, \Sigma^{-1}, \, \sigma^2 \sim N\left(D_i(\sigma^{-2}X_i^T \mathbf{y}_i + \Sigma^{-1}\theta_0), \, D_i\right)$$

for $i = 1, \ldots, k$, where $D_i^{-1} = \sigma^{-2}X_i^T X_i + \Sigma^{-1}$,

$$\mathbf{y}_i = \begin{pmatrix} y_{i1} \\ \vdots \\ y_{in_i} \end{pmatrix}, \quad \text{and} \quad X_i = \begin{pmatrix} 1 & x_{i1} \\ \vdots & \vdots \\ 1 & x_{in_i} \end{pmatrix}.$$

Similarly, the full conditional for θ_0 is

$$\theta_0 | \, \mathbf{y}, \, \{\theta_i\}, \, \Sigma^{-1}, \, \sigma^2 \sim N\left(V(k\Sigma^{-1}\bar{\theta} + C^{-1}\eta), \, V\right),$$

where $V = (k\Sigma^{-1} + C^{-1})^{-1}$ and $\bar{\theta} = \frac{1}{k}\sum_{i=1}^k \theta_i$.

For Σ^{-1}, under the parametrization in (7.19), the prior distribution (7.20) is proportional to

$$|\Sigma^{-1}|^{(\rho-3)/2} \exp\left[-\frac{1}{2} tr(\rho R \Sigma^{-1})\right].$$

Combining the normal likelihood for the random effects (7.18) with this prior distribution produces the updated Wishart distribution

$$\Sigma^{-1}\big|\, \mathbf{y}, \{\boldsymbol{\theta}_i\}, \boldsymbol{\theta}_0, \sigma^2$$
$$\sim W\left(\left[\textstyle\sum_{i=1}^{k}(\boldsymbol{\theta}_i - \boldsymbol{\theta}_0)(\boldsymbol{\theta}_i - \boldsymbol{\theta}_0)^T + \rho R\right]^{-1}, \; k + \rho\right).$$

Finally, the full conditional for σ^2 is the updated inverse gamma distribution

$$\sigma^2\big|\, \mathbf{y}, \{\boldsymbol{\theta}_i\}, \boldsymbol{\theta}_0, \Sigma$$
$$\sim IG\left(\tfrac{n}{2} + a, \; \left[\tfrac{1}{2}\textstyle\sum_{i=1}^{k}(\mathbf{y}_i - X_i\boldsymbol{\theta}_i)^T(\mathbf{y}_i - X_i\boldsymbol{\theta}_i) + b^{-1}\right]^{-1}\right),$$

where $n = \sum_{i=1}^{k} n_i$. Thus, the conditioning property of the Gibbs sampler enables closed form full conditional distributions for each parameter in the model, including the variance parameters – the sticking point in the original Lindley and Smith paper. Samples from all of these conditional distributions (except perhaps the Wishart) are easily obtained from any of a variety of statistical software packages. Wishart random deviates may be produced via an appropriate combination of normal and gamma random deviates. The reader is referred to the original paper on this topic by Odell and Feiveson (1966), or to the more specifically Bayesian treatment by Gelfand et al. (1990).

In our dataset, each rat was weighed once a week for five consecutive weeks. In fact, the rats were all the same age at each weighing: $x_{i1} = 8$, $x_{i2} = 15$, $x_{i3} = 22$, $x_{i4} = 29$, and $x_{i5} = 36$ for all i. As a result, we may simplify our computations by rewriting the likelihood (7.17) as

$$Y_{ij} \stackrel{ind}{\sim} N\left(\alpha_i + \beta_i(x_{ij} - \bar{x}), \, \sigma^2\right), \; i = 1, \ldots, k, \; j = 1, \ldots, n_i,$$

so that it is now reasonable to think of α_i and β_i as independent a priori. Thus, we may set $\Sigma = Diag(\sigma_\alpha^2, \sigma_\beta^2)$, and replace the Wishart prior (7.19) with a product of independent inverse gamma priors, say $IG(a_\alpha, b_\alpha)$ and $IG(a_\beta, b_\beta)$.

We complete the prior specification by choosing hyperprior values that determine very vague priors, namely, $C^{-1} = 0$ (so that $\boldsymbol{\eta}$ disappears entirely from the full conditionals), $a = a_\alpha = a_\beta = \epsilon$, and $b = b_\alpha = b_\beta = 1/\epsilon$ where $\epsilon = 0.001$. Using WinBUGS (Spiegelhalter et al., 2004), we ran three independent Gibbs sampling chains for 500 iterations each. These chains were started from three different points in the sample space that we believed to be overdispersed with respect to the posterior distribution, so that overlap among these chains would suggest convergence of the sampler. The first column of Figure 7.11 displays the Gibbs samples obtained for the population slope and intercept at time zero (birth), the latter now being given by $\mu = \alpha_0 - \beta_0\bar{x}$ due to our recentering of the x_{ij}s. As a satisfactory degree of convergence seems to have occurred by iteration 100, we use the output of all three chains over iterations 101 to 500 to obtain the posterior

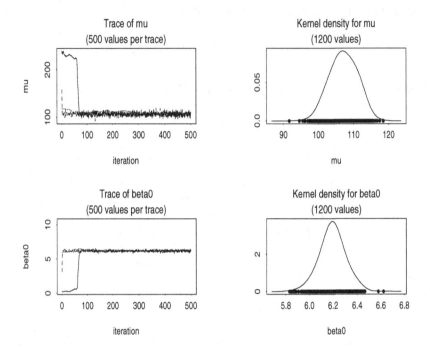

Figure 7.11 *Monitoring plots and kernel density estimates for μ and β_0, Gibbs sampling analysis of the rat data.*

density estimates shown in the second column of the figure. These are the kernel density estimates produced by WinBUGS; more accurate results could be obtained via the Rao-Blackwellization approach in equation (3.9). It appears that the average rat weighs about 106 grams at birth, and gains about 6.2 grams per day.

Gelfand et al. (1990) also discuss and illustrate the impact of missing data on the analysis. If Y_{ij} is missing, samples from its predictive distribution $p(y_{ij}|\mathbf{y})$ are easily generated, since its complete conditional distribution is nothing but the normal distribution given in the likelihood expression (7.17). The predictive could then be estimated using a kernel smooth of these samples as in Figure 7.11, or a mixture density estimate of the form (3.9), namely,

$$
\begin{aligned}
p(y_{ij}|\mathbf{y}) &= \int p(y_{ij}|\boldsymbol{\theta}_i, \sigma^2) p(\boldsymbol{\theta}_i, \sigma^2|\mathbf{y}) \\
&\approx \frac{1}{G} \sum_{g=1}^{G} p\left(y_{ij}|\boldsymbol{\theta}_i^{(g)}, \sigma^{2(g)}\right),
\end{aligned}
$$

where the superscript (g) indexes the postconvergence replications from the sampler. ∎

7.4 Continuous and categorical time series

Many datasets arising in statistics and biostatistics exhibit autocorrelation over time that cannot be ignored in their analysis. A useful model in such situations is the *state space model*, given in its most general form by

$$x_t = f_t(x_{t-1}) + u_t, \qquad \text{and}$$
$$y_t = h_t(x_t) + v_t, \qquad t = 1, \ldots, n, \tag{7.21}$$

where x_t is the (unobserved) $p \times 1$ state vector, y_t is the (observed) $q \times 1$ observation vector, and $f_t(\cdot)$ and $h_t(\cdot)$ are known, possibly nonlinear functions that depend on some unknown parameters. Typically, u_t and v_t are taken as independent and identically distributed, with $u_t \sim N_p(0, \Sigma)$ and $v_t \sim N_q(0, \Upsilon)$, where N_p denotes the p-dimensional normal distribution.

State space models are particularly amenable to a Bayesian approach because the time-ordered arrival of the data means that the notion of updating prior knowledge in the presence of new data arises quite naturally. Carlin, Polson, and Stoffer (1992) use Monte Carlo methods in an analysis of these models that does not resort to convenient normality or linearity assumptions at the expense of model adequacy. Writing the collection of unknown parameters simply as $\boldsymbol{\theta}$ for the moment, the likelihood for model (7.21) is given by

$$L(\boldsymbol{\theta}|x_0, \mathbf{x}, \mathbf{y}) = p(x_0) \prod_{t=1}^{n} g_u(x_t|x_{t-1}, \boldsymbol{\theta}) \prod_{t=1}^{n} g_v(y_t|x_t, \boldsymbol{\theta}) \tag{7.22}$$

for some densities g_u and g_v. Next, introducing the nuisance parameters $\boldsymbol{\lambda} = (\lambda_1, \ldots, \lambda_n)$ and $\boldsymbol{\omega} = (\omega_1, \ldots, \omega_n)$, they assume that

$$p(x_t|x_{t-1}, \lambda_t, \Sigma) = N_p(f_t(x_{t-1}), \lambda_t \Sigma), \qquad \text{and}$$
$$p(y_t|x_t, \omega_t, \Upsilon) = N_q(h_t(x_t), \omega_t \Upsilon), \qquad t = 1, \ldots, n, \tag{7.23}$$

where Σ and Υ are parameter matrices of sizes $p \times p$ and $q \times q$, respectively. Hence, unconditionally,

$$g_u(x_t|x_{t-1}, \Sigma) = \int_\Lambda p(x_t|x_{t-1}, \lambda_t, \Sigma) p_1(\lambda_t) d\lambda_t, \qquad \text{and}$$
$$g_v(y_t|x_t, \Upsilon) = \int_\Omega p(y_t|x_t, \omega_t, \Upsilon) p_2(\omega_t) d\omega_t, \qquad t = 1, \ldots, n. \tag{7.24}$$

Note that, by varying $p_1(\lambda_t)$ and $p_2(\omega_t)$, the distributions g_u and g_v are scale mixtures of multivariate normals for each t, thus enabling a wide variety of nonnormal error densities to emerge in (7.24). For example, in the univariate case, the distributions g_u and g_v can be double exponential, logistic, exponential power, or t densities.

As discussed just prior to Example 4.6, the conditioning feature of the

Gibbs sampler makes such normal scale mixtures very convenient computationally. Non-Gaussian likelihoods may be paired with non-Gaussian priors and Gaussian complete conditionals may still emerge for the state parameters (in the present case, the x_ts). Further, under conjugate priors the complete conditionals for Σ and Υ follow from standard Gaussian and Wishart distribution theory, again due to the conditioning on λ and ω. Finally, the complete conditionals for the nuisance parameters themselves are of known functional form. For example, for our state space model we have $p(\lambda_t | \lambda_{j \neq t}, \omega, \Sigma, \Upsilon, \mathbf{y}, \mathbf{x}) = p(\lambda_t | \Sigma, x_t, x_{t-1}) \propto p(x_t | x_{t-1}, \lambda_t, \Sigma) p_1(\lambda_t)$ by Bayes' Theorem. But the normalizing constant is known by construction: it is $g_u(x_t | x_{t-1}, \Sigma)$ from equation (7.24). Complete conditionals for the ω_t would of course arise similarly.

Thus, the sampling based methodology allows the solution of several previously intractable modeling problems, including the modeling of asymmetric densities on the positive real line (as might be appropriate for death rates), explicit incorporation of covariates (such as age and sex), heteroscedasticity of errors over time, multivariate analysis (including simultaneous modeling of both preliminary and final mortality estimates), and formal model choice criteria for selecting the best model from many. West (1992) uses importance sampling and adaptive kernel-type density estimation for estimation and forecasting in dynamic nonlinear state space models. In the special case of linear state space models, Carter and Kohn (1994) provide an algorithm for generating all of the elements in the state vector simultaneously, obtaining a far less autocorrelated sampling chain (and hence faster convergence) than would be possible using the fully conditional algorithm of Carlin, Polson, and Stoffer (1992). See also Carter and Kohn (1996) for further improvements in the context of conditionally Gaussian state space models. Chib and Greenberg (1995b) extend this framework to allow time invariant parameters in the state evolution equation, as well as multivariate observations in a vector autoregressive or moving average process of order 1. Marriott et al. (1996) use a hybrid Gibbs-Metropolis algorithm to analyze the exact ARMA likelihood (rather than that conditional on initial conditions), and incorporate the associated constraints on the model parameters to insure stationarity and invertibility of the process.

Carlin and Polson (1992) extend model (7.21) further, formulating it for categorical data by adding the condition $y_t = A(y_t^*)$ to the specification, where $A(\cdot)$ is a given many-to-one function. They also provide an example of biostatistical interest, applying the methodology to the infant sleep state study described by Stoffer et al. (1988). In this dataset, the experimenter observes only $y_t = 1$ or 0, corresponding to rapid eye movement (REM) or non-REM sleep, respectively, and the goal is to estimate the infant's underlying (continuous) sleep state x_t. Simple density estimates for the x_t are available both prior to the observation of y_t (prediction) and afterward (filtering). Albert and Chib (1993b) use a similar approach to handle underlying autoregressive structure of any order, and use the resulting hidden Markov model to advantage in analyzing U.S. GNP data. Chib (1996) provides a multivariate updating algorithm for a collection of discrete state variables, analogous to the result of Carter and Kohn (1994) for continuous variables in a state space model. See also Chib and Greenberg (1998) for MCMC Bayesian methods for analyzing *multivariate probit models*, even more general approaches for correlated binary response data.

7.5 Survival analysis and frailty models

Bayesian methods have been a subject of increasing interest among researchers engaged in the analysis of survival data. For example, in monitoring the results of a clinical trial, the Bayesian approach is attractive from the standpoint of allowing the input of additional relevant information concerning stopping or continuing the trial. Spiegelhalter et al. (2000) provide a remarkably comprehensive overview of and guide to the literature on Bayesian methods in clinical trials and other areas of health technology assessment (e.g., observational studies). Areas covered include prior determination, design, monitoring, subset analysis, and multicenter analysis; a guide to relevant web sites and software is also provided.

However, the Bayesian approach may be somewhat limited in its ability to persuade a broad audience (whose beliefs may or may not be characterized by the selected prior distribution) as to the validity of its results. In an attempt to remedy this, Chaloner et al. (1993) developed software for eliciting prior beliefs and subsequently demonstrate its application to a collection of P relevant parties (medical professionals, policy makers, patients, etc.) in an attempt to robustify the procedure against outlying prior opinions. Modern computational methods allow real-time evaluation of the trial's stopping rule, despite the P-fold increase in the computational burden.

7.5.1 Statistical models

We now outline the essential features of a survival model. Suppose that associated with individual i in our study, $i = 1, \ldots, n$, is a p-dimensional covariate vector $\mathbf{z}_i$ which does not change over time. We model the effect of these covariates using the familiar log-linear proportional hazards form, namely $h(t|\mathbf{z}) = h_0(t) \exp(\boldsymbol{\beta}^T \mathbf{z})$, where $\boldsymbol{\beta}$ is a p-vector of unknown parameters. Suppose that d failures are observed in the combined group over the course of the study. After listing the individuals in order of increasing time to failure or censoring, let $\mathbf{z}_i$ be the covariate vector associated with the ith individual in this list, $i = 1, \ldots, n$. Denote the jth *risk set*, or the collection of individuals still alive at the time of the jth death, by $\mathcal{R}_j$ for $j = 1, \ldots, d$. Then the Cox partial likelihood (Cox and Oakes, 1984, Chapter 7) for this model is given by

$$L(\boldsymbol{\beta}) = \prod_{j=1}^{d} \left(\frac{e^{\boldsymbol{\beta}^T \mathbf{z}_j}}{\sum_{k \in \mathcal{R}_j} e^{\boldsymbol{\beta}^T \mathbf{z}_k}} \right), \qquad (7.25)$$

where $\boldsymbol{\beta} = (\beta_1, \ldots, \beta_p)$ and $\mathbf{z}_j = (z_{1j}, \ldots, z_{pj})$. It is often convenient to treat this partial likelihood as a likelihood for the purpose of computing the posterior density. A justification for this is offered by Kalbfleisch (1978), who places a gamma process prior on the baseline hazard function (independent of the prior for $\boldsymbol{\beta}$), and shows that the marginal posterior density for $\boldsymbol{\beta}$ approaches a form proportional to the Cox partial likelihood as the baseline hazard prior becomes arbitrarily diffuse.

To complete the Bayesian model, we must specify a joint prior distribution $\pi(\boldsymbol{\beta})$ on the components of $\boldsymbol{\beta}$. Before doing so, however, we focus attention more clearly on the monitoring of treatment-control clinical trials. Suppose the variable of primary interest is the first component of $\mathbf{z}$, z_1, a binary variable indicating whether individual i has been randomly assigned to the control group ($z_1 = 0$) or the treatment group ($z_1 = 1$). Hence, interest focuses on the parameter β_1, and in particular on whether $\beta_1 = 0$. Now returning to the question of prior elicitation for $\boldsymbol{\beta}$, we make the simplifying assumption that $\pi(\boldsymbol{\beta}) = \pi_1(\beta_1)\pi_{-1}(\boldsymbol{\beta}_{-1})$, where $\boldsymbol{\beta}_{-1} = (\beta_2, \ldots, \beta_p)$. That is, β_1 is *a priori* independent of the covariate-effect parameters $\boldsymbol{\beta}_{-1}$. In the absence of compelling evidence to the contrary, this *a priori* lack of interaction between the treatment effect and covariates provides a reasonable initial model. Moreover, this prior specification does not force independence in the posterior, where the data determine the appropriate level of dependence. In addition, we assume a flat prior on the nuisance parameters, namely $\pi_{-1}(\boldsymbol{\beta}_{-1}) \equiv 1$. This eliminates specifying the exact shape of the prior, and produces an evaluation of covariate effects comparable to that produced by traditional frequentist methods. Of course, if objective evidence on covariate effects is available, it can be used. However, since "standard practice" in patient care changes rapidly over time, such a prior should be fairly vague or include a between-study variance component.

7.5.2 Treatment effect prior determination

It remains to specify the (univariate) prior distribution on the treatment effect, $\pi_1(\beta_1)$. An algorithm and corresponding software for eliciting such a prior from an expert was developed by Chaloner et al. (1993). While this method has been an operational boon to the Bayesian approach to this problem, its use of subjective expert opinion as the sole source of evidence in creating the prior distribution on β_1 may constitute an unnecessary risk in many trials. Objective development of a prior for the control-group failure rate may be possible using existing data, since this value depends only on the endpoint selected, and not the particular treatment under investigation. Even so, eligibility, diagnostic and other temporal drifts limit the precision of such a prior. Data relevant for the treatment group are less likely to exist, but may be available in the form of previous randomized trials of the treatment. Failing this, existing nonrandomized trial data might be employed, though prior uncertainty should be further inflated here to account for the inevitable biases in such data.

A sensible overall strategy for prior determination in clinical trials is recommended by Spiegelhalter, Freedman, and Parmar (1994). These authors suggest investigating and reporting the posterior distributions that arise over a *range* of priors, in the spirit of sensitivity analysis (see Subsection 4.2.1). For example, we might investigate the impact of a "clinical" prior, which represents the (typically optimistic) prior feelings of the trial's investigators, a "skeptical" prior, which reflects the opinion of a person or regulatory agency that doubts the treatment's effectiveness, and a "noninformative" prior, a neutral position that leads to posterior summaries formally equivalent to those produced by standard maximum likelihood techniques. The specific role of the prior and the uncertainty remaining in the results after the data have been observed can then be directly ascertained.

7.5.3 Computation and advanced models

For the above models and corresponding priors, we require efficient algorithms for computing posterior summaries (means, variances, density estimates, etc.) of the quantities of interest to the members of the clinical trial's data monitoring board. Naturally β_1 itself would be an example of such a quantity; the board might also be interested in the Bayes factor BF in favor of $H_0 : \beta_1 \neq 0$, or perhaps simply the posterior probability that the treatment is efficacious, given that this has been suitably defined. For simplicity of presentation, in what follows we concentrate solely on the posterior of β_1. The simplest method for summarizing β_1's posterior would be to use the first order (normal) approximation provided by Theorem 3.1. Such an approximation is extremely attractive in that it is computationally easy to produce quantiles of the posterior distribution, as well as a standardized Z-score for testing $H_0 : \beta_1 = \beta_1^{(0)}$ (for example, $Z = (E(\beta_1|\mathcal{R}) - \beta_1^{(0)})/\sqrt{V(\beta_1|\mathcal{R})}$). This statistic allows sequential graphical monitoring displays similar to traditional frequentist monitoring plots (e.g. Lan and DeMets, 1983).

If n is not sufficiently large, results appealing to asymptotic posterior normality will be inappropriate, and we must turn to more accurate methods such as Laplace's method or traditional numerical quadrature, such as the Gauss-Hermite rules. In fact, if the dimension of β is small, it will be very convenient to simply discretize the entire vector (instead of just β_1) and use a naive summation method, such as the trapezoidal rule. For high-dimensional β, Monte Carlo integration methods will likely provide the only feasible alternative.

These methods are especially convenient for an extension of model (7.25) known as a *frailty model*. Introduced by Vaupel, Manton, and Stallard (1979), this is a mixed model with random effects (called *frailties*) that correspond to an individual's health status. That is, we replace the Cox hazard function for subject i, $h(t|\mathbf{z}_i, \boldsymbol{\beta}) = h_0(t) \exp(\boldsymbol{\beta}^T \mathbf{z}_i)$, by

$$h(t|\mathbf{z}_i, \boldsymbol{\beta}, \gamma_i) = h_0(t) \, \gamma_i \, \exp(\boldsymbol{\beta}^T \mathbf{z}_i) \, ,$$

where the frailties γ_i are typically assumed to be i.i.d. variates from a specified distribution (e.g., gamma or lognormal). Clayton (1991) and Sinha (1993) show how the Gibbs sampler may be used to fit this model under an independent increment gamma process parametrization of the baseline hazard function. Gustafson (1994) and Sargent (1995) avoid this assumption by treating the Cox partial likelihood as a likelihood, and employing more sophisticated MCMC approaches.

A nonnormal posterior distribution for β_1 precludes traditional monitoring plots based on standardized Z-scores. However, Bayesian monitoring may still proceed by plotting the posterior probabilities that β_1 falls outside an "indifference zone" (or "range of equivalence"), $(\beta_{1,L}, \beta_{1,U})$. Recall from Subsection 4.2.2 that this approach, proposed by Freedman and Spiegelhalter (1983; 1989; 1992), assumes that for β_1 values within this prespecified range, we are indifferent as to the practical use of either the treatment or the control preparation. Since negative values of β_1 indicate an effective treatment, we can set $\beta_{1,L} = K < 0$ and $\beta_{1,U} = 0$, an additional benefit perhaps being required of the treatment in order to justify its higher cost in terms of resources, clinical effort, or toxicity. The Bayesian decision rule terminates the trial (or more realistically, provides a guideline for termination) when $P(\beta_1 > \beta_{1,U}|\mathcal{R})$ is sufficiently small (deciding in favor of the treatment), *or* when $P(\beta_1 < \beta_{1,L}|\mathcal{R})$ is sufficiently small (deciding in favor of the control). The case study in Section 8.2 provides an illustration.

7.6 Sequential analysis

As illustrated in equation (2.6), when data samples $\mathbf{y}_1, \mathbf{y}_2, \ldots, \mathbf{y}_K$ arrive over time, Bayes' Rule may be used sequentially simply by treating the current posterior as the prior for the next update, i.e.,

$$p(\boldsymbol{\theta}|\mathbf{y}_1, \ldots, \mathbf{y}_K) \propto f_K(\mathbf{y}_K|\boldsymbol{\theta}, \mathbf{y}_1, \ldots, \mathbf{y}_{K-1}) p(\boldsymbol{\theta}|\mathbf{y}_1, \ldots, \mathbf{y}_{K-1}) \, .$$

This formula is conceptually straightforward and simple to implement provided all that is required is an updated posterior for θ at each monitoring point. In many cases, however, the analyst faces a specific *decision* at each timepoint – say, a repeated hypothesis test regarding θ, the outcome of which may determine whether the experiment should be continued or not. For the frequentist, this raises the thorny issues of repeated significance testing and multiple comparisons. To protect the overall error rate α, many classical approaches resort to "α-spending functions," as in the aforementioned work by Lan and DeMets (1983).

In the Bayesian framework, the problem is typically handled by combining the formal decision theoretic tools of Appendix B with the usual Bayesian probability tools. The traditional method of attack for such problems is *backward induction* (see e.g. DeGroot, 1970, Chapter 12), a method which, while conceptually simple, is often difficult to implement when K is at all large. Subsection 7.6.1 describes a class of two-sided sequential decision problems (i.e., their likelihood, prior, and loss structures), while Subsection 7.6.2 outlines their solution via the traditional backward induction approach. Then in Subsection 7.6.3, we present a *forward* sampling algorithm which can produce the optimal stopping boundaries for a broad subclass of these problems at far less computational and analytical expense. In addition, this forward sampling method can *always* be used to find the best boundaries within a certain class of decision rules, even when backward induction is infeasible. We then offer a brief discussion, but defer exemplification of the methods until Subsection 8.2.4, as part of our second Chapter 8 case study.

7.6.1 Model and loss structure

Carlin, Kadane, and Gelfand (1998) present the sequential decision problem in the context of clinical trial monitoring. Since this setting is very common in practice, and is one we have already begun to develop in Sections 4.2.2 and 7.5, we adopt it here as well.

Suppose then that a clinical trial has been established to estimate the effect θ of a certain treatment. Information already available concerning θ is summarized by a prior distribution $p(\theta)$. As patient accrual and followup is getting underway, the trial's data safety and monitoring board (hereafter referred to simply as "the board") agrees to meet on K prespecified dates to discuss the trial's progress and determine whether or not it should be continued. At each monitoring point $k \in \{1, \ldots, K\}$, some cost A_k will have been invested in obtaining a new datapoint y_k having distribution $f(y_k|\theta)$, enabling computation of the updated posterior distribution $p(\theta|y_1, \ldots, y_k)$. It is convenient to think of $k = 0$ as an additional, preliminary monitoring point, allowing the board to determine whether the trial should be run at all, given only the prior and the set of costs $\{A_k\}$.

Suppose that for the treatment effect parameter $\theta \in \Re$, large negative values suggest superiority of the treatment, while large positive values suggest superiority of the placebo (inferiority of the treatment). Suppose further that in conjunction with substantive information and clinical expert opinion, an indifference zone (θ_L, θ_U) can be determined such that the treatment is preferred for $\theta < \theta_L$, the placebo for $\theta > \theta_U$, and either decision is acceptable for $\theta_L < \theta < \theta_U$.

At the final monitoring point (K), there are then two decisions available to the board: d_1 (stop and decide in favor of the treatment), or d_2 (stop and decide in favor of the placebo). Corresponding to these two decisions, the indifference zone (θ_L, θ_U) suggests the following loss functions for given true values of $\theta \in \Re$:

$$l_1^{(K)}(d_1, \theta) = s_1^{(K)}(\theta - \theta_U)^+ \text{ , and } l_2^{(K)}(d_2, \theta) = s_2^{(K)}(\theta_L - \theta)^+ \text{ ,} \quad (7.26)$$

where $s_1^{(K)}, s_2^{(K)} > 0$ and the "+" superscript denotes the "positive part" function, i.e., $y^+ \equiv \max(0, y)$. That is, incorrect decisions are penalized, with the penalty increasing linearly as θ moves further from the indifference zone. To facilitate the presentation of backward induction later in this section, for the moment we simplify the loss functions (7.26) by dropping the positive part aspect, resulting in linearly increasing *negative* losses (gains) for correct decisions.

Given the posterior distribution of the treatment parameter at this final time point, $p(\theta|y_K)$, the best decision is that corresponding to the smaller of the two posterior expected losses, $L_1^{(K)} \equiv E[l_1^{(K)}(d_1, \theta)|y_K] = s_1^{(K)}[E(\theta|y_K) - \theta_U]$, and $L_2^{(K)} \equiv E[l_2^{(K)}(d_2, \theta)|y_K] = s_2^{(K)}[\theta_L - E(\theta|y_K)]$. If $E(\theta|y_K)$ exists, then clearly $L_1^{(K)}$ increases in $E(\theta|y_K)$ while $L_2^{(K)}$ decreases in $E(\theta|y_K)$. The breakpoint γ_K between the optimality of the two decisions follows from equating these two expected losses. Specifically, decision d_1 (decide in favor of the treatment) will be optimal if and only if

$$E(\theta|y_K) \leq \gamma_K \equiv \frac{s_1^{(K)}\theta_U + s_2^{(K)}\theta_L}{s_1^{(K)} + s_2^{(K)}} \text{ ;} \quad (7.27)$$

otherwise (when $E(\theta|y_K) > \gamma_K$), decision d_2 is optimal. Therefore, when $s_1^{(K)} = s_2^{(K)}$, the breakpoint γ_K is simply the midpoint of the indifference zone. A similar result holds under the positive part loss functions in (7.26).

7.6.2 Backward induction

To obtain the optimal Bayesian decision in the sequential setting at any time point other than the last, one must consider the probability of data values that have not yet been observed. This requires "working the problem backward" (from time K back to time 0), or backward induction. To illustrate, consider time $K - 1$. There are now three available decisions: d_1, d_2, and d_3 (continue and decide after y_K is available for analysis). To simplify the notation, suppose for now that $K = 1$, so that no data have yet accumulated and hence the current posterior $p(\theta|y_{K-1})$ equals the prior

$p(\theta)$. The posterior expected losses are now $L_1^{(K-1)} = s_1^{(K-1)}[E(\theta) - \theta_U]$, $L_2^{(K-1)} = s_2^{(K-1)}[\theta_L - E(\theta)]$, and

$$L_3^{(K-1)} = A_K + \int_{R_1^{(K)}} \int_{-\infty}^{\infty} s_1^{(K)}(\theta - \theta_U)p(\theta|y_K)m(y_K)d\theta dy_K$$
$$+ \int_{R_2^{(K)}} \int_{-\infty}^{\infty} s_2^{(K)}(\theta_L - \theta)p(\theta|y_K)m(y_K)d\theta dy_K ,$$

(7.28)

where $R_1^{(K)} = \{y_K : E(\theta|y_K) \leq \gamma_K\}$, $R_2^{(K)} = \{y_K : E(\theta|y_K) > \gamma_K\}$, γ_K is as given in (7.27), and $m(y_K)$ is the marginal density of the (as yet unobserved) data value y_K. Once the prior $p(\theta)$ and the sampling cost A_K are chosen, the optimal decision is that corresponding to the minimum of $L_1^{(K-1)}$, $L_2^{(K-1)}$ and $L_3^{(K-1)}$. Alternatively, a class of priors (say, the normal distributions having fixed variance σ^2) might be chosen, and the resulting $L_i^{(K-1)}$ plotted as functions of the prior mean μ. The intersection points of these three curves would determine stopping regions as a function of μ, reminiscent of those found in traditional sequential monitoring displays.

While the integration involved in computing $L_3^{(K-1)}$ is a bit awkward, it is not particularly difficult or high-dimensional. Rather, the primary hurdle arising from the backward induction process is the explosion in bookkeeping complexity arising from the resulting alternating sequence of integrations and minimizations. The number of Monte Carlo generations required to implement a backward induction solution also increases exponentially in the number of backward steps.

7.6.3 Forward sampling

The backward induction method's geometric explosion in both organizational and computational complexity as K increases motivates a search for an alternative approach. We now describe a forward sampling method which permits finding the optimal sequential decision in a prespecified class of such decisions. We implement this method when the class consists of decision rules of the following form: at each step k,

- for $E(\theta|y_1, \ldots, y_k) \leq \gamma_{k,L}$, make decision d_1 (stop and choose the treatment);

- for $E(\theta|y_1, \ldots, y_k) > \gamma_{k,U}$, make decision d_2 (stop and choose placebo);

- for $\gamma_{k,L} < E(\theta|y_1, \ldots, y_k) \leq \gamma_{k,U}$, make decision d_3 (continue sampling).

Carlin, Kadane, and Gelfand (1998) provide sufficient conditions to ensure that the optimal strategy is of this form; customary normal mean, binomial, and Poisson models meet these conditions.

Note that $\gamma_{k,L} \leq \gamma_{k,U}$, with equality always holding for $k = K$ (i.e., $\gamma_{K,L} = \gamma_{K,U} \equiv \gamma_K$). This special structure motivates the following forward sampling algorithm, which for comparison with Section 7.6.2 is illustrated in the case $K = 2$. Suppose an independent sample $(\theta_j^*, y_{K-1,j}^*, y_{K,j}^*)$, $j =$

$1, \ldots, G$ is drawn via composition from the joint density $p(\theta, y_{K-1}, y_K) = p(\theta)f(y_{K-1}|\theta)f(y_K|\theta)$. When $K = 2$, decision rules can be indexed by the $2K-1 = 5$-dimensional vector $\gamma = (\gamma_{K-2,L}, \gamma_{K-2,U}, \gamma_{K-1,L}, \gamma_{K-1,U}, \gamma_K)'$. Suppose that at the ith iteration of the algorithm, a current estimate $\gamma^{(i)}$ of the optimal rule of this form is available. For each of the simulated parameter-data values, the loss incurred under the current rule $\gamma^{(i)}$ can be computed as follows:

$$
\begin{aligned}
\text{If } E(\theta) \leq \gamma^{(i)}_{K-2,L}, &\quad L_j = s_1^{(K-2)}(\theta_j^* - \theta_U) &\quad (7.29)\\
\text{else if } E(\theta) > \gamma^{(i)}_{K-2,U}, &\quad L_j = s_2^{(K-2)}(\theta_L - \theta_j^*)\\
\text{else if } E(\theta|y^*_{K-1,j}) \leq \gamma^{(i)}_{K-1,L}, &\quad L_j = s_1^{(K-1)}(\theta_j^* - \theta_U) + A_{K-1}\\
\text{else if } E(\theta|y^*_{K-1,j}) > \gamma^{(i)}_{K-1,U}, &\quad L_j = s_2^{(K-1)}(\theta_L - \theta_j^*) + A_{K-1}\\
\text{else if } E(\theta|y^*_{K-1,j}, y^*_{K,j}) \leq \gamma^{(i)}_K, &\quad L_j = s_1^{(K)}(\theta_j^* - \theta_U) + A_{K-1} + A_K\\
\text{else,} &\quad L_j = s_2^{(K)}(\theta_L - \theta_j^*) + A_{K-1} + A_K \, .
\end{aligned}
$$

Then $\bar{L} = \frac{1}{G}\sum_{j=1}^{G} L_j$ is a Monte Carlo estimate of the posterior expected loss under decision rule $\gamma^{(i)}$. If the current estimate is not the minimum value, it is adjusted to $\gamma^{(i+1)}$, and the above process is repeated. At convergence, the algorithm provides the desired optimal stopping boundaries.

Note that many of the $y^*_{K,j}$ values (and some of the $y^*_{K-1,j}$ values as well) will not actually figure in the determination of $\bar{L}$ because they contribute only to the later conditions in the decision tree (7.29). These cases correspond to simulated trials that stopped before these later data values were actually observed; ignoring them in a given replication creates the proper marginal distribution of the parameter and the observed data (if any). Indeed, note that when $E(\theta) \notin (\gamma_{K-2,L}, \gamma_{K-2,U}]$, all of the sampled values fall into one of the first two cases in (7.29). Second, $\bar{L}$ is not likely to be smooth enough as a function of $\gamma^{(i)}$ to allow its minimization via standard methods, such as a quasi-Newton method. Better results may be obtained by applying a more robust, univariate method (such as a golden section search) sequentially in each coordinate direction, i.e., within an iterated conditional modes algorithm (see Besag, 1986).

The main advantage of this approach over backward induction is clear: since the length of γ is $2K - 1$, the computational burden in forward sampling grows *linearly* in K, instead of exponentially. The associated bookkeeping and programming is also far simpler. Finally, the forward sampling algorithm is easily adapted to more advanced loss structures. For example, switching back to the positive part loss functions (7.26), a change that would cause substantial complications in the backward induction solution, leads only to a corresponding change in the scoring rule (7.29).

A potential disadvantage of the forward sampling algorithm is that it replaces a series of simple minimizations over discrete finite spaces (a choice among three alternatives) with a single, more difficult minimization over a continuous space ($\Re^{2K-1}$). One might well wonder whether this minimization is any more feasible for large K than backward induction. Fortunately, even when the exact minimum is difficult to find, it should be possible to find a rule that is nearly optimal via grid search over γ-space, meaning that approximate solutions will be available via forward sampling even when backward induction is infeasible.

It is critical to get a good estimate of *each* of the components of the optimizing γ, and not just the "components of interest" (typically, the early components of γ, near the current time point), for otherwise there is a risk of finding a local rather than a global minimum. As such, one may want to sample not from the joint distribution $p(\theta, y_1, \ldots, y_K)$, but from some modification of it designed to allow more realizations of the latter stages of the decision tree (7.29). To correct for this, one must then make the appropriate importance sampling adjustment to the L_j scores in (7.29). By deliberately straying from the true joint distribution, one is sacrificing overall precision in the estimated expected loss in order to improve the estimate of the later elements of γ. As such, the phrase "unimportance sampling" might be used to refer to this process.

Discussion

In the case of a one-parameter exponential family likelihood, forward sampling may be used to find the optimal rule based solely on the posterior mean. In a very general parametric case, the optimal sequential strategy could be a complex function of the data, leading to a very large class of strategies to be searched. But what of useful intermediate cases – say, a normal likelihood with both the mean and the variance unknown? Here, the class of possibly optimal sequential rules includes arbitrary functions of the sufficient statistic vector $(\bar{y}, s^2)$. In such a case, for the particular loss function in question, it may be possible to reduce the class of possible optimal strategies. But for practical purposes it may not be necessary to find the optimal sequential strategy. Forward sampling can be used to evaluate general sequential strategies of any specified form. If the class of sequential strategies offered to the algorithm is reasonably broad, the best strategy in that class is likely to be quite good, even if it is not optimal in the class of all sequential strategies. This constitutes a distinct advantage of the forward sampling algorithm. When backward induction becomes infeasible, as it quickly does even in simple cases, it offers no advice about what sequential strategy to employ. Forward sampling, however, can be applied in such a situation with a restricted class of sequential strategies to find a good practical one even if it is not the best possible.

Finally, forward sampling as a technique has been refined by several authors since its development. Brockwell and Kadane (2003) develop a gridding method that (when implemented with sufficient parallel computing power) allows forward sampling to work more efficiently, and thus tackle a far larger class of models. Forward sampling is also used by Müller et al. (2007), who (like us) reduce the allowable action space so that decisions can be made based on a low-dimensional summary statistic in a sequential dose ranging trial, and by Rossell et al. (2007), who use forward sampling to decide about sequential stopping one of many trials in a Phase II drug screening design.

7.7 Spatial and spatio-temporal models

Statisticians are increasingly faced with the task of analyzing data that are geographically referenced, and often presented in the form of maps. Echoing a theme already common in this book, this increase has been driven by advances in computing; in this case, the advent of *geographic information systems* (GISs), sophisticated computer software packages that allow the simultaneous graphical display and summary of spatial data. But while GISs such as `ARC/INFO`, `ArcView`, and `MapInfo` are extraordinarily powerful for data manipulation and display, as of the present writing, they have little capacity for statistical modeling and inference.

To establish a framework for spatial modeling, we follow that given in the classic textbook on the subject by Cressie (1993). Let $\mathbf{s} \in \Re^d$ be a location in d-dimensional space, and $X(\mathbf{s})$ be a (possibly vector-valued) data value observed at $\mathbf{s}$. The full dataset can then be modeled as the multivariate random process

$$\{X(\mathbf{s}) : \mathbf{s} \in D\}, \qquad (7.30)$$

where $\mathbf{s}$ varies over an index set $D \subset \Re^d$. Cressie (1993) then uses (7.30) to categorize spatial data into three cases:

- *Geostatistical data*, where $X(\mathbf{s})$ is a random vector at a location $\mathbf{s}$ that varies *continuously* over D, a fixed subset of $\Re^d$ that contains a d-dimensional rectangle of positive volume.

- *Lattice data*, where D is again a fixed subset (of regular or irregular shape), but now containing only countably many sites in $\Re^d$, normally supplemented by neighbor information.

- *Point pattern data*, where now D is itself random; its index set gives the locations of random events that are the spatial point pattern. $X(\mathbf{s})$ itself can simply equal 1 for all $\mathbf{s} \in D$ (indicating occurrence of the event) or possibly give some additional covariate information (producing a *marked point pattern process*).

An example of the first case might be a collection of measured oil or water reserves at various fixed *source* points $\mathbf{s}$, with the primary goal being to predict the reserve available at some unobserved *target* location **t**. In the second case, the observations might correspond to pixels on a computer screen, each of which has precisely four neighbors (up, down, left, and right), or perhaps average values of some variable over a geographic region (e.g., a county), each of whose neighbor sets contains those other counties that share a common border. Finally, point pattern data often arise as locations of certain kinds of trees in a forest, or residences of persons suffering from a particular disease (which in this case might be supplemented by age or other covariate information, producing a marked point pattern).

In the remainder of this section, we provide guidance on the Bayesian fitting of models for the first two cases above. While mining applications provided the impetus for much of the model development in spatial statistics, we find the term "geostatistical" somewhat misleading, since these methods apply much more broadly. Similarly, we dislike the term "lattice" because it connotes data corresponding to "corners" of a checkerboard-like grid. As such, we prefer to relabel these two cases more generically, according to the nature of the set D. Thus, in Subsection 7.7.1 we consider models for geostatistical data, or what we call *point source data models*, where the locations are fixed points in space that may occur anywhere in D. Then in Subsection 7.7.2, we turn to lattice data models, or what we term *regional summary data models*, where the source data are counts or averages over geographical regions.

The `maps` library in R has for some time allowed display of point source data and regional summary data at the state or county level in the United States; more recent R tools for spatial mappping and data analysis include `maptools`, `geoR`, `geoRglm`, `spatstat`, `spdep`, and, most recently, `spBayes`, which permits fully MCMC-based Bayesian analysis of point source data models in the multivariate normal case (including spatiotemporal models). Our treatment below is of necessity abbreviated; please see the textbook by Banerjee et al. (2004) for a full treatment of hierarchical modeling in the point source and regional summary data cases.

7.7.1 Point source data models

We begin with some general notation and terminology. Let $X(\mathbf{s}_i)$ denote the spatial process associated with a particular variable (say, environmental quality), observed at locations $\{\mathbf{s}_i : \mathbf{s}_i \in D \subset \Re^d\}$. A basic starting point is to assume that $X(\mathbf{s}_i)$ is a *second-order stationary* process, i.e., that $E[X(\mathbf{s})] = \mu$ and $Cov[X(\mathbf{s}_i), X(\mathbf{s}_j)] \equiv C(\mathbf{s}_i - \mathbf{s}_j) < \infty$. C is called the *covariogram*, and is analogous to the autocovariance function in time series analysis. It then follows that $Var[X(\mathbf{s}_i) - X(\mathbf{s}_j)] \equiv 2\gamma(\mathbf{s}_i - \mathbf{s}_j)$ is a function of the separation vector $\mathbf{s}_i - \mathbf{s}_j$ alone, and is called the *variogram*; $\gamma(\mathbf{s}_i - \mathbf{s}_j)$ itself is called the *semivariogram*. The variogram is said to be *isotropic* if it is a function solely of the Euclidean distance between $\mathbf{s}_i$ and $\mathbf{s}_j$, d_{ij}; in this case $2\gamma(\mathbf{s}_i - \mathbf{s}_j) = 2\gamma(d_{ij})$, so that $C(\mathbf{s}_i - \mathbf{s}_j) = C(d_{ij})$.

We define the *sill* of the variogram as $\lim_{d\to\infty} 2\gamma(d)$, and the *nugget* as $\lim_{d\to 0} 2\gamma(d)$. The latter need not be zero (i.e., the variogram may have a discontinuity at the origin) due to microscale variability or measurement error (Cressie, 1993, p. 59). If $C(d) \to 0$ as $d \to \infty$, then $2\gamma(d) \to 2C(\mathbf{0})$, the sill. For a monotonic variogram that reaches its sill exactly, the distance at which this sill is reached is called the *range*; observations farther apart than the range are uncorrelated. If the sill is reached only asymptotically, the distance such that $Corr[X(\mathbf{s}_i), X(\mathbf{s}_j)] = 0.05$ is called the *effective range*.

If we further assume a second-order stationary *Gaussian* process, our model can be written as

$$X(\mathbf{s}_i) = \mu + W(\mathbf{s}_i) + \epsilon(\mathbf{s}_i) \,, \tag{7.31}$$

where μ is again the grand mean, $W(\mathbf{s})$ accounts for spatial correlation, and $\epsilon(\mathbf{s})$ captures measurement error, $W(\mathbf{s})$ and $\epsilon(\mathbf{s})$ independent and normally distributed. To specify an isotropic spatial model, we might use a parametric covariance function, such as $Cov(W(\mathbf{s}_i), W(\mathbf{s}_j)) = \sigma^2 \rho(d_{ij}; \phi)$. For example, we might choose $\rho(d_{ij}; \phi) = \exp(-\phi d_{ij})$, $\phi > 0$, so that the covariance decreases with distance at constant exponential rate ϕ. The measurement error component is then often specified as $\epsilon(\mathbf{s}) \overset{iid}{\sim} N(0, \tau^2)$. From this model specification we may derive the variogram as

$$2\gamma(\tau^2, \sigma^2, \phi) = 2[\tau^2 + \sigma^2(1 - \rho(d_{ij}; \phi))] \,, \tag{7.32}$$

so that τ^2 is the nugget and $\tau^2 + \sigma^2$ is the sill.

Despite the occasionally complicated notation and terminology, equations (7.31) and (7.32) clarify that the preceding derivation determines only a four-parameter model. Bayesian model fitting thus proceeds once we specify the prior distribution for $\boldsymbol{\theta} = (\mu, \tau^2, \sigma^2, \phi)$; standard choices for these four components might be flat, inverse gamma, inverse gamma, and gamma, respectively (the latter since there is no hope for a conjugate specification for ϕ). Because this model is quite low-dimensional, any of the computational methods described in Chapter 3 might be appropriate; in particular, one could easily obtain draws $\{\boldsymbol{\theta}^{(g)}, g = 1, \ldots G\}$ from $p(\boldsymbol{\theta}|\mathbf{X})$.

Bayesian kriging

While marginal posterior distributions such as $p(\phi|\mathbf{x})$ might be of substantial interest, historically the primary goal of such models has been to predict the unobserved value $X(\mathbf{t})$ at some target location $\mathbf{t}$. The traditional minimum mean squared error approach to this spatial prediction problem is called *kriging*, a rather curious name owing to the influential early work of South African mining engineer D.G. Krige (see, e.g., Krige, 1951).

Handcock and Stein (1993) and Diggle et al. (1998) present the Bayesian implementation of kriging, which is most naturally based on the predictive distribution of the targets given the sources. More specifically, suppose we seek predictions at target locations $(\mathbf{t}_1, \ldots, \mathbf{t}_m)$ given (spatially correlated) response data at source locations $(\mathbf{s}_1, \ldots, \mathbf{s}_n)$. Following equations (7.31) and (7.32) above, the joint distribution of $\mathbf{X}_1 = (X(\mathbf{s}_1), \ldots, X(\mathbf{s}_n))'$ and $\mathbf{X}_2 = (X(\mathbf{t}_1), \ldots, X(\mathbf{t}_m))'$ is

$$[(\mathbf{X}_1, \mathbf{X}_2)' | \mu, \tau^2, \sigma^2, \phi] \sim MVN_{m+n}\left(\mu \mathbf{1}_{m+n}, \begin{pmatrix} \Sigma_{11} & \Sigma_{12} \\ \Sigma_{21} & \Sigma_{22} \end{pmatrix}\right), \quad (7.33)$$

where $\Sigma_{kl} \equiv Cov(\mathbf{X}_k, \mathbf{X}_l)$, $k, l = 1, 2$ based on our isotropic model (e.g., the ijth element of Σ_{11} is just $\sigma^2 \exp(-\phi d_{ij})$ where d_{ij} is the separation distance between $\mathbf{s}_i$ and $\mathbf{s}_j$, and similarly for the other blocks). Thus, using standard multivariate normal distribution theory (see e.g. Guttman, 1982, pp. 70-72), the conditional distribution of the target values $\mathbf{X}_2$ given the source values $\mathbf{X}_1$ and the model parameters is

$$[\mathbf{X}_2 | \mathbf{X}_1, \mu, \tau^2, \sigma^2, \phi] \sim MVN_m(\mu_{2.1}, \Sigma_{2.1}),$$

where

$$\mu_{2.1} = \mu \mathbf{1}_m + \Sigma_{21}\Sigma_{11}^{-1}(\mathbf{X}_1 - \mu \mathbf{1}_n), \text{ and } \Sigma_{2.1} = \Sigma_{22} - \Sigma_{21}\Sigma_{11}^{-1}\Sigma_{12}.$$

Therefore, marginalizing over the parameters, the predictive distribution we seek is given by

$$
\begin{aligned}
p(\mathbf{X}_2 | \mathbf{X}_1) &= \int p(\mathbf{X}_2 | \mathbf{X}_1, \mu, \tau^2, \sigma^2, \phi) p(\mu, \tau^2, \sigma^2, \phi | \mathbf{X}_1) d\mu d\tau^2 d\sigma^2 d\phi \\
&\approx \frac{1}{G} \sum_{g=1}^{G} p(\mathbf{X}_2 | \mathbf{X}_1, \mu^{(g)}, (\tau^2)^{(g)}, (\sigma^2)^{(g)}, \phi^{(g)}), \quad (7.34)
\end{aligned}
$$

where $(\mu^{(g)}, (\tau^2)^{(g)}, (\sigma^2)^{(g)}, \phi^{(g)})$ are the posterior draws obtained above. Point and interval estimates for the spatially interpolated values $X(\mathbf{t}_j)$ then arise from the appropriate marginal summaries of (7.34). Recall from Chapter 3 that we need not actually do the mixing in (7.34); merely drawing $\mathbf{X}_2^{(g)}$ values by composition from $p(\mathbf{X}_2 | \mathbf{X}_1, \mu^{(g)}, (\tau^2)^{(g)}, (\sigma^2)^{(g)}, \phi^{(g)})$ is sufficient to produce a histogram estimate of any component of $p(\mathbf{X}_2 | \mathbf{X}_1)$.

Anisotropy

If the spatial correlation between $X(\mathbf{s}_i)$ and $X(\mathbf{s}_j)$ depends not only on the length of the separation vector $\mathbf{h}_{ij} \equiv \mathbf{s}_i - \mathbf{s}_j$, but on its *direction* as well, the process $X(\mathbf{s})$ is said to be *anisotropic*. A commonly modeled form of anisotropy is *geometric range anisotropy*, where we assume there exists a positive definite matrix B such that the variogram takes the form

$$2\gamma(\mathbf{h}_{ij}) = 2\gamma((\mathbf{h}_{ij}' B \mathbf{h}_{ij})^{1/2}).$$

Taking $B = \phi^2 I$ produces $(\mathbf{h}'_{ij} B \mathbf{h}_{ij})^{1/2} = \phi d_{ij}$, hence isotropy. Thus, anisotropy replaces ϕ by the parameters in B; equation (7.32) generalizes to

$$2\gamma(\tau^2, \sigma^2, B) = 2[\tau^2 + \sigma^2(1 - \rho((\mathbf{h}'_{ij} B \mathbf{h}_{ij})^{1/2}))] \,, \qquad (7.35)$$

so that τ^2 and $\tau^2 + \sigma^2$ are again the nugget and sill, respectively.

In the most common spatial setting, $\Re^2$, the 2×2 symmetric matrix $B = \begin{pmatrix} \beta_{11} & \beta_{12} \\ \beta_{12} & \beta_{22} \end{pmatrix}$ completely defines the geometric anisotropy, and produces elliptical (instead of spherical) correlation contours in the plane having orientation ω defined implicitly by the relation

$$\cot(2\omega) = \frac{\beta_{11} - \beta_{22}}{2\beta_{12}} \,.$$

Also, the effective range in the direction η, r_η, is given by

$$r_\eta = \frac{c}{(\mathbf{h}'_\eta B \mathbf{h}_\eta)^{1/2}} \,,$$

where $\mathbf{h}_\eta = (\cos\eta, \sin\eta)$, the unit vector in the direction of η and c is a constant such that $\rho(c) = 0.05$. Thus, for our simple exponential ρ function above, we have $c = -\log(0.05) \approx 3$.

Ecker and Gelfand (1999) again recommend a simple product of independent priors for the components of $\boldsymbol{\theta} = (\mu, \tau^2, \sigma^2, B)$, suggesting a $Wishart(R, \nu)$ prior for B with small degrees of freedom ν. Again, while the notation may look a bit frightening, the result in $\Re^2$ is still only a 6-parameter model (up from 4 in the case of isotropy). Estimation and Bayesian kriging may now proceed as before; see Ecker and Gelfand (1999) for additional guidance on preanalytic exploratory plots, model fitting, and a real-data illustration. See also Handcock and Wallis (1994) for a spatio-temporal extension of Bayesian kriging, an application involving temperature shifts over time, and related discussion. Le, Sun, and Zidek (1997) consider Bayesian multivariate spatial interpolation in cases where certain data are missing by design.

7.7.2 Regional summary data models

In this subsection, we generally follow the outline of Waller et al. (1997), and consider spatial data (say, on disease incidence) that are available as summary counts or rates for a defined region, such as a county, district, or census tract. Within a region, suppose that counts or rates are observed for subgroups of the population defined by socio-demographic variables, such as gender, race, age, and ethnicity. Suppose further that for each subgroup within each region, counts or rates are collected within regular time intervals, e.g. annually, as in our example. Hence, in general notation, an observation can be denoted by $y_{i\ell t}$. Here $i = 1, ..., I$ indexes the regions,

$\ell = 1, ..., L$ indexes the subgroups, and $t = 1, ..., T$ indexes the time periods. In application, subgroups are defined through factor levels, so that if we used, say, three factors, subscript ℓ would be replaced by jkm.

We assume $y_{i\ell t}$ is a count arising from a probability model with a baseline rate and a region-, group-, and year-specific relative risk $\psi_{i\ell t}$. Based on a likelihood model for the vector of observed counts $\mathbf{y}$ given the vector of relative risks $\boldsymbol{\psi}$, and a prior model on the space of possible $\boldsymbol{\psi}$s, MCMC computational algorithms can estimate a posterior for $\boldsymbol{\psi}$ given $\mathbf{y}$, which is used to create a disease map. Typically, a map is constructed by color-coding the means, medians, or other summaries of the posterior distributions of the $\psi_{i\ell t}$.

Typically, the objective of the prior specification is to stabilize rate estimates by *smoothing* the crude map, though environmental equity concerns imply interest in *explaining* the rates. The crude map arises from the likelihood model alone using estimates (usually MLEs) of the $\psi_{i\ell t}$ based only on the respective $y_{i\ell t}$. Such crude maps often feature large outlying relative risks in sparsely populated regions, so that the map may be visually dominated by rates having very high uncertainty. Equally important, the maximum likelihood estimates fail to take advantage of the similarity of relative risks in adjacent or nearby regions. Hence, an appropriate prior for $\boldsymbol{\psi}$ will incorporate exchangeability and/or spatial assumptions which will enable the customary Bayesian "borrowing of strength" to achieve the desired smoothing.

The likelihood model assumes that, given $\boldsymbol{\psi}$, the counts $y_{i\ell t}$ are conditionally independent Poisson variables. The Poisson acts as an approximation to a binomial distribution, say, $y_{i\ell t} \sim Bin(n_{i\ell t}, p_{i\ell t})$ where $n_{i\ell t}$ is the known number of individuals at risk in county i within subgroup ℓ at time t, and $p_{i\ell t}$ is the associated disease rate. Often the counts are sufficiently large so that $y_{i\ell t}$ or perhaps $\sqrt{y_{i\ell t}}$ (or the Freeman-Tukey transform, $\sqrt{y_{i\ell t}} + \sqrt{y_{i\ell t} + 1}$) can be assumed to approximately follow a normal density. In our case, however, the partitioning into subgroups results in many small values of $y_{i\ell t}$ (including several 0s), so we confine ourselves to the Poisson model, i.e. $y_{i\ell t} \sim Po(n_{i\ell t}p_{i\ell t})$.

We reparametrize to relative risk by writing $E_{i\ell t}\psi_{i\ell t} = n_{i\ell t}p_{i\ell t}$, where $E_{i\ell t}$ is the expected count in region i for subgroup ℓ at time t. Thus our basic model takes the form

$$y_{i\ell t} \sim Po(E_{i\ell t}\psi_{i\ell t}) .$$

In particular, if p^* is an overall disease rate, then $\psi_{i\ell t} = p_{i\ell t}/p^*$ and $E_{i\ell t} = n_{i\ell t}p^*$. The ψs are said to be *externally standardized* if p^* is obtained from another data source, such as a standard reference table (see, e.g., Bernardinelli and Montomoli, 1992). The ψs are said to be *internally standardized* if p^* is obtained from the given dataset, e.g., if $p^* = \sum_{i\ell t} y_{i\ell t} / \sum_{i\ell t} n_{i\ell t}$. In our data example, we lack an appropriate standard reference table, and therefore rely on the latter approach. Under external standardizing a product Poisson likelihood arises, while under internal standardizing the joint distribution of the $y_{i\ell t}$ is multinomial. However, since likelihood inference is unaffected by whether or not we condition on $\sum_{i\ell t} y_{i\ell t}$ (see e.g. Agresti, 1990, p. 455-56), it is common practice to retain the product Poisson likelihood.

The interesting aspect in the modeling involves the prior specification for $\psi_{i\ell t}$, or equivalently for $\mu_{i\ell t} \equiv \log \psi_{i\ell t}$. Models for $\mu_{i\ell t}$ can incorporate a variety of *main effect* and *interaction* terms. Consider the main effect for subgroup membership. In our example, with two levels for sex and two levels for race, it takes the form of a sex effect plus a race effect plus a sex-race interaction effect. In general, we can write this main effect as $\varepsilon_\ell = \mathbf{x}_\ell^T \boldsymbol{\beta}$ for an appropriate design vector $\mathbf{x}_\ell$ and parameter vector $\boldsymbol{\beta}$. In the absence of prior information, we would take a vague, possibly flat, prior for $\boldsymbol{\beta}$. In many applications it would be sensible to assume additivity and write $\mu_{i\ell t} = \varepsilon_\ell + \varepsilon_{it}$. For instance, in our example this form expresses the belief that sex and race effects would not be affected by region and year. This is the only form we have investigated thus far though more complicated structures could, of course, be considered. Hence, it remains to elaborate ε_{it}.

Temporal modeling

As part of ε_{it}, consider a main effect for time δ_t. While we might specify δ_t as a parametric function (e.g., linear or quadratic), we prefer a qualitative form that lets the data reveal any presence of trend via the set of posterior densities for the δ_t. For instance, a plot of the posterior means or medians against time would be informative. But what sort of prior shall we assign to the set $\{\delta_t\}$? A simple choice is again a flat prior. An alternative is an autoregressive prior with say $AR(1)$ structure, that is, $\delta_t | \delta_{t-1} \sim N\left(\gamma_t + \rho(\delta_{t-1} - \gamma_{t-1}), \sigma_\delta^2\right)$. We might take $\gamma_t = \gamma t$, whence γ and σ_δ^2 are hyperparameters.

Spatial modeling

The main effect for regions, say η_i, offers many possibilities. For instance, we might have regional covariates collected in a vector $\mathbf{z}_i$, which contribute a component $h(\mathbf{z}_i)$. Typically h would be a specified parametric function $h(\mathbf{z}_i; \boldsymbol{\omega})$ possibly linear, i.e., $\mathbf{z}_i^T \boldsymbol{\omega}$. In any event, a flat prior for $\boldsymbol{\omega}$ would likely be assumed. In addition to or in the absence of covariate information we might include regional random effects θ_i to capture heterogeneity among the regions (Clayton and Bernardinelli, 1992; Bernardinelli and Montomoli, 1992). If so, since i arbitrarily indexes the regions, an exchangeable prior for the θ_is seems appropriate, i.e.,

$$\theta_i \overset{iid}{\sim} N\left(\kappa_\theta, \frac{1}{\tau}\right). \tag{7.36}$$

We then add a flat hyperprior for κ_θ, but require a proper hyperprior (typically a gamma distribution) for τ. Formally, with an improper prior such as $\tau^{(d-3)/2}$, $d = 0, 1, 2$, the resultant joint posterior need not be proper (Hobert and Casella, 1996). Pragmatically, an informative prior for τ insures well behaved MCMC model fitting; see Banerjee et al. (2004, Sec. 5.4) for further discussion.

Viewed as geographic locations, the regions naturally suggest the possibility of spatial modeling (Clayton and Kaldor, 1987; Cressie and Chan, 1989; Besag, York and Mollié, 1991; Clayton and Bernardinelli, 1992; Bernardinelli and Montomoli, 1992). As with exchangeability, such spatial modeling introduces association across regions. Of course, we should consider whether, after adjusting for covariate effects, it is appropriate for geographic proximity to be a factor in the correlation among observed counts. If so, we denote the spatial effect for region i by ϕ_i. In the spirit of Clayton and Kaldor (1987) and Cressie and Chan (1989) who extend Besag (1974), we model the ϕ_i using a conditional autoregressive (CAR) model. That is, we assume the conditional density of $\phi_i | \phi_{j \neq i}$ is proportional to $\exp\left[-\frac{\lambda}{2}(a_i\phi_i - \sum_{j \neq i} w_{ij}\phi_j)^2\right]$, where $w_{ij} \geq 0$ is a weight reflecting the influence of ϕ_j on the expectation of ϕ_i, and $a_i > 0$ is a "sample size" associated with region i. From Besag (1974), we may show that the joint density of the vector of spatial effects ϕ is proportional to $\exp(-\frac{\lambda}{2}\phi^T B\phi)$ where $B_{ii} = a_i$ and $B_{ij} = -a_iw_{ij}$. It is thus a proper multivariate normal density with mean $\mathbf{0}$ and covariance matrix B^{-1} provided B is symmetric and positive definite. Symmetry requires $a_iw_{ij} = a_jw_{ji}$. A proper gamma prior is typically assumed for λ.

Two special cases have been discussed in the literature. One, exemplified in Cressie and Chan (1989), Devine and Louis (1994), and Devine, Halloran and Louis (1994), requires a matrix of interregion distances d_{ij}. Then w_{ij} is set equal to $g(d_{ij})$ for a suitable decreasing function g. For example, Cressie and Chan (1989) choose g based on the estimated variogram of the observations. Because $w_{ij} = w_{ji}$, symmetry of B requires a_i constant. In fact, $a_i = g(0)$, which without loss of generality we can set equal to 1. A second approach, as in e.g. Besag, York, and Mollié (1991), defines a set ∂_i of *neighbors* of region i. Such neighbors can be defined as regions adjacent (contiguous) to region i, or perhaps as regions within a prescribed distance of region i. Let n_i be the number of neighbors of region i and let $w_{ij} = 1/n_i$ if $j \in \partial_i$, and 0 otherwise. Then, if $a_i = n_i$, B is symmetric. It is easy to establish that

$$\phi_i \mid \phi_{j \neq i} \sim N\left(\bar{\phi}_i, \frac{1}{\lambda n_i}\right), \qquad (7.37)$$

where $\bar{\phi}_i = n_i^{-1} \sum_{j \in \partial_i} \phi_j$. It is also clear that B is singular because the sum of all the rows or of all the columns is $\mathbf{0}$. Thus, the joint density is improper. In fact, it is straightforward to show that $\phi^T B \phi$ can be written as a sum of pairwise differences of neighbors. Hence, the joint density is invariant to translation of the ϕ_is, again demonstrating its impropriety. The notions of interregion distance and neighbor sets can be combined as in Cressie and Chan (1989) by setting $w_{ij} = g(d_{ij})$ for $j \in \partial_i$. In the exercises below, however, we confine ourselves to pairwise difference specification based upon adjacency, since this requires no inputs other than the regional map, and refer to such specifications as $CAR(\lambda)$ priors.

Typically the impropriety in the CAR prior is resolved by adding the sum-to-zero constraint $\sum_{i=1}^{I} \phi_i = 0$, which is awkward theoretically but simple to implement "on the fly" as part of an MCMC algorithm. That is, we simply subtract the current mean $\frac{1}{I} \sum_{i=1}^{I} \phi_i^{(g)}$ from all of the $\phi_i^{(g)}$ at the end of each iteration g. Again, adding a proper hyperprior for λ, proper posteriors for each of the ϕ_i ensue. The reader is referred to Eberly and Carlin (2000) for further insight into the relationship among identifiability, Bayesian learning, and MCMC convergence rates for spatial models using $CAR(\lambda)$ priors. For recent extensions of CAR models that permit smoothing using *two* neighborhood relations simultaneously (e.g., one for east-west neighbors and one for north-south neighbors over a regular lattice), see Reich et al. (2007).

Finally, note that we have chosen to model the spatial correlation in the prior specification using, in Besag's (1974) terminology, an auto-Gaussian model. One could envision introducing spatial correlation into the likelihood, i.e., directly amongst the $y_{i\ell t}$, resulting in Besag's auto-Poisson structure. However, Cressie (1993) points out that this model does *not* yield a Poisson structure marginally. Moreover, Cressie and Chan (1989) observe the intractability of this model working merely with y_is; for a set of $y_{i\ell t}$s, model fitting becomes infeasible.

A general form for the regional main effects is thus $\eta_i = h(\mathbf{z}_i; \boldsymbol{\omega}) + \theta_i + \phi_i$. This form obviously yields a likelihood in which θ_i and ϕ_i are not identifiable. However, the prior specification allows the possibility of separating these effects; the data can inform about each of them. A related issue is posterior integrability. One simple condition is that any effect in $\mu_{i\ell t}$ that is free of i and has a flat prior must have a fixed *baseline* level, such as 0. For example, $\mu_{i\ell t}$ cannot include an intercept. Equivalently, if both ϕ_i and θ_i are in the model then we must set $\kappa_\theta = 0$. Also, if δ_t terms are included we must include a constraint (say, $\delta_1 = 0$ or $\sum_t \delta_t = 0$), for otherwise we can add a constant to each ϕ_i and subtract it from each level

of the foregoing effect without affecting the likelihood times prior, hence
the posterior. It is also important to note that θ_i and ϕ_i may be viewed
as surrogates for unobserved regional covariates. That is, with additional
regional explanatory variables, they might not be needed in the model.
From a different perspective, for models employing the general form for η_i,
there may be strong collinearity between $\mathbf{z}_i$ and say ϕ_i, which again makes
the ϕ_i difficult to identify, hence the models difficult to fit.

Spatio-temporal interaction

To this point, $\mu_{i\ell t}$ consists of the main effects ε_ℓ, δ_t, and η_i. We now turn
to interactions, though as mentioned above the only ones we consider are
spatio-temporal. In general, defining a space-time component is thorny because it is not clear how to reconcile the different scales. Bernardinelli,
Clayton et al. (1995) assume the multiplicative form $(\nu_i^\theta + \nu_i^\phi)t$ for the
heterogeneity by time and clustering by time effects. We again prefer a
qualitative form and propose a *nested* definition, wherein heterogeneity
effects and spatial effects are nested within time. In this way, we can examine the evolution of spatial and heterogeneity patterns over time. We
write these effects as $\theta_i^{(t)}$ and $\phi_i^{(t)}$. That is, $\theta_i^{(t)}$ is a random effect for
the ith region in year t. Conditional exchangeability given t would lead
us to the prior $\theta_i^{(t)} \overset{iid}{\sim} N(\kappa_\theta^{(t)}, 1/\tau_t)$. Similarly, $\phi_i^{(t)}$ is viewed as a spatial effect for the ith region in year t. Again we adopt a CAR prior, but
now given t as well. Hence, we assume $\phi_i^{(t)}|\phi_{j\neq i}^{(t)}$ has density proportional
to $\exp\left[-\frac{\lambda_t}{2}(a_i\phi_i^{(t)} - \sum_{j\neq i} w_{ij}\phi_j^{(t)})^2\right]$. Again, proper gamma priors are assumed for the τ_t and the λ_t. Such definitions for $\theta_i^{(t)}$ and $\phi_i^{(t)}$ preclude the
inclusion of main effects θ_i and ϕ_i in the model. Also, if $\phi_i^{(t)}$ appears in
the model, then we cannot include δ_t terms that lack an informative prior,
since for any fixed t the likelihood could not identify both. Similarly, if both
$\theta_i^{(t)}$ and $\phi_i^{(t)}$ appear in the model, we must set $\kappa_\theta^{(t)} = 0$.

In summary, the most general model we envision for $\mu_{i\ell t}$ takes the form

$$\mu_{i\ell t} = \mathbf{x}_\ell^T\boldsymbol{\beta} + \mathbf{z}_i^T\boldsymbol{\omega} + \theta_i^{(t)} + \phi_i^{(t)} . \tag{7.38}$$

In practice, (7.38) might well include more effects than we need. In the
exercises below, we investigate more parsimonious reduced models.

7.8 Exercises

1. Let $G_k(t)$ be defined as in Subsection 7.1.2. Show that the posterior
   mode of $G_k(t)$, denoted $\tilde{G}_k(t)$, has equal mass at exactly k mass points
   that are derived from an expression which for each t is the likelihood of
   independent, non-identically distributed Bernoulli variables.

2. Let G_k^* be a discrete distribution with mass $1/k$ at mass points $(u_1 <$

$u_2 \ldots < u_k)$ and assume it is the edf of $(\theta_1, \ldots, \theta_k)$. Prove that if the Y_j are i.i.d., then under integrated squared-error loss the best estimates of the individual θ_j are the u-values with assignment according to the ordering of the Y_j. Also, show that the optimal ranks for the θ_j are the ranks of the Y_j.

3. Compute and compare the mean and variance induced by edfs formed from the coordinate-specific MLEs, the standard Bayes estimates, and $\bar{G}_k$ for the normal/normal, Poisson/gamma, and beta/binomial two-stage models.

4. Let $G_k^{(0.5)}(t)$ be the posterior median of $G_k(t)$.

   (a) Show that for each t, $G_k^{(0.5)}(t)$ minimizes expected, integrated absolute loss, $E\left[\int \mid G_k^{est}(t) - G_k(t) \mid dt\right]$.

   (b) Outline an algorithm to compute $G_k^{(0.5)}(t)$ using MCMC output.

   (c) Let $k = 100$ and assume that, a posteriori $(\theta_1, \ldots, \theta_k)$ are i.i.d. with distribution that is a 50/50 mixture of a N(−2, 1) and a N(2,1). Simulate from this distribution by drawing $B = 1000$ samples each of size 100 and compute $\bar{G}_k$ and $G_k^{(0.5)}(t)$ from the output. Graphically and numerically compare these estimates to each other and to the generating distribution.

5. Identify applications in which SEL in estimating ranks is the appropriate metric, and others in which (above γ)/(below γ) loss is appropriate. For the latter, discuss how you would select γ.

6. Table 7.1 shows that the AR(1) model "calms" longitudinal variation in the ranks. Discuss why this calming can be advantageous and why it might be disadvantageous.

7. Prove that if the target parameters $(\theta_1, \ldots, \theta_k)$ are independent and stochastically ordered, then for all j, $\tilde{P}_j(\gamma) = P_j$ (and therefore $\tilde{P}_j(\gamma)$ does not depend on γ). (Hint: Use the equivalence of $\tilde{P}$ and P^*.)

8. (Devroye, 1986, p. 38) Suppose X is a random variable having cdf F, and Y is a truncated version of this random variable with support restricted to the interval $[a, b]$. Then Y has cdf

$$G(y) = \begin{cases} 0, & y < a \\ \frac{F(y) - F(a)}{F(b) - F(a)}, & a \leq y \leq b \\ 1, & y > b \end{cases}.$$

Show that Y can be generated as $F^{-1}(F(a) + U[F(b) - F(a)])$, where U is a $Unif(0,1)$ random variate. (This result enables "one-for-one" generation from truncated distributions for which we can compute F and F^{-1}, either exactly or numerically.)

9. Consider the estimation of human mortality rates between ages x and $x + k$, where $x \geq 30$. Data available from a mortality study of a group of independent lives includes d_i, the number of deaths observed in the unit age interval $[x + i - 1, \, x + i]$, and e_i, the number of person-years the lives were under observation (*exposed*) in this interval. Thinking of the e_i as constants and under the simplifying assumption of a constant force of mortality θ_i over age interval i, one can show that

$$d_i | \theta_i \overset{iid}{\sim} Po(e_i \theta_i), \quad i = 1, \ldots, k \, .$$

An estimate of θ_i is thus provided by $r_i = d_i / e_i$, the unrestricted maximum likelihood estimate of θ_i, more commonly known as the *raw* mortality rate. We wish to produce a *graduated* (smoothed) sequence of θ_is, which conform to the *increasing* condition

$$\boldsymbol{\theta} \in S^{INC} = \{\boldsymbol{\theta} : \, 0 < \theta_1 < \cdots < \theta_k < B\}$$

for some fixed positive B.

Assume that the θ_is are an i.i.d. sample from a $G(\alpha, \beta)$ distribution *before* imposing the above constraints, where α is a known constant and β has an $IG(a, b)$ hyperprior distribution. Find the full conditionals for the θ_i and β needed to implement the Gibbs sampler in this problem. Also, describe how the result in the previous problem would be used in performing the required sampling.

10. Table 7.4 gives a dataset of male mortality experience originally presented and analyzed by Broffitt (1988). The rates are for one-year intervals, ages 35 to 64 inclusive (i.e., bracket i corresponds to age $34 + i$ at last birthday).

   (a) Use the results of the previous problem to obtain a sequence of graduated, strictly increasing mortality rate estimates. Set the upper bound $B = .025$, and use the data-based (empirical Bayes) hyperparameter values $\alpha = \hat{\alpha} \equiv \bar{r}^2 / (s_r^2 - \bar{r} \sum_{i=1}^{k} e_i^{-1} / k)$, $a = 3.0$, and $b = \hat{\alpha} / (2\bar{r})$, where $\bar{r} = \sum_{i=1}^{k} r_i / k$ and $s_r^2 = \sum_{i=1}^{k} (r_i - \bar{r})^2 / (k - 1)$. These values of a and b correspond to a rather vague hyperprior having mean and standard deviation both equal to $\hat{\beta} = \bar{r} / \hat{\alpha}$. (In case you've forgotten, $\hat{\alpha}$ and $\hat{\beta}$ are the method of moments EB estimates derived in Chapter 3, Problem 9.)

   (b) Do the graduated rates you obtained form a sequence that is not only increasing, but *convex* as well? What modifications to the model would guarantee such an outcome?

11. Consider again the model of Example 7.2. Suppose that instead of a straight-line growth curve, we wish to investigate the *exponential* growth

i	d_i	e_i	r_i	i	d_i	e_i	r_i
1	3	1771.5	0.0016935	16	4	1516.0	0.0026385
2	1	2126.5	0.0004703	17	7	1371.5	0.0051039
3	3	2743.5	0.0010935	18	4	1343.0	0.0029784
4	2	2766.0	0.0007231	19	4	1304.0	0.0030675
5	2	2463.0	0.0008120	20	11	1232.5	0.0089249
6	4	2368.0	0.0016892	21	11	1204.5	0.0091324
7	4	2310.0	0.0017316	22	13	1113.5	0.0116749
8	7	2306.5	0.0030349	23	12	1048.0	0.0114504
9	5	2059.5	0.0024278	24	12	1155.0	0.0103896
10	2	1917.0	0.0010433	25	19	1018.5	0.0186549
11	8	1931.0	0.0041429	26	12	945.0	0.0126984
12	13	1746.5	0.0074435	27	16	853.0	0.0187573
13	8	1580.0	0.0050633	28	12	750.0	0.0160000
14	2	1580.0	0.0012658	29	6	693.0	0.0086580
15	7	1467.5	0.0047700	30	10	594.0	0.0168350

Table 7.4 *Raw mortality data (from Broffitt, 1988).*

model,

$$Y_{ij} \overset{ind}{\sim} N\left(\alpha_i e^{\beta_i(x_{ij}-\bar{x})}, \sigma^2\right), \; i = 1,\ldots,k, \; j = 1,\ldots,n_i \,,$$

with the remaining priors and hyperpriors as previously specified.

(a) Of the full conditional distributions for $\{\alpha_i\}$, $\{\beta_i\}$, $\alpha_0, \beta_0, \Sigma^{-1}$ and σ^2, which now require modification?

(b) Of those distributions in your previous answer, which may still be derived in closed form as members of familiar families? Obtain expressions for these distributions.

(c) Of those full conditionals that lack a closed form, give a Metropolis or Hastings substep for generating the necessary MCMC samples.

(d) Write a program applying your results to the data in Table 7.3, and obtain estimated posterior distributions for α_0 and β_0. What does your fitted model suggest about the growth patterns of young rats?

12. Suppose we have a collection of binary responses $y_i \in \{0,1\}, i = 1,\ldots,n$, and associated k-dimensional predictor variables $\mathbf{x}_i$. Define the latent variables y_i^* as

$$y_i^* = \mathbf{x}_i^T \boldsymbol{\beta} + \epsilon_i, \; i = 1,\ldots,n \,,$$

where the ϵ_i are independent mean-zero errors having cumulative distribution function F, and $\boldsymbol{\beta}$ is a k-dimensional regression parameter.

i	x_{1i}	x_{2i}	y_i	i	x_{1i}	x_{2i}	y_i
1	3.7	0.825	1	21	0.4	2	0
2	3.5	1.09	1	22	0.95	1.36	0
3	1.25	2.5	1	23	1.35	1.35	0
4	0.75	1.5	1	24	1.5	1.36	0
5	0.8	3.2	1	25	1.6	1.78	1
6	0.7	3.5	1	26	0.6	1.5	0
7	0.6	0.75	0	27	1.8	1.5	1
8	1.1	1.7	0	28	0.95	1.9	0
9	0.9	0.75	0	29	1.9	0.95	1
10	0.9	0.45	0	30	1.6	0.4	0
11	0.8	0.57	0	31	2.7	0.75	1
12	0.55	2.75	0	32	2.35	0.03	0
13	0.6	3	0	33	1.1	1.83	0
14	1.4	2.33	1	34	1.1	2.2	1
15	0.75	3.75	1	35	1.2	2.0	1
16	2.3	1.64	1	36	0.8	3.33	1
17	3.2	1.6	1	37	0.95	1.9	0
18	0.85	1.415	1	38	0.75	1.9	0
19	1.7	1.06	0	39	1.3	1.625	1
20	1.8	1.8	1				

Table 7.5 *Finney's vasoconstriction data.*

Consider the model

$$Y_i = \begin{cases} 0, & \text{if } Y_i^* \le 0 \\ 1, & \text{if } Y_i^* > 0 \end{cases}.$$

(a) Show that if F is the standard normal distribution, this model is equivalent to the usual *probit* model for $p_i = P(Y_i = 1)$.

(b) Under a $N(\mu, \Sigma)$ prior for β, find the full conditional distributions for β and the Y_i^*, $i = 1, \ldots, n$.

(c) Use the results of the previous part to find the posterior distribution for β under a flat prior for the data in Table 7.5. Originally given by Finney (1947), these data give the presence or absence of vaso-constriction in the skin of the fingers following inhalation of a certain volume of air, x_1, at a certain average rate, x_2. (Include an intercept in your model, so $k = 3$.)

(d) How would you modify your computational approach if, instead of the probit model, we wished to fit the *logit* (logistic regression) model?

(e) Perform the actual MCMC fitting of the logistic and probit models using BRugs or WinBUGS. Do the fits differ significantly? (*Hint:* Follow the approach of Chapter 3, problem 4.)

13. Suppose that in the context of the previous problem, the data y_i are not merely binary but *ordinal*, i.e., $y_i \in \{0, 1, \ldots m\}$, $i = 1, \ldots, n$. We extend our previous model to

$$
Y_i = \begin{cases}
0, & \text{if } Y_i^* \leq \gamma_1 \\
1, & \text{if } \gamma_1 < Y_i^* \leq \gamma_2 \\
\vdots & \vdots \\
m-1, & \text{if } \gamma_{m-1} < Y_i^* \leq \gamma_m \\
m, & \text{if } Y_i^* > \gamma_m
\end{cases}.
$$

(a) Suppose we wish to specify G as the cdf of Y_i (e.g., $G \sim Bin(m, p_i)$). Find an expression for the appropriate choice of "bin boundaries" γ_j, $j = 1, \ldots m$, and derive the new full conditional distributions for β and the Y_i^* in this case.

(b) Now suppose instead that we wish to leave G unspecified. Can we simply add all of the γ_js into the sampling order and proceed? What are the appropriate full conditional distributions to use for the γ_j?

Station	Latitude	Longitude	Observed Data
i	s_{1i}	s_{2i}	$X(s_i)$
1	−18.9805	359.3649	0.085
2	−16.9316	359.1380	0.069
3	−15.1309	357.1999	0.084
4	−12.5863	356.9081	0.104
5	−12.0162	356.5347	0.111
6	−9.9322	355.4129	0.069
7	−9.7132	364.1625	0.049
8	−13.1500	350.8925	0.053
9	−11.2352	358.3001	0.082
10	−9.8524	359.5717	0.065
A	−13.1383	356.5103	−
B	−13.9711	356.0425	−

Table 7.6 *Ozone data at 10 monitoring stations, Atlanta metro area, 6/3/95.*

14. The data in Table 7.6 were originally analyzed by Tolbert et al. (2000), and record the horizontal (s_1) and vertical (s_2) coordinates (in meters $\times 10^4$) of 10 ozone monitoring stations in the greater Atlanta metro area. Figure 7.12 shows the station locations on a greater Atlanta zip code

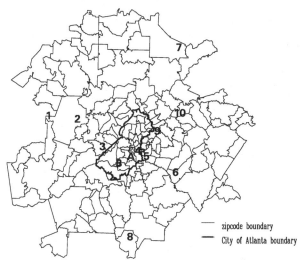

Figure 7.12 *Ten ozone monitoring stations (source locations) and two target locations, greater Atlanta.*

map. The table also gives the 10 corresponding ozone measurements $X(\mathbf{s}_i)$ (one-hour max) for June 3, 1995.

(a) Using the isotropic Bayesian kriging model of Section 7.7.1 applied to $Z(\mathbf{s}_i) = \log X(\mathbf{s}_i)$, find the posterior distribution of the spatial smoothing parameter ϕ and the true ozone concentrations at target locations A and B, whose coordinates are also given in Table 7.6. (As can be seen from Figure 7.12, these are two points on opposite sides of the same Atlanta city zip code.) Set $\tau^2 = 0$, and use a flat prior on μ, a vague $IG(3, 0.5)$ prior (mean 1, variance 1) on σ^2, and a vague $G(.001, 100)$ (mean .1, variance 10) prior on ϕ. Are your results consistent with the mapped source data?

(b) Why might an anisotropic model be more appropriate here? Fit this model, using a vague Wishart prior for the matrix B, and compare its fit to the isotropic model in part (a).

15. Table 7.7 presents observed (Y_i) and expected (E_i) cases of lip cancer in 56 counties in Scotland, which have been previously analyzed by Clayton and Kaldor (1987) and Breslow and Clayton (1993). The table also shows a spatial covariate, x_i, the percentage of each county's population engaged in agriculture, fishing, or forestry (AFF), the observed standardized mortality ratio (SMR) for the county, $100Y_i/E_i$, and the the spatial relation of each county to the others via the indices of those counties adjacent to it. The raw data and the AFF covariate are mapped in Figure 7.13 on page 370.

i	Y_i	E_i	x_i	SMR	Adjacent Counties (∂_i)
1	9	1.4	16	652.2	5, 9, 11, 19
2	39	8.7	16	450.3	7, 10
3	11	3.0	10	361.8	6, 12
4	9	2.5	24	355.7	18, 20, 28
5	15	4.3	10	352.1	1, 11, 12, 13, 19
6	8	2.4	24	333.3	3, 8
7	26	8.1	10	320.6	2, 10, 13, 16, 17
8	7	2.3	7	304.3	6
9	6	2.0	7	303.0	1, 11, 17, 19, 23, 29
10	20	6.6	16	301.7	2, 7, 16, 22
11	13	4.4	7	295.5	1, 5, 9, 12
12	5	1.8	16	279.3	3, 5, 11
13	3	1.1	10	277.8	5, 7, 17, 19
14	8	3.3	24	241.7	31, 32, 35
15	17	7.8	7	216.8	25, 29, 50
16	9	4.6	16	197.8	7, 10, 17, 21, 22, 29
17	2	1.1	10	186.9	7, 9, 13, 16, 19, 29
18	7	4.2	7	167.5	4, 20, 28, 33, 55, 56
19	9	5.5	7	162.7	1, 5, 9, 13, 17
20	7	4.4	10	157.7	4, 18, 55
21	16	10.5	7	153.0	16, 29, 50
22	31	22.7	16	136.7	10, 16
23	11	8.8	10	125.4	9, 29, 34, 36, 37, 39
24	7	5.6	7	124.6	27, 30, 31, 44, 47, 48, 55, 56
25	19	15.5	1	122.8	15, 26, 29
26	15	12.5	1	120.1	25, 29, 42, 43
27	7	6.0	7	115.9	24, 31, 32, 55
28	10	9.0	7	111.6	4, 18, 33, 45

Table 7.7 *Observed SMRs for Lip Cancer in 56 Scottish Counties.*

(a) Following Section 7.7.2, fit the model $Y_i \sim Po(E_i e^{\mu_i})$, where

$$\mu_i = \alpha x_i/10 + \theta_i$$

and the θ_i have the exchangeable prior (7.36). Estimate the posterior distributions of κ_θ and α, and comment on your findings.

(b) Repeat the previous analysis using the mean structure

$$\mu_i = \alpha x_i/10 + \phi_i$$

i	Y_i	E_i	x_i	SMR	Adjacent Counties (∂_i)
29	16	14.4	10	111.3	9, 15, 16, 17, 21, 23, 25, 26, 34, 43, 50
30	11	10.2	10	107.8	24, 38, 42, 44, 45, 56
31	5	4.8	7	105.3	14, 24, 27, 32, 35, 46, 47
32	3	2.9	24	104.2	14, 27, 31, 35
33	7	7.0	10	99.6	18, 28, 45, 56
34	8	8.5	7	93.8	23, 29, 39, 40, 42, 43, 51, 52, 54
35	11	12.3	7	89.3	14, 31, 32, 37, 46
36	9	10.1	0	89.1	23, 37, 39, 41
37	11	12.7	10	86.8	23, 35, 36, 41, 46
38	8	9.4	1	85.6	30, 42, 44, 49, 51, 54
39	6	7.2	16	83.3	23, 34, 36, 40, 41
40	4	5.3	0	75.9	34, 39, 41, 49, 52
41	10	18.8	1	53.3	36, 37, 39, 40, 46, 49, 53
42	8	15.8	16	50.7	26, 30, 34, 38, 43, 51
43	2	4.3	16	46.3	26, 29, 34, 42
44	6	14.6	0	41.0	24, 30, 38, 48, 49
45	19	50.7	1	37.5	28, 30, 33, 56
46	3	8.2	7	36.6	31, 35, 37, 41, 47, 53
47	2	5.6	1	35.8	24, 31, 46, 48, 49, 53
48	3	9.3	1	32.1	24, 44, 47, 49
49	28	88.7	0	31.6	38, 40, 41, 44, 47, 48, 52, 53, 54
50	6	19.6	1	30.6	15, 21, 29
51	1	3.4	1	29.1	34, 38, 42, 54
52	1	3.6	0	27.6	34, 40, 49, 54
53	1	5.7	1	17.4	41, 46, 47, 49
54	1	7.0	1	14.2	34, 38, 49, 51, 52
55	0	4.2	16	.0	18, 20, 24, 27, 56
56	0	1.8	10	.0	18, 24, 30, 33, 45, 55

Table 7.7 (continued)

where the ϕ_i have the CAR adjacency prior (7.37). How do your results change?

(c) Conduct an informal comparison of these two models using the DIC criterion (2.26); which model emerges as better? Also use equation (2.25) to compare the two effective model sizes p_D.

(d) How might we determine whether significant spatial structure remains in the data after accounting for the covariate?

Figure 7.13 *Scotland lip cancer data: (a) crude standardized mortality ratios (observed/expected × 100); (b) AFF covariate values.*

(e) Suppose we wanted to fit a model with *both* the heterogeneity terms (θ_i) and the clustering terms (ϕ_i). What constraints or modifications to the prior distributions would be required to insure model identifiability (and hence, MCMC sampler convergence)?

16. Fit the model described in Subsection 7.7.2 to the Ohio lung cancer data originally analyzed by Devine (1992, Chapter 4) and available online at http://www.biostat.umn.edu/~brad/data.html. Here, y_{ijkt} is the number of lung cancer deaths in county i during year t for gender j and race k in the state of Ohio, and n_{ijkt} is the corresponding exposed population count. Our subset of lung cancer data are recorded for $J = 2$ genders (male and female) and $K = 2$ races (white and nonwhite) for each of the $I = 88$ Ohio counties over an observation period of $T = 21$ years, namely 1968 to 1988 inclusive, yielding a total of 7392 observations. Internally standardized expected death counts are obtained in the usual way as $E_{ijkt} = n_{ijkt}\bar{y}$, where $\bar{y} = (\sum_{ijkt} y_{ijkt} / \sum_{ijkt} n_{ijkt})$, the average statewide death rate over the entire observation period.

We wish to fit the Poisson likelihood model

$$Y_{ijkt} \sim Po(E_{ijkt}e^{\mu_{ijkt}}) ,$$

where the log-relative risk μ_{ijkt} is of the form given in (7.38), namely,

$$\mu_{ijkt} = s_j\alpha + r_k\beta + s_jr_k\xi + \theta_i^{(t)} + \phi_i^{(t)} ,$$

where we adopt the gender and race scores

$$s_j = \begin{cases} 0 & \text{if male} \\ 1 & \text{if female} \end{cases} \quad \text{and} \quad r_k = \begin{cases} 0 & \text{if white} \\ 1 & \text{if nonwhite} \end{cases} .$$

Letting $\boldsymbol{\theta}^{(t)} = (\theta_1^{(t)},\ldots,\theta_I^{(t)})'$, $\boldsymbol{\phi}^{(t)} = (\phi_1^{(t)},\ldots,\phi_I^{(t)})'$, and denoting the I-dimensional identity matrix by $\mathbf{I}$, we adopt the prior structure

$$\boldsymbol{\theta}^{(t)} \mid \tau_t \stackrel{ind}{\sim} N\left(\mathbf{0},\frac{1}{\tau_t}\mathbf{I}\right) \quad \text{and} \quad \boldsymbol{\phi}^{(t)} \mid \lambda_t \stackrel{ind}{\sim} CAR(\lambda_t) , \quad t = 1,\ldots,T ,$$

so that heterogeneity and clustering may vary over time. Let us adopt a flat prior on α, β, and ξ, and conjugate, conditionally i.i.d. $Gamma(a,b)$ and $Gamma(c,d)$ priors on the τ_t and λ_t, respectively. Following Waller et al. (1997) (and the rough guidelines of Bernardinelli, Clayton, and Montomoli, 1995), we take $a = 1, b = 100$ (i.e., prior mean and standard deviation 100) for the nonspatial precisions τ_t, but $c = 1, d = 7$ (i.e., prior mean and standard deviation 7) for the spatial precisions λ_t.

(a) Fit this model in WinBUGS, and use the posterior of the main effects to estimate the risk of lung cancer death for white females, nonwhite males, and nonwhile females relative to white males (the baseline group). Also plot the posterior medians of the λ_t and the τ_t versus t, to check for evidence of change in the excess spatial or nonspatial variability over time.

(b) Use either WinBUGS or R to draw choropleth maps of the county-specific nonwhite female posterior mean relative risks for three years: 1968, 1978, and 1988. Also create three companion maps of some measure of the variability in these rates (say, interquartile range or 95% CI width). Discuss any interesting geographic or temporal patterns.

(c) Refit the model replacing the simple i.i.d. specifications for the τ_t and λ_t with AR(1) specifications. Compare the overall adequacy of these models and simplifications of them using DIC and other standard tools (e.g., residual plots). What is your favorite model for these data?

CHAPTER 8

Case studies

Most skills are best learned not through reading or quiet reflection, but through actual practice. This is certainly true of data analysis, an area with few hard-and-fast rules that is as much art as science. It is especially true of Bayes and empirical Bayes data analysis, with its substantial computing component and the challenging model options encouraged by the breadth of the methodology. The recent books edited by Gatsonis et al. (1993; 1995; 1997; 1999; 2002a,b), Gilks et al. (1996), and Berry and Stangl (1996; 2000) offer a broad range of challenging and interesting case studies in Bayesian data analysis and meta-analysis, most using modern MCMC computational methods. In addition, *Bayesian Analysis*, the official journal of the International Society for Bayesian Analysis (ISBA), publishes a great many applications of Bayesian methods to challenging datasets; see ba.stat.cmu.edu.

In this chapter, we provide three fully worked case studies arising from real-world problems and datasets. While all have a distinctly biomedical flavor, they cover a broad range of methods and data types that transfer over to applications in agriculture, economics, social science, and industry. All feature data with complex dependence structures (often introduced by an ordering in time, space, or both), which complicate classical analysis and suggest a hierarchical Bayes or EB approach as a more viable alternative.

Our first case study evaluates drug-induced response in a series of repeated measurements on a group of AIDS patients using both longitudinal modeling and probit regression. We then compare the survival times of patients in each drug and response group. The second study involves sequential analysis, in the context of monitoring a clinical trial. The results of using carefully elicited priors motivates a comparison with those obtained with vague priors, as well as the sort of robust analysis suggested in Subsection 4.2.2. Finally, our third case study concerns the modeling of infectious disease, extending standard Markov transition models and applying them to data from the 1918 flu pandemic.

8.1 Analysis of longitudinal AIDS data

8.1.1 Introduction and background

Large-scale clinical trials often require extended follow-up periods to evaluate the clinical efficacy of a new treatment, as measured for example by survival time. The resulting time lags in reporting the results of the trial to the clinical and patient communities are especially problematic in AIDS research, where the life-threatening nature of the disease heightens the need for rapid dissemination of knowledge. A common approach for dealing with this problem is to select an easily measured biological marker, known to be predictive of the clinical outcome, as a surrogate endpoint. For example, the number of CD4 lymphocytes per cubic millimeter of drawn blood has been used extensively as a surrogate marker for progression to AIDS and death in drug efficacy studies of HIV-infected persons. Indeed, an increase in the average CD4 count of patients receiving one of the drugs in our study (didanosine) in a clinical trial was the principal basis for its limited licensing in the United States.

Several investigators, however, have questioned the appropriateness of using CD4 count in this capacity. For example, Lin et al. (1993) showed that CD4 count was not an adequate surrogate for first opportunistic infection in data from the Burroughs Wellcome 02 trial, nor was it adequate for the development of AIDS or death in AIDS Clinical Trials Group (ACTG) Protocol 016 data. Choi et al. (1993) found CD4 count to be an incomplete surrogate marker for AIDS using results from ACTG Protocol 019 comparing progression to AIDS for two dose groups of zidovudine (AZT) and placebo in asymptomatic patients. Results of the joint Anglo-French "Concorde" trial (Concorde Coordinating Committee, 1994) showed that, while asymptomatic patients who began taking AZT immediately upon randomization did have significantly higher CD4 counts than those who deferred AZT therapy until the onset of AIDS-related complex (ARC) or AIDS itself, the survival patterns in the two groups were virtually identical.

We analyze data from a trial involving 467 persons with advanced HIV infection who were randomly assigned to treatment with the antiretroviral drugs didanosine (ddI) or zalcitabine (ddC). This study, funded by the National Institutes of Health (NIH) and conducted by the Community Programs for Clinical Research on AIDS (CPCRA), was a direct randomized comparison of these two drugs in patients intolerant to, or failures on, AZT. The details of the conduct of the ddI/ddC study are described elsewhere (Abrams et al., 1994; Goldman et al., 1996); only the main relevant points are given here. The trial enrolled HIV-infected patients with AIDS or two CD4 counts of 300 or less, and who fulfilled specific criteria for AZT intolerance or failure. CD4 counts were recorded at study entry and again at the 2, 6, 12, and 18 month visits (though some of these observations are missing for many individuals). The study measured several outcome variables; we consider only total survival time.

Our main goal is to analyze the association among CD4 count, survival time, drug group, and AIDS diagnosis at study entry (an indicator of disease progression status). Our population of late-stage patients is ideal for this purpose because it contains substantial information on both the actual (survival) and surrogate (CD4 count) endpoints.

8.1.2 Modeling of longitudinal CD4 counts

Let Y_{ij} denote the jth CD4 measurement on the ith individual in the study, $j = 1, \ldots, s_i$ and $i = 1, \ldots, n$. Arranging each individual's collection of observations in a vector $Y_i = (Y_{i1}, \ldots, Y_{is_i})^T$, we attempt to fit the mixed effects model (7.16), namely,

$$Y_i = \mathbf{X}_i \alpha + \mathbf{W}_i \beta_i + \epsilon_i \ ,$$

where as in Section 7.3, $\mathbf{X}_i$ is an $s_i \times p$ design matrix, α is a $p \times 1$ vector of fixed effects, $\mathbf{W}_i$ is an $s_i \times q$ design matrix (q typically less than p), and β_i is a $q \times 1$ vector of subject-specific random effects, usually assumed to be normally distributed with mean vector $\mathbf{0}$ and covariance matrix $\mathbf{V}$. We assume that given β_i, the components of ϵ_i are independent and normally distributed with mean 0 and variance σ^2 (marginalizing over β_i, the Y_i components are again correlated, as desired). The β_is capture subject-specific effects, as well as introduce additional variability into the model.

Because we wish to detect a possible increase in CD4 count two months after baseline, we fit a model that is linear, but with possibly different slopes before and after this time. Thus the subject-specific design matrix $\mathbf{W}_i$ for patient i in equation (7.16) has j^{th} row

$$\mathbf{w}_{ij} = (1 \ , \ t_{ij} \ , \ (t_{ij} - 2)^+) \ ,$$

where $t_{ij} \in \{0, 2, 6, 12, 18\}$ and $z^+ = \max(z, 0)$. Hence the three columns of $\mathbf{W}_i$ correspond to individual-level intercept, slope, and change in slope following the changepoint, respectively. We account for the effect of covariates by including them in the fixed effect design matrix $\mathbf{X}_i$. Specifically, we set

$$\mathbf{X}_i = (\mathbf{W}_i \mid d_i \mathbf{W}_i \mid a_i \mathbf{W}_i) \ , \tag{8.1}$$

where d_i is a binary variable indicating whether patient i received ddI ($d_i = 1$) or ddC ($d_i = 0$), and a_i is another binary variable telling whether the patient was diagnosed as having AIDS at baseline ($a_i = 1$) or not ($a_i = 0$). Notice from equation (8.1) that we have $p = 3q = 9$; the two covariates are being allowed to affect any or all of the intercept, slope, and change in slope of the overall population model. The corresponding elements of the α vector then quantify the effect of the covariate on the form

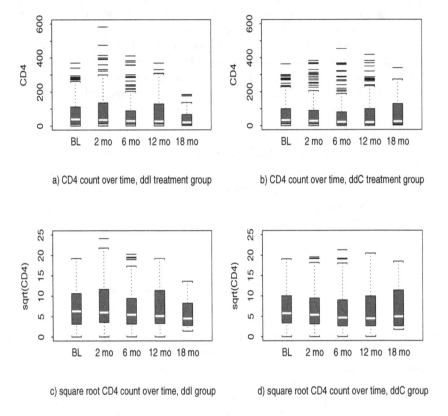

a) CD4 count over time, ddI treatment group

b) CD4 count over time, ddC treatment group

c) square root CD4 count over time, ddI group

d) square root CD4 count over time, ddC group

Figure 8.1 *Exploratory plots of CD4 count, ddI/ddC data.*

of the CD4 curve. In particular, our interest focuses on the α parameters corresponding to drug status, and whether they differ from 0.

Adopting the usual exchangeable normal model for the random effects, we obtain a likelihood of the form

$$\prod_{i=1}^{n} N_{s_i}(Y_i|\mathbf{X}_i\alpha + \mathbf{W}_i\beta_i, \sigma^2\mathbf{I}_{s_i}) \prod_{i=1}^{n} N_3(\beta_i|\mathbf{0}, \mathbf{V}), \qquad (8.2)$$

where $N_k(\cdot|\mu, \boldsymbol{\Sigma})$ denotes the k-dimensional normal distribution with mean vector μ and covariance matrix $\boldsymbol{\Sigma}$. To complete our Bayesian model specification, we adopt independent prior distributions $N_9(\alpha|c, \mathbf{D})$, $IG(\sigma^2|a, b)$, and $IW(\mathbf{V}|(\rho\mathbf{R})^{-1}, \rho)$, where IG and IW denote the inverse gamma and inverse Wishart distributions, respectively (see Appendix A for details).

Turning to the observed data in our study, boxplots of the individual CD4 counts for the two drug groups, shown in Figures 8.1(a) and (b), indicate a high degree of skewness toward high CD4 values. This, combined with the count nature of the data, suggests a square root transformation for each group. As Figures 8.1(c) and (d) show, this transformation improves matters considerably. The sample medians, shown as white horizontal bars on the boxplots, offer reasonable support for our assumption of a linear decline in square root CD4 after two months. The sample sizes at the five time points, namely (230, 182, 153, 102, 22) and (236, 186, 157, 123, 14) for the ddI and ddC groups, respectively, indicate an increasing degree of missing data. The total number of observations is 1405, an average of approximately 3 per study participant.

Prior determination

Values for the parameters of our priors (the hyperparameters) may sometimes be determined from past datasets (Lange, Carlin, and Gelfand, 1992), or elicited from clinical experts (Chaloner et al., 1993; see also Section 8.2). In our case, since little is reliably known concerning the CD4 trajectories of late-stage patients receiving ddI or ddC, we prefer to choose hyperparameter values that lead to fairly vague, minimally informative priors. Care must be taken, however, to ensure that this does not lead to an improper posterior distribution. For example, taking ρ extremely small and $\mathbf{D}$ extremely large would lead to confounding between the fixed and random effects. (To see this, consider the simple but analogous case of the oneway layout, where $Y_i \overset{iid}{\sim} N(\mu + \alpha_i, \sigma^2)$, $i = 1, \ldots, n$, and we place noninformative priors on *both* μ and the α_i.) To avoid this while still allowing the random effects a reasonable amount of freedom, we adopt the rule of thumb wherein $\rho = n/20$ and

$$\mathbf{R} = Diag((r_1/8)^2, (r_2/8)^2, (r_3/8)^2) \,,$$

where r_i is the total range of plausible parameter values across the individuals. Because $\mathbf{R}$ is roughly the prior mean of $\mathbf{V}$, this gives a ± 2 prior standard deviation range for β_i that is half of that plausible in the data. Since in our case Y_{ij} corresponds to square root CD4 count in late-stage HIV patients, we take $r_1 = 16$ and $r_2 = r_3 = 2$, and hence our rule produces $\rho = 24$ and $\mathbf{R} = Diag(2^2, (.25)^2, (.25)^2)$.

Turning to the prior on σ^2, we take $a = 3$ and $b = .005$, so that σ^2 has both mean and standard deviation equal to 10^2, a reasonably vague (but still proper) specification. Finally, for the prior on α, we set

$$c = (10, 0, 0, 0, 0, 0, -3, 0, 0) \,, \quad \text{and}$$
$$\mathbf{D} = Diag(2^2, 1^2, 1^2, (.1)^2, 1^2, 1^2, 1^2, 1^2, 1^2) \,,$$

a more informative prior, but strongly biased away from 0 only for the baseline intercept, α_1, and the intercept adjustment for a positive AIDS diagnosis, α_7. These values correspond to our expectation of mean baseline square root CD4 counts of 10 and 7 (100 and 49 on the original scale) for the AIDS-negative and AIDS-positive groups, respectively. This prior forces the drug group intercept (i.e., the effect at baseline) to be very small, since patients were assigned to drug group at random. Indeed, this randomization implies that $\alpha_4 \equiv 0$. Also, the prior allows the data to determine the degree to which CD4 trajectories depart from horizontal, both before and after the changepoint. Similarly, though α and the β_i are uncorrelated *a priori*, the priors are sufficiently vague to allow the data to determine the proper amount of correlation *a posteriori*.

Because the prior is fully conjugate with the likelihood, the Gibbs sampler is an easily implemented MCMC method, with full conditional distributions similar to those given for the rat data in Example 7.2; see also Section 2.4 of the hierarchical CD4 changepoint analysis in Lange, Carlin, and Gelfand (1992). Our model can also be described graphically and subsequentlty analyzed in WinBUGS; again we would follow the model of Example 7.2 (click Help, pull down to Examples Vol I, and click on Rats) or the more explicit help for our ddI/ddC dataset available online at www.mrc-bsu.cam.ac.uk/bugs/documentation/bugs06/addman06.html.

Results

We fit our model by running five parallel Gibbs sampler chains for 500 iterations each. A crudely overdispersed starting distribution for $\mathbf{V}^{-1}$ was obtained by drawing five samples from a Wishart distribution with the $\mathbf{R}$ matrix given above, but having $\rho = \dim(\beta_i) = 3$, the smallest value for which this distribution is proper. The gth chain for each of the other parameters was initialized at its prior mean plus $(g - 3)$ prior standard deviations, $g = 1, \ldots, 5$. While individually monitoring each of the 1417 parameters is not feasible, Figure 8.2 shows the resulting Gibbs chains for the first six components of α, the three components of β_8 (a typical random effect), the model standard deviation σ, and the (1,1) and (1,2) elements of $\mathbf{V}^{-1}$, respectively. Also included are the point estimate and 95th percentile of Gelman and Rubin's (1992) scale reduction factor (labeled "G&R" in the figure), which measures between-chain differences and should be close to 1 if the sampler is close to the target distribution, and the autocorrelation at lag 1 as estimated by the first of the five chains (labeled "lag 1 acf" in the figure). The figure suggests convergence for all parameters with the possible exception of the two $\mathbf{V}^{-1}$ components. While we have little interest in posterior inference on $\mathbf{V}^{-1}$, caution is required since a lack of convergence on its part could lead to false inferences concerning the other parameters that *do* appear to have converged (Cowles and Carlin, 1996; Kass et al., 1998). In this case, the culprit appears to be the large positive autocorrelations present in the chains, a frequent cause of degradation of the Gelman and Rubin statistic in otherwise well-behaved samplers. Since the plotted chains appear reasonably stable, we stop the sampler at this

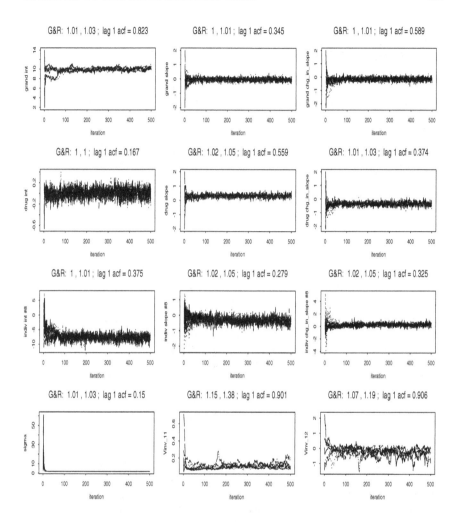

Figure 8.2 *Convergence monitoring plots, five independent chains. Horizontal axis is iteration number; vertical axes are* $\alpha_1, \alpha_2, \alpha_3, \alpha_4, \alpha_5, \alpha_6$, $\beta_{8,1}$, $\beta_{8,2}$, $\beta_{8,3}$, σ, $(\mathbf{V}^{-1})_{11}$, *and* $(\mathbf{V}^{-1})_{12}$.

point, concluding that an acceptable degree of convergence has been obtained.

Based on the visual point of overlap of the five chains in Figure 8.2, we discarded the first 100 iterations from each chain to complete sampler "burn-in." For posterior summarization, we then used all of the remaining 2000 sampled values despite their autocorrelations, in the spirit of Geyer (1992). For the model variance σ^2, these samples produce an estimated

			mode	95% interval	
Baseline	Intercept	α_1	9.938	9.319	10.733
	Slope	α_2	-0.041	-0.285	0.204
	Change in slope	α_3	-0.166	-0.450	0.118
Drug	Intercept	α_4	0.004	-0.190	0.198
	Slope	α_5	0.309	0.074	0.580
	Change in slope	α_6	-0.348	-0.671	-0.074
AIDS Dx	Intercept	α_7	-4.295	-5.087	-3.609
	Slope	α_8	-0.322	-0.588	-0.056
	Change in slope	α_9	0.351	0.056	0.711

Table 8.1 *Point and interval estimates, fixed effect parameters.*

posterior mean and standard deviation of 3.36 and 0.18, respectively. By contrast, the estimated posterior mean of V_{11} is 13.1, suggesting that the random effects account for much of the variability in the data. The dispersion evident in exploratory plots of the β_i posterior means (not shown) confirms the necessity of their inclusion in the model.

Posterior modes and 95% equal-tail credible intervals (based on posterior mixture density estimates) for all nine components of the fixed effect parameter vector α are given in Table 8.1. The posterior credible interval for α_4 is virtually unchanged from that specified in its prior, confirming that the randomization of patients to drug group has resulted in no additional information on α_4 in the data. The 95% credible sets of (.074, .580) and (−.671, −.074) for α_5 (pre-changepoint drug slope) and α_6 (post-changepoint drug slope), respectively, suggest that the CD4 trajectories of persons receiving ddI have slopes that are significantly different from those of the ddC patients, both before and after the two-month changepoint. AIDS diagnosis also emerges as a significant predictor of the two slopes and the intercept, as shown in the last three lines in the table.

Plots of the fitted population CD4 trajectories for each of the four possible drug-diagnosis combinations (not shown) suggest that an improvement in square root CD4 count typically occurs only for AIDS-negative patients receiving ddI. However, there is also some indication that the ddC trajectories "catch up" to the corresponding ddI trajectories by the end of the observation period. Moreover, while the trajectories of ddI patients may be somewhat higher, the difference is quite small, and probably insignificant clinically – especially compared to the improvement provided by a negative AIDS diagnosis at baseline.

The relatively small differences between the fitted pre- and post-changepoint slopes suggest that a simple linear decline model might well be suffi-

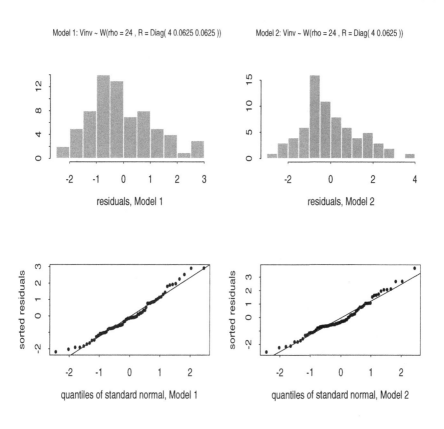

Figure 8.3 *Bayesian model choice based on 70 random observations from the 1405 total. Model 1: changepoint model; model 2: linear decay model.*

cient for these data. To check this possibility, we ran five sampling chains of 500 iterations each for this simplified model. We compared these two models using the cross-validation ideas described near the beginning of Section 4.6, using a 5% subsample of 70 observations, each from a different randomly selected individual. Figure 8.3 offers a side-by-side comparison of histograms and normal empirical quantile-quantile plots of the resulting residuals for Model 1 (changepoint) and Model 2 (linear). The q-q plots indicate a reasonable degree of normality in the residuals for both models. The absolute values of these residuals for the two models sum to 66.37 and 70.82, respectively, suggesting almost no degradation in fit using the simpler model, as anticipated. This similarity in quality of fit is confirmed by Figure 8.4(a), which plots the residuals versus their standard deviations. While there is some indication of smaller standard deviations for residuals

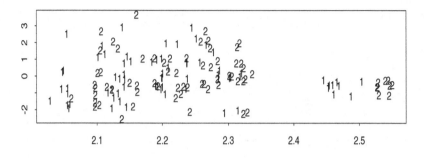

a) y_r - E(y_r) versus sd(y_r); plotting character indicates model number

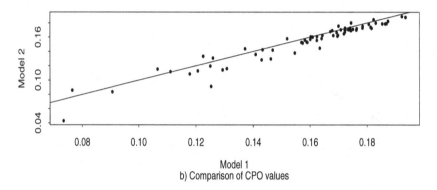

Model 1
b) Comparison of CPO values

Figure 8.4 *Bayesian model choice based on 70 random observations from the 1405 total. Model 1: changepoint model; model 2: linear decay model.*

in the full model (see, e.g., the two clumps of points on the right-hand side of the plot), the reduction appears negligible.

A pointwise comparison of CPO is given in Figure 8.4(b), along with a reference line marking where the values are equal for the two models. Because larger CPO values are indicative of better model fit, the predominance of points below the reference line implies a preference for Model 1 (the full model). Still, the log(CPO) values sum to LMPL scores of −130.434 and −132.807 for the two models, respectively, in agreement with our previous finding that the improvement offered by Model 1 is very slight. In summary, while the more complicated changepoint model was needed to address the specific clinical questions raised by the trial's designers concerning a CD4 boost at two months, the observed CD4 trajectories can be adequately explained using a simple linear model with our two covariates.

Finally, we provide the results of a more formal Bayesian model choice investigation using the marginal density estimation approach of Subsection 4.4.2 to compute a Bayes factor between the two models. We begin with a trick recommended by Chib and Carlin (1999): for Model 1, we write $p(\{\boldsymbol{\beta}_i\}, \boldsymbol{\alpha} | \sigma^2, \mathbf{V}^{-1}, \mathbf{y}) = p(\{\boldsymbol{\beta}_i\} | \boldsymbol{\alpha}, \sigma^2, \mathbf{V}^{-1}, \mathbf{y}) p(\boldsymbol{\alpha} | \sigma^2, \mathbf{V}^{-1}, \mathbf{y})$, and recognize that the reduced conditional $p(\boldsymbol{\alpha} | \sigma^2, \mathbf{V}^{-1}, \mathbf{y})$ is available in closed form. That is, we marginalize the likelihood over the random effects, obtaining

$$\mathbf{Y}_i | \boldsymbol{\alpha}, \sigma^2, \mathbf{V}, M = 1 \sim N\left(\mathbf{X}_i \boldsymbol{\alpha}, \ \sigma^2 \mathbf{I}_{s_i} + \mathbf{W}_i \mathbf{V} \mathbf{W}_i^\top\right) ,$$

from which we derive the reduced conditional for $\boldsymbol{\alpha}$. This trick somewhat reduces autocorrelation in the chains, and also allows us to work with the appropriate three-block extension of equation (4.22),

$$\begin{aligned}
\log f(\mathbf{Y} | M = 1) &= \log f(\mathbf{Y} | \boldsymbol{\alpha}', (\sigma^2)', (\mathbf{V}^{-1})', M = 1) \\
&+ \log p(\boldsymbol{\alpha}', (\sigma^2)', (\mathbf{V}^{-1})' | M = 1) - \log p(\boldsymbol{\alpha}' | \mathbf{Y}, (\sigma^2)', (\mathbf{V}^{-1})', M = 1) \\
&- \log \hat{p}((\sigma^2)' | \mathbf{Y}, (\mathbf{V}^{-1})', M = 1) - \log \hat{p}((\mathbf{V}^{-1})' | \mathbf{Y}, M = 1)
\end{aligned}$$

for Model 1, and similarly for Model 2.

Using 5 chains of 1000 post-burn-in iterations each, we estimated the log marginal likelihood as –3581.795 for Model 1 and –3577.578 for Model 2. These values are within .03 of values obtained independently for this problem by Chib (personal communication), and produce an estimated Bayes factor of $\exp[-3577.578 - (-3581.795)] = 67.83$, moderate to strong evidence in favor of Model 2 (the simpler model without a change point). This is of course consistent with our findings using the less formal model choice tools desribed above.

Summary and discussion

In modeling longitudinal data one must think carefully about the number of random effects to include per individual, and restrict their variability enough to produce reasonably precise estimates of the fixed effects. In our dataset, while there was some evidence that heterogeneous variance parameters σ_i^2 might be needed, including them would increase the number of subject-specific parameters per individual to four – one more than the average number of CD4 observations per person in our study. Fortunately, accidental model overparametrization can often be detected by ultrapoor performance of multichain convergence plots and diagnostics of the variety shown in Figure 8.2. In cases where reducing the total dimension of the model is not possible, Gelfand, Sahu, and Carlin (1995; 1996) show that the relative prior weight assigned to variability at the model's first stage (σ^2) versus its second stage $(\mathbf{V})$ can help suggest an appropriate transformation to improve algorithm convergence. More specifically, had our prior put more credence in small σ^2 values and larger random effects, a sampler operating on the transformed parameters $\alpha^* = \alpha$, $\beta_i^* = \beta_i + \alpha$ would have been recommended.

Missing data are inevitable, and consideration of the possible biasing effects of missingness is very important. Validity of inferences from our models depends on the data being observed at random (i.e., missingness cannot depend on what one would have observed) and an appropriate model structure. Little and Rubin (1987) refer to this combination as "ignorable missingness." In our application, missing CD4 counts are likely the result of patient refusal or a clinical decision that the patient was too ill to have CD4 measured. Thus, our data set is very likely biased toward the relatively higher CD4 counts in the later follow-up periods; we are essentially modeling CD4 progression among persons who continue to have CD4 measured. As a check on ignorability, one can explicitly model the missingness mechanism using Tobit-type or logistic models that link missingness to the unobserved data, and compare the resulting inferences. See Cowles et al. (1996) for an application of the Tobit approach in estimating compliance in a clinical trial.

8.1.3 CD4 response to treatment at two months

The longitudinal changepoint model of the previous section takes advantage of all the observations for each individual, but provides only an implicit model for drug-related CD4 response. For an explicit model, we reclassify each patient as "response" or "no response," depending on whether or not the CD4 count increased from its baseline value. That is, for $i = 1, \ldots, m = 367$, the number of patients who had CD4 measured at both the baseline and two month visits, we define $R_i = 1$ if $Y_{i2} - Y_{i1} \geq 0$, and $R_i = 0$ otherwise. Defining p_i as the probability that patient i responded, we fit the probit regression model

$$p_i = \Phi(\gamma_0 + \gamma_1 d_i + \gamma_2 a_i) , \tag{8.3}$$

where Φ denotes the cumulative distribution function of a standard normal random variable, the γs are unknown coefficients, and the covariates d_i and a_i describe treatment group and baseline AIDS diagnosis as in Subsection 8.1.2. We used probit rather than logistic regression since it is readily fit using the Gibbs sampler without need for Metropolis rejection (Carlin and Polson, 1992; Albert and Chib, 1993a).

Under a flat prior on $(\gamma_0, \gamma_1, \gamma_2)'$, the posterior distributions for each of the γ coefficients were used to derive point and interval estimates of the corresponding covariate effects. In addition, given particular values of the covariates, the posterior distributions were used with equation (8.3) to estimate the probability of a CD4 response in subgroups of patients by treatment and baseline prognosis. The point and 95% interval estimates for the regression coefficients obtained using the Gibbs sampler are shown in Table 8.2. Of the three intervals, only the one for prior AIDS diagnosis excludes 0, though the one for treatment group nearly does. The signs on the point estimates of these coefficients indicate that patients without a prior AIDS diagnosis and, to a lesser extent, those in the ddI group were more likely to experience a CD4 response, in agreement with our Subsection 8.1.2 results.

	Point Estimate	95% Confidence Limits	
		Lower	Upper
γ_0 (intercept)	.120	−.135	.378
γ_1 (treatment)	.226	−.040	.485
γ_2 (AIDS diagnosis)	−.339	−.610	−.068

Table 8.2 *Point and interval estimates, CD4 response model coefficients.*

Converting these results for the regression coefficients to the probability-of-response scale using equation (8.3), we compare results for a typical patient in each drug-diagnosis group. Table 8.3 shows that the posterior median probability of a response is larger by almost .09 for persons taking ddI, regardless of health status. If we compare across treatment groups instead of within them, we see that the patient without a baseline AIDS diagnosis has roughly a .135 larger chance of responding than the sick patient, regardless of drug status. Notice that for even the best response group considered (baseline AIDS-free patients taking ddI), there is a substantial estimated probability of *not* experiencing a CD4 response, confirming the rather weak evidence in favor of its existence under the full random effects model (7.16).

	ddI	ddC
AIDS diagnosis at baseline	.502	.413
No AIDS diagnosis at baseline	.637	.550

Table 8.3 *Point estimates, probability of response given treatment and prognosis.*

8.1.4 Survival analysis

In order to evaluate the clinical consequences of two-month changes in the CD4 lymphocyte count associated with the study drugs, we performed a parametric survival analysis. As in Subsection 8.1.3, the $m = 367$ patients with baseline and two-month CD4 counts were divided into responders and nonresponders. We wish to fit a proportional hazards model, wherein the hazard function h for a person with covariate values $\mathbf{z}$ and survival or censoring time t takes the form $h(t|\mathbf{z}, \boldsymbol{\beta}) = h_0(t) \exp(\mathbf{z}'\boldsymbol{\beta})$. In our case, we define four covariates as follows: $z_0 = 1$ for all patients; $z_1 = 1$ for ddI patients with a CD4 response, and 0 otherwise; $z_2 = 1$ for ddC patients without a CD4 response, and 0 otherwise; and $z_3 = 1$ for ddC patients with a CD4 response, and 0 otherwise. Writing $\boldsymbol{\beta} = (\beta_0, \beta_1, \beta_2, \beta_3)'$ and following Cox and Oakes (1984, Sec. 6.1), we obtain the loglikelihood

$$\log L(\boldsymbol{\beta}) = \sum_{i \in \mathcal{U}} \log h(t_i | \mathbf{z}_i, \boldsymbol{\beta}) + \sum_{i=1}^{m} \log S(t_i | \mathbf{z}_i, \boldsymbol{\beta}) , \tag{8.4}$$

where S denotes the survival function, $\mathcal{U}$ the collection of uncensored failure times, and $\mathbf{z}_i = (z_{0i}, z_{1i}, z_{2i}, z_{3i})'$. Our parametrization uses nonresponding ddI patients as a reference group; β_1, β_2, and β_3 capture the effect of being in one of the other three drug-response groups.

We followed Dellaportas and Smith (1993) by beginning with a Weibull model for the baseline hazard, $h_0(t) = \rho t^{\rho-1}$, but replaced their rejection sampling algorithm based on the concavity of the loglikelihood (8.4) with the easier-to-program Metropolis subchain approach (see Subsection 3.4.4). Our initial concern that the Weibull model might not be rich enough evaporated when we discovered extremely high posterior correlation between ρ and β_0. This suggests the baseline hazard may reasonably be thought of as exponential in our very ill population, and hence we fixed $\rho = 1$ in all subsequent calculations.

The resulting estimated marginal posterior distributions for the βs were fairly symmetric, and those for β_1 and β_2 were centered near 0. However, the 95% equal-tail posterior credible set for β_3, $(-1.10, 0.02)$, suggests predominantly negative values. To ease the interpretation of this finding, we transform the posterior $\boldsymbol{\beta}$ samples into corresponding ones from the survival function using the relation $S(t|\mathbf{z}, \boldsymbol{\beta}) = \exp\{-t \exp(\mathbf{z}'\boldsymbol{\beta})\}$. We do this for each drug-response group over a grid of nine equally-spaced t values from 0 to 800 days. Figure 8.5(a) gives a smoothed plot of the medians of the resulting samples, thus providing estimated posterior survival functions for the four groups. As expected, the group of ddC responders stands out, with substantially improved mortality. A more dramatic impression of this difference is conveyed by the estimated posterior distributions of median survival time, $\theta(\mathbf{z}) = (\log 2) \exp(-\mathbf{z}'\boldsymbol{\beta})$, shown in Figure 8.5(b). While a CD4 response translates into improved survival in both drug groups, the improvement is clinically significant only for ddC recipients.

8.1.5 Discussion

Analysis of our CD4 data using both a longitudinal changepoint model and a simpler probit regression model suggests that ddC is less successful than ddI in producing an initial CD4 boost in patients with advanced HIV infection. However, this superior short-term performance does not translate into improved survival for ddI patients. In fact, it is the patients

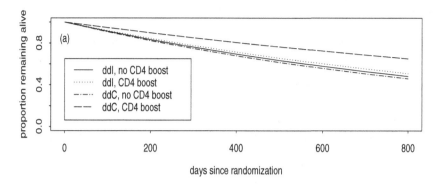

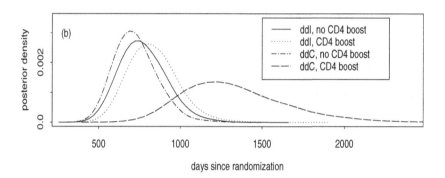

Figure 8.5 *Posterior summaries, ddI/ddC data: (a) estimated survival functions; (b) median survival time distributions.*

"responding" to ddC who have improved survival. It is quite possible that these ddC "responders"' are simply those persons whose CD4 would have increased in the absence of treatment, whereas the ddI "responders" reflect the short-term increase in CD4 produced by ddI. These results provide another example of the caution needed in evaluating treatments on the basis of potential surrogate markers (like CD4 count), and on the use of post-randomization information to explain results.

8.2 Robust analysis of clinical trials

8.2.1 Clinical background

In this section, we consider another case study related to AIDS, but in the context of the interim monitoring and final analysis of a clinical trial, namely, the CPCRA toxoplasmic encephalitis (TE) prophylaxis trial. Among the most ominous infections common in AIDS patients is encephalitis due to *Toxoplasma gondii*. This infection is the cause of death in approximately 50% of persons who develop it and the median survival is approximately six months. Additional clinical and immunological background concerning TE is provided in the review paper by Carlin, Chaloner, Louis, and Rhame (1995).

All patients entered into the study had either an AIDS defining illness or a CD4 count of less than 200. In addition all had a positive titre for *toxoplasma gondii*, and were therefore at risk for TE. The trial was originally designed with four treatment groups: active clindamycin, placebo clindamycin, active pyrimethamine, and placebo pyrimethamine with an allocation ratio of 2:1:2:1, respectively. It was anticipated that the two placebo groups would be pooled in a final analysis for a three treatment group comparison of placebo, active pyrimethamine, and active clindamycin.

The first patient was randomized in September 1990. The active and placebo clindamycin treatments were terminated soon after in March 1991, due to non-life-threatening toxicities that resulted in the discontinuation of medication for many patients in the active clindamycin group. People on either active or placebo clindamycin were offered rerandomization to either active or placebo pyrimethamine with a 2:1 allocation ratio. The trial then became a simple two treatment study of active pyrimethamine against placebo pyrimethamine.

8.2.2 Interim monitoring

In order to ensure scientific integrity and ethical conduct, decisions concerning whether or not to continue a clinical trial based on the accumulated data are made by an independent group of statisticians, clinicians and ethicists who form the trial's *data safety and monitoring board*, or DSMB. These boards have been a standard part of clinical trials practice (and NIH policy) since the early 1970s; see Fleming (1992) for a full discussion. As a result, the trial's statisticians require efficient algorithms for computing posterior summaries of quantities of interest to the DSMB.

For simplicity, we consider the case of $p = 2$ covariates: z_{1i} indicates whether individual i received pyrimethamine ($z_{1i} = 1$) or placebo ($z_{1i} = 0$), and z_{2i} denotes the baseline CD4 cell count of individual i at study entry. Using the method and software of Chaloner et al. (1993), a prior for p_1 was elicited from five AIDS experts, coded as A, B, C, D, and E. Each expert's prior was then discretized onto a 31-point grid, each support point having probability $1/31$. Finally, each prior was transformed to the β_1 scale and converted to a smooth curve suitable for graphical display using a histogram approach. In what follows, we consider only the prior of expert A, a physician at a university AIDS clinic; the paper by Carlin et al. (1993) provides a more complete analysis. While the trial protocol specified that only the onset of TE was to be treated as an endpoint, expert A was unable to separate the outcomes of death and TE. As such, this prior relates to the dual endpoint of TE or death.

As described in the report by Jacobson et al. (1994), the trial's data safety and monitoring board met on three occasions after the start of the trial in September of 1990 to assess its progress and determine whether it should continue or not. These three meetings analyzed the data available as of the file closing dates 1/15/91, 7/31/91, and 12/31/91, respectively. At its final meeting, the board recommended stopping the trial based on an informal stochastic curtailment rule: the pyrimethamine group had not shown significantly fewer TE events up to that time, and, due to the low TE rate, a significant difference was judged unlikely to emerge in the future. An increase in the number of deaths in the pyrimethamine group was also noted, but this was not a stated reason for the discontinuation of the trial (although subsequent follow-up confirmed this mortality increase). The recommendation to terminate the study was conditional on the agreement of the protocol chairperson after unblinding and review of the data. As a result, the trial did not actually stop until 3/30/92, when patients were instructed to discontinue their study medication.

We now create displays that form a Bayesian counterpart to standard monitoring boundaries. Using the Cox partial likelihood (7.25) as our likelihood, Figure 8.6 diplays the elicited prior, likelihood, exact posterior, and normal approximation to the posterior of β_1 for the four data monitoring dates listed above. The number of accumulated events is shown beneath each display. The exact posterior calculations were obtained by simple summation methods after discretizing β_2 onto a suitably fine grid. Notice that in the first frame the likelihood is flat and the prior equals the exact posterior because no information has yet accumulated. From this frame we see that expert A expects the pyrimethamine treatment to be effective at preventing TE/death; only one of the 31 prior support points are to the right of 0. The remaining frames show the excess of events (mostly deaths) accumulating in the treatment group, as the likelihood and posterior summaries move dramatically toward positive β_1 values. The extremely limited prior support for these values (and, in fact, no support in the region where the likelihood is maximized) causes the likelihood to appear one-tailed, and the exact posterior to assume a very odd shape. The normal approximation is inadequate when the sample size is small (due to the skew in the prior) and, contrary to the usual theory, also when it is large (due to the truncation effect of the prior on the likelihood).

Next, we turn to the monitoring plot associated with the posteriors in Figure 8.6. The trial protocol specifies a target reduction of 50% in hazard for the treatment relative to control. For a more sensitive analysis, we use a 25% reduction as the target, implying the indifference zone boundaries $\beta_{1,U} = 0$ and $\beta_{1,L} = \log(.75) = -.288$. Figure 8.7 plots these two tail

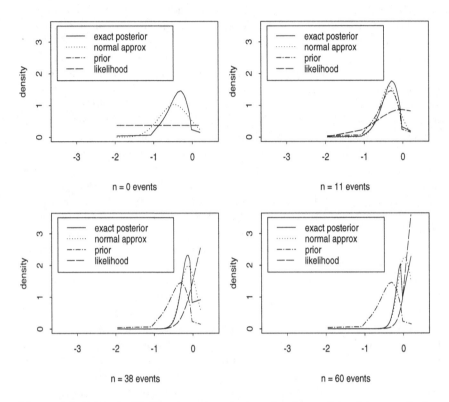

Figure 8.6 *Prior A, likelihood, and posteriors for β_1 on four data monitoring dates, TE trial data. Covariate = baseline CD4 count; monitoring dates are 1/15/91, 7/31/91, 12/31/91, and 3/30/92.*

areas versus accumulated information for the exact posterior, approximate normal posterior, and likelihood (or more accurately, the posterior under a prior on β_1 that is flat over the support range specified by our expert). This figure nicely captures the steady progression of the data's impact on expert A's prior. Using a stopping level of $p = .10$ (indicated by the horizontal reference line in the figure), the flat prior analysis recommends stopping the trial as of the penultimate monitoring point; by the final monitoring point the data converts the informative prior to this view as well. The inadequacy of the normal approximation is seen in its consistent overstatement of the upper tail probability.

Impact of subjective prior information

Figure 8.7 raises several questions and challenges for further research. For instance, notice that expert A's prior suggests stopping immediately (i.e.,

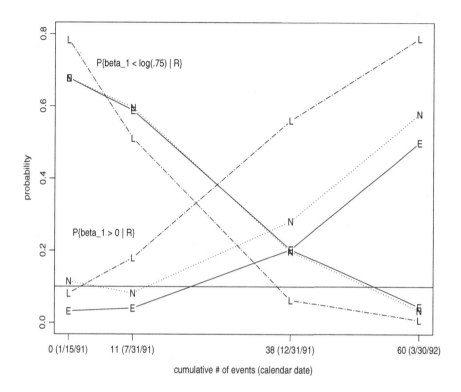

Figure 8.7 *Posterior monitoring plot for β_1, TE trial data. E = exact posterior; N = normal approximation; L = likelihood.*

not starting the trial) and deciding in favor of the treatment. Though none of the elicitees would have actually recommended such an action (recall the 50% reduction goal), it does highlight the issue of ethical priors for clinical trials, as alluded to earlier. In addition, the fact that expert A's clinically-derived prior bears almost no resemblance to the data, besides being medically interesting, is perhaps statistically troubling as well, since it suggests delaying the termination of an apparently harmful treatment. Perhaps expert A should have received more probabilistic training before elicitation began, or been presented with independent data on the issue.

Prior beliefs common to all five of the experts interviewed included TE being common in this patient population, and a substantial beneficial prophylactic effect from pyrimethamine. However, data from the trial were at odds with these prior beliefs. For instance, out of the 396 people in the trial, only 12 out of 264 on pyrimethamine and only 4 out of 132 on placebo actually got TE. The low rate is believed to be partly due to a change in the standard care of these patients during the trial. During the trial, information was released on the prophylactic effects of the antibiotic trimethoprim-sulphamethoxazole (TMP/SMX, or Bactrim). TE was extremely rare for patients receiving this drug, and many patients in our study were taking it concurrently with their study medication.

Finally, all five experts put high probability on a large beneficial effect of the treatment, but the rate of TE was actually higher in the pyrimethamine group than in the placebo group. In addition, a total of 46 out of 264 people died in the pyrimethamine group, compared to only 13 out of 132 in the placebo group. Apparently the assumption made at the design stage of the trial that there would be no difference in the non-TE death rates in the different trial arms was incorrect.

Still, the decision makers are protected by the simultaneous display of the flat prior (likelihood) results with those arising from the overly optimistic clinical prior; the results of a skeptical prior might also be included. Overall, this example clearly illustrates the importance of prior robustness testing in Bayesian clinical trial monitoring; this is the subject of Subsection 8.2.3.

Impact of proportional hazards

Our data indicate that the proportional hazards assumption may not hold in this trial. The relative hazard for death in the pyrimethamine group increased over follow-up compared to the placebo group (see Figure 1 of Jacobson et al., 1994). Specifically, while the hazard for the placebo group is essentially constant for the entire study period, the hazard in the pyrimethamine group changes level at approximately six months into the follow-up period. As such, we might reanalyze our data using a model that does not presume proportional hazards. Several authors have proposed models of this type (see, e.g., Louis, 1981b; Lin and Wei, 1989; Tsiatis, 1990). Other authors (e.g., Grambsch and Therneau, 1994) create diagnostic plots designed specifically to detect the presence of time-varying covariates in an otherwise proportional hazards setting.

The simplest approach might be to include a parameter in the Cox model (7.25) that explicitly accounts for any hazard rate change. In our case, this might mean including a new covariate z_{p+1} which equals 0 if the individual is in the placebo group ($z_1 = 0$), equals -1 if the individual is in the drug group ($z_1 = 1$) *and* the corresponding event (failure or censoring) occurs prior to time $t = 6$ months, and equals $+1$ if the individual is in the drug group and the corresponding event occurs after time $t = 6$. A posterior distribution for β_{p+1} having significant mass far from 0 would then confirm the assumed pattern of nonproportional hazards. Alternatively, we might replace $\beta_1 z_1$ in model (7.25) by $\beta_1 z_1 w(t; \beta_{p+1})$, where w is a weight function chosen to model the relative hazard in the drug group over time. For example, $w(t; \beta_{p+1}) = \exp(\beta_{p+1} t)$ could be used to model an increasing

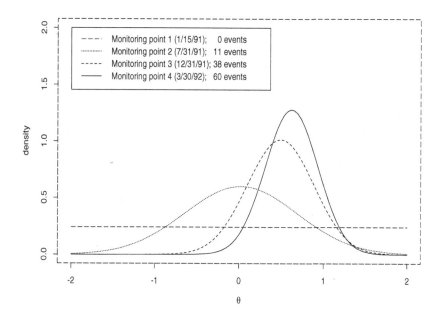

Figure 8.8 *Posterior for the treatment effect under a flat prior, TE trial data. Endpoint is TE or death; covariate is baseline CD4 count.*

relative hazard over time, with $\beta_1 = 0$ corresponding to the "null" case of proportional hazards.

8.2.3 Prior robustness and prior scoping

Phase III clinical trials (like ours) are large scale trials designed to identify patients who are best treated with a drug whose safety and effectiveness has already been reasonably well established. Despite the power and generality of the Bayesian approach, it is unlikely to generate much enthusiasm in monitoring such trials unless it can be demonstrated that the methods possess good frequentist properties. Figure 8.8 displays the posterior for θ under a flat prior (i.e., the standardized marginal likelihood) at each of the four dates mentioned above. Recalling that negative values of θ correspond to an efficacious treatment, the shift toward positive θ values evident in the third and fourth monitoring points reflects an excess of deaths in the treatment group. This emerging superiority of the placebo was quite surprising, both to those who designed the study and to those who provided the prior distributions.

Consider the percentile restriction approach of Subsection 4.2.2. We first integrate β_2 out of the Cox model likelihood $L(\beta)$ by a simple summation method after discretizing β_2 onto a suitably fine grid. This produces the marginal likelihood for β_1, which we write in the notation of Section 4.2.2 simply as $f(x|\theta)$. At the fourth monitoring point, stopping probability $p = 0.10$ and weight $\pi = 0.25$ on the null hypothesis produce the values $f(x|\hat{\theta}_1) = 1.28$ and $c = 3f(x|0) = 0.48$, where $\hat{\theta} = .600$ is the marginal MLE of θ.

Suppose we seek the **a** for which $\mathcal{H}_c$ is nonempty given a specific indifference zone (θ_L, θ_U). Because negative values of θ are indicative of an efficacious treatment, in practice one often takes $\theta_U = 0$ and $\theta_L < 0$, the additional benefit being required of the treatment in order to justify its higher cost in terms of resources, clinical effort, or toxicity. We adopt this strategy and, following the previous subsection, choose $\theta_L = \log(.75) = -.288$, so that a reduction in hazard for the treatment relative to control of at least 25% is required to conclude treatment superiority.

At monitoring point four we have $\theta_U < \hat{\theta} = .600$, so from equation (4.14) the **a** that satisfy the condition $\sup_G \int f(x|\theta)dG(\theta) \geq c$ are such that

$$a_U \geq \frac{[f(x|\theta_U) - f(x|\theta_L)]a_L + [c - f(x|\theta_U)]}{f(x|\hat{\theta}) - f(x|\theta_U)}. \tag{8.5}$$

Under the Cox model, we obtain $f(x|\theta_L) = .02$ and $f(x|\theta_U) = .18$, so again taking $p = .10$ and $\pi = .25$, equation (8.5) simplifies to $a_U \geq .145a_L + .273$. For the (a_L, a_U) pairs satisfying this equation, there exists at least one conditional (over H_1) prior G that has these tail areas and permits stopping and rejecting H_0, given the data collected so far. Conversely, for conditional priors G featuring an (a_L, a_U) pair located in the lower region, stopping and rejecting H_0 is not yet possible. Points lying on the boundary between these two regions correspond to the maximum amount of mass the prior may allocate to the indifference zone before stopping becomes impossible.

Next, consider the case of stopping and rejecting H_1 (i.e., accepting H_0) at monitoring point four. This occurs if $P(\theta \neq 0|x) < p$, that is, when $\int f(x|\theta)dG(\theta) \leq c^* = \left(\frac{p}{1-p}\right)\left(\frac{\pi}{1-\pi}\right)f(x|0)$. Hence, we are now interested in the infimum of $\int f(x|\theta)dG(\theta)$. Since $\min\{f(x|\theta_L), f(x|\theta_U)\} = f(x|\theta_L) = .020$ for our dataset, we have from (4.13) that the **a** satisfying the condition $\inf_G \int f(x|\theta)dG(\theta) \leq c^*$ are such that

$$a_U \geq (1 - 50c^*) - a_L = .700 - a_L, \tag{8.6}$$

where we have again let $p = .10$ and $\pi = .25$. Here, increasing π (equivalently c) serves to decrease the intercept of the boundary, and vice versa. In this regard, note that $\pi = 0$ is required to make stopping impossible for all **a**. On the other hand, if π is at least .529, then there is at least one prior for which it is possible to stop and accept the null regardless of **a**.

Figure 8.9 shows the four prior tail area regions determined by the two

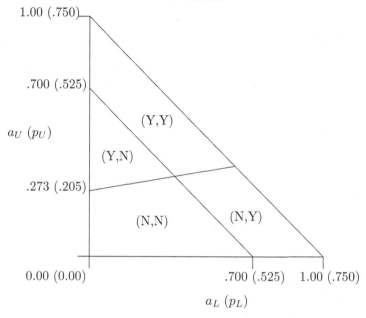

Figure 8.9 *Conditional prior tail area regions, with ordered pair indicating whether it is possible to stop and reject H_0 and H_1, respectively (Y = yes, N = no).*

stopping boundaries discussed above. Each region is labeled with an ordered pair, the components of which indicate whether stopping to reject is possible for the prior tail areas in this region under the null and alternative hypotheses, respectively. The relatively large size of the lower left region indicates the mild nature of the evidence provided by these data; there are many prior tail area combinations for which it is not yet possible to stop and reject *either* H_0 or H_1.

Because it may be difficult to think solely in terms of the conditional prior G, we have also labeled the axes of Figure 8.9 with the corresponding unconditional tail areas, which we denote by p_L and p_U. That is, $p_L \equiv P_{unc}(\text{treatment superior}) = a_L(1 - \pi)$, and $p_U \equiv P_{unc}(\text{control superior}) = a_U(1 - \pi)$; recall that $\pi = .25$ in our example. Increasing π (equivalently c) increases only the a_U-intercept in boundary (8.5), so that checking the impact of this assumption is straightforward. In particular, for this dataset we remark that $\pi \leq .014$ makes stopping and rejecting H_0 possible for *all* **a** (boundary (8.5) has a_L-intercept 1), while $\pi \geq .410$ makes such stopping impossible *regardless* of **a** (boundary (8.5) has a_U-intercept 1).

We now turn to the semiparametric approach of Subsection 4.2.2. We use a $N(\theta|\hat{\theta}_k, \hat{\sigma}_k^2)$ approximation to the likelihood at monitoring point k, $k =$

$2, 3, 4$, as suggested by Figure 8.8. Since our analysis is targeted toward the members of the data safety and monitoring board, we restrict attention to "skeptical" conditional priors g by setting $\mu = 0$. We may then compute the Bayes factor BF in favor of H_0 that arises from equations (4.16) and (4.17) as a function of the conditional prior standard deviation τ for $0 < \tau < 4$. To facilitate comparison with previous figures, Figure 8.10(a) plots B not versus τ, but, instead, versus the conditional prior upper tail probability a_U, a one-to-one function of τ. Assigning a mass $\pi = .25$ to the null region (the indifference zone) and retaining our rejection threshold of $p = .1$, we would reject H_0 for $B < \frac{1}{3}$; this value is marked on the figure with a dashed horizontal reference line. The message from this figure is clear: no uniform-normal mixture priors of the form (4.15) allow rejection of H_0 at monitoring points 2 or 3, but by monitoring point 4, rejection of H_0 is favored by all but the most extreme priors (i.e., those that are either extremely vague or essentially point masses at 0).

Next, suppose we wish to investigate robustness as π varies. We reject H_0 if and only if $B(\tau) \le \left(\frac{p}{1-p} \right) \left(\frac{1-\pi}{\pi} \right)$, or equivalently,

$$\pi \le \left[\frac{(1-p)B(\tau)}{p} + 1 \right]^{-1} .$$

Again using the one-to-one relationship between a_U and τ, Figure 8.10(b) plots this boundary in (π, a_U)-space for the final three monitoring points. Conditional on the data, all priors that correspond to points lying to the left of a given curve lead to rejection of H_0, while all those to the right do not. (Note the difference in interpretation between this graph and Figure 8.9, where each plotted point corresponded to infinitely many priors.) We see the mounting evidence against H_0 in the boundaries' gradual shift to the right, increasing the number of priors that result in rejection.

Finally, we turn to the fully parametric approach outlined in Subsection 4.2.2. Fixing the prior mean μ at 0 again, the prior support for the indifference zone, π, is now determined by τ. This loss of one degree of freedom in the prior means that a bivariate plot like Figure 8.10(b) is no longer possible, but a univariate plot as in Figure 8.10(a) is still sensible. To obtain the strongest possible results for our dataset, in Figure 8.11(a) we plot the posterior probability of $H_L : \theta < \theta_L$, rather than H_0, versus τ. We see that rejection of this hypothesis is possible for all skeptical normal priors by the third monitoring point; by the fourth monitoring point, the evidence against this hypothesis is overwhelming. While this analysis considers a far smaller class of priors than considered in previous figures, it still provides compelling evidence by the third monitoring point that the treatment is not superior in terms of preventing TE or death.

It is of interest to compare our results with those obtained using the ϵ-contamination method of Greenhouse and Wasserman (1995). As men-

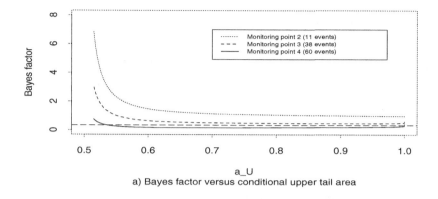

a) Bayes factor versus conditional upper tail area

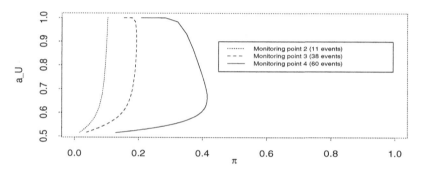

b) Conditional upper tail area versus prior mass on indifference zone;
for combinations to the left of each curve,
all priors result in stopping to reject Ho

Figure 8.10 *Semiparametric prior partitions, TE trial data.*

tioned in Subsection 4.2.2, this method enables computation of upper and lower bounds on posterior expectations for priors of the form $(1-\epsilon)g_0 + \epsilon g$, where g_0 is some baseline distribution and g can range across the set of all possible priors. Reasonably tight bounds for moderately large ϵ suggest robustness of the original result to changes in the prior. Figure 8.11(b) plots the Greenhouse and Wasserman bounds for the posterior probability of H_L for the final three monitoring points, where we have taken g_0 as our $N(\theta|\mu, \tau^2)$ prior with $\mu = 0$ and $\tau = 1$. At the third monitoring point, the posterior probability is bounded below the rejection threshold of .1 only for ϵ roughly less than .17, suggesting that we cannot stray too far from the baseline normal prior and maintain total confidence in our conclusion. By the fourth monitoring point, however, the upper bound stays below the threshold for ϵ values as large as .65, indicating a substantial degree of robustness to prior specification in our stopping decision.

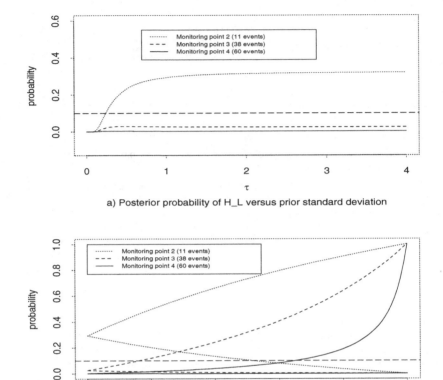

a) Posterior probability of H_L versus prior standard deviation

b) Bounds on posterior probability of H_L versus prior contamination factor

Figure 8.11 *Fully parametric prior partitions, TE trial data.*

8.2.4 *Sequential decision analysis*

In this subsection, we apply the fully Bayesian decision theoretic approach of Subsections 7.6.2 and 7.6.3 to our dataset. The posteriors pictured in Figure 8.8 are nothing but standardized versions of the marginal partial likelihoods for the treatment effect θ given all the data observed up to the given date, and so provide the necessary inputs for these approaches.

Consider first the backward induction approach of Subsection 7.6.2 in this setting. The very near-normal appearance of the curves at the latter three time points justifies an assumption of normality for the prior and likelihood terms. In fact, our dataset provides a good test for the $K = 2$ calculations in Subsection 7.6.2, because information that would suggest stopping accumulates only at the final two monitoring points. Thus, we take the appropriate normal approximation to the 7/31/91 likelihood, a

$N(\mu, \sigma^2_{K-2})$ distribution with $\mu = .021$ and $\sigma^2_{K-2} = .664^2$, as our prior $p(\theta)$, and assume independent $N(\theta, \sigma^2_{K-1})$ and $N(\theta, \sigma^2_K)$ distributions for $f(y_{K-1}|\theta)$ and $f(y_K|\theta)$, respectively. Following our earlier discussion we assume the likelihood variance parameters to be known, using the data-based values $\sigma^2_{K-1} = .488^2$ and $\sigma^2_K = .515^2$. Coupled with the normal-normal conjugacy of our prior-likelihood combination, this means that the marginal distributions $m(y_{K-1})$ and $m(y_K|y_{K-1})$ emerge as normal distributions as well. Specifically, the samples required to estimate $L_3^{(K-2)}$ may be obtained as

$$y^*_{K-1,i} \overset{iid}{\sim} N(\mu, \ \sigma^2_{K-2} + \sigma^2_{K-1}), \ i = 1, \ldots, B,$$

and

$$y^*_{K,ij}|y^*_{K-1,i} \overset{iid}{\sim} N\left(\frac{\sigma^2_{K-1}\mu + \sigma^2_{K-2}y^*_{K-1,i}}{\sigma^2_{K-2} + \sigma^2_{K-1}} \ , \ \frac{\sigma^2_{K-1}\sigma^2_{K-2}}{\sigma^2_{K-1} + \sigma^2_{K-2}} + \sigma^2_K \right),$$

where $j = 1, \ldots, B$ for each i.

We use the same indifference zone as above; in the notation of Subsection 7.6.2 these are $\theta_U = 0$ and $\theta_L = \log(.75) = -.288$. We adopt a very simple loss and cost structure, namely, $s_1^{(k)} = s_2^{(k)} = 1$ and $A_k = .1$, both for all k. Performing the backward induction calculations using $B = 5000$ Monte Carlo draws, we obtain $L_1^{(K-2)} = .021$, $L_2^{(K-2)} = -.309$, and $L_3^{(K-2)} = -.489$, implying that for this cost, loss, and distributional structure, it is optimal to continue sampling. However, a one-step-back analysis at time $K - 1$ using $p(\theta|y_{K-1})$ as the prior finds that $L_2^{(K-1)}$ is now the smallest of the three values by far, meaning that the trial should now be stopped and the treatment discarded. This confirms the board's actual recommendation, albeit for a slightly different reason (excess deaths in the drug group, rather than a lack of excess TE cases in the placebo group).

Figure 8.12 expands our time $K-2$ analysis by replacing the observed 7/31/91 mean $\mu = .021$ with a grid of points centered around this value. We can obtain the intersection points of $L_3^{(K-2)}(\mu)$ with $L_1^{(K-2)}(\mu)$ and $L_2^{(K-2)}(\mu)$ via interpolation as $\gamma_{K-2,L} = -.64$ and $\gamma_{K-2,U} = .36$, respectively. Notice that these values determine a continuation region that contains the observed 7/31/91 mean $\mu = .021$, and is centered around the midpoint of the indifference zone, $(\theta_U + \theta_L)/2 = -.144$.

Because our loss functions are monotonic and we have normal likelihoods, the forward sampling method of Subsection 7.6.3 is applicable and we may compare its results with those above. Using the same loss and cost structure, setting $\mu = .021$, and using $G = 10^6$ Monte Carlo samples, the algorithm quickly deduces that $\gamma_{K-2,L} < \mu < \gamma_{K-2,U}$, so that continuation of the trial is indeed warranted. The algorithm also finds $\gamma_{K-1,L} = -0.260$ and $\gamma_{K-1,U} = -0.044$, implying that the trial should be continued beyond the penultimate decision point (12/31/91) only if the posterior mean at

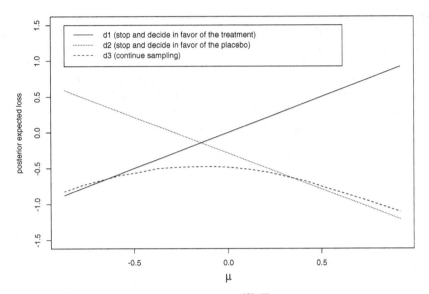

Figure 8.12 *Plot of posterior expected loss $L_i^{(K-2)}$ versus μ for the three decisions ($i = 1, 2, 3$) computed via backward induction, using the normal/normal model with variances estimated from TE trial data. Choosing the treatment is optimal for $\mu < -.65$, while choosing the placebo is optimal for $\mu > .36$; otherwise, it is optimal to continue the trial to time $K - 1$.*

this point lies within this interval. Note from Figure 8.8 that the observed mean, $E(\theta|y_{K-1}) = .491$, lies above this interval, again in agreement with the board's decision to stop at this point and reject the treatment.

As with the backward induction method, obtaining exact values via forward sampling for the initial cutoffs $\gamma_{K-2,L}$ and $\gamma_{K-2,U}$ is more involved, since it is again necessary to resort to a grid search of μ values near the solution. Still, using $G = 10^6$ forward Monte Carlo draws we were able to confirm the values found above (−.64 and .36) via backward induction. While the continued need for grid search is a nuisance, we remark that for forward sampling, the *same* G Monte Carlo draws may be used for each candidate μ value.

Using forward sampling we may extend our analysis in three significant ways. First, we extend to the case $K = 3$, so that the 1/15/91 likelihood now plays the role of the prior, with the remaining three monitoring points representing accumulating data. Second, we return to the "positive part" loss structure (7.26), simply by adding this modification to our definition of the Monte Carlo losses L_j, as in (7.29). Note that this is still a monotone loss specification (hence, suitable for forward sampling in the restricted class that depends only on the posterior mean), but one which leads to substantial complications in the backward induction solution. Finally, the positive part loss allows us to order the losses so that incorrect decisions are penalized more heavily the later they are made. Specifically, we take $s_1^{(k)} = s_2^{(k)} = k + 3$, for $k = 0, 1, 2, 3$.

Because forward sampling cannot be implemented with the improper prior suggested by the 1/15/91 likelihood in Figure 8.8, we instead retain a normal specification, i.e., a $N(\mu, \sigma_0^2)$ with $\mu = 0$ and $\sigma_0^2 = 5$. This prior is quite vague relative to our rather precise likelihood terms. We also retain our uniform cost structure, $A_k = .1$ for all k, and simply set $\sigma_1^2 = \sigma_2^2 = \sigma_3^2 = .560^2$, the sample mean of the three corresponding values estimated from our data, again giving our procedure something of an empirical Bayes flavor. Again using $G = 10^6$ Monte Carlo samples, the forward algorithm quickly determines $\gamma_{0,L} << \mu << \gamma_{0,U}$ (so that the trial should indeed be carried out), and goes on to find precise values for the remainder of the γ vector. In particular, we obtain $\gamma_{1,L} = -0.530$ and $\gamma_{1,U} = 0.224$, an interval that easily contains the observed 7/31/91 mean of .021, confirming the board's decision that the trial should not be stopped this early.

Finally, we investigate the robustness of our conclusions by replacing our normal likelihood with a much heavier-tailed Student's t with $\nu = 3$ degrees of freedom. Since there is now no longer a single sufficient statistic for θ, the optimal rule need not be solely a function of the posterior mean, but we may still use forward sampling to find the best member in this class of posterior mean-based rules. We retain our previous prior, cost, and loss structure, and set $\sigma_k^2 = .560^2/3$, $k = 1, 2, 3$, so that $Var(Y_k|\theta) = \nu\sigma_k^2/(\nu - 2) = .560^2$, as before. Again using $G = 10^6$ Monte Carlo samples, our forward algorithm now requires univariate numerical integration to find the posterior means to compare to the elements of the γ vector, but is otherwise unchanged. We now obtain $\gamma_{1,L} = -0.503$ and $\gamma_{1,U} = 0.205$, a slightly narrower interval than before (probably because the t_3 has a more concentrated central 95% region than the normal for a fixed variance), but one which still easily contains the observed 7/31/91 mean of .021.

8.2.5 Discussion

As seen in the progression of displays in Subsection 8.2.3, the practical usefulness of a prior partitioning analysis often depends on the breadth of the class of priors considered. One must select the prior class carefully; large classes may lead to overly broad posterior bounds, while narrow classes may eliminate many plausible prior candidates. The quasiunimodal, semiparametric, and fully parametric classes considered here constitute only a small fraction of those discussed in the Bayesian robustness literature. Other possibilities include density bounded classes, density ratio classes, and total variation classes; see Wasserman and Kadane (1992) for a discussion and associated computational strategies.

For computational convenience, we have frequently assumed normal distributions in the above discussion. Using Monte Carlo integration methods, however, prior partitioning can be implemented for any combination of likelihood and prior. Similarly, one could apply the method in settings where θ is a multivariate parameter, such as a multi-arm trial or a simultaneous study of effectiveness and toxicity. As such, prior partitioning offers an attractive complement to traditional prior elicitation and robustness methods in a wide array of clinical trial settings.

Armed with the forward sampling computational framework, many challenges in the practical application of fully Bayesian methods to clincial trial monitoring may now be tackled. Better guidelines for the proper choice of loss functions $l_i^{(k)}(d_i, \theta)$ and corresponding sampling costs $\{A_k\}$, expressed in either dollars earned or lives saved, are needed. For example, if an incorrect decision favoring the treatment were judged C times more egregious than an equivalent one favoring the placebo (i.e., one where the true value lay equally far from the indifference zone, but in the opposite direction), we might choose $s_1^{(k)} = C \cdot s_2^{(k)}$.

Finally, though we stress the importance of Bayesian robustness in the monitoring of Phase III clinical trials, more traditional Bayesian analyses are proving effective in the design and analysis of Phase I studies (dose-ranging, toxicity testing) and Phase II studies (indications of potential clinical benefit). In these settings, the ethical consequences of incorrect decisions are less acute than in Phase III studies, and decisions primarily involve the design of the next phase in drug development (including whether there should *be* another phase!). See Berry (1991; 1993) for discussions of these issues, and Berry et al. (2002) for a case study.

8.3 Modeling of infectious diseases

8.3.1 Introduction and data

Mathematical models have been widely used to study infectious disease epidemics. Among the classics is the chain-binomial model introduced by Greenwood (1931) and by Reed and Frost (see Abbey, 1952). It extends the Susceptible-Infective-Removal (SIR) model to the discrete time, stochastic transition context. Disease incidence and mortality are usually reported on a daily basis and great fluctuations are often observed, a situation suited for chain-binomial models of such population-level, time-series information.

The SIR model is built on compartments representing disease status (e.g, susceptibles, infectives, symptomatic cases, cures, deaths). Usually, only some of the transitions among compartments are observed, for example new symptomatic cases and deaths. Therefore, much of the information needed for direct estimation of compartment holding times is missing, making development and implementation of valid estimation methods such as likelihood-based approaches difficult. However, formal Bayesian analysis implemented by MCMC is an effective approach for dealing with this data structure. It is relatively easy to implement, allows incorporation of prior information, and produces the full joint posterior distribution of the parameters (Becker, 1999).

To demonstrate the effectiveness of the Bayesian approach and to highlight cautions, we model influenza incidence and death data from the 1918 pandemic in Baltimore, MD, and Newark, NJ. We use information on a single wave of incidence during the late summer and fall of 1918, apply our modeling strategy separately to data from each city, then compare and discuss results. Our approach to parameterization, model building and model evaluation are applicable to similar data sets and to multiple waves. On the methodologic front, our analyses illustrate assessing model identifiability by algorithm convergence, and assessing model goodness of fit by external simulation. Our approach is a Bayesian generalization of Lekone and Finkenstädt (2006), who analyze an ebola outbreak and include a control strategy.

The Baltimore dataset was collected by Wade Hampton Frost and Edgar Sydenstricker of the U.S. Public Health Service in conjunction with Departments of Health (Frost and Sydenstricker, 1919) using a household survey shortly after the epidemic. Surveyed household members were asked if anyone in the household had been ill with or had died with influenza-like symptoms between 25 August and 10 December 1918. Originally, the dataset contained individual-level linkage between symptom onset and death, but this information has never been published and is missing. Only daily incidence of disease onset and daily mortality from illness consistent with influenza are reported. In total, there were 7498 influenza cases and 145 influenza-related deaths among 33,776 individuals surveyed. The top panels of Figure 8.13 below provide incidence and mortality curves.

The Newark data cover the period 27 September through 12 November, 1918. These data were collected through daily reports of cases and deaths to the city's department of health (Galishoff, 1969). Among the 435,000 residents, 26,235 influenza cases and 1720 influenza-related deaths were reported (see the bottom panels of Figure 8.13).

8.3.2 Stochastic compartmental model

Model Specification

Our analysis is structured by a compartment model wherein on day t each individual is in one of the five disease states listed at the top of Table 8.4. The bottom of the table and Figure 8.14 give information on state transitions. Each day, a susceptible individual (in S) has a chance of becoming infected through contact with infectious asymptomatic (E) or symptomatic

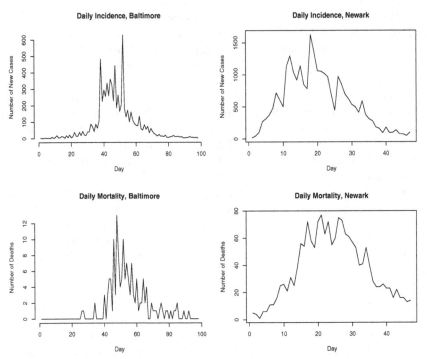

Figure 8.13 *Influenza incidence and mortality curves. Top panels: Baltimore, MD, 25 August 1918 → 10 December 1918; bottom panels: Newark, NJ, 27 September 1918 → 12 November 1918.*

(I) individuals. Infected individuals are first asymptomatic (in NE) and after a stochastic holding time become symptomatic by "surfacing" (entering I).

After an additional, stochastic holding time the individual either recovers (enters C) or dies (enters D). Death counts are observed, but recoveries are not and we pose a deterministic relation between them. For $0 \leq \gamma \leq 1$, let $ND_t = (1 - \gamma)\{ND_t + NC_t\}$. Table 8.5 gives this and other transitions.

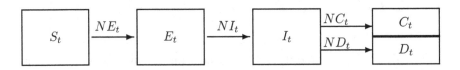

Figure 8.14 *Disease states and daily transitions.*

Compartment/ Transition Counts at Day t	State	Probability Distribution
S_t	Susceptible	
E_t	Asymptomatic Infectives (Exposed)	
I_t	Symptomatic Infectives	
C_t	Recovered (Cured) – Lifetime Immunity	
D_t	Deceased	
NE_t	New Infected	P_{NE}
NI_t	New Symptomatic Infectives	P_{NI}
NC_t	New Recoveries	$\propto ND_t$
ND_t	New Deaths	P_{ND}

Table 8.4 *Disease states, transition counts, and probability distributions.*

$$NE_t = S_{t-1} - S_t$$
$$S_t + E_t = S_{t-1} + E_{t-1} - NI_t$$
$$I_t = I_{t-1} + NI_t - (NC_t + ND_t)$$
$$NC_t = \left(\frac{\gamma}{1-\gamma}\right) ND_t$$
$$D_t = \sum_{s=1}^{t} ND_s$$

Table 8.5 *Transition relations.*

Stochastic transitions

We use an extended chain-binomial model, which is a time-indexed, population-based Markov chain. Each day, flows among states follow probability distributions that depend only on the numbers of individuals currently in the compartments, denoted by $\mathbf{V}_t = (S_t, E_t, I_t, C_t, D_t)$, and not on compartment residence times Bailey (1975). Conditional on $\mathbf{V}_{t-1}$, transition counts are independent, governed by the following probability distributions:

$$NE_t|\mathbf{V}_{t-1} \sim P_{NE} = \text{Poisson}\left[\lambda_t(\rho, \delta, r_t, \varepsilon)\right]$$
$$NI_t|\mathbf{V}_{t-1} \sim P_{NI} = \text{Binomial}(\theta, E_t) \tag{8.7}$$
$$ND_t|\mathbf{V}_{t-1} \sim P_{ND} = \text{Binomial}(\eta, I_t)$$

where

$$\lambda_t(\rho, \delta, r_t, \varepsilon) \;=\; r_t \times \left(\frac{\varepsilon E_{t-1} + \rho\, \delta I_{t-1}}{S_{t-1} + E_{t-1} + \delta I_{t-1} + C_{t-1}} \right) \times S_{t-1} \qquad (8.8)$$

$$\text{and} \;\; \log(r_t) \;=\; \beta_0 + \beta_1(t - t_{mid}) + \beta_2(t - t_{mid})^2 + \beta_3(t - t_{mid})^3,$$

with t_{mid} being the middle day of the epidemic. On the first day of the epidemic, NI_1 diagnosed cases surface from the E_0 asymptomatics, and then the transition model takes over.

Disease transmissibility (r_t) is the composite of infectivity and the social contact rate. Most previous models assume time-constant transmissibility $(r_t \equiv r)$, but we find that time-dependence greatly improves model fit. The log cubic parameterization of r_t provides sufficient flexibility.

Time-varying reproductive number

The basic reproductive number $(\tilde{R})$ is the expected number of individuals infected by an infective on the first day of the epidemic (Dietz, 1993). It is used to characterize an epidemic, but only applies to the front end of the epidemic. A time-varying reproductive number (R_t) can be computed from equation (8.8). It reports the expected number of new infections that an individual who becomes infectious at day t will induce. Time-dependence results from a combination of the time-dependence of r_t and changes in population composition. With DOI denoting the typical duration of infectivity, we have:

$$R_t = \frac{S_t}{S_t + E_t + \delta I_t + C_t} \left[r_t + (DOI - 1) \cdot \delta \cdot r_t \right] \qquad (8.9)$$

Using this notation, $\tilde{R} = R_1$ or if that is unstable, R_2.

8.3.3 Parameter estimation and model building

If daily counts of all state transitions were observed, maximum likelihood estimation would be straightforward. However, we only have information on the total reference population size, daily incidence of symptomatic flu-like illness (NI_t) and daily mortality (ND_t). Though, in principle, a missing data likelihood can be constructed and EM algorithm implemented, use of a formal Bayesian model implemented by MCMC substantially simplifies the programming. Furthermore, it confers the additional benefits of producing posterior distributions for all unknowns (all parameters and all daily state membership tallies) and allows use of informative prior distributions that bring into the system empirical evidence from other investigations (for example, blood titer studies) and expert understanding of infectious disease natural history. The Bayesian methodology has been previously used in infectious disease modeling (O'Neill, 2002). We use WinBUGS to implement our Bayesian approach; see the book's web site for our code. For each model considered, we used a burn-in of 4000 iterations, followed by 6000 production iterations.

Groups of parameters

The parameters, $\{E_0, \varepsilon, \theta, \gamma, \eta, \rho, \delta, r_t\}$, can be classified into four groups:

Directly estimable parameters: Both γ and η can be estimated directly. The ratio $(1 - \gamma) = \{\text{deaths}/(\text{deaths} + \text{recoveries})\}$ is estimated by,

$$1 - \hat{\gamma} = \frac{\text{total flu caused deaths}}{\text{total influenza cases}} = \frac{\sum_t ND_t}{\sum_t NI_t}. \qquad (8.10)$$

For individuals in the symptomatic state we assume a constant daily probability of death, η. Associated with each death are $\left(\frac{\gamma}{1-\gamma}\right)$ cures. Therefore, for each individual in the symptomatic state, the time to removal follows a geometric distribution with daily event probability (removal rate) $\left(\frac{\eta}{1-\gamma}\right)$ and so the expected duration in the symptomatic state (EDS) is $\left(\frac{1-\gamma}{\eta}\right)$. To estimate η, divide total deaths by total person-time at risk for influenza caused death,

$$\text{Person-time at risk} = \sum_t tNI_t - \frac{1}{1-\gamma}\sum_t tND_t$$

$$\hat{\eta} = \frac{\sum_t ND_t}{\text{Person-time at risk}} \qquad (8.11)$$

$$\text{EDS} = \frac{1-\hat{\gamma}}{\hat{\eta}}. \qquad (8.12)$$

Though γ and η can be directly estimated, we include them in the MCMC so we can obtain full posterior distributions.

Parameters with point priors: We use point priors for $(\theta, E_0, \varepsilon, \rho)$. From (8.7), the incubation period is geometric with mean $1/\theta$. Reported incubation length is around two days (Longini et al., 2004), which implies $\hat{\theta} = 0.5$. We use this value in our primary analyses, but include sensitivity analyses. We use an initial number of latent infectives that is compatible with $\theta = 0.5$, setting $\hat{E}_0 = 2 \cdot NI_1$.

We denote the infectious proportion among the latent cases by ε. Infectiousness usually starts 24 hours before symptoms, which implies that on average $\hat{\varepsilon} = 0.5$. Similarly, we use ρ to denote the infectious proportion among the symptomatic infectives. The commonly observed duration of infectiousness (DOI) is four days (Longini et al., 2004). Using expected duration of symptomatic disease stage (EDS) directly estimated from the data, we can set the infectious percentage of the symptomatic infectives to

$$\hat{\rho} = \frac{DOI - 1}{EDS}. \qquad (8.13)$$

Furthermore, we require that at day $t = 0$ (one day before the first cases are identified) there are $E_0 = 2 \cdot NI_1$ asymptomatics, $I_0 = 0$ symptomatics, and no individuals who have yet experienced illness or death due to pandemic influenza, $(C_0 + D_0 = 0)$.

Not all of these parameters need to be fixed; some can be estimated from daily incidence and death and we explore various combinations of fixing some parameters and estimating others via MCMC.

Parameters with low information priors: We use low information priors for $(\gamma, \eta, \beta_0, \beta_1, \beta_2, \beta_3)$, namely

$$\gamma \sim Beta(1,1),$$
$$\eta \sim Beta(1,1), \qquad (8.14)$$
$$\text{and } (\beta_0, \beta_1, \beta_2, \beta_3) \overset{iid}{\sim} Gaussian(0, 10000)$$

Constrained parameters: The parameter δ is the mixing rate of symptomatic infectives relative to susceptibles and latent symptomatics. Influenza symptoms, hospitalization and quarantine reduce social contact and so we require $\delta < 1$. We evaluate δ using sensitivity analysis (see Subsection 8.3.4).

Assessing fit

We assess model fit by comparing day-specific percentiles of the ensemble of simulated incidence and mortality curves to observed disease and death incidence. The appropriate simulation must be conducted outside of the MCMC iterations and not extracted from within it. To construct valid comparisons, we sample at random 100 vectors from the joint posterior distribution of $(\beta_0, \beta_1, \beta_2, \beta_3, \eta, \gamma)$, but use fixed values for the remaining parameters. We use R (R Development Core Team, 2006) to simulate one epidemic per parameter set. Day t disease and death incidence are generated by sampling from a Binomial(θ, E_t) and a Binomial(η, I_t), with E_t and I_t generated by a stochastic process that conditions only on model parameters and not on the observed data.

Figure 8.15 displays fit assessments for Baltimore and Newark based on external simulations. The day-specific median curves are compatible with observed data and, and the interval between the 2.5^{th} and 97.5^{th} percentiles contains most of the observed values.

Within MCMC, simulated disease and death incidence curves are also generated using imputed values for E_t and I_t from each pass through the chain to generate daily disease incidence via a Binomial(θ, E_t) and death incidence via a Binomial(η, I_t). However, comparison of percentiles of these curves with observed incidence is overly optimistic because the imputed E_t and I_t are generated by a stochastic process that conditions on the most recent updating of model parameters and these updates depend on the observed disease and death incidence. To illustrate this over-optimism,

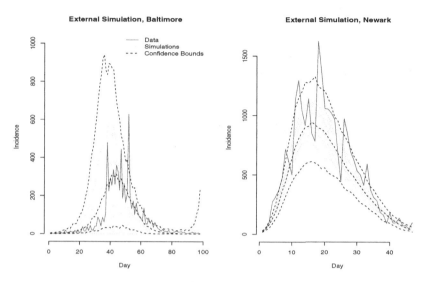

Figure 8.15 *Simulations from outside of the* WinBUGS *program. Day-specific* $2.5^{th}, 50^{th},$ *and* 97.5^{th} *percentiles from the simulation are connected. The left panel shows Baltimore's daily incidence, and Newark's is on the right.*

Figure 8.16 plots day-specific medians, 2.5th, and 97.5th percentiles of 6000 posterior-sampled epidemics, along with observed incidence and mortality. The extracted curves match the observed curves almost perfectly.

Need for a time-varying r_t

The fit of the classic SIR model, e.g., a time-constant $r_t \equiv r$ $(\beta_1 = \beta_2 = \beta_3 = 0)$, is poor (see Yin et al., 2007). For example, many generated Baltimore epidemics grow slowly and fall substantially short of producing the the observed incidence. For 100 simulated epidemics, the median total disease incidence is approximately 5000, but the observed value is 7498. As such, we employ a time-varying r_t. Several formulations were tried, including use of a single change point at the peak of the epidemic and two change points, one immediately before and one after the peak of the epidemic. Both of these models allow for different rate of increase or decrease in incidence, but model fit was still poor.

8.3.4 Results

In our primary analyses, $\theta = 0.5$ and $\varepsilon = 0.5$ are used for both Baltimore and Newark; E_0, ρ and δ are city-specific.

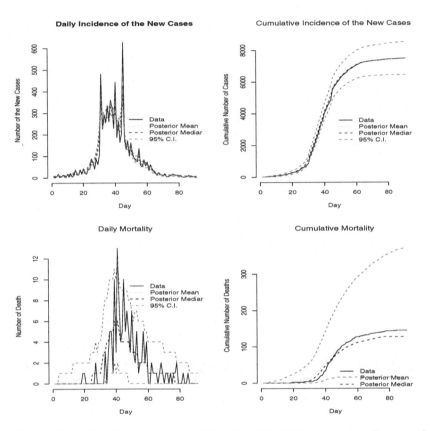

Figure 8.16 *Simulations from within the* WinBUGS *program. Day-specific* $2.5^{th}, 50^{th}, 97.5^{th}$ *percentiles from the simulation are connected. Plots on the right are cumulative incidence and mortality.*

Baltimore

The top panel of Figure 8.17 plots the best-fitting Baltimore SIR, along with observed and smoothed Baltimore incidence. The two epidemics are highly discrepant. Relative to the SIR curve, Baltimore incidence rises slowly followed by a rapid, almost vertical, increase and then a relatively rapid decrease. This discrepancy makes it essentially impossible for time-constant transmissibility models to fit the data. They cannot simultaneously produce the observed total number of cases and a shape that is close to the observed curve.

Calculated from the Baltimore data, $E_0 = 2 \cdot NI_1 = 4$ and $\rho = 0.42$; see equation (8.13). Initially, we gave δ a flat, $Beta(1,1)$, prior, but diagnostics show that the model is only weakly identified (δ and β_0 are strongly

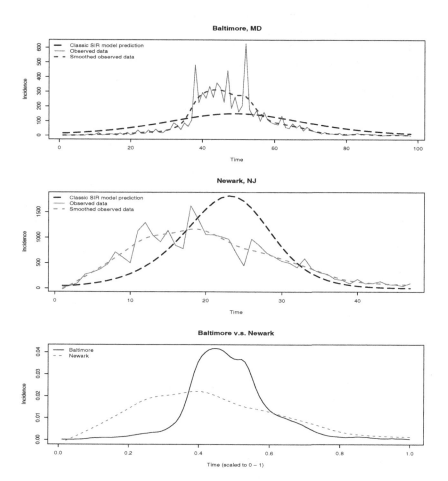

Figure 8.17 *Comparison of the shapes of the incidence curves. Top panel: Baltimore incidence and the best fitting nonstochastic, SIR curve. The SIR model is parameterized to match Baltimore's cumulative incidence and duration. Middle panel: Newark incidence and the best fitting, nonstochastic, SIR model curve. The SIR model is parameterized to match Newark's cumulative incidence and duration. Bottom panel: Comparison of Baltimore and Newark incidence. Curves are smoothed and then standardized to have the same cumulative incidence (area under the curve) and duration.*

negatively correlated); MCMC chains have persistent autocorrelation and do not mix well even after 20,000 iterations, resembling random walks. The negative correlation reflects, for example, the need for transmissibility r_t to increase and compensate for a small δ in order to produce the observed disease incidence.

Either δ or β_0 must have a highly informative prior and we evaluated $\delta = (1.0, 0.5, 0.1, 0.01, 0.001, 0.0001)$. When $\delta = 1.0$, the median of the total number of incidence cases is around 200, and $\delta = 0.001$ produces approximately the correct total number of influenza cases (7498). Therefore, most of the infections are produced by asymptomatic infectives.

We use MCMC output to estimate r_t and R_t. The top left panel of Figure 8.18 shows that r_t starts from slightly below 1.0, increases to 1.3 at around 20 days and then decreases. The reproductive number curve, R_t (the top right panel of Figure 8.18) has a shape similar to r_t and is close to it because δ is small; see (8.9). For this model, $\tilde{R} = R_1 = 0.9$ with R_t increasing to a maximum of 1.4 at day 21, then decreasing until near the end of the epidemic. The increase at the end is induced by the log-cubic parameterization, has essentially no impact on model fit and should not be over-interpreted. A log-quadratic parameterization avoids this increase, but the log-cubic form provides a better fit.

We also ran models with weakly informative priors on θ. Results are very sensitive to the prior and the model is poorly identified. For example, with a Beta(1,1) prior on θ and flat priors on ρ and A_0, posterior percentiles for θ are $_{2.5^{th}}50^{th}{}_{97.5^{th}} \equiv_{0.10} 0.10_{0.11}$, implying an expected duration of 10 days in the asymptomatic state. This expectation is five times longer than the typical duration of 2 days ($\theta = 0.5$).

Newark

The second panel of Figure 8.17 compares the Newark data to the classic SIR model matched for total incidence and duration. Newark incidence is quite compatible with the SIR model, but early in the epidemic it increases more rapidly than predicted (for small t, r_t is higher than the SIR-constant value). However, the day of maximum observed incidence is considerably earlier in the epidemic than for the SIR model.

Calculated from the Newark data, $E_0 = 2 \cdot NI_1 = 32$ and $\hat{\rho} = 0.66$; equation (8.13). We use a method similar to the Baltimore analysis to explore δ's impact on the Newark model. The δ estimated from the MCMC chain with a low-information prior produces $_{2.5^{th}}50^{th}{}_{97.5^{th}} \equiv_{0.001} 0.020_{0.050}$. Assessed by external simulation, this model fits well, though MCMC diagnostics indicate that the chain has not mixed well. Therefore, we evaluated $\delta = (1.0, 0.5, 0.25, 0.1, 0)$. Unlike the Baltimore results, all values produce a good fit as assessed by external simulation, but very small values of δ generate epidemics that slightly underestimate the total number of cases (likely due to a too small r_t).

When δ is estimated from the chain, the posterior percentiles turn out to be $_{2.5^{th}}50^{th}{}_{97.5^{th}} \equiv_{0.001} 0.02_{0.05}$, and so we ran MCMC chains using $\delta = 0$ and 0.1 and estimate r_t and R_t. The bottom left panel of Figure 8.18 displays the two r_t; the bottom right panel displays the two R_t. Both r_t

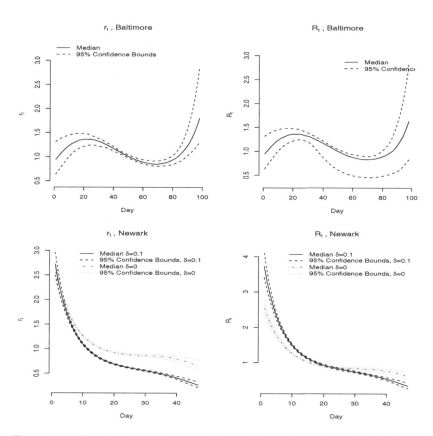

Figure 8.18 *Results of Baltimore model and Newark analyses. Top left panel: Baltimore, r_t for the model with $\delta \equiv 0.001$; top right panel: Baltimore, R_t, with $\delta \equiv 0.001$; bottom left panel: Newark, r_t, with $\delta \equiv 0.1$ (lower curves), and with $\delta \equiv 0$ (upper curves); bottom right panel: Newark, R_t with $\delta \equiv 0.1$ (lower curves), and with $\delta \equiv 0$ (upper curves). Day-specific $2.5^{th}, 50^{th}, 97.5^{th}$ percentiles are connected.*

and R_t are strictly decreasing. For the two δ-values, $\tilde{R} = R_1 = 2.4$ and 4.0, then R_t decreases rapidly, dropping below 1 after 20 days.

We also ran the model with several weakly informative priors on θ. Though there is high posterior correlation among parameters and there is persistent autocorrelation in the chains, each chain mixes well and results are insensitive to moderate changes in priors. Indeed, with a flat prior, the posterior distribution for θ is compatible with the external estimate of 0.5, with percentiles $_{2.5^{th}}50^{th}{}_{97.5^{th}} \equiv {}_{0.35} 0.43 {}_{0.51}$.

Comparison of Baltimore and Newark

To compare the shape of the Baltimore and Newark incidence curves, we plotted standardized versions in the third panel of Figure 8.17. Each incidence curve is smoothed and then scaled to produce the same total incidence (the same area under the curve) in the interval $[0, 1]$. Conceptually, we have two epidemics of the same duration and the same magnitude. This comparison shows a substantial shape difference between the two cities. The the Newark curve quite compatible with an SIR model; the Baltimore curve departs dramatically from SIR-form, displaying a "head wall" of accelerating incidence. Compatibility with the SIR form makes results for Newark far less sensitive to prior specification and generally considerably more stable than those for Baltimore. The Newark model produces reasonable estimates of key parameters including θ and ρ and results depend only weakly on the value of δ. None of these attributes apply to Baltimore and highly informative priors are necessary to obtain sensible results.

8.3.5 Discussion

The influenza pandemic of 1918 caused a large number of deaths worldwide. Though it has been the subject of many thorough epidemiological studies, several features remain unexplained, for example the appearance of multiple temporal waves in incidence. In both Baltimore and Newark resurgences in deaths attributed to influenza were observed, though not as dramatic as in other cities of the United States (Collins et al., 1930). Several hypotheses have been offered, including the impact of public health countermeasures, of individual behavioral responses to the epidemic, of seasonality on transmission and severity, of genetic changes in the virus over time (Ferguson et al., 2005; Barry, 2003).

We have investigated the temporal dynamics of the first of multiple waves through a stochastic compartment model with time-dependent disease transmissibility. Though our model deals with only the autumn wave, our approach can be generalized to model multiple waves. We find that models incorporating a time-dependent transmission rate fit morbidity and mortality data in Baltimore and Newark better than time-constant models, suggesting that the transmissibility of influenza was modified during the course of the outbreak. Available data are not informative on the cause or causes of time-dependence, but we speculate that either public health measures enacted during the course of the epidemic or responses by the general public to the epidemic produced a reduction in contact rates. Using a time-varying expected duration in the asymptomatic, infectious state $(1/\theta_t)$ is an alternative to use a time-varying r_t, but it is unlikely that this duration would change appreciably over the short duration of the pandemic.

As the third plot in Figure 8.17 reveals, there were fundamental differences between the Baltimore and Newark epidemics. Also, transmissibility's time-dependence is more complex in Baltimore (top panels of Figure 8.18) and estimates for Baltimore are sensitive to prior specification. The Newark model is substantially more stable and robust. We have identified several candidate explanations for incompatibility of the Baltimore data with an SIR model. Cases could have been unevenly distributed, overall data could be an aggregation of several, local, SIR-compatible epidemics. There could have been a strongly time-varying contact process due to social awareness or quarantine, or infectivity of the virus could have changed (very unlikely).

To produce a good fit in Baltimore, a small relative contact rate for diagnosed cases (δ) is needed, whereas a wide range of δ's, including small values, produces a good fit for Newark. A small value is consistent with the acute and severe symptoms of the 1918 pandemic (Barry, 2003) and is consistent with studies that indicate viral shedding is highest during the first few days of illness while symptoms can persist for several days after viral loads decline (Hayden et al., 1998). The small value of δ indicates that relatively few infections are induced by those with symptoms. Therefore, the disease incidence and death time-series are separable into two, nearly independent processes: a process that generates new infections and a process that removes individuals from the system through either recovery or death. The death time-series and its relation to the incidence series directly identify the mortality rate, η, and the case fatality rate $(1 - \gamma)$. However, we do estimate them within the MCMC system to obtain the full, joint posterior distribution of all unknowns.

Our results are compatible with some features of analyses from other locations. Mills et al. (2004) analyzed mortality data from the fall wave in 45 U.S. cities. The 25th, 50th and 75th percentiles of the estimated basic reproductive numbers are 1.7, 2.0 and 2.3. Chowell et al. (2006) analyzed spring and fall wave data from Geneva, Switzerland. Estimated basic reproductive numbers were $_{1.45}1.49_{1.53}$ and $_{3.57}3.75_{3.93}$ respectively. Massad et al. (2006) estimated the basic reproductive number in São Paulo, Brazil to be 2.68. Sertsou et al. (2006) modeled data from an institutional outbreak in New Zealand in the fall of 1918, estimating basic reproductive numbers ranging from 1.3 to 3.1. The maximums of our time-varying reproductive numbers, 1.4 for Baltimore and 2.6 to 3.8 for Newark, are consistence with these values.

The Bayesian formalism, implemented by MCMC, is very effective in analyzing this type of time-series data. Most state transitions and occupancy counts are not observed. Writing down the incomplete data likelihood and implementing an EM or other algorithm to find MLEs, while "straightforward" involves a series of nested summations (98 for Baltimore) with summation limits depending on other indices. The WinBUGS program is straightforward (see www.biostat.umn.edu/~brad/data.html) and supports stabilization or importation of external information by use of informative priors.

One needs to be aware of the pitfalls and the limitations of either the Bayesian or frequentist approach. For both, parameter identifiability and model fit must be checked. In Bayesian terms, validity depends on chains converging to a single ergodic distribution irrespective of the starting point and they must be run sufficiently long to mix. Autocorrelation must attenuate. These are the Bayesian manifestations of parameter identifiability. If they do not hold, identifiability must be established via informative priors (an option not available to the frequentist).

Model fit must be assessed by external simulation and sufficiently vigorous sensitivity analyses must be conducted. As shown by analysis of the Baltimore data, highly sensitive estimates can be produced by the incompatibility between the model assumptions and the actual process which generated the data. Similar cautions apply to a frequentist analysis.

Available data are not sufficient to identify all aspects of the infection-generating process and informative priors are needed. For some parameters we use a degenerate prior, which can lead to underestimation of posterior uncertainty for other parameters. However, we find that, in general, uncertainty is essentially unchanged when degenerate priors are replaced by moderate-information priors. Of course, a degenerate prior does reduce the dimension of the joint posterior distribution.

Use of a stochastic model, whether Bayesian or frequentist, aids in identifying parameters and we extract a great deal of structure from the daily incidence curves. Generally, identification depends on more than first moments. For example, assuming that daily exits from the asymptomatic infective state (E) are binomial, as in (8.7), provides information over and above assuming a fixed holding time. Assuming "binomial" and other distributional forms carries information and we have not evaluated robustness to distributional form. See Yin et al. (2007) for details of simulation-based studies of identifiability and bias.

Finally, additional information will be very valuable in identifying parameters and in model assessment, especially individual-level information on symptom onset, recovery or death; and group-level information on spatio-temporal social contacts among susceptibles, symptomatics, and those who have recovered. Collecting this and similar information will require development of an effective infrastructure and rapid implementation.

Appendices

Distributional catalog

This appendix provides a summary of the statistical distributions used in this book. For each distribution, we list its abbreviation, its support, constraints on its parameters, its density function, and its mean, variance, and other moments as appropriate. We also give additional information helpful in implementing a Bayes or EB analysis (e.g., relation to a particular sampling model or conjugate prior). We denote density functions generically by the letter p, and the random variable for the distribution in question by a Roman letter: x for scalars, $\mathbf{x}$ for vectors, and $\mathbf{V}$ for matrices. Parameter values are assigned Greek letters. The reader should keep in mind that many of these distributions are most commonly used as prior distributions, so that the random variable would itself be denoted as a Greek character.

In modern applied Bayesian work, a key issue is the generation of random draws from the distributions in this catalog. Since whole textbooks (Devroye, 1986; Ripley, 1987; Gentle, 1998) have been devoted to this subject, we make no attempt at a comprehensive treatment here, but only provide a few remarks and pointers. First, many modern Bayesian software packages (such as WinBUGS) essentially "hide" the requisite sampling from the user altogether. High-level (but still general purpose) programming environments (such as *R, S-plus, Gauss,* or *Matlab*) feature functions for generating samples from most of the standard distributions below. Lower-level languages (such as Fortran or C) often include only a $Uniform(0,1)$ generator, since samples from all other distributions can be built from these. In what follows, we provide a limited number of generation hints, and refer the reader to the aforementioned textbooks for full detail.

A.1 Discrete

A.1.1 Univariate

- **Binomial:** $X \sim Bin(n, \theta)$, $x = 0, 1, 2, \ldots, n$, $0 \leq \theta \leq 1$, n any positive integer, and

$$p(x|n, \theta) = \binom{n}{x} \theta^x (1 - \theta)^{n-x} ,$$

where

$$\binom{n}{x} = \frac{n!}{x!(n-x)!} .$$

$E(X) = n\theta$ and $Var(X) = n\theta(1 - \theta)$.

The binomial is commonly assumed in problems involving a series of independent, success/failure trials all having probability of success θ (*Bernoulli trials*): X is the number of successes in n such trials. The binomial is also the appropriate model for the number of "good" units obtained in n independent draws with replacement from a finite population of size N having G good units, and where $\theta = G/N$, the proportion of good units.

- **Poisson:** $X \sim Po(\theta)$, $x = 0, 1, 2, \ldots$, $\theta > 0$, and

$$p(x|\theta) = \frac{e^{-\theta} \theta^x}{x!} .$$

$E(X) = \theta$ and $Var(X) = \theta$.

The Poisson is commonly assumed in problems involving counts of "rare events" occurring in a given time period, and for other discrete data settings having countable support. It is also the limiting distribution for the binomial when n goes to infinity and the expected number of events converges to θ.

- **Negative Binomial:** $X \sim NegBin(r, \theta)$, $x = 0, 1, 2, \ldots$, $0 \leq \theta \leq 1$, r any positive integer, and

$$p(x|r, \theta) = \binom{x + r - 1}{x} \theta^r (1 - \theta)^x .$$

$E(X) = r(1 - \theta)/\theta$ and $Var(X) = r(1 - \theta)/\theta^2$.

The name "negative binomial" is somewhat misleading; "inverse binomial" might be a better name. This is because it arises from a series of Bernoulli trials where the number of *successes*, rather than the total number of trials, is fixed at the outset. In this case, X is the number of failures preceding the rth success. This distribution is also sometimes used to model countably infinite random variables having variance substantially larger than the mean (so that the Poisson model would be inappropriate). In fact, it is the marginal distribution of a Poisson random variable whose rate θ follows a gamma distribution (see Section A.2 below).

- **Geometric:** $X \sim Geom(\theta) \equiv NegBin(1, \theta)$, $x = 0, 1, 2, \ldots$, $0 \le \theta \le 1$, and

$$p(x|\theta) = \theta(1 - \theta)^x .$$

$E(X) = (1 - \theta)/\theta$ and $Var(X) = (1 - \theta)/\theta^2$.

The geometric is the special case of the negative binomial having $r = 1$; it is the number of failures preceding the first success in a series of Bernoulli trials.

A.1.2 Multivariate

- **Multinomial:** $\mathbf{X} \sim Mult(n, \boldsymbol{\theta})$ for $\mathbf{X} = (x_1, \ldots, x_k)'$, where $x_i \in \{0, 1, 2, \ldots, n\}$ and $\sum_{i=1}^{k} x_i = n$, $\boldsymbol{\theta} = (\theta_1, \ldots, \theta_k)'$ where $0 \le \theta_i \le 1$ and $\sum_{i=1}^{k} \theta_i = 1$, and

$$p(\mathbf{x}|n, \boldsymbol{\theta}) = \frac{n!}{\prod_{i=1}^{k} x_i!} \prod_{i=1}^{k} \theta_i^{x_i} .$$

$E(X_i) = n\theta_i$, $Var(X_i) = n\theta_i(1 - \theta_i)$, and $Cov(X_i, X_j) = -n\theta_i\theta_j$.

The multinomial is the multivariate generalization of the binomial; note that for $k = 2$, $Mult(n, \boldsymbol{\theta}) = Bin(n, \theta)$ with $\theta = \theta_1 = 1 - \theta_2$. For a population having $k > 2$ mutually exclusive and exhaustive classes, X_i is the number of elements of class i that would be obtained in n independent draws with replacement from the population. Note well that this is really only $(k - 1)$-dimensional distribution, because $x_k = n - \sum_{i=1}^{k-1} x_i$ and $\theta_k = 1 - \sum_{i=1}^{k-1} \theta_i$.

A.2 Continuous

A.2.1 Univariate

- **Beta:** $X \sim Beta(\alpha, \beta)$, $x \in [0, 1]$, $\alpha > 0$, $\beta > 0$, and

$$p(x|\alpha, \beta) = \frac{\Gamma(\alpha + \beta)}{\Gamma(\alpha)\Gamma(\beta)} x^{\alpha-1}(1 - x)^{\beta-1} ,$$

where $\Gamma(\cdot)$ denotes the *gamma function*, defined by the integral equation

$$\Gamma(\alpha) \equiv \int_0^\infty y^{\alpha-1} e^{-y} dy , \quad \alpha > 0 .$$

Using integration by parts, we have that $\Gamma(\alpha) = (\alpha - 1)\Gamma(\alpha - 1)$, so since $\Gamma(1) = 1$, we have that for integer α,

$$\Gamma(\alpha) = (\alpha - 1)! \quad .$$

One can also show that $\Gamma(1/2) = \sqrt{\pi}$.

For brevity, the pdf is sometimes written as

$$\frac{1}{B(\alpha, \beta)} x^{\alpha-1}(1 - x)^{\beta-1} \ ,$$

where $B(\cdot, \cdot)$ denotes the *beta function*,

$$B(\alpha, \beta) \equiv \frac{\Gamma(\alpha)\Gamma(\beta)}{\Gamma(\alpha + \beta)} \ .$$

Returning to the distribution itself, we have $E(X) = \alpha/(\alpha + \beta)$ and $Var(X) = \alpha\beta/[(\alpha + \beta)^2(\alpha + \beta + 1)]$; it is also true that $E(X^2) = \alpha(\alpha + 1)/[(\alpha + \beta)(\alpha + \beta + 1)]$. The beta is a common and flexible distribution for continuous random variables defined on the unit interval (so that rescaled and/or recentered versions may apply to any continuous variable having finite range). For $\alpha < 1$, the distribution has an infinite peak at 0; for $\beta < 1$, the distribution has an infinite peak at 1. For $\alpha = \beta = 1$, the distribution is flat (see the next bullet). For α and β both greater than 1, the distribution is concave down, and increasingly concentrated around its mean as $\alpha + \beta$ increases. The beta is the conjugate prior for the binomial likelihood; for an example, see Subsection 2.11.

- **Uniform:** $X \sim Unif(\theta_L, \theta_U)$, $x \in [\theta_L, \theta_U]$, $-\infty < \theta_L < \theta_U < \infty$, and

$$p(x|\theta_L, \theta_U) = \frac{1}{\theta_U - \theta_L} \ .$$

$E(X) = (\theta_L + \theta_U)/2$ and $Var(X) = (\theta_U - \theta_L)^2/12$. Clearly $Unif(0, 1) \equiv Beta(1, 1)$.

- **Normal** (or **Gaussian**): $X \sim N(\mu, \sigma^2)$, $x \in \Re$, $-\infty < \mu < \infty$, $\sigma^2 > 0$, and

$$p(x|\mu, \sigma^2) = \frac{1}{\sqrt{2\pi}\sigma} \exp\left[-\frac{1}{2}\left(\frac{x - \mu}{\sigma}\right)^2\right] \ .$$

$E(X) = \mu$ and $Var(X) = \sigma^2$. The normal is the single most common distribution in statistics, due in large measure to the Central Limit Theorem. It is symmetric about μ, which is not only its mean, but its median and mode as well.

A $N(0, 1)$ variate Z can be generated as $\sqrt{-2\log(U_1)}cos(2\pi U_2)$, where U_1 and U_2 are independent $Unif(0, 1)$ random variates; in fact, the quantity $\sqrt{-2\log(U_1)}sin(2\pi U_2)$ produces a second, independent $N(0, 1)$ variate. This is the celebrated *Box-Muller* (1958) method. A $N(\mu, \sigma^2)$ variate can then be generated as $\mu + \sigma Z$.

- **Double Exponential** (or *Laplace*): $X \sim DE(\mu, \sigma^2)$, $x \in \Re$, $-\infty < \mu < \infty$, $\sigma^2 > 0$, and

$$p(x|\mu, \sigma^2) = \frac{1}{2\sigma} \exp\left(-\frac{|x-\mu|}{\sigma}\right).$$

$E(X) = \mu$ and $Var(X) = 2\sigma^2$.

Like the normal, this distribution is symmetric and unimodal, but has heavier tails and a somewhat different shape, being strictly concave up on both sides of μ.

- **Logistic:** $X \sim Logistic(\mu, \sigma^2)$, $x \in \Re$, $-\infty < \mu < \infty$, $\sigma^2 > 0$, and

$$p(x|\mu, \sigma^2) = \frac{\exp\left(-\frac{x-\mu}{\sigma}\right)}{\sigma\left[1 + \exp\left(-\frac{x-\mu}{\sigma}\right)\right]^2}.$$

$E(X) = \mu$ and $Var(X) = (\pi^2/3)\sigma^2$.

The logistic is another symmetric and unimodal distribution, more similar to the normal in appearance than the double exponential, but with even heavier tails.

- **t** (or *Student's t*): $X \sim t(\nu, \mu, \sigma^2)$, $x \in \Re$, $\nu > 0$, $-\infty < \mu < \infty$, $\sigma^2 > 0$, and

$$p(x|\mu, \sigma^2, \nu) = \frac{\Gamma[(\nu+1)/2]}{\sigma\sqrt{\nu\pi}\Gamma(\nu/2)}\left[1 + \frac{1}{\nu}\left(\frac{x-\mu}{\sigma}\right)^2\right]^{-(\nu+1)/2}.$$

$E(X) = \mu$ (if $\nu > 1$) and $Var(X) = \nu\sigma^2/(\nu-2)$ (if $\nu > 2$). The parameter ν is referred to as the *degrees of freedom* and is usually taken to be a positive integer, though the distribution is proper for any positive real number ν. The t is a common heavy-tailed (but still symmetric and unimodal) alternative to the normal distribution. The leading term in the pdf can be rewritten

$$\frac{\Gamma[(\nu+1)/2]}{\sigma\sqrt{\nu\pi}\Gamma(\nu/2)} = \frac{1}{\sigma\sqrt{\nu}B(\frac{1}{2}, \frac{\nu}{2})},$$

where $B(\cdot, \cdot)$ again denotes the beta function.

- **Cauchy:** $X \sim Cau(\mu, \sigma^2) \equiv t(1, \mu, \sigma^2)$, $x \in \Re$, $-\infty < \mu < \infty$, $\sigma^2 > 0$, and

$$p(x|\mu, \sigma^2) = \frac{1}{\sigma\pi\left[1 + \left(\frac{x-\mu}{\sigma}\right)^2\right]}.$$

$E(X)$ and $Var(X)$ do not exist, though μ is the median of this distribution. This special case of the t distribution has the heaviest possible tails and provides a wealth of counterexamples in probability theory.

- **Gamma:** $X \sim G(\alpha, \beta)$, $x > 0$, $\alpha > 0$, $\beta > 0$, and

$$p(x|\alpha, \beta) = \frac{x^{\alpha-1}e^{-x/\beta}}{\Gamma(\alpha)\beta^\alpha} .$$

$E(X) = \alpha\beta$ and $Var(X) = \alpha\beta^2$. Note also that if $X \sim G(\alpha, \beta)$, then $Y = cX \sim G(\alpha, c\beta)$.

The gamma is a flexible family for continuous random variables defined on the positive real line. For $\alpha < 1$, the distribution has an infinite peak at 0 and is strictly decreasing. For $\alpha = 1$, the distribution intersects the vertical axis at $1/\beta$, and again is strictly decreasing (see the Exponential distribution below). For $\alpha > 1$, the distribution starts at the origin, increases for a time and then decreases; its appearance is roughly normal for large α. The gamma is the conjugate prior distribution for a Poisson rate parameter θ.

- **Exponential:** $X \sim Expo(\beta) \equiv G(1, \beta)$, $x > 0$, $\beta > 0$, and

$$p(x|\beta) = \frac{1}{\beta}e^{-x/\beta} .$$

$E(X) = \beta$ and $Var(X) = \beta^2$.

- **Chi-square:** $X \sim \chi^2(\nu) \equiv G(\nu/2, 2)$, $x > 0$, $\nu > 0$, and

$$p(x|\nu) = \frac{x^{\nu/2-1}e^{-x/2}}{\Gamma(\nu/2)2^{\nu/2}} .$$

$E(X) = \nu$ and $Var(X) = 2\nu$. As with the t distribution, the parameter ν is referred to as the *degrees of freedom* and is usually taken to be a positive integer, though a proper distribution results for any positive real number ν.

- **Inverse Gamma:** $X \sim IG(\alpha, \beta)$, $x > 0$, $\alpha > 0$, $\beta > 0$, and

$$p(x|\alpha, \beta) = \frac{e^{-1/(\beta x)}}{\Gamma(\alpha)\beta^\alpha x^{\alpha+1}} .$$

$E(X) = 1/[\beta(\alpha-1)]$ (provided $\alpha > 1$) and $Var(X) = 1/[\beta^2(\alpha-1)^2(\alpha-2)]$ (provided $\alpha > 2$). Note also that if $X \sim IG(\alpha, \beta)$, then $Y = cX \sim IG(\alpha, \beta/c)$.

A better name for the inverse gamma might be the *reciprocal gamma*, because $X = 1/Y$ where $Y \sim G(\alpha, \beta)$ (or in distributional shorthand, $IG(\alpha, \beta) \equiv 1/G(\alpha, \beta)$).

Despite its poorly behaved moments and somewhat odd, heavy-tailed appearance, the inverse gamma is very commonly used in Bayesian statistics as the conjugate prior for a variance parameter σ^2 arising in a normal likelihood function. Choosing α and β appropriately for such a prior can be aided by solving the above equations for $\mu \equiv E(X)$ and $\tau^2 \equiv Var(X)$ for α and β. This results in

$$\alpha = (\mu/\tau)^2 + 2 \quad \text{and} \quad \beta = \frac{1}{\mu\left[(\mu/\tau)^2 + 1\right]} \;.$$

Setting the prior mean and standard deviation both equal to μ (a reasonably vague specification) thus produces $\alpha = 3$ and $\beta = 1/(2\mu)$.

- **Inverse Gaussian** (or *Wald*): $X \sim InvGau(\mu, \lambda)$, $x > 0$, $\mu > 0$, $\lambda > 0$, and

$$p(x|\mu, \lambda) = \sqrt{\frac{\lambda}{2\pi x^3}} \exp\left[-\frac{\lambda(x - \mu)^2}{2\mu^2 x}\right] \;.$$

$E(X) = \mu$ and $Var(X) = \mu^3/\lambda$.

The inverse Gaussian has densities that resemble those of the gamma distribution. Despite its somewhat formidable form, all its positive and negative moments exist.

A.2.2 Multivariate

- **Dirichlet: X** $\sim D(\boldsymbol{\alpha})$, $\mathbf{X} = (x_1, \ldots, x_k)'$ where $0 \le x_i \le 1$ and $\sum_{i=1}^k x_i = 1$, $\boldsymbol{\alpha} = (\alpha_1, \ldots, \alpha_k)'$ where $\alpha_i \ge 0$, and

$$p(\mathbf{x}|\boldsymbol{\alpha}) = \frac{\Gamma(\alpha_0)}{\prod_{i=1}^k \Gamma(\alpha_i)} \prod_{i=1}^k x_i^{\alpha_i - 1} \;.$$

$E(X_i) = \alpha_i/\alpha_0$, $Var(X_i) = [\alpha_i(\alpha_0 - \alpha_i)]/[\alpha_0^2(\alpha_0 + 1)]$, and finally $Cov(X_i, X_j) = -(\alpha_i\alpha_j)/[\alpha_0^2(\alpha_0 + 1)]$, where $\alpha_0 \equiv \sum_{i=1}^k \alpha_i$.

The Dirichlet is the multivariate generalization of the beta; note that for $k = 2$, $D(\boldsymbol{\alpha}) = Beta(\alpha_1, \alpha_2)$. Note that this is really only $(k - 1)$-dimensional distribution, since $x_k = 1 - \sum_{i=1}^{k-1} x_i$. The Dirichlet is the conjugate prior for the multinomial likelihood. It also forms the foundation for the *Dirichlet process prior*, the basis of most nonparametric Bayesian inference (see Section 2.6).

- **Multivariate Normal** (or *Multinormal*, or *Multivariate Gaussian*): $\mathbf{X} \sim N_k(\boldsymbol{\mu}, \boldsymbol{\Sigma})$, $\mathbf{x} \in \Re^k$, $\boldsymbol{\mu} = (\mu_1, \ldots, \mu_k)'$ where $-\infty < \mu_i < \infty$, $\boldsymbol{\Sigma}$ a $k \times k$ positive definite matrix, and

$$p(\mathbf{x}|\boldsymbol{\mu}, \boldsymbol{\Sigma}) = \frac{|\boldsymbol{\Sigma}^{-1}|^{1/2}}{(2\pi)^{k/2}} \exp\left[-\frac{1}{2}(\mathbf{x} - \boldsymbol{\mu})'\boldsymbol{\Sigma}^{-1}(\mathbf{x} - \boldsymbol{\mu})\right] \;,$$

where $|\boldsymbol{\Sigma}^{-1}|$ denotes the determinant of $\boldsymbol{\Sigma}^{-1}$. $E(\mathbf{X}) = \boldsymbol{\mu}$ (i.e., $E(X_i) = \mu_i$ for all i), and $Var(\mathbf{X}) = \boldsymbol{\Sigma}$ (i.e., $Var(X_i) = \sigma_{ii}$ and $Cov(X_i, X_j) = \sigma_{ij}$ where $\boldsymbol{\Sigma} = (\sigma_{ij})$). The multivariate normal forms the basis of the likelihood in most common linear models, and also serves as the conjugate prior for mean and regression parameters in such likelihoods.

A $N_k(\mathbf{0}, \mathbf{I})$ variate $\mathbf{z}$ can be generated simply as a vector of k independent, univariate $N(0, 1)$ random variates. For general covariance matrices $\boldsymbol{\Sigma}$, we first factor the matrix as $\boldsymbol{\Sigma} = \mathbf{LL}'$, where $\mathbf{L}$ is a lower-triangular matrix (this is often called the *Cholesky factorization*; see, e.g., Thisted, 1988, pp. 81–83). A $N_k(\boldsymbol{\mu}, \boldsymbol{\Sigma})$ variate $\mathbf{z}$ can then be generated as $\boldsymbol{\mu} + \mathbf{Lz}$.

- **Multivariate t** (or *Multi-t*): $\mathbf{X} \sim t_k(\nu, \boldsymbol{\mu}, \boldsymbol{\Sigma})$, $\mathbf{x} \in \Re^k$, $\nu > 0$, $\boldsymbol{\mu} = (\mu_1, \dots, \mu_k)'$ where $-\infty < \mu_i < \infty$, $\boldsymbol{\Sigma}$ a $k \times k$ positive definite matrix, and $p(\mathbf{x}|\nu, \boldsymbol{\mu}, \boldsymbol{\Sigma})$ is

$$\frac{|\boldsymbol{\Sigma}^{-1}|^{1/2}\Gamma[(\nu + k)/2]}{(\nu\pi)^{k/2}\Gamma(\nu/2)} \left[1 + \frac{1}{\nu}(\mathbf{x} - \boldsymbol{\mu})'\boldsymbol{\Sigma}^{-1}(\mathbf{x} - \boldsymbol{\mu})\right]^{-(\nu+k)/2}.$$

$E(\mathbf{X}) = \boldsymbol{\mu}$ (if $\nu > 1$), and $Var(\mathbf{X}) = \nu\boldsymbol{\Sigma}/(\nu - 2)$ (if $\nu > 2$). The multivariate t provides a heavy-tailed alternative to the multivariate normal while still accounting for correlation among the elements of $\mathbf{x}$. Typically, $\boldsymbol{\Sigma}$ is called a *scale matrix*, and is approximately equal to the variance matrix of $\mathbf{X}$ for large ν.

- **Wishart:** $\mathbf{V} \sim W(\boldsymbol{\Omega}, \nu)$, $\mathbf{V}$ a $k \times k$ symmetric and positive definite matrix, $\boldsymbol{\Omega}$ a $k \times k$ symmetric and positive definite parameter matrix, $\nu > 0$, and

$$p(\mathbf{V}|\nu, \boldsymbol{\Omega}) = c\frac{|\mathbf{V}|^{(\nu-k-1)/2}}{|\boldsymbol{\Omega}|^{\nu/2}} \exp\left[-\frac{1}{2}tr(\boldsymbol{\Omega}^{-1}\mathbf{V})\right],$$

provided the shape parameter (or "degrees of freedom") $\nu \geq k$. Here, the proportionality constant c takes the awkward form

$$c = \left[2^{(\nu k)/2}\pi^{k(k-1)/4}\prod_{j=1}^{k}\Gamma\left(\frac{\nu + 1 - j}{2}\right)\right]^{-1}.$$

$E(V_{ij}) = \nu\Omega_{ij}$, $Var(V_{ij}) = \nu(\Omega_{ij}^2 + \Omega_{ii}\Omega_{jj})$, and finally $Cov(V_{ij}, V_{kl}) = \nu(\Omega_{ik}\Omega_{jl} + \Omega_{il}\Omega_{jk})$.

The Wishart is a multivariate generalization of the gamma, originally derived as the sampling distribution of the sum of squares and crossproducts matrix, $\sum_{i=1}^{n}(\mathbf{X}_i - \bar{\mathbf{X}})(\mathbf{X}_i - \bar{\mathbf{X}})'$, where $\mathbf{X}_i \stackrel{iid}{\sim} N_k(\boldsymbol{\mu}, \mathbf{V})$. In Bayesian analysis, just as the reciprocal of the gamma (the inverse gamma) is often used as the conjugate prior for a variance parameter σ^2 in a normal likelihood, the "reciprocal" (inverse) of the Wishart (the *inverse Wishart*) is often used as the conjugate prior for a variance-covariance *matrix* $\boldsymbol{\Sigma}$ in a multivariate normal likelihood (see Example 7.2).

A random draw from a Wishart with ν an integer can be obtained using the idea in the preceding paragraph: If $\mathbf{x}_1, \ldots, \mathbf{x}_\nu$ are independent draws from a $N_k(\mathbf{0}, \mathbf{\Omega})$ distribution, then $\mathbf{V} = \sum_{i=1}^\nu \mathbf{x}_i \mathbf{x}_i'$ is a $W(\mathbf{\Omega}, \nu)$ random variate. Non-integral ν requires the general algorithm originally given by Odell and Feiveson (1966); see also the textbook treatment by Gentle (1998, p. 107), or the more specifically Bayesian treatment by Gelfand et al. (1990).

- **Inverse Wishart:** $\mathbf{V} \sim IW(\mathbf{\Omega}, \nu)$, $\mathbf{V}$ a $k \times k$ symmetric and positive definite matrix, $\mathbf{\Omega}$ a $k \times k$ symmetric and positive definite parameter matrix, $\nu > 0$, and

$$
p(\mathbf{V}|\nu, \mathbf{\Omega}) = c \frac{|\mathbf{V}|^{-(\nu+k+1)/2}}{|\mathbf{\Omega}|^{\nu/2}} \exp\left[-\frac{1}{2} tr(\mathbf{\Omega}^{-1}\mathbf{V}^{-1})\right],
$$

provided the shape parameter (or "degrees of freedom") $\nu \geq k$, and where proportionality constant c again takes the awkward form

$$
c = \left[2^{(\nu k)/2} \pi^{k(k-1)/4} \prod_{j=1}^k \Gamma\left(\frac{\nu+1-j}{2}\right)\right]^{-1}.
$$

$E(V_{ij}) = \frac{1}{\nu-k-1}\Omega_{ij}^{-1}$, provided $\nu > k + 1$. The variance and covariance expressions are complex; see e.g. Dreze and Richard (1983, pp. 587–588).

As already mentioned in the previous entry in this list, in Bayesian analysis the inverse Wishart is often used as the conjugate prior for a variance-covariance matrix $\mathbf{V} = \mathbf{\Sigma}$ in a multivariate normal likelihood (see Example 7.2). To simplify the notation somewhat, the $\mathbf{\Omega}$ matrix is often written as $\mathbf{S}^{-1}$, so that $E(\mathbf{\Sigma}) = \frac{1}{\nu-k-1}\mathbf{S}$ (again for $\nu > k + 1$). Also, if $k = 1$, then S is scalar and $IW(S^{-1}, \nu) \equiv IG\left(\frac{\nu}{2}, \frac{2}{S}\right)$.

Decision theory

B.1 Introduction

As discussed in Chapters 1 and 5, the Bayesian approach with low-information priors strikes an effective balance between frequentist robustness and Bayesian efficiency. It occupies a middle ground between the two and, when properly tuned, can generate procedures with excellent frequentist properties, often "beating the frequentist approach at its own game." As an added benefit, the Bayesian approach more easily structures complicated models and inferential goals.

To make comparisons among approaches, we need a method of keeping score, commonly referred to as a *loss structure*. Loss structures may be explicit or implicit; here we discuss the structure adopted in a *decision-theoretic* framework. We outline the general decision framework, highlighting those features that suit the present purpose. We also provide examples of how the theory can help the practitioner. Textbooks such as Ferguson (1967) and DeGroot (1970), as well as the more recent essay by Brown (2000), provide a more comprehensive treatment of the theory.

Setting up the general decision problem requires a prior distribution, a sampling distribution, a class of allowable actions and decision rules, and a loss function. Specific loss function forms correspond to point estimation, interval estimation, and hypothesis testing. We use the notation

$$
\begin{aligned}
\textit{prior distribution:} \quad & G(\theta),\ \theta \in \Theta & \text{(B.1)} \\
\textit{sampling distribution:} \quad & f(\mathbf{x}|\theta) \\
\textit{allowable actions:} \quad & a \in \mathcal{A} \\
\textit{decision rules:} \quad & d \in \mathcal{D} : \mathcal{X} \to \mathcal{A} \\
\textit{loss function:} \quad & l(\theta, a)\ .
\end{aligned}
$$

The loss function $l(\theta, a)$ computes the loss incurred when θ is the true state of nature and we take action a. Thus, for point estimation of a parameter θ, we might use *squared error loss* (SEL),

$$
l(\theta, a) = (\theta - a)^2\ ,
$$

or *weighted squared error loss* (WSEL),

$$l(\theta, a) = w(\theta)(\theta - a)^2 \ ,$$

or *absolute error loss*,

$$l(\theta, a) = |\theta - a| \ ,$$

or, for discrete parameter spaces, *0–1 loss*,

$$l(\theta, a) = \begin{cases} 0 & \text{if } \theta = a \\ 1 & \text{if } \theta \neq a \end{cases} \ . \tag{B.2}$$

The decision rule d maps the observed data $\mathbf{x}$ into an action a; in the case of point estimation, this action is a proposed value for θ (e.g., $d(\mathbf{x}) = \bar{x}$).

The Bayesian outlook on the problem of selecting a decision rule is as follows. In light of the data $\mathbf{x}$, our opinion as to the state of nature is summarized by the *posterior* distribution of θ, which is given by Bayes' Theorem as

$$dG(\theta|\mathbf{x}) = \frac{f(\mathbf{x}|\theta)dG(\theta)}{m_G(\mathbf{x})} \ ,$$

where

$$m_G(\mathbf{x}) = \int f(\mathbf{x}|u)dG(u) \ .$$

The Bayes rule minimizes the *posterior risk*,

$$\rho(G, d(\mathbf{x})) = E_{\theta|\mathbf{x}}[l(\theta, d(\mathbf{x}))] = \int l(\theta, d(\mathbf{x}))dG(\theta|\mathbf{x}) \ . \tag{B.3}$$

Note that posterior risk is a single number regardless of the dimension of θ, so choosing d to minimize $\rho(G, d(\mathbf{x}))$ is well defined.

B.1.1 Risk and admissibility

In the frequentist approach, we have available neither a prior nor a posterior distribution. For a decision rule d, define its *frequentist risk* (or simply *risk*) for a given true value of θ as

$$R(\theta, d) = E_{\mathbf{x}|\theta}[l(\theta, d(\mathbf{x}))] = \int l(\theta, d(\mathbf{x}))f(\mathbf{x}|\theta)d\mathbf{x} \ , \tag{B.4}$$

the average loss, integrated over the distribution of $\mathbf{X}$ conditional on θ. Note that frequentist risk is a *function of* θ, not a single number like the posterior risk (B.3).

Example B.1 In the interval estimation setting of Example 1.1, consider the loss function

$$l(\theta, a) = \begin{cases} 0 & \text{if } \theta \in \delta(\mathbf{x}) \\ 1 & \text{if } \theta \notin \delta(\mathbf{x}) \end{cases} \ .$$

Then the risk function is

$$
\begin{aligned}
R(\theta, d) &= E_{\mathbf{x}|\theta,\sigma^2}[l(\theta, \delta(\mathbf{X}))] \\
&= P_{\mathbf{x}|\theta,\sigma^2}[\theta \notin \delta(\mathbf{X})] \\
&= .05 ,
\end{aligned}
$$

which is constant over all possible values of θ and σ^2. ∎

This controlled level of risk across all parameter values is one of the main selling points of frequentist confidence intervals. Indeed, plotting R versus θ for various candidate rules can be very informative. For example, if d_1 and d_2 are two rules such that

$$
R(\theta, d_1) \leq R(\theta, d_2) \text{ for all } \theta ,
$$

with strict inequality for at least one value of θ, then under the given loss function we would never choose d_2 because its risk is never smaller that d_1's and can be larger. In such a case, the rule d_2 is said to be *inadmissible*, and *dominated* by d_1. Note that d_1 may itself be inadmissible, since there may be yet another rule which uniformly beats d_1. If no such rule exists, d_1 is called *admissible*.

Admissibility is a sensible and time-honored criterion for comparing decision rules, but its utility in practice is rather limited. While inadmissible rules often need not be considered further, for most problems there will be many admissible rules, creating the problem of which one to select. Moreover, admissibility is not a guarantee of sensible performance; the example in Subsection 5.6.1 provides an illustration. Therefore, additional criteria are required to select a frequentist rule. We consider the three most important such criteria in turn.

B.1.2 Unbiased rules

Unbiasedness is a popular method for reducing the number of candidate decision rules and allowing selection of a "best" rule within the reduced class. A decision rule $d(x)$ is unbiased if

$$
E_{x|\theta}[l(\theta', d(x))] \geq E_{x|\theta}[l(\theta, d(x))] \text{ for all } \theta \text{ and } \theta' . \tag{B.5}
$$

Under squared error loss, (B.5) is equivalent to requiring $E_{x|\theta}[d(x)] = \theta$. For interval estimation, unbiasedness requires that the interval have a greater chance of covering the true parameter than any other parameter. For hypothesis testing, unbiasedness is equivalent to the statistical power being greater under an alternative hypothesis than under the null.

Though unbiasedness has the attractive property of reducing the number of candidate rules and allowing the frequentist to find an optimal rule, it is often a very high price to pay, even for frequentist evaluations. Indeed, a principal thesis of this book is that a little bias can go a long way in improving frequentist performance. In Section B.2 below, we provide the basic example based on squared error loss. Here we present three estimation examples of how unbiasedness can work against logical and effective inference.

Example B.2 : Estimating P(no events). Ferguson (1967) provides the following compelling example of a problem with unbiased estimates. Assume a Poisson distribution and consider the goal of estimating the probability that there will be no events in a time period of length $2t$ based on the observed number of events, X, in a time period of length t. Therefore, with λ the Poisson parameter we want to estimate $e^{-2t\lambda}$, based on X which is distributed Poisson($t\lambda$). The MLE of this probability is e^{-2X}, while the best (indeed, the only) unbiased estimate is $(-1)^X$. This latter result is derived by matching terms in two convergent power series.

This unbiased estimate is patently absurd; it is either $+1$ or -1 depending on whether an even or an odd number of events has occurred in the first t time units. Using almost any criterion other than unbiasedness, the MLE is preferred. For example, it will have a substantially smaller mean squared error (MSE). ∎

Example B.3 : Constrained parameter space. In many models the unbiased estimate can lie outside the allowable parameter space. A common example is the components of variance models (i.e., the Type II ANOVA model), where

$$Y_{ij} = \mu + b_i + \epsilon_{ij}, \ i = 1, \ldots, k, \ j = 1, \ldots, n_i,$$

where $\mu \in \Re$, $b_i \overset{iid}{\sim} N(0, \tau^2)$, and $\epsilon_{ij} \overset{iid}{\sim} N(0, \sigma^2)$. The unbiased estimate of τ^2, computed using the within and between MSEs, can sometimes take negative values (as can the MLE). Clearly, performance of these estimates is improved by constraining them to be nonnegative. ∎

Example B.4 : Dependence on the stopping rule. As with all frequentist evaluations, unbiasedness requires attention to the sampling plan and, if applicable, to the stopping rule used in collecting the data. Consider again Example 1.2, where now we wish to estimate the unknown success probability based on x observed successes in n trials. This information unambiguously determines the MLE as x/n, but the unbiased estimate depends on how the data were obtained. If n is fixed, then the MLE is unbiased. But if the sampling were "inverse" with data gathered until r failures were obtained, then the number of observed successes, x, is negative binomial, and the unbiased estimate is $\frac{x}{x+r-1} = \frac{x}{n-1}$. If both n and x are random, it may be impossible to identify an unbiased estimate. ∎

B.1.3 Bayes rules

While many frequentists would not even admit to the existence of a subjective prior distribution $G(\theta)$, a possibly attractive frequentist approach to choosing a best decision rule does depend on the formal use of a prior. For a prior G, define the *Bayes risk* as

$$r(G, d) = E_\theta E_{\mathbf{x}|\theta} l(\theta, d(\mathbf{x})) = E_\theta R(\theta, d) , \qquad (B.6)$$

the expected frequentist risk with respect to the chosen prior G. Bayes risk is alternatively known as *empirical Bayes risk* because it averages over the variability in both θ and the data. Reversing the order of integration in (B.6), we obtain the alternate computational form

$$r(G, d) = E_{\mathbf{x}} E_{\theta|\mathbf{x}} l(\theta, d(\mathbf{x})) = E_{\mathbf{x}} \rho(G, d(\mathbf{x})) ,$$

the expected posterior risk with respect to the marginal distribution of the data $\mathbf{X}$. Because this is the posterior loss one expects *before* having seen the data, the Bayes risk is sometimes referred to as the *preposterior risk*. Thus $r(G, d)$ is directly relevant for Bayesian experimental design (see Section 6.1).

Since $r(G, d)$ is a scalar quantity, we can choose the rule d_G that minimizes the Bayes risk, i.e.,

$$d_G(\mathbf{x}) = \arg\min_{d \in \mathcal{D}} r(G, d) . \qquad (B.7)$$

This minimizing rule is called the *Bayes rule*. This name is somewhat confusing, because a subjective Bayesian would not average over the data (as $r(G, d)$ does) when choosing a decision rule, but instead choose one that minimized the posterior risk (B.3) given the observed data. Fortunately, it turns out that these two operations are virtually equivalent; under very broad conditions, minimizing the Bayes risk (B.6) is equivalent to minimizing the posterior risk (B.3) for all $\mathbf{x}$ such that $m_G(\mathbf{x}) > 0$. For further discussion of this result and related references, see Berger (1985, p. 159).

Example B.5 : Point estimation. Under SEL, the Bayes rule is found by minimizing the posterior risk,

$$\rho(G, a) = \int (\theta - a)^2 g(\theta|\mathbf{x}) d\theta .$$

Taking the derivative with respect to a, we have

$$\frac{\partial}{\partial a}[\rho(G, a)] = \int 2(\theta - a)(-1) g(\theta|\mathbf{x}) d\theta .$$

Setting this expression equal to zero and solving for a, we obtain

$$a = \int \theta \, g(\theta|\mathbf{x}) d\theta = E(\theta|\mathbf{x}) ,$$

the posterior mean, as a solution. Because

$$\frac{\partial^2}{\partial a^2}[\rho(G,a)] = 2\int g(\theta|\mathbf{x})d\theta = 2 > 0 \,,$$

the second derivative test implies our solution is indeed a minimum, confirming that the posterior mean is the Bayes estimate of θ under SEL.

We remark that under absolute error loss, the Bayes estimate is the posterior median, while under 0–1 loss, it is the posterior mode. ∎

Example B.6 : Interval estimation. Consider the loss function

$$l(\theta,a) = I_{\{\theta\notin a\}} + c \times \text{volume(a)},$$

where I is the indicator function of the event in curly brackets and a is a subset of the parameter space Θ. Then the Bayes rule will be the region (or regions) for θ having highest posterior density, with c controlling the tradeoff between the volume of a and the posterior probability of coverage. Subsection 2.3.2 gives a careful description of Bayesian confidence intervals; for our present purpose we note (and Section 5.6 demonstrates) that under noninformative prior distributions, these regions produce confidence intervals with excellent frequentist coverage and modest volume. ∎

Example B.7 : Hypothesis testing. Consider the comparison of two simple hypotheses, $H_0 : \theta = \theta_0$ and $H_1 : \theta = \theta_1$, so that we have a two-point parameter space $\Theta = \{\theta_0, \theta_1\}$. We use the "0–$l_i$" loss function

$$l(\theta,a) = al_0 + (1-a)l_1,$$

where both l_0 and l_1 are nonnegative, and $a \in \{0,1\}$ gives the index of the accepted hypothesis. Then it can be shown that the Bayes rule with respect to a prior on $\pi = P(\theta = \theta_0)$ is a likelihood ratio test; in fact, many authors use this formulation to prove the Neyman-Pearson lemma. This model is easily generalized to a countable parameter space. ∎

B.1.4 Minimax rules

A final (and very conservative) approach to choosing a best decision rule in a frequentist framework is to control the worst that can happen; that is, to minimize the maximum risk over the parameter space. Mathematically, we choose the rule d^* satisfying the relation

$$\sup_{\theta\in\Theta} R(\theta, d^*) = \inf_{d\in\mathcal{D}} \sup_{\theta\in\Theta} R(\theta, d) \,.$$

This d^* may sacrifice a great deal for this control, since its risk may be unacceptably high in certain regions of the parameter space. The minimax rule is attempting to produce controlled risk behavior for all θ *and* $\mathbf{x}$, but is making no evaluation of the relative likelihood of the θ values.

The Bayesian approach is helpful in finding minimax rules via a theorem which states that a Bayes rule (or limit of Bayes rules) that has constant risk over the parameter space is minimax. Generally, the minimax rule is not unique (indeed, in Chapter 5 we find more than one), and failure to find a constant risk Bayes rule does not imply that the absence of a minimax rule. Of course, as with all other decision rules, the minimax rule depends on the loss function. For the Gaussian sampling distribution, the sample mean is minimax (it is constant risk and the limit of Bayes rules). Finding minimax rules for the binomial and the exponential distributions have been left as exercises.

B.2 Procedure evaluation and other unifying concepts

B.2.1 Mean squared error (MSE)

To see the connection between the Bayesian and frequentist approaches, consider again the problem of estimating a parameter under squared error loss (SEL). With the sample $\mathbf{X} = (X_1, X_2, \ldots, X_n)$, let $d(\mathbf{X})$ be our estimate of θ. Its frequentist risk is

$$E_{\mathbf{x}|\theta}[l(\theta, d(\mathbf{x}))] = E_{\mathbf{x}|\theta}[(\theta - d(\mathbf{x}))^2] \equiv MSE_d(\theta) ,$$

the *mean squared error* of d given the true θ. Its posterior risk with respect to some prior distribution G is

$$E_{\theta|\mathbf{x}}[l(\theta, d(\mathbf{x}))] = E_{\theta|\mathbf{x}}[(\theta - d(\mathbf{x}))^2] \equiv MSE_{d,G}(\mathbf{x}) ,$$

and its preposterior risk is

$$E_{\theta,\mathbf{x}}[l(\theta, d(\mathbf{x}))] = E_{\theta,\mathbf{x}}[(\theta - d(\mathbf{x}))^2] \equiv MSE_{d,G} ,$$

which equals $E_\theta[MSE_d(\theta)]$ and $E_{\mathbf{x}}[MSE_{d,G}(\mathbf{x})]$. We have not labeled d as "frequentist" or "Bayesian"; it is simply a function of the data. Its properties can be evaluated using any frequentist or Bayesian criteria.

B.2.2 The variance-bias tradeoff

We now indicate how choosing estimators having minimum MSE requires a tradeoff between reducing variance and bias. Assume that conditional on θ, the data are i.i.d. with mean θ and variance $\sigma^2(\theta)$. Let $d(\mathbf{x})$ be the sample mean, $\bar{X} = \frac{1}{n}\sum_{i=1}^n X_i$, and consider estimators of θ of the form

$$d_c(\mathbf{x}) = c\bar{X} . \tag{B.8}$$

For $0 < c < 1$, this estimator shrinks the sample mean toward 0 and has the following risks:

$$\text{frequentist:} \quad c^2 \frac{\sigma^2(\theta)}{n} + (1 - c)^2\theta^2$$

$$\text{posterior:} \quad Var(\theta|\mathbf{x}) + [c\bar{x} - E(\theta|\mathbf{x})]^2,$$

preposterior (Bayes): $c^2 \dfrac{E[\sigma^2(\theta)]}{n} + (1-c)^2\{Var(\theta) + [E(\theta)]^2\},$

where in the last expression the expectations and variances are with respect to the prior distribution G. Notice that all three expressions are of the form "variance + bias squared."

Focusing on the frequentist risk, notice that the first term decreases like $\frac{1}{n}$ and the second term is constant in n. If $\theta = 0$ or is close to 0, then a c near 0 will produce small risk. More generally, for relatively small n, it will be advantageous to set c to a value betweeen 0 and 1 (reducing the variance at the expense of increasing the bias), but, as n increases, c should converge to 1 (for large n, bias reduction is most important). This phenomenon underlies virtually all of statistical modeling, where the standard way to reduce bias is to add terms (degrees of freedom, hence more variability) to the model.

Taking $c = 1$ produces the usual unbiased estimate with MSE equal to the sampling variance. Comparing $c = 1$ to the general case shows that if

$$\frac{n\theta^2 - \sigma^2(\theta)}{n\theta^2 + \sigma^2(\theta)} < c \le 1,$$

then it will be advantageous to use the biased estimator. This relation can be turned around to find the interval of θs around 0 for which the estimator beats $\bar{X}$. This interval has endpoints that shrink to 0 like $\frac{1}{\sqrt{n}}$.

Turning to the Bayesian posterior risk, the first term also decreases like $\frac{1}{n}$, and the Bayes rule (not restricted to the current form) will be $E(\theta|\mathbf{x})$. The c that minimizes the preposterior risk is

$$c = c_n = \frac{n\{Var(\theta) + [E(\theta)]^2\}}{n\{Var(\theta) + [E(\theta)]^2\} + E[\sigma^2(\theta)]}.$$

As with the frequentist risk, for relatively small n it is advantageous to use $c \neq 1$ and have c_n increase to 1 as n increases.

Similar results to the foregoing hold when an estimator (B.8) is augmented by an additive offset (i.e., $d_{a,c}(\mathbf{x}) = a + c\bar{X}$). We have considered SEL and MSE for analytic simplicity; qualitatively similar conclusions and relations hold for a broad class of convex loss functions.

Shrinkage estimators of the form (B.8) arise naturally in the Bayesian context. The foregoing discussion indicates that they can have attractive frequentist properties (i.e., lower MSE than the standard estimator, $\bar{X}$). Under squared error loss, one wants to strike a tradeoff between variance and bias, and rules that effectively do so will have good properties.

B.3   Other loss functions

We now evaluate the Bayes rules for three loss functions that show the effectiveness of Bayesian structuring and how intuition must be aided by the rules of probability. Chapter 5 and Section 7.1 present additional examples of the importance of Bayesian structuring.

B.3.1 Generalized absolute loss

Consider the loss function for $0 \leq p \leq 1$,

$$p|\theta - a| \quad \text{if} \quad a < \theta$$
$$(1 - p)|\theta - a| \quad \text{if} \quad a \geq \theta .$$

It is straightforward to show that the optimal estimate is the pth percentile of the posterior distribution. Therefore, $p = .5$ gives the posterior median. Other values of p are of more than theoretical interest, as illustrated below in Problem 3.

B.3.2 Testing with a distance penalty

Consider a hypothesis testing situation wherein there is a penalty for making an incorrect decision and, if the null hypothesis is rejected, an additional penalty that depends on the distance of the parameter from its null hypothesis value. The loss function

$$l(\theta, a) = \begin{cases} 0 & , \quad a = 0, \theta < 0 \text{ or } a = 1, \theta > 0 \\ \theta^2 & , \quad a = 0, \theta > 0 \text{ or } a = 1, \theta < 0 \end{cases}$$

computes this score. Assume the posterior distribution is $N(\mu, \tau^2)$. Then

$$R(\theta, a = 1) - R(\theta, a = 0) = \tau^2[1 - 2\Phi(\mu/\tau)] + \frac{\mu}{\tau}\phi(\mu/\tau) ,$$

where the first term corresponds to 0–1 loss, while the second term corresponds to the distance adjustment. If $\mu > 0$, 0–1 loss implies $a = 1$, but with the distance adjustment, if $\tau < 1$, then for small $\mu > 0$, $a = 0$. This decision rule is nonintuitive, but best.

B.3.3 A threshold loss function

Through its Small Area Income and Poverty Estimates (SAIPE) project, the U.S. Census Bureau is improving estimated poverty counts and rates used in allocating Title I funds (National Research Council, 2000). Allocations must satisfy a "hold-harmless" condition (an area-specific limit on the reduction in funding from one year to the next) and a threshold rule (e.g., concentration grants kick in when estimated poverty exceeds 15%). National Research Council (2000) demonstrates that statistical uncertainty induces unintended consequences of a hold-harmless condition, because allocations cannot appropriately adjust to changing poverty estimates. Similarly, statistical uncertainty induces unintended consequences of a threshold

Condition	Eligible for Concentration Funds?	Loss
$\theta \geq T$	yes	$(\theta - a)^2$
$\theta < T$	no	a^2

Table B.1 *A threshold loss function.*

rule. Unintended consequences can be ameliorated by optimizing loss functions that reflect societal goals associated with hold-harmless, threshold, and other constraints. However, amelioration will be limited and societal goals may be better met through replacing these constraints by ones designed to operate in the context of statistical uncertainty.

Example B.8 To show the quite surprising influence of thresholds on optimal allocation numbers, we consider a greatly simplified, mathematically tractable example. Let θ be the true poverty rate for a single area (e.g., a county), a be the amount allocated to the area per child in the population base, and $\mathbf{Y}$ denote all data. For a threshold T, consider the societal loss function in Table B.1. The optimal per-child allocation value is then

$$a_T(\mathbf{Y}) = E(\theta \mid \theta \geq T, \mathbf{Y}) \times \text{pr}(\theta \geq T \mid \mathbf{Y}) .$$

Though the allocation formula as a function of the true poverty rate (θ) has a threshold, the optimal allocation value has no threshold. Furthermore, for $T > 0$, $a_T(\mathbf{Y})$ is not the center of the the posterior distribution (neither the mean, median, nor mode). It is computed by multiplying the posterior mean (conditional on $\theta > T$) by a posterior tail area. This computation always produces an allocation value smaller than the posterior mean; for example, if the posterior distribution is exponential with mean μ, $a_T = \mu(1 + r)e^{-r}$ where $r = T/\mu$. But, this reduction is compensated by the lack of a threshold in a_T. It is hard to imagine coming up with an effective procedure in this setting without Bayesian structuring.

Actual societal loss should include multiplication by population size, incorporate a dollar cap on the total allocation over all candidate areas, and "keep score" over multiple years. Absolute error should replace squared-error in comparing a to θ. ∎

B.4 Multiplicity

There are several opportunities for taking advantage of multiple analyses, including controlling error rates, multiple outcomes, multiple treatments, interim analyses, repeated outcomes, subgroups, and multiple studies. Each of these resides in a domain for control of statistical properties, including

α	$Z_{\alpha/2}$	$Z_{\alpha_{bf}/2}$	$\sqrt{-2\log\alpha}$
.01	2.57	2.81	3.03
.02	2.33	2.57	2.80
.05	1.96	2.24	2.45
.10	1.65	1.95	2.15

Table B.2 *Comparison of upper univariate confidence limits, multiple comparisons setting with $K = 2$.*

control for a single analysis, a set of endpoints, a single study, a collection of studies, or even an entire research career! It is commonly thought that the Bayesian approach pays no attention to multiplicity. For example, as shown in Example 1.2, the posterior distribution does not depend on a data monitoring plan or other similar aspects of an experimental design. However, a properly constructed prior distribution and loss function can acknowledge and control for multiplicity.

Example B.9 When a single parameter is of interest in a K-variate parameter structure, the Bayesian formulation automaticallly acknowledges multiplicity. For the $K = 2$ case, consider producing a confidence interval for θ_1 when: $\boldsymbol{\theta} = (\theta_1, \theta_2) \sim N_2(\mathbf{0}, I_2)$. The $(1 - \alpha)$ HPD region is

$$\| \boldsymbol{\theta} \|^2 \leq -2\log(\alpha) .$$

Projecting this region to the first coordinate gives

$$-\sqrt{-2\log(\alpha)} \leq \theta_1 \leq \sqrt{-2\log(\alpha)} .$$

Table B.2 shows that for $K = 2$ the Bayesian HPD interval is longer than either the unadjusted or the Bonferroni adjusted intervals (α_{bf} is the Bonferroni non-coverage probability). That is, the Bayesian interval is actually more conservative. However, as K increases, the Bayesian HPD interval becomes narrower than the Bonferroni interval. ■

B.5 Multiple testing

B.5.1 Additive loss

Under additive, component-specific loss, Bayesian inference with components that are independent *a priori* separately optimizes inference for each component with no accounting for the number of comparisons. However, use of a hyperprior Bayes (or empirical Bayes) approach links the components because the posterior "borrows information" among components. The k-ratio t-test is a notable example. With F denoting the F-test for a one-way ANOVA,

$$F = (\hat{\sigma}^2 + K\hat{\tau}^2)/\hat{\sigma}^2$$
$$(1 - \hat{B}) = (F - 1)/F$$
$$\text{and } Z_{12} = \left(\frac{F - 1}{F}\right)^{1/2} \frac{Y_1 - Y_2}{\sqrt{2}\sigma}.$$

The magnitude of F adjusts the test statistic. For large K, under the global null hypothesis $H_0 : P(\text{all } Z_{ij} = 0) \geq 0.5$, the rejection rate is much smaller than 0.5. Note that the procedure depends on the estimated prior mean ($\hat{\mu}$) and shrinkage ($\hat{B}$) and the number of candidate coordinates can influence these values. If the candidate coordinates indicate high heterogeneity, then $\hat{B}$ is small as is the Bayesian advantage. Therefore, it is important to consider what components to include in an analysis to control heterogeneity while retaining the opportunity for discovery.

B.5.2 Non-additive loss

If one is concerned about multiplicity, the loss function should reflect this concern. Consider a testing problem with a loss function that penalizes for individual coordinate errors and adds an extra penalty for making two errors:

$$\begin{aligned}
\text{parameters:} \quad & \theta_1, \theta_2 \in \{0, 1\} \\
\text{decisions:} \quad & a_1, a_2 \in \{0, 1\} \\
l(a, \boldsymbol{\theta}) = \quad & a_1(1 - \theta_1)\ell_0 + (1 - a_1)\theta_1\ell_1 \\
& + a_2(1 - \theta_2)\ell_0 + (1 - a_2)\theta_2\ell_1 \\
& + \gamma(1 - \theta_1)(1 - \theta_2)a_1 a_2.
\end{aligned}$$

With $\ell_0 = \ell_1 = 1$ and $G_{ij} = P(\theta_1 = i, \theta_2 = j|data)$, a straightforward computation shows that the risk is

$$a_1(1 - 2G_{1+}) + a_2(1 - 2G_{+1}) + \gamma a_1 a_2 G_{00} + \{G_{1+} + G_{+1}\}.$$

The four possible values of (a_1, a_2) then produce the risks in Table B.3, and the the Bayes decision rule (the (a_1, a_2) that minimize risk) is given in Table B.4. Note that in order to declare the second component "$a_2 = 1$," we require more compelling evidence than if $\gamma = 0$.

Discussion

Statistical decision rules can be generated by any philosophy under any set of assumptions. They can then be evaluated by any criteria, even those arising from an utterly different philosophy. We contend (and much of this book shows) that the Bayesian approach is an excellent "procedure generator," even if one's evaluation criteria are frequentist. This somewhat

a_1	a_2	Risk $- \{G_{1+} + G_{+1}\}$
0	0	0
1	0	$1 - 2G_{1+}$
0	1	$1 - 2G_{+1}$
1	1	$(1 - 2G_{1+}) + (1 - 2G_{1+}) + \gamma G_{00}$

Table B.3 *Risks for the possible decision rules.*

Condition	Optimal Rule
$G_{1+} \leq .5, G_{+1} \leq .5$	$a_1 = 0, a_2 = 0$
$G_{1+} \leq .5, G_{+1} > .5$	$a_1 = 0, a_2 = 1$
$G_{1+} > .5, G_{+1} \leq .5$	$a_1 = 1, a_2 = 0$
$G_{1+} > G_{+1} > .5$	$a_1 = 1, a_2 = \begin{cases} 0, & \text{if } (2G_{+1} - 1) < \gamma G_{00} \\ 1, & \text{if } (2G_{+1} - 1) \geq \gamma G_{00} \end{cases}$

Table B.4 *Optimal decision rule for non-additive loss.*

agnostic view considers the parameters of the prior, or perhaps the entire prior, as "tuning parameters" that can be used to produce a decision rule with broad validity. Of course, no approach automatically produces broadly valid inferences, even in the context of the Bayesian models. A procedure generated assuming a overconfident prior that turns out to be far from the truth will perform poorly in both the Bayesian and frequentist senses.

B.6 Exercises

1. Show that the Bayes estimate of a parameter θ under weighted squared error loss, $l(\theta, a) = w(\theta)(\theta - a)^2$, is given by
$$d_\pi(\mathbf{x}) = \frac{E[\theta w(\theta)|\mathbf{x}]}{E[w(\theta)|\mathbf{x}]}.$$

2. Suppose that the posterior distribution of θ, $p(\theta|\mathbf{x})$, is discrete with support points $\{\theta_1, \theta_2, \ldots\}$. Show that the Bayes rule under 0–1 loss (B.2) is the posterior mode.

3. Consider the following loss function:
$$l(\theta, a) = \begin{cases} p|\theta - a|, & \theta > a \\ (1 - p)|\theta - a|, & \theta \leq a \end{cases}$$

Show that the Bayes estimate is the $100 \times p^{th}$ percentile of the posterior distribution.

(Note: The posterior median is the Bayes estimate for $p = .5$. This loss function does have practical application for other values of p. In many organizations, employees are allowed to set aside pre-tax dollars for healthcare expenses. Federal rules require that one cannot get back funds not used in a 12-month period, so one should not put away too much. However, putting away too little forces some healthcare expenses to come from post-tax dollars. If losses are linear in dollars (or other currency), then this loss function applies with p equal to the individual's marginal tax rate.)

4. For the binomial distribution with n trials, find the minimax rule for squared error loss and normalized squared error loss, $L(\theta, a) = \frac{(\theta - a)^2}{\theta(1-\theta)}$.
   (*Hint:* The rules are Bayes rules or limits of Bayes rules.)

   For each of these estimates, compute the frequentist risk (the expected loss with respect to $f(x|\theta)$), and plot this risk as a function of θ (the *risk plot*). Use these risk plots to choose and justify an estimator when θ is

   (a) The head probability of a randomly selected coin.
   (b) The failure probability for high-quality computer chips.
   (c) The probability that there is life on other planets.

5. For the exponential distribution based on a sample of size n, find the minimax rule for squared error loss.

6. Find the frequentist risk under squared error loss (i.e., the MSE) for the estimator $d_{a,c}(\mathbf{x}) = a + c\bar{X}$. What true values of θ favor choosing a small value of c (high degree of shrinkage)? Does your answer change if n is large?

7. Let X be a random variable with mean μ and variance σ^2. You want to estimate μ under SEL, and propose an estimate of the form $(1 - b)X$.

   (a) Find b^*, the b that minimizes the MSE.
       (*Hint:* Use the "variance + (bias)^2" representation of MSE.)
   (b) Discuss the dependence of b^* on μ and σ^2 and its implications on the role of shrinkage in estimation.

8. Consider a linear regression having true model of the form $Y_i = \alpha + \beta x_i + \epsilon_i$, where the ϵ_i are i.i.d. with mean 0 and variance σ^2. Suppose you have a sample of size n, use least-squares estimates (LSEs), and want to minimize the *prediction error*:

$$\mathrm{PE} = E[Y_{new} - \hat{Y}(X_{new})^2] \, .$$

Here, $\hat{Y}(x)$ is the prediction based on the estimated regression equation and X_{new} is randomly selected from a distribution with mass $\frac{1}{n}$ at each of the $x_1, \ldots, x_n$.

(a) Assume that $\sum_i x_i = 0$ and show that even though you may *know* that $\beta \neq 0$, if $\sum_i x_i^2$ is sufficiently small, it is better to force $\hat{\beta} = 0$ than to use the LSE for it.

   (Note: This result underlies the use of the C_p statistic and PRESS residuals in selecting regression variables; see the documentation for SAS Proc REG.)

(b) *(More difficult)* As in the previous exercise, assume you plan to use a prediction of the form $\hat{Y}(x) = \hat{\alpha} + (1 - b)\hat{\beta}x$. Find b^*, the optimal value of b.

9. In an errors-in-variables simple regression model, the least squares estimate of the regression slope (β) is biased toward 0, an example of *attenuation*. Specifically, if the true regression (through the origin) is $Y = x\beta + \epsilon$, but Y is regressed on X, with $X = x + \delta$, then the least squares estimate ($\hat{\beta}$) has expectation: $E[\hat{\beta}] \approx \rho\beta$, with $\rho = \frac{\sigma_x^2}{\sigma_x^2 + \sigma_\delta^2} \leq 1$.

   If ρ is known or well-estimated, one can correct for attenuation and produce an unbiased estimate by using $\hat{\beta}/\rho$ to estimate β. However, this estimate may have poor MSE properties. To minimize MSE, consider an estimator of the form: $\hat{\beta}_c = c(\hat{\beta}/\rho)$.

(a) Let $\sigma^2 = Var[\hat{\beta}]$ and find the c that minimizes MSE.

(b) Discuss the solution's implications on the variance/bias tradeoff and the role of shrinkage in estimation.

10. Compute the HPD confidence interval in Example B.9 for a general K, and compare its length to the unadjusted and Bonferroni adjusted intervals.

Answers to selected exercises

Chapter 1

1. Though we may have no prior understanding of male hypertension, the structure of the problem dictates that $\theta \in [0, 1]$, providing some information. In addition, values close to each end of this interval seem less likely, since we know that while not *all* males are hypertensive, at least *some* of them are.

   Suppose we start with $\hat{\theta} = .3$. Given the knowledge that four of the first five men we see are in fact hypertensive, we would likely either (a) keep our estimate unchanged because too few data have yet accumulated to sway us, or (b) revise our initial estimate upward somewhat, say, to .35. (It certainly would not seem plausible to revise our estimate *downward* at this point.)

   Once we have observed 400 hypertensive men out of a sample of 1000, however, the data's pull on our thinking is hard to resist. Indeed, we might completely discard our poorly informed initial estimate in favor of the data-based estimate, $\hat{\theta} = .4$.

   All of the above characteristics of sensible data analysis (plausible starting estimates, gradual revision, and eventual convergence to the data-based value) are features we would want our inferential system to possess. These features *are* possessed by Bayes and EB systems.

5.(a) p-value $= P(X \geq 13 \text{ or } X \leq 4) = .049$ under H_0. So, using the usual Type I error rate $\alpha = .05$, we would stop and reject H_0.

  (b) For the two-stage design, we now have

$$
\begin{aligned}
p - \text{value} \quad &= \quad P(X_1 \geq 13 \text{ or } X_1 \leq 4) \\
&\quad + P(X_1 + X_2 \geq 29 \text{ and } 4 < X_1 < 13) \\
&\quad + P(X_1 + X_2 \leq 15 \text{ and } 4 < X_1 < 13) \\
&= \quad .085 ,
\end{aligned}
$$

   which is no longer significant at $\alpha = .05$! Thus, even though the observed data was exactly the same, the mere *contemplation* of a second stage to the experiment (having no effect on the data) changes the answer in the classical hypothesis testing framework.

(c) The impact of adding imaginary stages to the design could be that the p-value approaches 1 even as x_1 remains fixed at 13.

(d) Strictly speaking, since this aspect of the design (early stopping due to ill effects on the patients) wasn't anticipated, the p-value is not computable.

(e) To us it suggests that p-values are less than objective evidence because not only the form of the likelihood, but also extra information (like the design of the experiment), is critical to their computation and interpretation.

6.(a) $B = .5$, so $\theta|y \sim N(.5\mu + .5y, .5(2)) = N(2,1)$.

(b) Now $B = .1$, so $\theta|y \sim N(.1\mu + .9y, .9(2)) = N(3.6, 1.8)$. The increased uncertainty in the prior produces a posterior that is much more strongly dominated by the likelihood; the posterior mean shrinks only 10% of the way toward the prior mean μ. The overall precision in the posterior is also much lower, reflecting our higher overall uncertainty regarding θ. Despite this decreased precision, a frequentist statistician might well prefer this prior because it allows the data to drive the analysis. It is also likely to be more robust, i.e., to perform well over a broader class of θ-estimation settings.

Chapter 2

3. From Appendix Section A.2, since $\chi^2(\nu) \equiv G(\nu/2, 2)$, we have that $G(\alpha, 2) \equiv \chi^2(2\alpha)$. But if $X \sim G(\alpha, 2)$, a simple Jacobian transformation shows that $Y = \beta X/2 \sim G(\alpha, \beta)$. Finally, $Z = 1/Y \sim IG(\alpha, \beta)$ by definition, so we have that the necessary samples can be generated as

$$\sigma^2 = \frac{2}{\beta X}, \text{ where } X \sim \chi^2(2\alpha)$$

and

$$\alpha = \frac{1}{2} + a , \quad \beta = \left[\frac{1}{2}(x - \theta)^2 + \frac{1}{b}\right]^{-1} .$$

4. Because $\Theta = \Re^+$ in this example, a difficulty with the histogram approach is its finite support (though in this example, terminating the histogram at, say, 10 feet would certainly seem safe). Matching a parametric form avoids this problem and also streamlines the specification process, but an appropriate shape for a population maximum may not be immediately obvious, nor may the usual families be broad enough to truly capture your prior beliefs.

5.(a) $\theta \sim Beta(\alpha, \beta)$

(b) $\theta \sim Beta(\alpha, \beta)$ (likelihood same as in (a))

(c) $\boldsymbol{\theta} \sim D(\boldsymbol{\alpha})$

    (d) $\beta \sim IG(a, b)$

    (e) none available

    (f) $\boldsymbol{\theta} \sim N_k(\boldsymbol{\mu}, \mathbf{V})$

    (g) $\boldsymbol{\Sigma} \sim IW(\boldsymbol{\Omega}, \nu)$

6.(a) scale; $p(\sigma) \propto 1/\sigma$

    (b) location-scale; $p(\theta, a) \propto 1/a$

    (c) location; $p(\mu) \propto 1$

    (d) scale; $p(\beta) \propto 1/\beta$

    (e) neither (α is often called a *shape* parameter)

7. The Jeffreys prior for γ is

$$I(\gamma) = -E_{\mathbf{x}|\gamma}\left[\frac{d^2}{d\gamma^2}\log p(\mathbf{x}|\gamma)\right].$$

But, using the chain rule, the product rule, and the chain rule again, we can show that

$$\frac{d^2}{d\gamma^2}\log p(\mathbf{x}|\gamma) = \frac{d\log p(\mathbf{x}|\theta)}{d\theta} \cdot \frac{d^2\theta}{d\gamma^2} + \frac{d^2\log p(\mathbf{x}|\theta)}{d\theta^2} \cdot \left(\frac{d\theta}{d\gamma}\right)^2.$$

Because the second terms in these two products are constant with respect to $\mathbf{x}$ and the score statistic has expectation 0, we have

$$\begin{aligned}
I(\gamma) &= -E_{\mathbf{x}|\theta}\left[\frac{d^2\log p(\mathbf{x}|\theta)}{d\theta^2}\right] \cdot \left(\frac{d\theta}{d\gamma}\right)^2 \\
&= I(\theta) \cdot \left(\frac{d\theta}{d\gamma}\right)^2.
\end{aligned}$$

Hence, $[I(\gamma)]^{1/2} = [I(\theta)]^{1/2} \cdot |d\theta/d\gamma|$, as required.

10.(a) For the joint posterior of θ and σ^2, we have

$$p(\theta, \sigma^2|\mathbf{y}) \propto (\sigma^2)^{-n/2-1} \exp\{\frac{1}{2\sigma^2}\sum_{i=1}^{n}(y_i - \theta)^2\}. \qquad (\text{C.1})$$

This expression is proportional to an inverse gamma density, so integrating it with respect to σ^2 produces

$$p(\theta|\mathbf{y}) \propto \left[\sum_{i=1}^{n}(y_i - \theta)^2\right]^{-n/2}.$$

However, using the identity $\sum_{i=1}^{n}(y_i - \theta)^2 = C + n(\bar{y} - \theta)^2$, where $C = \sum_{i=1}^{n}(y_i - \bar{y})^2$, we have that

$$p(\theta|\mathbf{y}) \propto [C + n(\bar{y} - \theta)^2]^{-n/2} \propto [1 + t^2/(n-1)]^{-n/2}.$$

Node	Mean	SD	MC Error	2.5%	Median	97.5%
β_1	4.441	0.4198	0.004398	3.634	4.44	5.262
β_2	-3.354	0.3977	0.00376	-4.136	-3.358	-2.578
β_3	3.250	0.3964	0.004212	2.479	3.252	4.023
β_4	1.561	0.4513	0.004016	0.671	1.559	2.445
σ	0.7394	0.0266	2.762E-4	0.69	0.7386	0.7943

Table C.1 WinBUGS *posterior summaries (5000 samples after 100 burn-in).*

(b) Here, the above identity must be employed right away, in order to integrate θ out of the joint posterior (C.1) as proportional to a normal form. The result for $p(\sigma^2|\mathbf{y})$ is then virtually immediate.

12.(a) Our WinBUGS code (also available on the book's webpage) is

BUGS code
```
model {
  for (i in 1:389) {
    Y[i] ~ dnorm(mu[i], tau)
    mu[i] <- beta[1] + beta[2]*X[i,1] +
      beta[3]*X[i,2] + beta[4]*X[i,3]
  }
  for (i in 1:4) {
    beta[i] ~ dflat()
  }
  tau ~ dgamma(3,b)
  b <- 1/(1/(2*(1/(0.73*0.73))))
  sigma <- 1/sqrt(tau)
}
## Inits (2 choices):
  list(beta = c(0,0,0,0), tau = 1.0)
  list(beta = c(0.5,0.5,0.5,0.5), tau = 4.0)
```

Note that we have used $\hat{\sigma} = 0.73$ from OLS regression. History plots reveal essentially immediate convergence; the summaries from 5000 samples following 100 burn-in samples are shown in Table C.1. Because 95% credible intervals that exclude 0 indicate "significant" parameters, here all the βs are significant. Locations farther from the airports are more desirable (β_3 and β_4 are positive), while land value is inversely related to distance from Lake Michigan (β_2 is negative).

(b) We proceed by two steps, first updating σ^2 and then drawing $\boldsymbol{\beta}$ as a block given σ^2. Under the given prior, the full conditional for $\boldsymbol{\beta}$ is

$$\boldsymbol{\beta}^{(g)} \sim N((\mathbf{X}^T\mathbf{X})^{-1}\mathbf{X}^T\mathbf{Y}, (\sigma^2)^{(g)}(\mathbf{X}^T\mathbf{X})^{-1}) \qquad \text{(C.2)}$$

Next, the marginal posterior of σ^2 is given by

$$P(\sigma^2|\mathbf{y}) \propto \frac{1}{(\sigma^2)^{\frac{(n-p)}{2}+1}} \exp\left(-\frac{(n-p)s^2}{2\sigma^2}\right), \qquad \text{(C.3)}$$

which we recognize as the kernel of an $IG(\frac{N-p}{2}, \frac{(N-p)s^2}{2})$ distribution, where $s^2 = \hat{\sigma}^2 = (\mathbf{y} - \mathbf{X}\hat{\boldsymbol{\beta}})^T(\mathbf{y} - \mathbf{X}\hat{\boldsymbol{\beta}})/(N-p)$. While the marginal posterior distribution of $\boldsymbol{\beta}$ could be derived as a noncentral t distribution, samples may be obtained more easily via composition sampling from the two conditional distributions, which we coded in R as follows:

R code

```
data <- read.table(file="land_data.txt",header=T,sep="")
library(MASS)

exact <- function(data, NITER) {
  Y <- data$Y
  X <- as.matrix(data[,2:4])
  X <- cbind(rep(1,times=length(Y)),X)
  N <- length(Y)
  p <- dim(X)[2]

  tXX.inv <- solve(t(X) %*% X)
  beta.hat <- tXX.inv %*% t(X) %*% Y
  s.sq <- t(Y - X%*%beta.hat)%*%(Y - X%*%beta.hat)/(N - p)

  sigma.sq <- rep(0, times = NITER)
  beta <- matrix(0, nrow = NITER, ncol = p)

  for (i in 1:NITER) {
    sigma.sq[i] <- 1/rgamma(1, (N-p)/2, (N-p)*s.sq/2)
    sigma.sq.of.beta <- (tXX.inv) * sigma.sq[i]
    beta[i,] <- mvrnorm(1, beta.hat, sigma.sq.of.beta)
  }
  sigma <- sqrt(sigma.sq)
  cbind(beta,sigma)
}

samples <- exact(data,NITER=5000)
```

Results in Table C.2 are again very similar to those seen earlier. Note that because sampling is exact (not MCMC), there is no need for burn-in.

(c) For each Monte Carlo draw $\boldsymbol{\theta}_j$, we generate a replicate data set $\mathbf{y}_{rep,j}$. We then calculate a discrepancy measure for each replicate, $T(\mathbf{y}_{rep,j}, \boldsymbol{\theta}_j)$, as well as for the original data, $T(\mathbf{y}, \boldsymbol{\theta}_j)$. Here, we adopt the sum of squared residuals as our discrepancy measure (omitting standardization by the posterior variance because it is the same in

Node	Mean	SD	2.5%	Median	97.5%
β_1	4.4371	0.4192	3.6105	4.4382	5.2527
β_2	−3.3565	0.4017	−4.1252	−3.3621	−2.5613
β_3	3.2482	0.3944	2.4823	3.2433	4.0256
β_4	1.5707	0.4519	0.6960	1.5776	2.4423
σ	0.7384	0.0265	0.6896	0.7381	0.7921

Table C.2 R *exact sampler posterior summaries (5000 samples).*

both statistics):

$$T(\mathbf{y}, \boldsymbol{\theta}) = (\mathbf{y} - \mathbf{X}\boldsymbol{\beta})^T (\mathbf{y} - \mathbf{X}\boldsymbol{\beta}) \, ,$$

where $\mathbf{X}$ is the $n \times p$ matrix of predictors, $\boldsymbol{\beta}$ is the $p \times 1$ vector of parameter values, and $\mathbf{y}$ is the $n \times 1$ data vector. Comparing the distributions of the discrepancy statistic for the replicate and original datasets yields the Bayesian p-value,

$$p_{Bayes} = Pr\left(T(\mathbf{y}_{rep}, \boldsymbol{\theta}) > T(\mathbf{y}, \boldsymbol{\theta}) \mid \mathbf{y}\right) \, .$$

Sample R code follows:

R code
```
Tstats <- function(data, NITER) {
    Y <- data$Y
    X <- as.matrix(data[,2:4])
    X <- cbind(rep(1,times=length(Y)),X)
    N <- length(Y)
    p <- dim(X)[2]

    tXX.inv <- solve(t(X) %*% X)
    beta.hat <- tXX.inv %*% t(X) %*% Y
    s.sq <- t(Y - X%*%beta.hat)%*%(Y - X%*%beta.hat)/(N-p)

    sigma.sq <- rep(0, times = NITER)
    beta <- matrix(0, nrow = NITER, ncol = p)
    y.rep <- matrix(0, nrow = N, ncol = NITER)
    T.rep <- rep(0, times=NITER)
    T.orig <- rep(0, times = NITER)

    for (i in 1:NITER) {
      sigma.sq[i] <- 1/rgamma(1, (N-p)/2, (N-p)*s.sq/2)
      sigma.sq.of.beta <- (tXX.inv) * sigma.sq[i]
      beta[i,] <- mvrnorm(1, beta.hat, sigma.sq.of.beta)
      y.rep[,i] <- mvrnorm(1, X%*%beta[i,], sigma.sq[i]*diag(N))
      T.rep[i] <- t(y.rep[,i] - X%*%beta[i,]) %*%
        (y.rep[,i] - X%*%beta[i,])
```

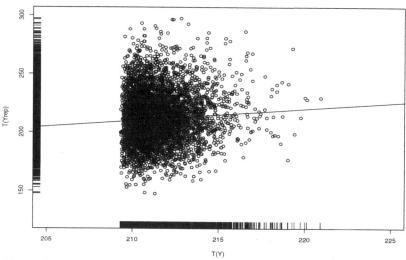

Figure C.1 *Discrepancy statistics, components of Bayesian p-value calculation.*

```
      T.orig[i] <- t(Y - X%*%beta[i,])%*%(Y - X%*%beta[i,])
   }
   diff <- (T.rep - T.orig)
   cbind(T.rep,T.orig,diff)
}

Trep <- Tstats(data,5000)
p.val <- length(Trep[Trep[,3]>0,3]) / 5000
```

Using 5000 Monte Carlo posterior samples, obtain the scatterplot of $T(\mathbf{y}, \boldsymbol{\theta})$ versus $T(\mathbf{y}_{rep}, \boldsymbol{\theta})$ shown in Figure C.1. The Bayesian p-value is the total proportion of values above the line of equality in the figure, which turns out to be 0.51, indicating no problem with model inadequacy here.

13.(a) Using classical methods, we would calculate:

$$W = \frac{\overline{y}_A - \overline{y}_B}{\sqrt{\frac{\hat{\sigma}_A^2}{n_A} + \frac{\hat{\sigma}_B^2}{n_B}}} = 2.97$$

Comparing this to a Student's t with $32 + 36 - 2 = 66$ degrees of freedom, we obtain a highly significant two-sided p-value of .004. The allegedly more appropriate Welch-Satterthwaite degrees of freedom approximation gives

$$df = \frac{\left(\frac{s_A^2}{n_A} + \frac{s_B^2}{n_B}\right)^2}{\frac{s_A^4}{n_A^2(n_A-1)} + \frac{s_B^4}{n_B^2(n_B-1)}} = \frac{\left(\frac{0.20^2}{36} + \frac{0.24^2}{32}\right)^2}{\frac{0.20^4}{36^2(36-1)} + \frac{0.24^4}{32^2(32-1)}} = 60.6,$$

and thus a p-value that is again .004 (to three decimal places).

(b) The control and experimental groups may be sampled separately because their parameters are independent both in the likelihood and
the priors. For both groups, the two-step composition sampling pro-
cess first draws σ_j^2, $j = 1, \ldots, N$ from $[\sigma^2|\mathbf{y}] = IG\left(\frac{n-1}{2}, \frac{(n-1)s^2}{2}\right)$,
where $s^2 = \hat{\sigma}^2$. For each sampled σ_j^2, we then draw from $[\mu|\sigma_j^2, \mathbf{y}] = N\left(\bar{y}, \frac{\sigma_j^2}{n}\right)$. This produces a sample from the joint posterior $[\mu_A, \mu_B|\mathbf{y}_A, \mathbf{y}_B]$,
from which we can create the mean differences $\Delta_j = \mu_{A,j} - \mu_{B,j}$. The
empirical relative proportions of Δ_j on either side of 0 gives a good
idea of the "significance" of the data. Alternatively (and more informally), the exclusion of 0 from the 95% credible interval for Δ might
also lead us to reject H_0.

Sample R code is as follows:

R code
```
BFsamples <- function(A,B,NITER) {
    sigma.sq.A <- 1/rgamma(NITER, (A$N - 1) / 2,
      (A$N - 1) * A$sd^2 / 2)
    mu.A <- rnorm(NITER, A$y.bar, sqrt(sigma.sq.A/A$N))
    sigma.sq.B <- 1/rgamma(NITER, (B$N - 1) / 2,
      (B$N - 1) * B$sd^2 / 2)
    mu.B <- rnorm(NITER, B$y.bar, sqrt(sigma.sq.B/B$N))
    diff <- mu.A- mu.B
    diff
}

ctrl <- list(N=32, y.bar = 1.013, sd = 0.24)
expt <- list(N=36, y.bar = 1.173, sd = 0.20)
post.samples <- BFsamples(A=expt,B=ctrl,NITER=5000)
```

Using the quantile command on post.samples, we obtained a 95%
equal-tail credible interval of (0.055 , 0.268), suggesting a significant difference between the two groups. Using the sort command,
we found there were just 10 negative Δ_j in our sample, leading to
an empirical Bayes factor of $4990/10 = 499$, strong evidence of an
increased response in the experimental group. Note however that because the priors are improper, this is really a *pseudo* Bayes factor
(PBF) and, in fact, it is a PBF for slightly different hypotheses than
asked for in the problem, namely $H_0 : \mu_A > \mu_B$ versus $H_a : \mu_A < \mu_B$.

14. Using the DIC tool in WinBUGS, under the $G(.1,.1)$ prior we obtained

an effective model size of $p_D = 3.10$, very close to the actual parameter count of 3, and a DIC score of -43.2. Under the $U(.01, 100)$ prior, we instead obtained $p_D = 2.90$ and $DIC = -49.3$, the "smaller" (more negative) DIC score indicating superiority of this model. This DIC difference suggests that the uniform prior leads to better overall model fit (since the p_D scores are virtually identical), which here we interpret as a prior that permits maximal consistency with the observed data.

In general, the use of DIC to choose among competing, informative prior specifications does *not* appear sensible because any data-based model choice statistic will always select the prior that is most congruent with the observed data. We would in effect be "using the same data twice," once to select the prior, and again to compute the posterior under this prior. As such, the safest approach would appear to require an honest assessment of any existing prior information *before* the data are studied.

21. Boxplots of the random effects and p_D scores across models indicate high degrees of shrinkage and little benefit in fit arising from most of the random effects. Even the "no random effects" model is surprisingly competitive in terms of DIC score.

22. The model can be specified very similarly to that in Example 2.16 as

BUGS code
```
model
{
## Standardise x's and coefficients##
for (j in 1 : p) {
   b[j] <- beta[j] / sd(x[ , j ])
   for (i in 1 : N) {
   z[i,j] <- (x[i,j]-mean(x[,j]))/sd(x[,j])
   }
}
b0 <- beta0 - b[1]*mean(x[,1])-b[2]*mean(x[,2])-b[3]*mean(x[,3])

# statistical model
for (i in 1 : N) {
   Y[i] ~ dnorm(mu[i], tau)
   mu[i] <- beta0+beta[1]*z[i,1]+beta[2]*z[i,2]+beta[3]*z[i,3]

#    ultimately we'll keep only one element of
#    each stat for "leave-one-out" validation.
   sresid[i] <- (Ytrue[i]-mu[i])/sigma
   outlier[i] <- step(sresid[i]-1.5)+step(-(sresid[i]+1.5))
   CPO[i]<-sqrt(tau)*exp(-tau/2*(Ytrue[i]-mu[i])*(Ytrue[i]-mu[i]))
   }

# Priors
beta0 ~ dnorm(0, 0.00001)
```

```
for (j in 1 : p) {beta[j] ~ dnorm(0, 0.00001)}
sigma ~ dunif(0.01,100)   #  vague Gelman prior for sigma
tau<- 1/(sigma*sigma)
}
```

The BRugs code is then

BRugs code
```
Y<-c(42,37,37,28,18,18,19,20,15,14,14,13,11,12,8,7,8,8,9,15,15)
N<-length(Y)
p<-3
for(i in 1:N){Ytemp<-Y; Ytemp[i]<-NA;
  dput(pairlist(p=p,N=N,Y=Ytemp, Ytrue=Y),
  paste("data",i,".txt",sep=""))}

mysresid<-rep(0,N);myoutlier<-rep(0,N);myCPO<-rep(0,N)

library(BRugs)

for(i in 1:N){

modelCheck("model.txt")
modelData( paste("data",i,".txt",sep=""))
modelData( paste("dataX.txt",sep=""))
modelCompile(numChains=2)
modelInits(c("Inits1.txt","Inits2.txt"))
modelGenInits()

modelUpdate(5000)
samplesSet(c("beta0","beta","sigma","tau"))
summarySet(c("sresid","outlier","CPO"))
#    running mean and quantiles calculated; samples not stored
modelUpdate(5000)

#samplesHistory("*", mfrow = c(4, 2))
#samplesBgr("*")
#samplesStats(c("beta0","beta","sigma","tau"))
#dput(samplesStats(c("beta0","beta","sigma","tau")),
#   paste("stats",i,".txt",sep=""))

#record the posterior means, keeping only the ith elements:
mysresid[i]<-summaryStats("sresid")[i,1]
myoutlier[i]<-summaryStats("outlier")[i,1]
myCPO[i]<-summaryStats("CPO")[i,1]

samplesClear(c("beta0","beta","sigma","tau"))
summaryClear(c("sresid","outlier","CPO"))
}
```

```
            mysresid
            myoutlier
            myCPO
```

This then produces the results:

```
   > mysresid
    [1]   1.39300 -0.80710   1.70200   2.08100 -0.52180 -0.95910
    [7]  -0.89350 -0.49820 -1.09400   0.46320   0.91670   1.03600
   [13]  -0.48750 -0.01554   0.83210   0.29690 -0.74590 -0.15710
   [19]  -0.20590   0.44210 -3.76500
   > myoutlier
    [1]  0.4307 0.1618 0.6351 0.8569 0.0001 0.0559 0.1356 0.0357
    [9]  0.1859 0.0202 0.1016 0.2052 0.0101 0.0033 0.1020 0.0010
   [17]  0.1863 0.0011 0.0027 0.0000 0.9927
   > myCPO
    [1]  0.122700 0.188400 0.082620 0.047180 0.244200 0.179400
    [7]  0.184900 0.229700 0.158700 0.233700 0.183300 0.164600
   [13]  0.236400 0.254300 0.192000 0.255500 0.185500 0.259300
   [19]  0.255000 0.249800 0.004631
```

23.(a) Here is the WinBUGS code for the solution. Note that several lines needed for the solution to part (b) are commented out ("#" sign).

BUGS code
```
model
{
# define pi's; standardise x's and coefficients
for (j in 1 : p) {
   pi[j] ~ dbern(theta[1])
#   pi[ j ] <- inout[ j ]-1
#   inout[ j ]~ dcat(theta[])

   b[j] <- beta[j] / sd(x[ , j ])
   for (i in 1 : N) {z[i,j] <- (x[i,j]-mean(x[,j]))/sd(x[,j])}
   }

theta[1] ~ dbeta(1,1)          # prior for unif-bernoulli model
# theta[1:2] ~ ddirch(a[])     # prior for dirichlet-dcat model

# statistical model
for (i in 1 : N) {
   Y[i] ~ dnorm(mu[i], tau)
   mu[i] <- beta0 + pi[1]*beta[1]*z[i,1] + pi[2]*beta[2]*z[i,2]
                   + pi[3]*beta[3]*z[i,3]
   }
# priors
beta0 ~ dnorm(0, 0.00001)
for (j in 1 : p) {beta[j] ~ dnorm(0, 0.00001) }
sigma ~ dunif(0.01,100)        #  vague Gelman prior for sigma
```

```
tau<- 1/(sigma*sigma)
}
```

The posterior for θ resembles a $Beta(2,2)$ distribution, so many inclusion probabilities other than $1/2$ are encouraged (recall the three model-specific estimates are roughly 0, 0.3, and 1).

(b) Commenting out the code associated with **dbern** and **dbeta** and commenting back in the **inout**, **dcat** and **ddirch** code produces the solution to this part. No significant differences since this is essentially the same model, provided we use the line

BUGS code   `a = c(1,1),`

in the **data** loading, since then then Dirichlet(1,1) prior is, like the Beta(1,1), equivalent to the Uniform(0,1).

Chapter 3

3. The probability p of accepting a rejection candidate is

$$
\begin{aligned}
p &= P\left(L(\theta)\pi(\theta) > MUg(\theta)\right) \\
&= \int P\left(U < \frac{L(\theta)\pi(\theta)}{Mg(\theta)}\right) g(\theta)d\theta \\
&= \int \frac{L(\theta)\pi(\theta)}{Mg(\theta)} \cdot g(\theta)d\theta \\
&= \frac{1}{M} \int L(\theta)\pi(\theta)d\theta \\
&= \frac{c}{M} .
\end{aligned}
$$

4. The new posterior is

$$
p_2(\theta|\mathbf{y}) \propto \frac{\pi_2(\theta)}{\pi_1(\theta)} p_1(\theta|\mathbf{y}) .
$$

In the notation of the weighted bootstrap, we have $g(\theta) = p_1(\theta|\mathbf{y})$, and we seek samples from

$$
L(\theta)\pi(\theta) = \frac{\pi_2(\theta)}{\pi_1(\theta)} p_1(\theta|\mathbf{y}) \quad \Rightarrow \quad w(\theta) = \frac{\pi_2(\theta)}{\pi_1(\theta)} .
$$

Thus, if we compute

$$
q_i = \frac{\pi_2(\theta_i)/\pi_1(\theta_i)}{\sum_{j=1}^{N} \pi_2(\theta_j)/\pi_1(\theta_j)} ,
$$

then resampling θ_i^*s with probability q_i provides a sample from $p_2(\theta|\mathbf{y})$. This procedure is likely to work well only if π_2 has lighter tails than π_1.

14. The Hastings acceptance ratio r is

$$
\begin{aligned}
r &= \frac{p(v)q(u, v)}{p(u)q(v, u)} \\
&= \frac{p(v)p(u_i|u_{j\neq i})}{p(u)p(v_i|u_{j\neq i})} \\
&= \frac{p(v_i|u_{j\neq i})p(u_{j\neq i})p(u_i|u_{j\neq i})}{p(u_i|u_{j\neq i})p(u_{j\neq i})p(v_i|u_{j\neq i})} \\
&\equiv 1 \,,
\end{aligned}
$$

since v and u differ only in their i^{th} components.

16.(a) Because $\log p_Z(z) \propto (\delta - 1)\log z - \beta z + k\log(1 - e^{-z})$, we have

$$
\frac{\partial \log p_Z(z)}{\partial z} \propto \frac{\delta - 1}{z} - \beta + \frac{k}{e^z - 1}
$$

and

$$
\frac{\partial^2 \log p_Z(z)}{\partial z^2} \propto \frac{-(\delta - 1)}{z^2} - \frac{ke^z}{(e^z - 1)^2} < 0 \text{ for } \delta \geq 1 \,.
$$

Thus, $p_Z(z)$ is guaranteed to be log-concave provided $\delta \geq 1$.

(b) Marginalizing the joint density, we have

$$
\begin{aligned}
\int p_{Z,U}(z, u)du &\propto \int \cdots \int z^{\delta-1}e^{-\beta z}\prod_{i=1}^{k} I_{(e^{-z},1)}(u_i)du_1\cdots du_k \\
&= z^{\delta-1}e^{-\beta z}\prod_{i=1}^{k}\int I_{(e^{-z},1)}(u_i)du_i \\
&= z^{\delta-1}e^{-\beta z}\prod_{i=1}^{k}(1 - e^{-z}) \\
&= z^{\delta-1}e^{-\beta z}(1 - e^{-z})^k \quad \propto \quad p_Z(z)
\end{aligned}
$$

(c) The full conditional for U is

$$
p_{U|Z}(u|z) \propto p_{U,Z}(u, z) \propto \prod_{i=1}^{k} I_{(e^{-z},1)}(u_i) = \prod_{i=1}^{k} Unif(e^{-z}, 1) \,,
$$

i.e., conditionally independent uniform distributions. For Z, we have

$$
p_{Z|U}(z|u) \propto z^{\delta-1}e^{-\beta z}I_{(e^{-z},1)}(u_1)\cdots I_{(e^{-z},1)}(u_k) \,.
$$

Therefore, $e^{-z} < u_i$ for all i implies $z > -\log u_i$ for all i, which in turn implies $z > -\log u_{min}$ where $u_{min} = \min(u_1, \ldots, u_k)$. Hence,

$$
p_{Z|U}(z|u) \propto z^{\delta-1}e^{-\beta z}I_{(-\log u_{min}, \infty)}(z) \,,
$$

a *truncated gamma* distribution. Alternately sampling from $p_{U|Z}$ and

$p_{Z|U}$ produces a Gibbs sampler which converges to $p_{U,Z}$; peeling off the $Z^{(g)}$ draws after convergence yields the desired sample from the D-distribution p_Z. Special subroutines required to implement this algorithm would include a $Unif(0,1)$ generator, so that the u_i draws could be computed as $e^{-z} + (1 - e^{-z}) \cdot Unif(0,1)$, and a gamma cdf and inverse cdf routine, so that the z draws could be found using the truncated distribution formula given in Chapter 7, problem 8.

19.(b) Analytically, we have

$$f_x(x) = \int_0^\infty \left[\int_0^\infty y e^{-yx} t e^{-ty} dy \right] f_x(t) dt$$

$$= \int_0^\infty \left[\frac{t}{(x+t)^2} \right] f_x(t) dt .$$

$f_x(t) = 1/t$ is a solution to this equation. But no Gibbs convergence is possible here because this is not a density function; $\int_0^\infty f_x(t) dt = +\infty$. The trouble is that the complete conditionals must determine a *proper* joint density.

21.(a) $p(\theta_1|\theta_2, y)$ and $p(\theta_2|\theta_2, y)$ are

$$N \left(\frac{b_1^2(y - \theta_2) + a_1}{1 + b_1^2} , \frac{b_1^2}{1 + b_1^2} \right)$$

and

$$N \left(\frac{b_2^2(y - \theta_1) + a_2}{1 + b_2^2} , \frac{b_2^2}{1 + b_2^2} \right) ,$$

respectively.

(b) $p(\theta_1|y)$ and $p(\theta_2|y)$ are

$$N \left(\frac{b_1^2(y - a_2) + (1 + b_2^2)a_1}{1 + b_1^2 + b_2^2} , \frac{b_1^2(1 + b_2^2)}{1 + b_1^2 + b_2^2} \right)$$

and

$$N \left(\frac{b_2^2(y - a_1) + (1 + b_1^2)a_2}{1 + b_1^2 + b_2^2} , \frac{b_2^2(1 + b_1^2)}{1 + b_1^2 + b_2^2} \right) ,$$

respectively, so the data do indeed inform (i.e., there is "posterior learning").

(c) Answer provided in Figure C.2. Convergence is obtained immediately for μ, but never for θ_1 or θ_2. Note that the estimates of $E(\mu|y)$ at 100 and 1000 iterations are numerically identical.

(d) Answer provided in Figure C.3. Now we see very slow convergence for μ, due to the very slow convergence for the (barely identified) θ_1 and θ_2. The estimates of $E(\mu|y)$ at 100 and 1000 iterations are now appreciably different (with only the latter being correct), but there is no indication in the 100 iteration μ plot, G&R statistic, or lag 1 sample autocorrelation that stopping this early is unsafe.

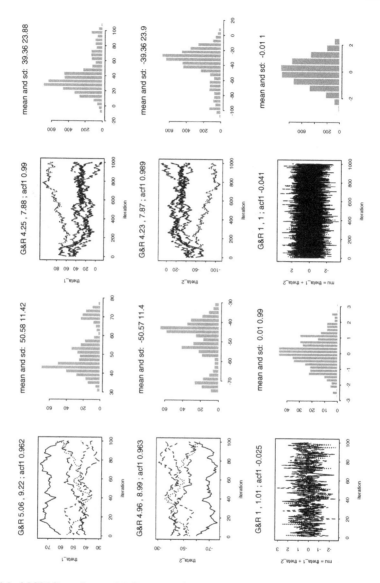

Figure C.2 *MCMC analysis of $N(\theta_1 + \theta_2, 1)$ likelihood, Case 1: Both parameters have prior mean 50 and prior standard deviation 1000.*

(e) This example provides evidence against the recommendation of several previous authors that when only a subset of a model's parameters are of interest, convergence of the remaining parameters need not be monitored. While completely unidentified parameters in an

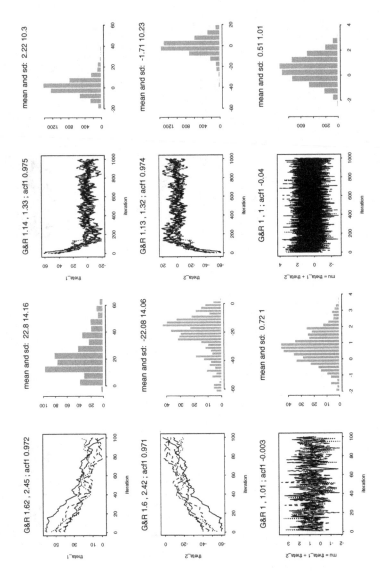

Figure C.3 *MCMC analysis of* $N(\theta_1 + \theta_2, 1)$ *likelihood, Case 2: Both parameters have prior mean 50 and prior standard deviation 10.*

overparametrized model may be safely ignored, those that are identified only by the prior must still be monitored, and slow convergence thereof must be remedied. As a result, monitoring a representative subset of *all* the parameters present is recommended, along with the use of simple reparametrizations where appropriate to speed overall convergence.

Chapter 4

1.(a) Sample WinBUGS code follows.

BUGS code
```
model
{
  for( i in 1:N ) {
    for( j in 1:T ) {
      HGB[i , j] ~ dnorm(mu[i , j],tau)
      mu[i , j] <- beta0[i] + beta1[i]*(X[j] - meanX)
    }
    beta0[i] ~ dnorm(mu0,tau0)
    beta1[i] ~ dnorm(mu1,tau1)
    junk[i] <- newarm[i]
  }
  sigma ~ dunif(0.001,1000)
  tau <- 1 /(sigma*sigma)
  mu0 ~ dnorm(0.0,1.0E-6)
  mu1 ~ dnorm(0.0,1.0E-6)
  sigma0 ~ dunif(0,100)
  sigma1 ~ dunif(0,100)
  tau0 <- 1/(sigma0*sigma0)
  tau1 <- 1/(sigma1*sigma1)
}
```

The apparently silly introduction of the junk variable keeps WinBUGS from objecting to the presence of newarm in the datafile. History plots show essentially immediate convergence; after discarding 100 burn-in samples, a further 5000 samples from posterior distributions of μ_0, μ_1, σ, σ_0, and σ_1 are shown in Table C.3.

The estimate for μ_1 is significant and positive and, in fact, most of the participants' fitted slopes $\mu_1 + \beta_{1i}$ are positive as well. However, as revealed by a caterpillar plot of the β_{1i} (not shown), many participants have significant probabilities of negative slopes. Turning to the intercept parameters, because the X variable is centered, they are interpreted as the height of trajectory in the *middle* of the observation period. Although all patients started the study with clinically defined anemia (HGB < 12 for males, < 11 for females), there is considerable variation in the posterior distributions for the β_{0i}, with posterior means ranging from approximately 9 (participant 115) to 14 (participant 287). Interestingly, neither of these participants had data past week 11, so these "intercepts" are already extrapolations to some extent.

Node	Mean	SD	MC error	2.5%	Median	97.5%
μ_0	11.6	0.0541	7.297E−04	11.49	11.6	11.7
μ_1	0.0907	0.0049	9.194E−05	0.0811	0.0907	0.1004
σ	0.9789	0.0111	1.377E−04	0.9576	0.9789	1.001
σ_0	0.9340	0.0402	6.103E−04	0.8568	0.9332	1.017
σ_1	0.0736	0.0042	8.844E−05	0.0656	0.0735	0.0819
HGB[10,5]	12.57	1.03	0.01	10.57	12.6	14.6
HGB[10,16]	14.3	1.06	0.0114	12.2	14.3	16.3
HGB[10,17]	14.4	1.08	0.0123	12.3	14.4	16.5
HGB[10,18]	14.6	1.08	0.0112	12.4	14.5	16.7
HGB[10,19]	14.7	1.12	0.0126	12.6	14.7	16.9
HGB[10,20]	14.9	1.13	0.0120	12.7	14.9	17.1
HGB[10,21]	15.0	1.16	0.0141	12.8	15.0	17.3
HGB[10,22]	15.2	1.18	0.0135	12.8	15.2	17.5
$\beta_{0,10}$	13.6	0.289	0.00375	13.0	13.6	14.1
$\beta_{1,10}$	0.152	0.0457	5.65E-4	0.0637	0.152	0.242

Table C.3 *Posterior summaries, basic model, HGB data.*

(b) Table C.3 also shows results for Patient 10, namely his fitted intercept $\beta_{0,10}$, slope $\beta_{1,10}$, and posterior predictive imputations for his missing HGB values. Note that the fitted HGB values are increasing slowly over time, consistent with this patient's positive fitted slope $\mu_1 + \beta_{1,10}$. The estimated standard deviations of the imputed values increase for the later weeks, after the participant was lost to follow-up; uncertainty increases as we move further from the bulk of the data.

(c) In the dataset, subjects with `newarm[i]` = 2 were assigned to the experimental protocol, while subjects with `newarm[i]` = 1 received the control treatment. In the `WinBUGS` code below, we subtract 1 from `newarm[i]` to ensure that treatment coefficients enter the model only for subjects in the experimental group. The revised models partition the intercept into a shared μ_0, a random effect centered around 0 ($\beta_{0,i}$), and a treatment component (γ_0). Similarly, the slope has grand component (μ_1), a random effect ($\beta_{1,i}$), and a treatment effect (γ_1).

BUGS code
```
model
{
  for( i in 1:N ) {
    for( j in 1:T ) {
      HGB[i , j] ~ dnorm(mu[i , j],tau)
      # Model 1
      mu[i , j] <- mu0 +beta0[i] +(mu1+beta1[i])*(X[j]-meanX)
```

```
                # Model 2
                + gamma0*(newarm[i]-1)
                # Model 3
                + gamma1*(newarm[i]-1)*(X[j] - meanX)
                # Model 4 contains both of the lines above
                }
                junk[i] <- newarm[i] + gamma0 + gamma1
                beta0[i] ~ dnorm(0, tau0)
                beta1[i] ~ dnorm(0, tau1)
            }
        sigma ~ dunif(0.001,1000)
        tau <- 1 /(sigma*sigma)
        mu0 ~ dflat()
        mu1 ~ dflat()
        gamma0 ~ dnorm(0, .0001)
        gamma1 ~ dnorm(0, .0001)
        sigma0 ~ dunif(0,100)
        sigma1 ~ dunif(0,100)
        tau0 <- 1/(sigma0*sigma0)
        tau1 <- 1/(sigma1*sigma1)
    }
```

Table C.4 shows results (again based on 5000 post-convergence samples) for Models 1-4, constructed by adding successively adding the treatment intercept (Model 2), treatment slope (Model 3), or both (Model 4) to the base model (Model 1). Both treatment intercept (γ_0) and slope (γ_1) emerge as significantly negative. A difference in intercept would seem to indicate differences in the treatment groups at randomization, but again, the centering of X means only a difference in the middle of the study. The plot of observed and fitted (fixed effects only) HGB trajectories for the two treatment groups in Figure C.4 clarifies this distinction: the differences arising at week 11.5 are indeed due to the difference in slopes; there is no appreciable difference in HGB at baseline. Finally, DIC can be used to compare these models, but the results are not particularly enlightening; neither the fit ($\overline{D}$) nor the effective size (p_D) is much affected by the inclusion or elimination of the treatment group parameters.

2.(a) For the nonlinear model with independent a_i and b_i, we obtained slightly smaller point estimates of $Y_{2,7}$ and $Y_{2,8}$, but the differences were no more than 0.02.

(b) One possible WinBUGS solution is given on the book's homepage, www.biostat.umn.edu/~brad/data.html, and reproduced here:

BUGS code
```
model
{
    for (i in 1:N) {
```

Node	Mean	SD	MC error	2.5%	Median	97.5%
Model 1: No treatment effect						
μ_0	11.6	0.0553	2.444E−03	11.49	11.6	11.71
μ_1	0.0909	0.0049	1.291E−04	0.0816	0.0908	0.1006
σ	0.979	0.0112	1.312E−04	0.9571	0.9785	1.001
σ_0	0.934	0.0407	5.765E−04	0.8573	0.9335	1.016
σ_1	0.0737	0.0042	8.468E−05	0.0658	0.0736	0.0820
Model 2: Treatment intercept						
γ_0	−0.252	0.112	5.431E−03	−0.4717	−0.2538	−0.0364
μ_0	11.7	0.0757	3.670E−03	11.58	11.72	11.87
μ_1	0.0906	0.0050	1.435E−04	0.0808	0.0906	0.1004
σ	0.979	0.0114	1.338E−04	0.9569	0.9789	1.002
σ_0	0.926	0.0411	6.149E−04	0.8497	0.9251	1.011
σ_1	0.0737	0.0042	8.936E−05	0.0660	0.0736	0.0822
Model 3: Treatment slope						
γ_1	−0.0196	0.0096	3.000E−04	−0.0386	−0.0195	−0.0009
μ_0	11.6	0.0527	2.296E−03	11.49	11.6	11.7
μ_1	0.1	0.0068	2.239E−04	0.0870	0.1003	0.1135
σ	0.979	0.0112	1.381E−04	0.9569	0.9789	1.002
σ_0	0.936	0.041	5.371E−04	0.8576	0.9352	1.02
σ_1	0.073	0.0042	9.671E−05	0.0652	0.0730	0.0815
Model 4: Treatment slope and intercept						
γ_0	−0.294	0.106	5.017E−03	−0.5018	−0.2961	−0.0900
γ_1	−0.0234	0.0096	2.499E−04	−0.0423	−0.0234	−0.0044
μ_0	11.7	0.0736	3.480E−03	11.59	11.74	11.88
μ_1	0.102	0.0066	1.777E−04	0.0890	0.1019	0.1151
σ	0.979	0.0111	1.366E−04	0.958	0.9788	1.001
σ_0	0.925	0.0405	6.056E−04	0.8492	0.9244	1.008
σ_1	0.0731	0.0042	8.702E−05	0.0652	0.0730	0.0817

Table C.4 *Posterior summaries, enhanced models, HGB data.*

```
for (j in 1:T) {
  Z[i,j] ~ dnorm(mu[i,j],tau[i])
  Y[i,j] <- exp(Z[i,j])
  mu[i,j] <- log(30) - theta[i,1]
                - exp(theta[i,2]-theta[i,1])*X[j]
  } # end of j loop
```

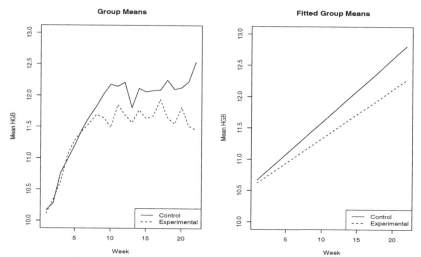

Figure C.4 *Observed and fitted group means, HGB data.*

```
theta[i , 1:2]  ~ dmnorm (thetamu[1:2] , Omega[1:2, 1:2])
beta[i,1] <- log(30) - theta[i,1]
beta[i,2] <- -exp(theta[i,2] - theta[i,1])

tau[i]   <- 1/ ( sigma[i] * sigma[i])
sigma[i] ~dunif(0.001,100)
} # end of i loop

thetamu[1:2]  ~ dmnorm(meen[1:2], prec[1:2, 1:2])
Omega[1:2 , 1:2]  ~ dwish(R[1:2, 1:2], 2)
beta.s2[1:2, 1:2] <- inverse(Omega[1:2, 1:2])
for (i in 1 : 2) {beta.s[i] <- sqrt(beta.s2[i, i]) }

} # end of nonlinear PK program
```

The data are the same as given in Example 2.13. We use three different sets of starting values,

BUGS code
```
list(sigma = c(.5,.5,.5,.5,.5,.5,.5,.5,.5,.5),
     thetamu=c(-5,-5),
     Omega=structure(.Data = c(0.1, 0, 0, 0.1), .Dim=c(2,2)))
list(sigma= c(1,1,1,1,1,1,1,1,1,1), thetamu=c(0,0),
     Omega=structure(.Data = c(0.1, 0, 0, 0.1), .Dim=c(2,2)))
list(sigma= c(2,2,2,2,2,2,2,2,2,2), thetamu=c(5,5),
     Omega=structure(.Data = c(1, 0, 0, 1), .Dim=c(2,2)))
```

For this model, we obtained model choice statistics of $\overline{D} = -66.1, p_D =$

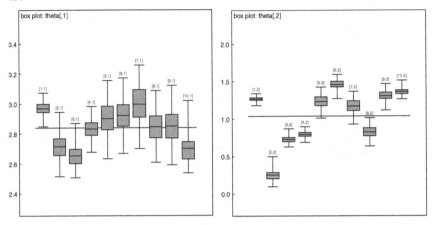

Figure C.5 *Boxplots of individual-specific log-volume parameters θ_{0i} (left) and log-clearance parameters θ_{1i} (right), nonlinear PK model, Cadralazine data.*

14.7, and DIC $= -51.4$. By contrast, for the linear model we obtained $\overline{D} = -59.3, p_D = 18.0$, and DIC $= -41.3$. Thus, the nonlinear model appears to fit better and uses slightly fewer effective parameters, translating to a fairly significant victory in terms of DIC. (Note that the new model's use of individual-specific error variances σ_i^2 may also be playing a role in the model's improved fit.) A bivariate scatterplot of the μ_i samples (obtained by applying the `Correlation` tool to the `thetamu` node above) reveals a weak positive correlation near 0.4, empirical evidence that the log-volume and log-clearance parameters are *not* independent.

Finally, boxplots of the θ_{0i} and θ_{1i} samples across the patients are shown in Figure C.5. The former samples do not suggest large differences across patients, while the latter do, with Patient 2 again emerging with a significantly slower drug clearance than the other patients. Boxplots of the β_{0i} and β_{1i} produced by this code (obtained by applying the `Comparison` tool to the `beta[,1]` and `beta[,2]` nodes above) are similar but slightly narrower than those produced by the linearized model and previously shown in Figure 2.20.

5.(a) Sample `WinBUGS` code is as follows:

BUGS code
```
model {
    for (i in 1:N) {
      Y[i] ~ dbern(p[i])
      logit(p[i]) <- beta0 + beta1 * (X[i] - mean(X[]))
    }
    beta0 ~ dflat()
    beta1 ~ dflat()
```

Node	Mean	SD	MC error	2.5%	Median	97.5%
β_0	-0.4046	0.09048	2.978E-4	-0.5821	-0.4043	-0.2275
β_1	-1.720	0.1975	6.831E-4	-2.115	-1.716	-1.341

Table C.5 *WinBUGS posterior summaries, logit model, forest co-presence data.*

```
}

# Inits
list(beta0 = 0.0, beta1 = -1)
list(beta0 = -0.5, beta1 = -2))
```

Results from 50,000 samples after a 500 sample burn-in period are shown in Table C.5. We can interpret the β coefficients on the log odds scale; exponentiating yields odds ratios. The 95% posterior credible interval for β_1 is strongly separated from 0, and indicates that with increasing distance from the edge, the odds of co-presence decrease.

(b) This Metropolis-Hastings algorithm uses a (bivariate) normal proposal distribution centered at the current value of the chain, and flat priors for the β parameters. The likelihood is simply binomial, and we may ignore the contribution of the proposal to the r ratio, because it is symmetric (standard Metropolis). Sample R code follows:

R code
```
mh <- function(data,NITER,init,nu) {
  Y <- data[,'Copresence']
  X <- data[,'LogEdgeDistance']-mean(data[,'LogEdgeDistance'])
  X <- cbind(rep(1,times=length(Y)),X)

  Beta <- matrix(NA,NITER,3)
  Beta[1,] <- c(init,0)

  ## Unnormalized log binomial likelihood
  lpost <- function(Y,pi) {
    return ( sum(Y*log(pi)+(1-Y)*log(1-pi)) )
  }

  for (i in 2:NITER) {
    Beta.star <- mvrnorm(1,Beta[i-1,1:2], nu^2*diag(2))
    pi.star <- exp(X%*%Beta.star) /
      (1 + exp(X%*%Beta.star) )
    pi.last <- exp(X%*%Beta[i-1,1:2]) /
      (1 + exp(X%*%Beta[i-1,1:2]))
    log.P.Beta.star <- lpost(Y,pi.star)
    log.P.Beta.last <- lpost(Y,pi.last)
```

Node	Mean	SD	2.5%	Median	97.5%
β_0	−0.4042	0.0907	−0.5827	−0.4040	−0.2265
β_1	−1.7197	0.1976	−2.1160	−1.7169	−1.3379

Table C.6 *R Metropolis-Hastings posterior summaries, logit model, forest co-presence data.*

```
r <- min(exp(log.P.Beta.star - log.P.Beta.last),1)
Beta[i,3] <- r

if (r >= 1){
  Beta[i,1:2] <- Beta.star
  } else {
  u <- runif(1,0,1)
  if (u <= r) {
    Beta[i,1:2] <- Beta.star
  } else {
    Beta[i,1:2] <- Beta[i-1,1:2]
  }
 }
}
  return (Beta)
}

chain1 <- mh(data,NITER=10000, init=c(0.0,-1),nu=0.19)
chain2 <- mh(data,NITER=10000, init=c(-0.5,-2),nu=0.19)
```

Clearly, the estimates and credible intervals arising from this algorithm (and shown in Table C.6) are very similar to the WinBUGS results in Table C.5. Recall that rejection ratios near 0.20 for block updating or 0.40 for scalar parameters often lead to a reasonably good performance; here we obtained 0.39 as the mean of r.

(c) The WinBUGS code simply replaces logit with cloglog in the code for part (a). Results are shown in Table C.7; β_1 has changed due to the new model, but is still significant and negative. The p_D and DIC values in Table C.8 are very similar, indicating that the choice of link function is not crucial for these data. This is confirmed by the very similar visual appearance of the two fitted models in Figure C.6.

6. Standard hierarchical Bayes calculations yield the following complete conditional distributions:

$$\theta|\boldsymbol{\beta}, \boldsymbol{\lambda}, \sigma^2, \mathbf{y} \quad \propto \quad \exp\left\{-\frac{1}{2\sigma^2}(\mathbf{y} - F_\theta\boldsymbol{\beta})^T\Sigma_i^{-1}(\mathbf{y} - F_\theta\boldsymbol{\beta})\right\}p(\theta)$$

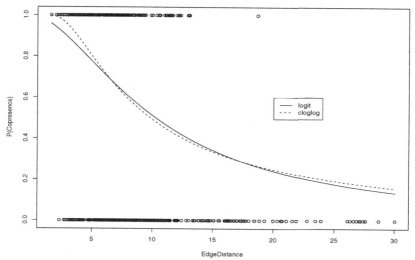

Figure C.6 *Fitted models, forest co-presence data.*

Node	Mean	SD	MC error	2.5%	Median	97.5%
β_0	−0.7149	0.0713	2.859E-4	−0.8569	−0.7139	−0.5776
β_1	−1.25	0.1357	5.32E-4	−1.517	−1.249	−0.9852

Table C.7 *WinBUGS posterior summaries, cloglog model, forest co-presence data.*

$$
\begin{aligned}
\boldsymbol{\beta}|\theta,\boldsymbol{\lambda},\sigma^2,\mathbf{y} \quad &\sim \quad N(Bb, B), \\
\text{where } B^{-1} \quad &= \quad \frac{1}{\sigma^2}F_\theta^T\Sigma_i^{-1}F_\theta + \Sigma_0^{-1} \\
\text{and } b \quad &= \quad \frac{1}{\sigma^2}F_\theta^T\Sigma_i^{-1}\mathbf{y} + \Sigma_0^{-1}\boldsymbol{\beta}_0 \\
\sigma^2|\theta,\boldsymbol{\beta},\boldsymbol{\lambda},\mathbf{y} \quad &\sim \quad IG\left(a_0 + \frac{n}{2},\right.\\
&\qquad \left.\left\{b_0^{-1} + \frac{1}{2}\sum_{i=1}^{n}(y_i - f(\mathbf{x}_i,\theta)\boldsymbol{\beta})^2/\lambda_i\right\}^{-1}\right).
\end{aligned}
$$

Including the λ_i complete conditionals given in Example 4.6, all of these are available in closed form with the exception of that of θ, for which one can use a rejection sampling method (e.g., Metropolis-Hastings).

Model	$\overline{D}$	$\hat{D}$	p_D	DIC
logit	726.414	724.418	1.995	728.409
cloglog	727.567	725.561	2.006	729.574

Table C.8 *DIC comparison of logit and cloglog models, forest co-presence data.*

8. The posterior probability of H_0, $P_G(\theta \in [\theta_L, \theta_U]|x)$, is given by

$$\frac{\int_{\theta_L}^{\theta_U} f(x|\theta) \left[\frac{\pi}{\theta_U - \theta_L} I_{[\theta_L, \theta_U]}(\theta) + (1 - \pi)g(\theta) \right] d\theta}{\int f(x|u) \left[\frac{\pi}{\theta_U - \theta_L} I_{[\theta_L, \theta_U]}(u) + (1 - \pi)g(u) \right] du}.$$

Setting this expression $\leq p$ and keeping in mind that $g(\theta)$ has no support on the interval $[\theta_L, \theta_U]$, we obtain $\mathcal{H}_c$ as the set of all G such that

$$\int f(x|\theta)dG(\theta) \geq c = \frac{1-p}{p} \frac{\pi}{1-\pi} \frac{1}{\theta_U - \theta_L} \int_{\theta_L}^{\theta_U} f(x|\theta)d\theta.$$

9. While the faculty member's solution is mathematically correct, it is *not* practical because we lack a closed form for the ratio in the expression (i.e., $p(\boldsymbol{\theta}|\mathbf{y}_{(i)})$ and $p(\boldsymbol{\theta}|\mathbf{y})$ have different and unknown normalizing constants). A conceptually simple alternative would be to run the MCMC algorithm again after deleting y_i, obtaining $\boldsymbol{\theta}^{(g)}$ samples from $p(\boldsymbol{\theta}|\mathbf{y}_{(i)})$ and subsequently obtaining the conditional predictive mean as

$$E(y_i|\mathbf{y}_{(i)}) = \int E(y_i|\boldsymbol{\theta})p(\boldsymbol{\theta}|\mathbf{y}_{(i)})d\boldsymbol{\theta}$$

$$\approx \frac{1}{G} \sum_{g=1}^{G} E(y_i|\boldsymbol{\theta}^{(g)}).$$

A more sophisticated approach (suggested by a more sophisticated faculty member, Prof. Hal Stern of the University of California, Irvine) would be to write

$$\int E(y_i|\boldsymbol{\theta})p(\boldsymbol{\theta}|\mathbf{y}_{(i)})d\boldsymbol{\theta} = \frac{\int E(y_i|\boldsymbol{\theta})f(\mathbf{y}_{(i)}|\boldsymbol{\theta})\pi(\boldsymbol{\theta})d\boldsymbol{\theta}}{\int f(\mathbf{y}_{(i)}|\boldsymbol{\theta})\pi(\boldsymbol{\theta})d\boldsymbol{\theta}}$$

$$= \frac{\int E(y_i|\boldsymbol{\theta}) \frac{f(\mathbf{y}_{(i)}|\boldsymbol{\theta})m(\mathbf{y})}{f(\mathbf{y}|\boldsymbol{\theta})\pi(\boldsymbol{\theta})}\pi(\boldsymbol{\theta})p(\boldsymbol{\theta}|\mathbf{y})d\boldsymbol{\theta}}{\int \frac{f(\mathbf{y}_{(i)}|\boldsymbol{\theta})m(\mathbf{y})}{f(\mathbf{y}|\boldsymbol{\theta})\pi(\boldsymbol{\theta})}\pi(\boldsymbol{\theta})p(\boldsymbol{\theta}|\mathbf{y})d\boldsymbol{\theta}}$$

$$\approx \frac{\sum_{g=1}^{G} E(y_i|\boldsymbol{\theta}^{(g)}) \frac{f(\mathbf{y}_{(i)}|\boldsymbol{\theta}^{(g)})}{f(\mathbf{y}|\boldsymbol{\theta}^{(g)})}}{\sum_{g=1}^{G} \frac{f(\mathbf{y}_{(i)}|\boldsymbol{\theta}^{(g)})}{f(\mathbf{y}|\boldsymbol{\theta}^{(g)})}},$$

where the $\boldsymbol{\theta}^{(g)}$ samples now come directly from the full posterior $p(\boldsymbol{\theta}|\mathbf{y})$, which plays the role of the importance sampling density g in (3.4). In the second line above, the prior terms $\pi(\boldsymbol{\theta})$ cancel in both the numerator and denominator, while the marginal density terms $m(\mathbf{y})$ can be pulled through both integrals and cancelled, since they are free of $\boldsymbol{\theta}$. Note that this solution will be accurate provided $p(\boldsymbol{\theta}|\mathbf{y}) \approx p(\boldsymbol{\theta}|\mathbf{y}_{(i)})$, which should hold unless the dataset is quite small or y_i is an extreme outlier.

11. We have

$$
\begin{aligned}
f(y_i|\mathbf{y}_{(i)}) &= \int f(y_i|\boldsymbol{\theta}, \mathbf{y}_{(i)})p(\boldsymbol{\theta}|\mathbf{y}_{(i)})d\boldsymbol{\theta} \\
&= \int f(y_i|\boldsymbol{\theta})p(\boldsymbol{\theta}|\mathbf{y}_{(i)})d\boldsymbol{\theta} \\
&\approx \int f(y_i|\boldsymbol{\theta})p(\boldsymbol{\theta}|\mathbf{y})d\boldsymbol{\theta} \\
&\approx \frac{1}{G}\sum_{g=1}^{G} f(y_i|\boldsymbol{\theta}^{(g)}) \,,
\end{aligned}
$$

where $\boldsymbol{\theta}^{(g)}$ are the MCMC samples from $p(\boldsymbol{\theta}|\mathbf{y})$. Because the CPO is the conditional likelihood, small values indicate poor model fit. Thus, we might classify those datapoints having CPO values in the bottom 5 to 10% as potential outliers meriting further investigation.

12. We have

$$
\begin{aligned}
p'_D &= \int P[D(\mathbf{y}^*, \boldsymbol{\theta}) > D(\mathbf{y}, \boldsymbol{\theta})|\boldsymbol{\theta}]\, p(\boldsymbol{\theta}|\mathbf{z})d\boldsymbol{\theta} \\
&= \int\int I_{\{D(\mathbf{y}^*,\boldsymbol{\theta})>D(\mathbf{y},\boldsymbol{\theta})\}}(\mathbf{y}^*, \boldsymbol{\theta})f(\mathbf{y}|\boldsymbol{\theta})p(\boldsymbol{\theta}|\mathbf{z})d\mathbf{y}d\boldsymbol{\theta} \\
&\approx \frac{1}{G}\sum_{g=1}^{G} I_{\{D(\mathbf{y}^{*(g)},\boldsymbol{\theta}^{(g)})>D(\mathbf{y},\boldsymbol{\theta}^{(g)})\}}(\mathbf{y}^{*(g)}, \boldsymbol{\theta}^{(g)}) \,,
\end{aligned}
$$

where $(\mathbf{y}^{*(g)}, \boldsymbol{\theta}^{(g)}) \sim p(\mathbf{y}^*, \boldsymbol{\theta}|\mathbf{z}) = f(\mathbf{y}^*|\boldsymbol{\theta})p(\boldsymbol{\theta}|\mathbf{z})$, and I_S denotes the indicator function of the set S. That is,

$$
\hat{p}'_D = \frac{\text{number of } D(\mathbf{y}^{*(g)}, \boldsymbol{\theta}^{(g)})s > D(\mathbf{y}, \boldsymbol{\theta}^{(g)})}{G}.
$$

The sampler may not have been designed to produce predictive samples $\mathbf{y}^{*(g)}$, but such samples are readily available at iteration g by drawing from $f(\mathbf{y}^*|\boldsymbol{\theta}^{(g)})$, a density that is available in closed form by construction.

14. For each method, we run five independent chains for 50,000 iterations following a 10,000 iteration burn-in period. Table C.9 summarizes our results, reporting the estimated posterior probability for model 2, a

Method	$\hat{P}(M=2|\mathbf{y})$	SD	$\widehat{BF}_{21}$	$\hat{\rho}(1)$	Time
CC	.70806	.001721	4848.4	.567	22.8"
RJ-M	.70861	.004058	4861.3	.589	18.7"
RJ-G	.70906	.002394	4871.9	.593	7.9"
RJ-R	.70750	.002004	4835.1	.660	6.7"
Chib-1			4860.7		13.6"
Chib-2			4860.3		14.0"
target	.70865		4862		

Table C.9 *Comparison of different methods for the simple linear regression example. Here, CC = Carlin and Chib's method; RJ-M = reversible jump using Metropolis steps if the current model is proposed; RJ-G = reversible jump using Gibbs steps if the current model is proposed; RJ-R = reversible jump on the reduced model (i.e., with the regression coefficients integrated out); Chib-1 = Chib's method evaluated at posterior means; and Chib-2 = Chib's method evaluated at frequentist LS solutions.*

batched standard deviation estimate for this probability (using 2500 batches of 100 consecutive iterations), the Bayes factor in favor of model 2, $\widehat{BF}_{21}$, the lag 1 sample autocorrelation for the model indicator, $\hat{\rho}(1)$, and the execution time for the FORTRAN program in seconds. Model 2 is clearly preferred, as indicated by the huge Bayes factor. The "target" probability and Bayes factor are as computed using traditional (non-Monte Carlo) numerical integration by Green and O'Hagan (1998). Programming notes for the various methods are as follows:

(a) For Chib's (1995) marginal density method, we may use a two-block Gibbs sampler that treats σ^2 or τ^2 as θ_1, and $(\alpha, \beta)'$ or $(\gamma, \delta)'$ as θ_2. We experimented with two different evaluation points (θ_1', θ_2'); see the table caption.

(b) For Carlin and Chib's (1995) product space search, we have $\theta_1 = (\alpha, \beta, \sigma)$ and $\theta_2 = (\gamma, \delta, \tau)$. For the regression parameters, one may use independent univariate normal pseudopriors that roughly equal the corresponding first-order approximation to the posterior. Acceptable choices are $\alpha|(M=2) \sim N(3000, 52^2)$, $\beta|(M=2) \sim N(185, 12^2)$, $\gamma|(M=1) \sim N(3000, 43^2)$, and $\delta|(M=1) \sim N(185, 9^2)$. For σ^2 and τ^2, taking the pseudopriors equal to the IG priors produces acceptable results. A bit of tinkering reveals that $\pi_1 = .9995$, $\pi_2 = .0005$ produces $M^{(g)}$ iterates of 1 and 2 in roughly equal proportion.

(c) For Green's (1995) reversible jump algorithm, we use log transforms of the error variances, i.e., $\lambda = \log \sigma^2$, $\omega = \log \tau^2$, to simplify the choice of proposal density. The probabilities of proposing new models are $h(1,1) = h(1,2) = h(2,1) = h(2,2) = 0.5$. The dimension matching requirement is automatically satisfied without generating an additional random vector; moreover, the the Jacobian term in the acceptance probability is equal to 1. Because the two regression models are similar in interpretation, when a move between models is proposed, we simply set $(\alpha, \beta, \lambda)' = (\gamma, \delta, \omega)'$ or $(\gamma, \delta, \omega)' = (\alpha, \beta, \lambda)'$. The acceptance probabilities are then given by $\alpha_{1\to2} = \min\left\{1, \frac{f(\mathbf{y}|\gamma,\delta,\omega,M=2)\pi_2}{f(\mathbf{y}|\alpha,\beta,\lambda,M=1)\pi_1}\right\}$, and $\alpha_{2\to1} = \min\left\{1, \frac{f(\mathbf{y}|\alpha,\beta,\lambda,M=1)\pi_1}{f(\mathbf{y}|\gamma,\delta,\omega,M=2)\pi_2}\right\}$.

When the proposed model is the same as the current model, we update using a standard current-point Metropolis step. That is, for model 1, we draw $(\alpha^*, \beta^*, \lambda^*)' \sim N\left((\alpha^{(k)}, \beta^{(k)}, \lambda^{(k)})', Diag(5000, 250, 1)\right)$ and set $\alpha^{(k+1)}, \beta^{(k+1)}, \lambda^{(k+1)})' = (\alpha^*, \beta^*, \lambda^*)'$ with probability $r = p(\alpha^*, \beta^*, \lambda^*)/p(\alpha^{(k)}, \beta^{(k)}, \lambda^{(k)})$. Similar results hold for model 2. Alternatively, we may perform the "within model" updates using a standard Gibbs step, thus avoiding the log transform and the trivariate normal proposal density, and simply using instead the corresponding full conditional distributions.

Finally, we may also use the reversible jump method on a somewhat reduced model, i.e., we analytically integrate the slopes and intercepts (α, β, γ, and δ) out of the model and use proposal densities for σ^2 and τ^2 that are identical to the corresponding priors. The acceptance probabilities are $\alpha_{1\to2} = \min\left\{1, \frac{\pi_2}{\pi_1}\frac{f(\mathbf{y}|\tau^2,M=2)}{f(\mathbf{y}|\sigma^2,M=1)}\right\}$ and $\alpha_{2\to1} = \min\left\{1, \frac{\pi_1}{\pi_2}\frac{f(\mathbf{y}|\sigma^2,M=1)}{f(\mathbf{y}|\tau^2,M=2)}\right\}$.

Summarizing Table C.9, the reversible jump algorithm operating on the reduced model (with the regression coefficients integrated out) appears to be slightly more accurate and faster than the corresponding algorithms operating on the full model. Of course, some extra effort is required to do the integration before programming, and posterior samples are obviously not produced for any parameters no longer appearing in the sampling order. The marginal density (Chib) method does not require preliminary runs (only a point of high posterior density, $\boldsymbol{\theta}'$), and only a rearrangement of existing computer code. The accuracy of results produced by this method is more difficult to assess, but is apparently higher for the given runsize, since repeated runs produced estimates for $\widehat{BF}_{21}$ consistently closer to the target than any of the other methods.

Chapter 5

1.(a) For $f(y|\theta) = N(y|\theta, 1)$, we have that $\theta = y + f'(x|\theta)/f(x|\theta)$. The result follows from substituting this expression for θ_i in the numerator of the basic posterior mean formula,

$$E(\theta_i|y_i) = \frac{\int \theta_i f(x_i|\theta_i)dG(\theta_i)}{\int f(x_i|\theta_i)dG(\theta_i)} \ ,$$

and simplifying.

(*Note:* The solution to a more general version of this problem is given by Maritz and Lwin, 1989, p. 73.)

2. The computational significance of this result is that we may find the joint marginal density of $\mathbf{y}$ (formerly a k-dimensional integral with respect to $\boldsymbol{\theta}$) as the product over i of $m(y_i)$, the result of a one-dimensional integral. Because $m(\mathbf{y}|\eta)$ must emerge in closed form for computation of a marginal MLE $\hat\eta$ to be feasible, this reduction in dimensionality plays an important role in PEB analysis.

3.(a) $m(y_i|\eta) = \Gamma(\alpha + \eta)y_i^{\alpha-1}/[\Gamma(\alpha)\Gamma(\eta)(y_i + 1)^{\alpha+\eta}]$

(b) With $\alpha = 2$,

$$L(\eta|\mathbf{y}) \propto \prod_{i=1}^{k} \frac{\eta(\eta + 1)y_i}{(y_i + 1)^{\eta+2}} \ ,$$

Taking the derivative of the log-likelihood with respect to η and setting the result equal to 0, we have that the marginal MLE $\hat\eta$ is a positive root of the quadratic equation

$$S\eta^2 + (S - 2k)\eta - k = 0 \ ,$$

where $S = \sum_{i=1}^{k} \log(y_i + 1)$.

4.(a) The marginal loglikelihood for B is proportional to

$$\frac{k}{2} \log B - \frac{B}{2} \sum_{i=1}^{k} y_i^2 \ .$$

Taking the derivative with respect to B, setting equal to 0, solving, and remembering the restriction that $0 \le B \le 1$, we have that

$$\widehat{B}_{MLE} = \min\left(\frac{k}{\sum_{i=1}^{k} y_i^2} , 1\right) \ .$$

The resulting PEB point estimates are $\hat\theta_i^{EB} = (1 - \widehat{B}_{MLE})Y_i$.

(b) Since $\sum_{i=1}^{k}(\sqrt{B}Y_i)^2 = B||\mathbf{Y}||^2 \sim \chi^2(k) = G(k/2, 2)$, it follows that $||\mathbf{Y}||^2 \sim G(k/2, 2/B)$, and hence that $1/(||\mathbf{Y}||^2) \sim IG(k/2, 2/B)$. Thus, we can compute

$$E(\widehat{B}) = E\left(\frac{k - 2}{||\mathbf{Y}||^2}\right) = \frac{k - 2}{\frac{2}{B}\left(\frac{k}{2} - 1\right)} = \frac{B(k - 2)}{k - 2} = B \ ,$$

and so $\widehat{B}$ is indeed unbiased for B. But then

$$\hat{\theta}_i^{EB} = (1 - \widehat{B})Y_i = \left(1 - \frac{k-2}{||\mathbf{Y}||^2}\right) Y_i = \hat{\theta}_i^{JS} .$$

6. We need to compute

$$E_{\mathbf{Y}|\theta}\left[\frac{1}{||\mathbf{Y}||^2}\right] .$$

Consider the case where $\sigma^2 = 1$ and replace $\mathbf{Y}$ by $\mathbf{Z}$; the general case follows by writing $\mathbf{Y} = \sigma\mathbf{Z}$. Now $||\mathbf{Z}||^2$ is distributed as a noncentral chi-square on k df with noncentrality parameter $\gamma = ||\boldsymbol{\theta}||^2 /2$. But a noncentral chi-square can be represented as a discrete mixture of central chi-squares, i.e., $\chi^2(k + 2J)$ where $J \sim Po(\gamma)$. Therefore,

$$E_{\mathbf{Z}|\theta}\left[\frac{1}{||\mathbf{Z}||^2}\right] = E_{\mathbf{Z}|\gamma}\left[\frac{1}{||\mathbf{Z}||^2}\right]$$

$$= \sum_{j=0}^{\infty} \frac{\gamma^j}{j!}e^{-\gamma}\frac{1}{k+2j-2} ,$$

because $1/\chi^2(k + 2J) \equiv IG((k + 2j)/2, 2)$ (see Appendix Section A.2).

9. The population moments in the marginal family $m(r_i|\alpha, \beta)$ are

$$E(r_i) = \frac{1}{t_i}E(Y_i) = \frac{1}{t_i}E(E(Y_i|\theta_i)) = \frac{1}{t_i}E(t_i\theta_i) = E(\theta_i) = \alpha\beta$$

and

$$\begin{aligned}
Var(r_i) &= \tfrac{1}{t_i^2}Var(Y_i) \\
&= \tfrac{1}{t_i^2}[Var(E(Y_i|\theta_i)) + E(Var(Y_i|\theta_i))] \\
&= \tfrac{1}{t_i^2}[Var(t_i\theta_i) + E(t_i\theta_i)] \\
&= \tfrac{1}{t_i^2}[t_i^2(\alpha\beta^2) + t_i(\alpha\beta)] = \alpha\beta^2 + t_i^{-1}\alpha\beta .
\end{aligned}$$

This results in the system of two equations and two unknowns

$$\bar{r} = \alpha\beta$$
$$s_r^2 = \alpha\beta^2 + \alpha\beta \sum_{i=1}^{k} t_i^{-1}/k$$

Solving, we obtain $\hat{\alpha} = \bar{r}^2/(s_r^2 - \bar{r}\sum_{i=1}^{k} t_i^{-1}/k)$ and $\hat{\beta} = \bar{r}/\hat{\alpha}$.

14.(a) The MLE is equal to the Bayes rule when $M = 0$, and carries a risk (MSE) of $E_{X|\theta}(\frac{X}{n} - \theta)^2 = Var_{X|\theta}(\frac{X}{n}) = \theta(1 - \theta)/n$. Subtracting the risk of the Bayes estimate given in equation (5.41) and setting $\mu = \frac{1}{2}$, we have risk improvement by the Bayes estimate if and only if

$$\frac{\theta(1-\theta)}{n}\left[1 - \left(\frac{n}{M+n}\right)^2\right] - \left(\frac{M}{M+n}\right)^2\left(\theta - \frac{1}{2}\right)^2 \geq 0 . \quad \text{(C.4)}$$

The two θ values for which this expression equals 0 (available from the quadratic formula) are the lower and upper bounds of the risk improvement interval. Clearly this interval will be symmetric about $\theta = \frac{1}{2}$, since there the first term in (C.4) is maximized while the second is minimized (equal to 0).

(b) Because $\theta|x \sim Beta(a + x, b + n - x)$, we have

$$
\begin{aligned}
\hat{\theta}_{Bayes} &= \frac{\int \frac{\theta}{\theta(1-\theta)} \theta^{a+x-1}(1-\theta)^{b+n-x-1} d\theta}{\int \frac{1}{\theta(1-\theta)} \theta^{a+x-1}(1-\theta)^{b+n-x-1} d\theta} \\
&= \frac{\int \theta^{a+x-1}(1-\theta)^{b+n-x-2} d\theta}{\int \theta^{a+x-2}(1-\theta)^{b+n-x-2} d\theta} \\
&= \frac{\frac{\Gamma(a+x)\Gamma(b+n-x-1)}{\Gamma(a+b+n-1)}}{\frac{\Gamma(a+x-1)\Gamma(b+n-x-1)}{\Gamma(a+b+n-2)}} \\
&= \frac{a + x - 1}{a + b + n - 1}
\end{aligned}
$$

from the result of Appendix B, problem 1.

Chapter 6

1.(b) This is actually the prior used by Müller and Parmigiani (1995). Using the "shortcut" method (sampling both θ_j^* and y_j^* with $N = 1$), they obtain $\tilde{n} = 29$.

Even this mixture prior *is* amenable to a fully analytical solution, since the posterior distribution is still available as a mixture of beta densities. The authors then obtain the refined estimate $\tilde{n} = 34$.

4. Following the notation of Subsection 6.2.2, we let $\mathbf{a}^T = (1, -1)$, $\boldsymbol{\xi}^T = (\mu_C, \mu_X)$, and $\beta = \mathbf{a}^T \boldsymbol{\xi}$. Because we are considering only efficacy and not cost, we have $\bar{\mathbf{y}} = (\bar{e}_C, \bar{e}_X)'$, and the sampling variance matrix is

$$
S = \begin{pmatrix} \frac{\sigma^2}{n} & 0 \\ 0 & \frac{\sigma^2}{n} \end{pmatrix}.
$$

(a) Solving expression (6.14) for n, we obtain

$$
n \geq \frac{(1.645 + 0.84)^2 2(4.5^2)}{(\beta^*)^2},
$$

where $\beta_d = \mathbf{a}'\mathbf{m}_d$ is the proposed true treatment effect. For the values of β_d given in the problem, we obtain the sample sizes shown in Table C.10.

Alternatively, we can adopt a simulation-based approach. First, given the parameter value specified by the design (point) prior, we draw from the likelihood to generate M sets of simulated data. We then analyze each set using the appropriate analysis prior and tally the

β_d	n
5	11
10	3
15	2
20	1

Table C.10 *Sample sizes n required for fixed values of β_d.*

number of times that the posterior credible interval excludes the null value. Finally, we use the proportion of null rejections divided by M as the estimated assurance, $\hat{\delta}$, shown for the relevant n and β^* values in Table C.11.

			n			
β_d	1	2	3	5	10	11
5	0.19	0.29	0.40	0.54	0.82	0.82
10	0.49	0.72	0.88	0.97	1.00	1.00
15	0.76	0.96	0.99	1.00	1.00	1.00
20	0.93	0.99	1.00	1.00	1.00	1.00

Table C.11 *Estimated Bayesian assurance $\hat{\delta}$ for fixed values of β_d and various samples sizes n.*

Sample R code for this calculation is as follows:

R code
```
calc.assure <- function(NREP,NITER,N,var,
    Ebeta.d,Ebeta.a,Vbeta.d,Vbeta.a) {

  assurance <- matrix(0,length(Ebeta.d),length(N),
    dimnames=list(Ebeta.d,N))
  CI <- data.frame(cbind("lower"=rep(0,NITER),
    "upper"=rep(0,NITER)))

  for (k in 1:length(N)) {
    for (j in 1:length(Ebeta.d)) {

      sim.data <- matrix(0,NITER,N[k])
      posterior <- matrix(0,NREP,NITER)

      for (i in 1:NREP) {
      ## Simulate data
```

```
         sim.data[i,] <- rnorm(N[k],Ebeta.d[j],sqrt(var))

         ## Simulate posterior distribution
         posterior[i,] <- rnorm(NITER,
             mean = var/(var+N[k]*Vbeta.a)*Ebeta.a +
               N[k]*Vbeta.a/(var+ N[k]*Vbeta.a)*mean(sim.data[i,]),
             sd = sqrt( 1/((N[k]/var) + (1/Vbeta.a)) ))

         CI[i,] <- quantile(posterior[i,],
           probs=c(0.05,0.95))
         }

       ## Calculate assurance using one-sided 95% CIs
       exclude.zero <- CI[(CI$lower > 0),]
       assurance[j,k] <- dim(exclude.zero)[1] / NREP
       }
     }
     assurance
     }

     point.prior <- calc.assure(NREP=1000,NITER=5000,
       N=c(1,2,3,5,10,11), var=2*(4.5^2),
       Ebeta.d=c(5,10,15,20), Vbeta.d = 0,
       Ebeta.a=0, Vbeta.a=100000)

     xtable(point.prior)
```

(b) We proceed as in the simulation approach, but replace the point design prior $V_d = 0$ for β with the $Normal(7.5, 2^2)$. The R code is:

R code

```
     calc.assure <- function(NREP,NITER,N,var,
         Ebeta.d,Vbeta.d,Ebeta.a,Vbeta.a) {

       beta <- rnorm(NREP,Ebeta.d,sqrt(Vbeta.d))
       assurance <- matrix(0,1,length(N),
         dimnames=list("Assurance",N))
       CI <- data.frame(cbind("lower"=rep(0,NREP),
         "upper"=rep(0,NREP)))
       posterior <- matrix(0,NREP,NITER)

       for (k in 1:length(N)) {
         for (j in 1:NREP) {

           ## Simulate data
           sim.data <- rnorm(N[k],beta[j],sqrt(var))

           ## Simulate posterior distribution
           posterior[j,] <- rnorm(NITER,
```

n	5	10	15	20
$\hat{\delta}$	0.72	0.94	0.97	0.98

Table C.12 *Estimated Bayesian assurance $\hat{\delta}$ for various n assuming a $N(7.5, 2^2)$ design prior and a vague analysis prior for β.*

```
            mean = var/(var + N[k]*Vbeta.a)*Ebeta.a +
              N[k]*Vbeta.a/(var + N[k]*Vbeta.a)*mean(sim.data),
            sd = sqrt( 1/((N[k]/var) + (1/Vbeta.a)) )
          )
          CI[j,] <- quantile(posterior[j,],
            probs=c(0.05,0.95))
        }

        ## Calculate assurance using one-sided 95% CIs
        exclude.zero <- CI[(CI$lower > 0),]
        assurance[1,k] <- dim(exclude.zero)[1] / NREP
      }
      assurance
    }

    norm.prior <- calc.assure(NREP=100,
      NITER=5000,N=c(5,10,15,20),var=2*(4.5^2),
      Ebeta.d=7.5,Vbeta.d=4,Ebeta.a=0,Vbeta.a=10000)

    xtable(norm.prior)
```

Results are given in Table C.12; they indicate that in order to achieve assurance of $\delta = 0.80$, we need a sample size between 5 and 10. This is larger than in most of the point prior scenarios above because we have incorporated uncertainty about the true value of β.

(c) Here we let the observed successes be $Y_C \sim Bin(n, p_C)$ and $Y_X \sim Bin(n, p_X)$ in the control and experimental groups, respectively. The parameter of interest is now $\beta = p_X - p_C$, and the relevant hypotheses are $H_0 : \beta = 0$ and $H_a : \beta > 0$.

We continue to assume that the samples are independent. Thus, we can simulate as above after specifying a design prior. The difference here is that this prior must generate values for p_C *and* p_X, not merely for the treatment difference β_d. Given these design prior values, we simulate 2M sets of binomial data and analyze each using conjugate priors to obtain posteriors distributions for p_X and p_C. We then subtract these to obtain a posterior distribution for β. Finally, the assurance is estimated as the proportion of posterior credible intervals for β that exclude the null value of 0.

The R code under point design priors for p_X and p_C is as follows:

R code
```
calc.assure <- function(NREP,NITER,N,
  prior.X,prior.C,a,b) {

    assurance <- matrix(0,length(prior.X),length(N),
      dimnames=list(
      paste(prior.C," vs ",prior.X,sep=""),N)
    CI <- data.frame(cbind("lower"=rep(0,NITER),
      "upper"=rep(0,NITER)))
    posterior.Beta <- matrix(0,NREP,NITER)
    for (k in 1:length(N)) {
      for (j in 1:length(prior.X)) {
        sim.data.X <- matrix(0,NITER,N[k])
        sim.data.C <- matrix(0,NITER,N[k])
        posterior.X <- matrix(0,NREP,NITER)
        posterior.C <- matrix(0,NREP,NITER)
        for (i in 1:NREP) {
          sim.data.X[i,] <- rbinom(N[k],1,prior.X[j])
          sim.data.C[i,] <- rbinom(N[k],1,prior.C[j])
          posterior.X[i,] <- rbeta(NITER,
            a + sum(sim.data.X[i,]),
            b + N[k] - sum(sim.data.X[i,])
            )
          posterior.C[i,] <- rbeta(NITER,
            a + sum(sim.data.C[i,]),
            b + N[k] - sum(sim.data.C[i,])
            )
          posterior.Beta[i,] <- posterior.X[i,]
            - posterior.C[i,]
          CI[i,] <- quantile(posterior.Beta[i,],
            probs=c(0.05,0.95))
        }
        exclude.zero <- CI[(CI$lower > 0),]
        assurance[j,k] <- dim(exclude.zero)[1] / NREP
      }
    }
    assurance
}

beta.point <- calc.assure(NREP=1000,
  NITER=5000,N=c(10,20,50,100,150),
prior.X = c(.51,.55,.60,.70),
prior.C = c(0.50,0.50,0.50,0.50),
a = 1, b = 1)
```

| | | | n | | |
p_C vs. p_X	10	20	50	100	150
0.50 vs. 0.51	0.064	0.057	0.047	0.084	0.050
0.50 vs. 0.55	0.084	0.079	0.125	0.177	0.220
0.50 vs. 0.60	0.132	0.139	0.247	0.421	0.521
0.50 vs. 0.70	0.247	0.356	0.620	0.901	0.970

Table C.13 *Estimated Bayesian assurance for fixed values of n, p_C, and p_X.*

```
xtable(beta.point,digits=rep(3,6),
  display=rep('fg',6))
```

Results using a noninformative Beta(1,1) (i.e., Uniform(0,1)) analysis prior for both the control and experimental groups are given in Table C.13. Thus, if we want a posterior probability of $\omega = 0.95$ and an assurance of $\delta = 0.90$ to detect a difference of 0.20 in the success proportions of the two drugs (with control succeeding half the time), we need 100 patients per group. The values in the first row of this table appear numerically unreliable and essentially measure Type I error because $\beta_d \approx 0$.

Chapter 7

9. The Bayesian model is

$$p(\mathbf{y}, \boldsymbol{\theta}|\boldsymbol{\lambda}) \propto \prod_{i=1}^{k} \theta_i^{d_i} \exp(-e_i\theta_i) \prod_{i=1}^{k} \theta_i^{\alpha-1} \beta^{-\alpha} \exp(-\theta_i/\beta) I_{S_i}(\theta_i) .$$

Hence, using the results of Gelfand, Smith, and Lee (1992), the full conditionals for the θ_i are given by

$$p(\theta_i|\mathbf{y}, \beta, \theta_{j\neq i}) \propto G\left(\theta_i \mid \alpha^*, \ \beta^*\right) I_{(\theta_{i-1},\theta_{i+1})}(\theta_i) \qquad (C.5)$$

for $i = 1, \dots, k$, where $\alpha^* = \alpha + d_i$, $\beta^* = (\beta^{-1} + e_i)^{-1}$, $\theta_0 \equiv 0$, and $\theta_{k+1} \equiv B$. The complete conditional for β is readily available as the untruncated form

$$p(\beta|\mathbf{y}, \boldsymbol{\theta}) = IG\left(a + k\alpha, \ \left\{b^{-1} + \sum_{i=1}^{k} \theta_i\right\}^{-1}\right) .$$

The result in the previous problem could be used in generating from the truncated gamma distributions in (C.5), provided subroutines for the gamma cdf (incomplete gamma function) and inverse cdf were available.

10.(b) The simplest solution here is to restrict $\boldsymbol{\theta}$ to lie in an *increasing convex* constraint set,

$$\{\boldsymbol{\theta}: \ \theta_1 > 0 \ , \ \theta_k < B \ , \ 0 < \theta_2 - \theta_1 < \cdots < \theta_k - \theta_{k-1}\} \ .$$

However, Gelman (1996) points out that this is a very strong set of constraints, and one which becomes stronger as k increases (e.g., as we graduate more years of data, or even subdivide our yearly data into monthly intervals). It is akin to fitting a quadratic to the observed data.

13.(a) This is the approach taken by Carlin and Polson (1992).

  (b) This is the approach taken by Albert and Chib (1993a) and (in a multivariate response setting) by Cowles et al. (1996).

14.(a) Running 3 chains of our Gibbs-Metropolis sampler for 1000 iterations produced the following posterior medians and 95% equal-tail credible intervals: for ϕ, 1.58 and (0.28, 4.12); for $X(\mathbf{s}_A)$, .085 and (.028, .239); and for $X(\mathbf{s}_B)$, .077 and (.021, .247). The spatially smoothed predictions at points A and B are consistent with the high values at monitoring stations 4 and 5 to the east, the moderate value at station 3 to the west, and the low value at station 8 to the far south.

  (b) Ozone is a wind-borne pollutant, so if a prevailing wind (say, from west to east) dominated Atlanta's summer weather pattern, it would be reasonable to expect some anisotropy (in this hypothetical case, more spatial similarity west to east than north to south).

A credible interval for β_{12} that excluded 0 would provide evidence that the anisotropic model is warranted in this case. One might also compare the two models formally (assuming proper priors) via Bayes factors (see Subsections 4.4 and 4.5) or informally via predictive or penalized likelihood criteria (see Subsection 4.6).

15. Based on 1000 Gibbs samples from single chain generated in the BUGS language (and collected after a 500-iteration burn-in period), Spiegelhalter et al. (1995b) report the following posterior mean and standard error estimates:

  (a) α, .73 ± .14; κ_θ, −.54 ± .16. Thus, the observed rates are slightly lower than expected from the standard table, but higher for those counties having a greater percentage of their population engaged in agriculture, fishing, and forestry (AFF).

  (b) α, .39 ± .12 (an overall intercept term cannot be fit in the presence of our translation-invariant CAR prior). While the covariate remains significant in this model, its reduced importance would appear due to this model's ability to account for spatial similarity. (We would expect some confounding of these two effects because counties where AFF employment is high would often be adjacent.)

(c) Spiegelhalter et al. (2002) obtain $p_D = 39.7, DIC = 101.9$ for the part (a) model (covariate plus exchangeble random effects), and $p_D = 29.4, DIC = 89.0$ for the part (b) model (covariate plus spatial random effects). Thus, DIC prefers the spatial model, largely due to its slightly smaller effective sample size.

(d) The fact that the spatial plus covariate model fits slightly better than the heterogeneity plus covariate model suggests that there is some excess spatial variability in the data that is not explained by the covariate. To get a better idea of this variability, we could map the fitted values $\hat{\phi}_i = E(\phi_i|\mathbf{y})$, thinking of them as spatial residuals. A pattern in this map might then indicate the presence of a lingering spatially-varying covariate not yet accounted for.

(e) Suppose we begin with proper hyperpriors for τ and λ. The translation invariance of the CAR prior forces to set $\kappa_\theta = 0$. Alternatively, we could estimate an overall level κ_θ after imposing the constraint $\sum_i \theta_i = 0$ and/or $\sum_i \phi_i = 0$.

16.(a) Running 5 parallel, initially overdispersed MCMC chains for 500 iterations, we found an acceptable degree of convergence by around the 100th iteration. Using the final 400 iterations from all 5 chains, we obtained the 95% posterior credible sets $(-1.10, -1.06)$, $(0.00, 0.05)$, and $(-0.27, -0.17)$ for α, β, and ξ, respectively. The corresponding point estimates are translated into the fitted relative risks for the four subgroups in Table C.14. It is interesting that the fitted sex-race interaction ξ reverses the slight advantage white men hold over non-white men, making nonwhite females the healthiest subgroup, with a relative risk nearly four times smaller than either of the male groups. Many Ohio counties have very small nonwhite populations, so this result could be an artifact of our inability to model covariate-region interactions. It could also be due to our failure to age-standardize the rates (which the online dataset *will* permit).

Demographic Subgroup	Contribution to μ_{ijkt}	Fitted Log-Relative Risk	Fitted Relative Risk
White males	0	0	1
White females	α	−1.08	0.34
Nonwhite males	β	0.02	1.02
Nonwhite females	$\alpha + \beta + \xi$	−1.28	0.28

Table C.14 *Fitted relative risks for the four socio-demographic subgroups in the Ohio lung cancer data.*

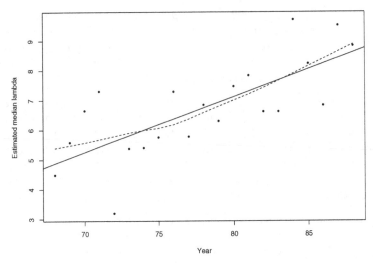

Figure C.7 *Estimated posterior medians for λ_t versus t, full model.*

Turning to the spatio-temporal parameters, histograms of the sampled values (not shown) showed $\theta_i^{(t)}$ distributions centered near 0 in most cases, but $\phi_i^{(t)}$ distributions typically removed from 0. This suggests some degree of clustering in the data, but no significant additional heterogeneity beyond that explained by the CAR prior. Figures C.7 and C.8 check for differential heterogeneity and clustering effects over time. The solid line in both plots is the least squares regression fit, while the dashed line is the result of a Tukey-type running median smoother (smooth in R). The λ_t plot shows a clear, almost linear increase, suggesting that the spatial similarity of lung cancer cases is increasing over the 21-year time period. On the other hand, the posterior medians for τ_t plotted versus t in Figure C.8 are all quite near the prior mean of 100 (again suggesting very little excess heterogeneity) and provide less of an indication of trend. What trend there is appears to be downward, suggesting that heterogeneity is also increasing over time (recall that τ_t is the precision in a mean-zero prior for $\theta_i^{(t)}$).

(b) Because under our model the expected number of deaths for a given subgroup in county i during year t is $E_{ijkt}\exp(\mu_{ijkt})$, we have that the (internally standardized) expected death rate per thousand is $1000\bar{y}\exp(\mu_{ijkt})$. The first row of Figure C.9 maps point estimates of these fitted rates for nonwhite females for the three years, greyscale-coded from lowest (white) to highest (black) into seven intervals: less than .08, .08 to .13, .13 to .18, .18 to .23, .23 to .28, .28 to .33, and

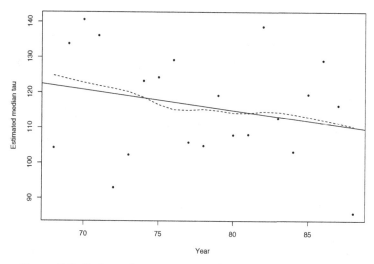

Figure C.8 *Estimated posterior medians for τ_t versus t, full model.*

greater than .33. The second row of the figure shows the interquartile ranges (IQRs) of these rates, also greyscale-coded into seven intervals: less than .01, .01 to .02, .02 to .03, .03 to .04, .04 to .05, .05 to .06, and greater than .06.

Figure C.9 reveals lung cancer death rates are increasing over time, as indicated by the gradual darkening of the counties in the figure's first row. But their variability is also increasing somewhat, as we would expect given our Poisson likelihood. This variability is smallest for high-population counties, such as those containing the cities of Cleveland (northern border, third from the right), Toledo (northern border, third from the left), and Cincinnati (southwestern corner). Lung cancer rates are high in these industrialized areas, but there is also a pattern of generally increasing rates as we move from west to east across the state for a given year, possibly due to a lower incidence of smoking in the predominantly agricultural west, as compared to those in the more mining and manufacturing-oriented east.

Appendix B

1. This proof is similar to that given in Example B.5.
2. For action a, the posterior risk is

$$
\begin{aligned}
\rho(\pi, a) &= \sum_i l(\theta_i, a) \pi(\theta_i | \mathbf{x}) \\
&= 1 - \pi(a | \mathbf{x}) ,
\end{aligned}
$$

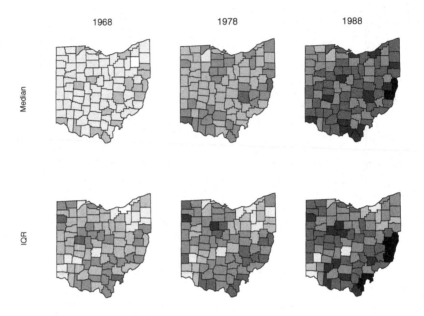

Figure C.9 *Posterior median and interquartile range (IQR) by county and year, nonwhite female lung cancer death rate per 1000 population.*

because $l(\theta_i, a) = 1$ in the sum unless $\theta_i = a$ and $\sum_i \pi(\theta_i|\mathbf{x}) = 1$. To minimize this quantity, we take a to maximize $\pi(a|\mathbf{x})$, hence the Bayes action is the posterior mode.

9. The optimal value of c is

$$c = \frac{(\rho\beta)^2}{\sigma^2 + (\rho\beta)^2} \ .$$

Because $c \leq 1$, the optimal estimator shrinks the unbiased estimator toward 0 by an amount that depends on the relative size of σ and $(\rho\beta)$. As $\sigma \to \infty$, $c \to 0$ (variance reduction is all important). When $\sigma \to 0$, $c \to 1$ (when the variance is small, bias reduction is the dominant goal).

References

Abbey, H. (1952). An examination of the Reed-Frost theory of epidemics. *Human Biology*, **24**, 201–233.

Abrams, D.I., Goldman, A.I., Launer, C., Korvick, J.A., Neaton, J.D., Crane, L.R., Grodesky, M., Wakefield, S., Muth, K., Kornegay, S., Cohn, D.L., Harris, A., Luskin-Hawk, R., Markowitz, N., Sampson, J.H., Thompson, M., Deyton, L., and the Terry Beirn Community Programs for Clinical Research on AIDS (1994). Comparative trial of didanosine and zalcitabine in patients with human immunodeficiency virus infection who are intolerant of or have failed zidovudine therapy. *New England Journal of Medicine*, **330**, 657–662.

Adler, S.L (1981). Over-relaxation method for the Monte Carlo evaluation of the partition function for multiquadratic actions. *Physical Review D*, **23**, 2901–2904.

Agresti, A. (1990). *Categorical Data Analysis*. New York: John Wiley & Sons.

Albert, J.H. (1993). Teaching Bayesian statistics using sampling methods and MINITAB. *The American Statistician*, **47**, 182–191.

Albert, J.H. (1996). *Bayesian Computation Using Minitab*. Belmont, CA: Wadsworth.

Albert, J.H. (2007). *Bayesian Computation with R*. New York: Springer.

Albert, J.H. and Chib, S. (1993a). Bayesian analysis of binary and polychotomous response data. *J. Amer. Statist. Assoc.*, **88**, 669–679.

Albert, J.H. and Chib, S. (1993b). Bayes inference via Gibbs sampling of autoregressive time series subject to Markov mean and variance shifts. *Journal of Business and Economic Statistics*, **11**, 1–15.

Almond, R.G. (1995). *Graphical Belief Modeling*. London: Chapman and Hall.

Ames, B.N., Lee, F.D., and Durston, W.E. (1973). An improved bacterial test system for the detection and classification of mutagens and carcinogens. *Proceedings of the National Academy of Sciences*, **70**, 782–786.

Andersen, P.K. and Gill, R.D. (1982). Cox's regression model for counting process: a large sample study. *Annals of Statistics*, **10**, 1100–20.

Andrews, D.F. and Mallows, C.L. (1974). Scale mixtures of normality. *J. Roy. Statist. Soc., Ser. B*, **36**, 99–102.

Antoniak, C.E. (1974). Mixtures of Dirichlet processes with applications to non-parametric problems. *Ann. Statist.*, **2**, 1152–1174.

Bailey, N.T.J. (1975). *The Mathematical Theory of Infectious Diseases and its Applications*. London: Charles Griffin.

Banerjee, S., Carlin, B.P. and Gelfand, A.E. (2004). *Hierarchical Modeling and Analysis for Spatial Data*. Boca Raton, FL: Chapman and Hall/CRC Press.

Barry, J.M. (2004). *The Great Influenza: The Story of the Deadliest Pandemic in History*. New York: Penguin Group.

Basu, S. and Chib, S. (2003). Marginal likelihood and Bayes factors for Dirichlet process mixture models. *J. Amer. Statist. Assoc.*, **98**, 224–235.

Bates, D.M. and Watts, D.G. (1988). *Nonlinear Regression Analysis and its Applications*. New York: John Wiley & Sons.

Bayarri, M.J. and Castellanos, M.E. (2007). Bayesian checking of the second levels of hierarchical models. *Statistical Science*, **22**, 322–343.

Bayes, T. (1763). An essay towards solving a problem in the doctrine of chances. *Philos. Trans. Roy. Soc. London*, **53**, 370–418. Reprinted, with an introduction by George Barnard, in 1958 in *Biometrika*, **45**, 293–315.

Becker, N. G. (1999). Statistical studies of infectious disease incidence. *Journal of Royal Statistical Society, Ser. B*, **61**, 287–307.

Becker, R.A., Chambers, J.M., and Wilks, A.R. (1988). *The New S Language: A Programming Environment for Data Analysis and Graphics*. Pacific Grove, CA: Wadsworth and Brooks/Cole.

Benjamini, Y. and Hochberg, Y. (1995). Controlling the false discovery rate: A practical and powerful approach to multiple testing. *J. Roy. Statist. Soc., Ser. B*, **57**, 289–300.

Berg-Wolf, M., Peng, G., Xiang, Y., Huppler-Hullsiek, K., MacArthur, R.D., Novak, R.M., Kozal, M., Schmetter, B., Henely, C., and Dehlinger, M. (2006). Long-term comparison of Nevirapine versus Efavirenz when combined with other antiretroviral drugs in HIV-1 positive antiretroviral-naive persons: The NNRTI substudy of the CPCRA 058 FIRST Study. Poster presented at the XVI International AIDS Conference, August 17, 2006, Toronto, Canada.

Berger, J.O. (1984). The robust Bayesian viewpoint (with discussion). In *Robustness in Bayesian Statistics*, ed. J. Kadane, Amsterdam: North Holland.

Berger, J.O. (1985). *Statistical Decision Theory and Bayesian Analysis*, 2nd ed. New York: Springer-Verlag.

Berger, J.O. (1994). An overview of robust Bayesian analysis (with discussion). *Test*, **3**, 5–124.

Berger, J.O. (2000). Bayesian analysis: A look at today and thoughts of tomorrow. *J. Amer. Statist. Assoc.*, **95**, 1269–1276.

Berger, J.O. and Berliner, L.M. (1986). Robust Bayes and empirical Bayes analysis with ϵ-contaminated priors. *Annals of Statistics*, **14**, 461–486.

Berger, J.O. and Berry, D.A. (1988). Statistical analysis and the illusion of objectivity. *American Scientist*, **76**, 159–165.

Berger, J.O. and Delampady, M. (1987). Testing precise hypotheses (with discussion). *Statistical Science*, **2**, 317–52.

Berger, J.O. and Pericchi, L.R. (1996). The intrinsic Bayes factor for linear models. In *Bayesian Statistics 5*, J.M. Bernardo, J.O. Berger, A.P. Dawid, and A.F.M. Smith, eds., Oxford: Oxford University Press, pp. 25–44.

Berger, J.O. and Sellke, T. (1987). Testing a point null hypothesis: The irreconcilability of p values and evidence (with discussion). *J. Amer. Statist. Assoc.*, **82**, 112–122.

Berger, J.O. and Wolpert, R. (1984). *The Likelihood Principle*. Hayward, CA: Institute of Mathematical Statistics Monograph Series.

Berger, R.L. and Hsu, J.C. (1996). Bioequivalence trials, intersection-union tests and equivalence confidence sets (with discussion). *Statistical Science*, **11**, 283–319.

Bernardinelli, L., Clayton, D.G., and Montomoli, C. (1995). Bayesian estimates of disease maps: How important are priors? *Statistics in Medicine*, **14**, 2411–2431.

Bernardinelli, L., Clayton, D.G., Pascutto, C., Montomoli, C., Ghislandi, M., and Songini, M. (1995). Bayesian analysis of space-time variation in disease risk. *Statistics in Medicine*, **14**, 2433–2443.

Bernardinelli, L. and Montomoli, C. (1992). Empirical Bayes versus fully Bayesian analysis of geographical variation in disease risk. *Statistics in Medicine*, **11**, 903–1007.

Bernardo, J.M. (1979). Reference posterior distributions for Bayesian inference (with discussion). *J. Roy. Statist. Soc. B*, **41**, 113–147.

Bernardo, J.M., Bayarri, M.J., Berger, J.O., Dawid, A.P., Heckerman, D., Smith, A.F.M., and West, M. (eds). (2003). *Bayesian Statistics 7*. Oxford: Oxford University Press.

Bernardo, J.M., Bayarri, M.J., Berger, J.O., Dawid, A.P., Heckerman, D., Smith, A.F.M., and West, M. (eds). (2007). *Bayesian Statistics 8*. Oxford: Oxford University Press.

Bernardo, J.M., Berger, J.O., Dawid, A.P., and Smith, A.F.M., eds. (1992). *Bayesian Statistics 4*. Oxford: Oxford University Press.

Bernardo, J.M., Berger, J.O., Dawid, A.P., and Smith, A.F.M., eds. (1996). *Bayesian Statistics 5*. Oxford: Oxford University Press.

Bernardo, J.M., Berger, J.O., Dawid, A.P., and Smith, A.F.M., eds. (1999). *Bayesian Statistics 6*. Oxford: Oxford University Press.

Bernardo, J.M., DeGroot, M.H., Lindley, D.V., and Smith, A.F.M., eds. (1980). *Bayesian Statistics*. Valencia: University Press.

Bernardo, J.M., DeGroot, M.H., Lindley, D.V., and Smith, A.F.M., eds. (1985). *Bayesian Statistics 2*. Amsterdam: North Holland.

Bernardo, J.M., DeGroot, M.H., Lindley, D.V., and Smith, A.F.M, eds. (1988). *Bayesian Statistics 3*. Oxford: Oxford University Press.

Bernardo, J.M. and Smith, A.F.M. (1994). *Bayesian Theory*. New York: John Wiley & Sons.

Berry, D.A. (1991). Bayesian methods in phase III trials. *Drug Information Journal*, **25**, 345–368.

Berry, D.A. (1993). A case for Bayesianism in clinical trials (with discussion). *Statistics in Medicine*, **12**, 1377–1404.

Berry, D.A. (1996). *Statistics: A Bayesian Perspective*. Belmont, CA: Duxbury.

Berry, D.A. (2006). Bayesian clinical trials. *Nature Reviews Drug Discovery*, **5**, 27–36.

Berry, D.A., Müller, P., Grieve, A.P., Smith, M., Parke, T., Blazek, R., Mitchard, N., and Krams, M. (2002). Adaptive Bayesian designs for dose-ranging drug trials (with discussion). In *Case Studies in Bayesian Statistics, Volume V*, C. Gatsonis et al., eds., New York: Springer-Verlag, pp.99–181.

Berry, D.A. and Stangl, D.K., eds. (1996). *Bayesian Biostatistics*. New York: Marcel Dekker.

Berry, D.A. and Stangl, D.K., eds. (2000). *Meta-Analysis in Medicine and Health Policy*. Boca Raton, FL: Chapman and Hall/CRC Press.

Besag, J. (1974). Spatial interaction and the statistical analysis of lattice systems (with discussion). *J. Roy. Statist. Soc., Ser. B*, **36**, 192-236.

Besag, J. (1986). On the statistical analysis of dirty pictures (with discussion). *J. Roy. Statist. Soc., Ser. B* **48**, 259-302.

Besag, J. (1994). Comment on "Representation of knowledge in complex systems," by U. Grenander and M.I. Miller, *J. Roy. Statist. Soc., Ser. B*, **56**, 591-592.

Besag, J. and Green, P.J. (1993). Spatial statistics and Bayesian computation (with discussion). *J. Roy. Statist. Soc., Ser. B*, **55**, 25-37.

Besag, J., Green, P., Higdon, D., and Mengersen, K. (1995). Bayesian computation and stochastic systems (with discussion). *Statistical Science*, **10**, 3-66.

Besag, J., York, J.C., and Mollié, A. (1991). Bayesian image restoration, with two applications in spatial statistics (with discussion). *Annals of the Institute of Statistical Mathematics*, **43**, 1-59.

Best, N.G., Cowles, M.K., and Vines, K. (1995). CODA: Convergence diagnosis and output analysis software for Gibbs sampling output, Version 0.30. Technical report, Medical Research Council Biostatistics Unit, Institute of Public Health, Cambridge University.

Best, N.G., Ickstadt, K., Wolpert, R.L., Cockings, S., Elliott, P., Bennett, J., Bottle, A., and Reed, S. (2002). Modeling the impact of traffic-related air pollution on childhood respiratory illness (with discussion). In *Case Studies in Bayesian Statistics, Volume V*, C. Gatsonis et al., eds., New York: Springer-Verlag, pp. 183-259.

Birnbaum, A. (1962). On the foundations of statistical inference (with discussion). *J. Amer. Statist. Assoc.*, **57**, 269-326.

Bliss, C.I. (1935). The calculation of the dosage-mortality curve. *Annals of Applied Biology*, **22**, 134-167.

Box, G.E.P. (1980). Sampling and Bayes inference in scientific modelling and robustness. *J. Roy. Statist. Soc., Ser. A*, **143**, 383-430.

Box, G.E.P. and Tiao, G. (1973). *Bayesian Inference in Statistical Analysis*. London: Addison-Wesley.

Box, G.E.P. and Muller, M.E. (1958). A note on the generation of random normal deviates. *Ann. Math. Statist.*, **29**, 610-611.

Brandwein, A.C. and Strawderman, W.E. (1990). Stein estimation: The spherically symmetric case. *Statistical Science*, **5**, 356-369.

Breslow, N.E. (1984). Extra-Poisson variation in log-linear models. *Appl. Statist.*, **3**, 38-44.

Breslow, N.E. (1990). Biostatistics and Bayes (with discussion). *Statistical Science*, **5**, 269-298.

Breslow, N.E. and Clayton, D.G. (1993). Approximate inference in generalized linear mixed models. *J. Amer. Statist. Assoc.*, **88**, 9-25.

Brockwell, A.E. and Kadane, J.B. (2003). A gridding method for Bayesian sequential decision problems. *Journal of Computational and Graphical Statistics*,

12, 566–584.

Broemeling, L.D. (2007). *Bayesian Biostatistics and Diagnostic Medicine*. Boca Raton, FL: Chapman and Hall/CRC Press.

Broffitt, J.D. (1988). Increasing and increasing convex Bayesian graduation (with discussion). *Trans. Soc. Actuaries*, **40**, 115–148.

Brooks, S.P. and Gelman, A. (1998). General methods for monitoring convergence of iterative simulations. *J. Comp. Graph. Statist.*, **7**, 434–455.

Brooks, S.P. and Giudici, P. (1999). Convergence assessment for reversible jump MCMC simulations. In *Bayesian Statistics 6*, eds. J.M. Bernardo, J.O. Berger, A.P. Dawid, and A.F.M. Smith. Oxford: Oxford University Press, pp. 733–742.

Brooks, S.P. and Giudici, P. (2000). MCMC convergence assessment via two-way ANOVA. *J. Comp. Graph. Statist.*, **9**, 266–285.

Brooks, S.P. and Roberts, G.O. (1998). Convergence assessment techniques for Markov chain Monte Carlo. *Statistics and Computing*, **8**, 319–335.

Brown, L.D. (2000). An essay on statistical decision theory. *J. Amer. Statist. Assoc.*, **95**, 1277–1281.

Brownlee, K.A. (1965). *Statistical Theory and Methodology in Science and Engineering*. New York: John Wiley & Sons.

Burden, R.L., Faires, J.D., and Reynolds, A.C. (1981). *Numerical Analysis*, 2nd ed. Boston: Prindle, Weber and Schmidt.

Bureau, A., Dupuis, J., Falls, K., Lunetta, K.L., Hayward, B., Keith, T.P., and Van Eerdewegh, P. (2005). Identifying SNPs predictive of phenotype using random forests. *Genet. Epidemiol.*, **28**, 171–82.

Butler, S.M. and Louis, T.A. (1992). Random effects models with non-parametric priors. *Statistics in Medicine*, **11**, 1981–2000.

Carlin, B.P. (1992). A simple Monte Carlo approach to Bayesian graduation. *Trans. Soc. Actuaries*, **44**, 55–76.

Carlin, B.P. (1996). Hierarchical longitudinal modeling. In *Markov Chain Monte Carlo in Practice*, eds. W.R. Gilks, S. Richardson, and D.J. Spiegelhalter, London: Chapman and Hall, pp. 303–319.

Carlin, B.P., Chaloner, K., Church, T., Louis, T.A., and Matts, J.P. (1993). Bayesian approaches for monitoring clinical trials with an application to toxoplasmic encephalitis prophylaxis. *The Statistician*, **42**, 355–367.

Carlin, B.P., Chaloner, K.M., Louis, T.A., and Rhame, F.S. (1995). Elicitation, monitoring, and analysis for an AIDS clinical trial (with discussion). In *Case Studies in Bayesian Statistics, Volume II*, eds. C. Gatsonis, J.S. Hodges, R.E. Kass and N.D. Singpurwalla, New York: Springer-Verlag, pp. 48–89.

Carlin, B.P. and Chib, S. (1995). Bayesian model choice via Markov chain Monte Carlo methods. *J. Roy. Statist. Soc., Ser. B*, **57**, 473–484.

Carlin, B.P. and Gelfand, A.E. (1990). Approaches for empirical Bayes confidence intervals. *J. Amer. Statist. Assoc.*, **85**, 105–114.

Carlin, B.P. and Gelfand, A.E. (1991a). A sample reuse method for accurate parametric empirical Bayes confidence intervals. *J. Roy. Statist. Soc., Ser. B*, **53**, 189–200.

Carlin, B.P. and Gelfand, A.E. (1991b). An iterative Monte Carlo method for nonconjugate Bayesian analysis. *Statistics and Computing*, **1**, 119–128.

Carlin, B.P. and Hodges, J.S. (1999). Hierarchical proportional hazards regression models for highly stratified data. *Biometrics*, **55**, 1162–1170.

Carlin, B.P., Kadane, J.B., and Gelfand, A.E. (1998). Approaches for optimal sequential decision analysis in clinical trials. *Biometrics*, **54**, 964–975.

Carlin, B.P., Kass, R.E., Lerch, F.J., and Huguenard, B.R. (1992). Predicting working memory failure: A subjective Bayesian approach to model selection. *J. Amer. Statist. Assoc.*, **87**, 319–327.

Carlin, B.P. and Klugman, S.A. (1993). Hierarchical Bayesian Whittaker graduation. *Scandinavian Actuarial Journal*, **1993.2**, 183–196.

Carlin, B.P. and Louis, T.A. (1996). Identifying prior distributions that produce specific decisions, with application to monitoring clinical trials. In *Bayesian Analysis in Statistics and Econometrics: Essays in Honor of Arnold Zellner*, eds. D. Berry, K. Chaloner, and J. Geweke, New York: John Wiley & Sons, pp. 493–503.

Carlin, B.P. and Louis, T.A. (2000). Empirical Bayes: past, present and future. *J. Amer. Statist. Assoc.*, **95**, 1286–1289.

Carlin, B.P. and Polson, N.G. (1991). Inference for nonconjugate Bayesian models using the Gibbs sampler. *Canadian Journal of Statistics*, **19**, 399–405.

Carlin, B.P. and Polson, N.G. (1992). Monte Carlo Bayesian methods for discrete regression models and categorical time series. In *Bayesian Statistics 4*, J.M. Bernardo, J.O. Berger, A.P. Dawid, and A.F.M. Smith, eds., Oxford: Oxford University Press, pp. 577–586.

Carlin, B.P., Polson, N.G., and Stoffer, D.S. (1992). A Monte Carlo approach to nonnormal and nonlinear state-space modeling. *J. Amer. Statist. Assoc.*, **87**, 493–500.

Carlin, B.P. and Sargent, D.J. (1996). Robust Bayesian approaches for clinical trial monitoring. *Statistics in Medicine*, **15**, 1093–1106.

Carlin, J.B. and Louis, T.A. (1985). Controlling error rates by using conditional expected power to select tumor sites. *Proc. Biopharmaceutical Section of the Amer. Statist. Assoc.*, 11–18.

Carter, C.K. and Kohn, R. (1994). On Gibbs sampling for state space models. *Biometrika*, **81**, 541–553.

Carter, C.K. and Kohn, R. (1996). Markov chain Monte Carlo in conditionally Gaussian state space models. *Biometrika*, **83**, 589–601.

Carvalho, B., Bengtsson, H., Speed, T., and Irizarry, R. (2007). Exploration, normalization, and genotype calls of high-density oligonucleotide SNP array data. *Biostatistics*, **8**, 485–499.

Casella, G. (1985). An introduction to empirical Bayes data analysis. *The American Statistician*, **39**, 83–87.

Casella, G. and Berger, R.L. (1990). *Statistical Inference*. Pacific Grove, CA: Wadsworth & Brooks-Cole.

Casella, G. and George, E. (1992). Explaining the Gibbs sampler. *The American Statistician*, **46**, 167–174.

Casella, G. and Hwang, J. (1983). Empirical Bayes confidence sets for the mean of a multivariate normal distribution. *J. Amer. Statist. Assoc.*, **78**, 688–698.

Celeux, G., Forbes, F., Robert, C.P., and Titterington, D.M. (2006). Deviance information criteria for missing data models (with discussion). *Bayesian Analysis*, **1**, 651–706.

Centers for Disease Control and Prevention, National Center for Health Statistics (1988). *Public Use Data Tape Documentation Compressed Mortality File, 1968–1985*. Hyattsvile, MD: U.S. Department of Health and Human Services.

Chaloner, K. (1996). The elicitation of prior distributions. In *Bayesian Biostatistics*, eds. D. Berry and D. Stangl, New York: Marcel Dekker, pp. 141–156.

Chaloner, K., Church, T., Louis, T.A., and Matts, J.P. (1993). Graphical elicitation of a prior distribution for a clinical trial. *The Statistician*, **42**, 341–353.

Chen, D.-G., Carter, E.M., Hubert, J.J., and Kim, P.T. (1999). Empirical Bayes estimation for combinations of multivariate bioassays. *Biometrics*, **55**, 1038–1043.

Chen, M.-H. (1994). Importance-weighted marginal Bayesian posterior density estimation. *J. Amer. Statist. Assoc.*, **89**, 818–824.

Chen, M.-H., Shao, Q.-M., and Ibrahim, J.G. (2000). *Monte Carlo Methods in Bayesian Computation*. New York: Springer-Verlag.

Chib, S. (1995). Marginal likelihood from the Gibbs output. *J. Amer. Statist. Assoc.*, **90**, 1313–1321.

Chib, S. (1996). Calculating posterior distributions and modal estimates in Markov mixture models. *Journal of Econometrics*, **75**, 79–97.

Chib, S. and Carlin, B.P. (1999). On MCMC sampling in hierarchical longitudinal models. *Statistics and Computing*, **9**, 17–26.

Chib, S. and Greenberg, E. (1995a). Understanding the Metropolis-Hastings algorithm. *The American Statistician*, **49**, 327–335.

Chib, S. and Greenberg, E. (1995b). Hierarchical analysis of SUR models with extensions to correlated serial errors and time-varying parameter models. *Journal of Econometrics*, **68**, 339–360.

Chib, S. and Greenberg, E. (1998). Analysis of multivariate probit models. *Biometrika*, **85**, 347–361.

Chib, S. and Jeliazkov, I. (2001). Marginal likelihood from the Metropolis-Hastings output. *J. Amer. Statist. Assoc.*, **96**, 270–281.

Choi, S., Lagakos, S.W., Schooley, R.T., and Volberding, P.A. (1993) CD4+ lymphocytes are an incomplete surrogate marker for clinical progession in persons with asymptomatic HIV infection taking zidovudine. *Ann. Internal Med.*, **118**, 674–680.

Chowell, G., Ammon, C., Hengartner, N. and Hyman, J. (2006). Estimation of the reproductive number of the Spanish flu epidemic in Geneva, Switzerland. *Vaccine*, **24**, 6747–6750.

Christensen, O.F., Roberts, G.O., and Skold, M. (2006). Robust Markov chain Monte Carlo methods for spatial generalized linear mixed models. *J. Comp. Graph. Statist.*, **15**, 1–17.

Christiansen, C.L. and Morris, C.N. (1997a). Hierarchical Poisson regression modeling. *J. Amer. Statist. Assoc.*, **92**, 618–632.

Christiansen, C.L. and Morris, C.N. (1997b). Improving the statistical approach to health care provider profiling. *Annals of Internal Medicine*, **127**, 764–768.

Clayton, D.G. (1991). A Monte Carlo method for Bayesian inference in frailty models. *Biometrics*, **47**, 467–485.

Clayton, D. (1994). Bayesian analysis of frailty models. Technical report, Medical Research Council Biostatistics Unit, Cambridge, UK.

Clayton, D.G. and Bernardinelli, L. (1992). Bayesian methods for mapping disease risk. In *Geographical and Environmental Epidemiology: Methods for Small-Area Studies*, P. Elliott, J. Cuzick, D. English, and R. Stern, eds., Oxford: Oxford University Press.

Clayton, D.G. and Kaldor, J.M. (1987). Empirical Bayes estimates of age-standardized relative risks for use in disease mapping. *Biometrics*, **43**, 671-681.

Clyde, M., DeSimone, H., and Parmigiani, G. (1996). Prediction via orthogonalized model mixing. *J. Amer. Statist. Assoc.*, **91**, 1197-1208.

Collins, S.D., Frost, W.H., Gover, M., and Sydenstricker, E. (1930). Mortality from influenza and pneumonia in 50 large cities of the United States 1910–1929. *Public Health Reports*, **45**, 2277-2328.

Concorde Coordinating Committee (1994). Concorde: MRC/ANRS randomized double-blind controlled trial of immediate and deferred zidovudine in symptom-free HIV infection. *Lancet*, **343**, 871-881.

Congdon, P. (2003). *Applied Bayesian Modelling*. New York: John Wiley & Sons.

Congdon, P. (2006). Bayesian model choice based on Monte Carlo estimates of posterior model probabilities. *Computational Statistics and Data Analysis*, **50**, 346-357.

Congdon, P. (2007a). *Bayesian Statistical Modelling*, 2nd ed. New York: John Wiley & Sons.

Congdon, P. (2007b). Model weights for model choice and averaging. *Statistical Methodology*, **4**, 143-157.

Conlon, E.M. and Louis, T.A. (1999). Addressing multiple goals in evaluating region-specific risk using Bayesian methods. In *Disease Mapping and Risk Assessment for Public Health*, A. Lawson, A. Biggeri, D. Böhning, E. Lesaffre, J.-F. Viel, and R. Bertollini, eds., New York: John Wiley & Sons, pp. 31-47.

Cook, T.D. and DeMets, D.L., eds. (2008). Introduction to Statistical Methods for Clinical Trials. Boca Raton, FL: Chapman and Hall/CRC Press.

Cooper, H. and Hedges, L. eds. (1994). *The Handbook of Research Synthesis*. New York: Russell Sage Foundation.

Cornfield, J. (1966a). Sequential trials, sequential analysis and the likelihood principle. *The American Statistician*, **20**, 18-23.

Cornfield, J. (1966b). A Bayesian test of some classical hypotheses – with applications to sequential clinical trials. *J. Amer. Statist. Assoc.*, **61**, 577-594.

Cornfield, J. (1969). The Bayesian outlook and its applications. *Biometrics*, **25**, 617-657.

Cowles, M.K. (1994). Practical issues in Gibbs sampler implementation with application to Bayesian hierarchical modeling of clinical trial data. Unpublished Ph.D. disssertation, Division of Biostatistics, University of Minnesota.

Cowles, M.K. and Carlin, B.P. (1996). Markov chain Monte Carlo convergence diagnostics: A comparative review. *J. Amer. Statist. Assoc.*, **91**, 883-904.

Cowles, M.K., Carlin, B.P., and Connett, J.E. (1996). Bayesian Tobit modeling of longitudinal ordinal clinical trial compliance data with nonignorable missingness. *J. Amer. Statist. Assoc.*, **91**, 86-98.

Cowles, M.K. and Rosenthal, J.S. (1998). A simulation approach to convergence rates for Markov chain Monte Carlo algorithms. *Statistics and Computing*, **8**, 115-124.

Cox, D.R. and Hinkley, D.V. (1974). *Theoretical Statistics*. London: Chapman and Hall.

Cox, D.R. and Oakes, D. (1984). *Analysis of Survival Data*. London: Chapman and Hall.

Creasy, M.A. (1954). Limits for the ratio of means. *J. Roy. Statist. Soc., Ser. B*, **16**, 186–194.

Cressie, N. (1989). Empirical Bayes estimates of undercount in the decennial census. *J. Amer. Statist. Assoc.*, **84**, 1033–1044.

Cressie, N.A.C. (1993). *Statistics for Spatial Data*, revised ed. New York: John Wiley & Sons.

Cressie, N. and Chan, N.H. (1989). Spatial modeling of regional variables. *J. Amer. Statist. Assoc.*, **84**, 393–401.

Damien, P., Laud, P.W., and Smith, A.F.M. (1995). Approximate random variate generation from infinitely divisible distributions with applications to Bayesian inference. *J. Roy. Statist. Soc.*, Ser. B, **57**, 547–563.

Damien, P., Wakefield, J., and Walker, S. (1999). Gibbs sampling for Bayesian non-conjugate and hierarchical models by using auxiliary variables. *J. Roy. Statist. Soc.*, Ser. B, **61**, 331–344.

Daniels, M.J. and Kass, R.E. (1999). Nonconjugate Bayesian estimation of covariance matrices and its use in hierarchical models. *J. Amer. Statist. Assoc.*, **94**, 1254–1263.

Daniels, M. and Normand, S.T. (2006). Longitudinal profiling of health care units based on continuous and discrete patient outcomes. *Biostatistics*, **7**, 1–15.

Deely, J.J. and Lindley, D.V. (1981). Bayes empirical Bayes. *J. Amer. Statist. Assoc.*, **76**, 833–841.

DeGroot, M.H. (1970). *Optimal Statistical Decisions*. New York: McGraw-Hill.

Dellaportas, P. and Smith, A.F.M. (1993). Bayesian inference for generalised linear and proportional hazards models via Gibbs sampling. *J. Roy. Statist. Assoc., Ser. C (Applied Statistics)*, **42**, 443–459.

Dellaportas, P., Forster, J.J., and Ntzoufras, I. (2002). On Bayesian model and variable selection using MCMC. *Statistics and Computing*, **12**, 27–36.

Dempster, A.P., Laird, N.M., and Rubin, D.B. (1977). Maximum likelihood estimation from incomplete data via the EM algorithm (with discussion). *J. Roy. Statist. Soc., Ser. B*, **39**, 1–38.

Dempster, A.P., Selwyn, M.R., and Weeks, B.J. (1983). Combining historical and randomized controls for assessing trends in proportions. *J. Amer. Statist. Assoc.*, **78**, 221–227.

DerSimonian, R. and Laird, N.M. (1986). Meta-analysis in clinical trials. *Controlled Clinical Trials*, **7**, 177–188.

DeSantis, F. and Spezzaferri, F. (1999). Methods for default and robust Bayesian model comparison: the fractional Bayes factor approach. *International Statistical Review*, **67**, 267–286.

DeSouza, C.M. (1991). An empirical Bayes formulation of cohort models in cancer epidemiology. *Statistics in Medicine*, **10**, 1241-1256.

Devine, O.J. (1992). Empirical Bayes and constrained empirical Bayes methods for estimating incidence rates in spatially aligned areas. Unpublished Ph.D. dissertation, Division of Biostatistics, Emory University.

Devine, O.J., Halloran, M.E., and Louis, T.A. (1994). Empirical Bayes methods for stabilizing incidence rates prior to mapping. *Epidemiology*, **5**, 622–630.

Devine, O.J. and Louis, T.A. (1994). A constrained empirical Bayes estimator for incidence rates in areas with small populations. *Statistics in Medicine*, **13**, 1119–1133.

Devine, O.J., Louis, T.A., and Halloran, M.E. (1994). Empirical Bayes estimators for spatially correlated incidence rates. *Environmetrics*, **5**, 381–398.

Devine, O.J., Louis, T.A., and Halloran, M.E. (1996). Identifying areas with elevated incidence rates using empirical Bayes estimators. *Geographical Analysis*, **28**, 187–199.

Devroye, L. (1986). *Non-Uniform Random Variate Generation*. New York: Springer-Verlag.

Dey, D.K. and Berger, J.O. (1983). On truncation of shrinkage estimators in simultaneous estimation of normal means. *J. Amer. Statist. Assoc.*, **78**, 865–869.

Dey, D.K., Gelfand, A.E., Swartz, T.B., and Vlachos, P.K. (1998). A simulation-intensive approach for checking hierarchical models. *Test*, **7**, 325–346.

Diebolt, J. and Robert, C.P. (1994). Estimation of finite mixture distributions through Bayesian sampling. *J. Roy. Statist. Soc., Ser. B*, **56**, 363–375.

Dietz, K. (1993). The estimation of the basic reproduction number for infectious diseases. *Statist. Meth. Med. Res.*, **2(1)**, 23–41.

Diggle, P.J. (1983). *Statistical Analysis of Spatial Point Patterns*. London: Academic Press.

Diggle, P.J., Tawn, J.A., and Moyeed, R.A. (1998). Model-based geostatistics (with discussion). *J. Roy. Statist. Soc., Ser. C*, **47**, 299–350.

Doss, H. and Narasimhan, B. (1994). Bayesian Poisson regression using the Gibbs sampler: Sensitivity analysis through dynamic graphics. Technical report, Department of Statistics, The Ohio State University.

Draper, D. (1995). Assessment and propagation of model uncertainty (with discussion). *J. Roy. Statist. Soc., Ser. B*, **57**, 45–97.

Draper, D. and Gittoes, M. (2004). Statistical analysis of performance indicators in UK higher education. *J. Roy. Statist. Soc., Ser. A*, **167**, 449–474.

Draper, D. and Krnjajic, M. (2007). Bayesian model specification. Technical report, Department of Applied Mathematics and Statistics, University of California – Santa Cruz.

Dreze, J.H. and Richard, J.F. (1983). Bayesian analysis of simultaneous equations systems. In *Handbook of Econometrics, Vol 1*, Z. Griliches and M.D. Intriligator, eds., Amsterdam: North Holland, pp. 517–598.

DuMouchel, W.H. (1988). A Bayesian model and a graphical elicitation procedure for multiple comparisons. In *Bayesian Statistics 3*, eds. J.M. Bernardo, M.H. DeGroot, D.V. Lindley, and A.F.M. Smith, Oxford: Oxford University Press, pp. 127–145.

DuMouchel, W. (1999). Bayesian data mining in large frequency tables, with an application to the FDA spontaneous reporting system (with discussion). *The American Statistician*, **53**, 177–190.

DuMouchel, W.H. and Harris, J.E. (1983). Bayes methods for combining the results of cancer studies in humans and other species (with discussion). *J.*

Amer. Statist. Assoc., **78**, 293–315.

Eberly, L.E. and Carlin, B.P. (2000). Identifiability and convergence issues for Markov chain Monte Carlo fitting of spatial models. *Statistics in Medicine*, **19**, 2279–2294.

Ecker, M.D. and Gelfand, A.E. (1999). Bayesian modeling and inference for geometrically anisotropic spatial data. *Mathematical Geology*, **31**, 67–83.

Edwards, R.G. and Sokal, A.D. (1988). Generalization of the Fortium-Kasteleyn-Swendsen-Wang representation and Monte Carlo algorithm. *Physical Review D*, **38**, 2009–2012.

Edwards, W., Lindman, H., and Savage, L.J. (1963). Bayesian statistical inference for psychological research. *Psych. Rev.*, **70**, 193–242.

Efron, B. (1986). Why isn't everyone a Bayesian? (with discussion). *Amer. Statistician*, **40**, 1–11.

Efron, B. (1996). Empirical Bayes methods for combining likelihoods (with discussion). *J. Amer. Statist. Assoc.*, **91**, 538–565.

Efron, B. (2005). Bayesians, frequentists, and scientists. *J. Amer. Statist. Assoc.*, **100**, 1–5.

Efron, B. and Morris, C.N. (1971). Limiting the risk of Bayes and empirical Bayes estimators – Part I: The Bayes case. *J. Amer. Statist. Assoc.*, **66**, 807–815.

Efron, B. and Morris, C.N. (1972a). Limiting the risk of Bayes and empirical Bayes estimators – Part II: The empirical Bayes case. *J. Amer. Statist. Assoc.*, **67**, 130–139.

Efron, B. and Morris, C.N. (1972b). Empirical Bayes on vector observations: An extension of Stein's method. *Biometrika*, **59**, 335–347.

Efron, B. and Morris, C. (1973a). Stein's estimation rule and its competitors – An empirical Bayes approach. *J. Amer. Statist. Assoc.*, **68**, 117–130.

Efron, B. and Morris, C. (1973b). Combining possibly related estimation problems (with discussion). *J. Roy. Statist. Soc., Ser. B*, **35**, 379–421.

Efron, B. and Morris, C. (1975). Data analysis using Stein's estimator and its generalizations. *J. Amer. Statist. Assoc.*, **70**, 311–319.

Efron, B. and Morris, C. (1977). Stein's paradox in statistics. *Scientific American*, **236**, 119–127.

Efron, B. and Tibshirani, R. (2002). Empirical Bayes methods and false discovery rates for microarrays. *Genet. Epidemiol.*, **23**, 70–86.

Efron, B., Tibshirani, R., Storey, J.D., and Tusher, V. (2001). Empirical Bayes analysis of a microarray experiment. *J. Amer. Statist. Assoc.*, **96**, 1151–1160.

Erkanli, A., Soyer, R., and Costello, E.J. (1999). Bayesian inference for prevalence in longitudinal two-phase studies. *Biometrics*, **55**, 1145–1150.

Escobar, M.D. (1994). Estimating normal means with a Dirichlet process prior. *J. Amer. Statist. Assoc.*, **89**, 268–277.

Escobar, M.D. and West, M. (1992). Computing Bayesian nonparametric hierarchical models. Discussion Paper 92-A20, Institute for Statistics and Decision Sciences, Duke University.

Escobar, M.D. and West, M. (1995). Bayesian density estimation and inference using mixtures. *J. Amer. Statist. Assoc.*, **90**, 577–588.

Etzioni, R. and Carlin, B.P. (1993). Bayesian analysis of the Ames *Salmonella*/microsome assay. In *Case Studies in Bayesian Statistics (Lecture*

Notes in Statistics, Vol. 83), eds. C. Gatsonis, J.S. Hodges, R.E. Kass, and N.D. Singpurwalla, New York: Springer-Verlag, pp. 311–323.

Evans, M. and Swartz, T. (1995). Methods for approximating integrals in statistics with special emphasis on Bayesian integration problems. *Statistical Science*, **10**, 254–272; discussion and rejoinder: **11**, 54–64.

Fay R.E., and Herriot, R.A. (1979). Estimates of income for small places: An application of James-Stein procedures to census data. *J. Amer. Statist. Assoc.*, **74**, 269–277.

Ferguson, N.M., Cummings, D.A., Cauchemez, S., Fraser, C., Riley, S., Meeyai, A., Iamsirithaworn, S., and Burke, D.S. (2005). Strategies for containing an emerging influenza pandemic in Southeast Asia. *Nature*, **437**, 209–214.

Ferguson, T.S. (1967). *Mathematical Statistics: A Decision Theory Approach.* New York: Academic Press.

Ferguson, T.S. (1973). A Bayesian analysis of some nonparametric problems. *Ann. Statist.*, **1**, 209–230.

Fieller, E.C. (1954). Some problems in interval estimation. *J. Roy. Statist. Soc., Ser. B*, **16**, 175–185.

Finney, D.J. (1947). The estimation from individual records of the relationship between dose and quantal response. *Biometrika*, **34**, 320–334.

Fisher, R.A. (1922). On the mathematical foundations of theoretical statistics. *Philos. Trans. Roy. Soc. London, Ser. A*, **222**, 309–368.

Fleming, T.R. (1992). Evaluating therapeutic interventions: Some issues and experiences (with discussion). *Statistical Science*, **7**, 428–456.

Freedman, L.S., Lowe, D., and Macaskill, P. (1984). Stopping rules for clinical trials incorporating clinical opinion. *Biometrics*, **40**, 575–586.

Freedman, L.S. and Spiegelhalter, D.J. (1983). The assessment of subjective opinion and its use in relation to stopping rules for clinical trials. *The Statistician*, **33**, 153–160.

Freedman, L.S. and Spiegelhalter, D.J. (1989). Comparison of Bayesian with group sequential methods for monitoring clinical trials. *Controlled Clinical Trials*, **10**, 357–367.

Freedman, L.S. and Spiegelhalter, D.J. (1992). Application of Bayesian statistics to decision making during a clinical trial. *Statistics in Medicine*, **11**, 23–35.

Frost, W.H. and Sydenstricker, E. (1919). Influenza in Maryland. *Public Health Reports*, Washington, DC: Government Printing Office, **34(510)**, 491–504.

Fuller, W.A. (1987). *Measurement Error Models.* New York: John Wiley & Sons.

Galishoff, S. (1969). *Public health in Newark, 1832-1918.* Unpublished Ph.D. thesis, New York University.

Gamerman, D. and Lopes, H.F. (2006). *Markov Chain Monte Carlo: Stochastic Simulation for Bayesian Inference*, 2nd ed. London: Chapman and Hall/CRC Press.

Garthwaite, P.H., Kadane, J.B., and O'Hagan, A. (2005). Statistical methods for eliciting probability distributions. *J. Amer. Statist. Assoc.*, **100**, 680–701.

Gatsonis, C., Hodges, J.S., Kass, R.E., and Singpurwalla, N.D., eds. (1993). *Case Studies in Bayesian Statistics.* New York: Springer-Verlag.

Gatsonis, C., Hodges, J.S., Kass, R.E., and Singpurwalla, N.D., eds. (1995). *Case Studies in Bayesian Statistics, Volume II.* New York: Springer-Verlag.

Gatsonis, C., Hodges, J.S., Kass, R.E., McCulloch, R.E., Rossi, P., and Singpur-walla, N.D., eds. (1997). *Case Studies in Bayesian Statistics, Volume III*. New York: Springer-Verlag.

Gatsonis, C., Kass, R.E., Carlin, B.P., Carriquiry, A.L., Gelman, A., Verdinelli, I., and West, M., eds. (1999). *Case Studies in Bayesian Statistics, Volume IV*. New York: Springer-Verlag.

Gatsonis, C., Kass, R.E., Carlin, B.P., Carriquiry, A.L., Gelman, A., Verdinelli, I., and West, M., eds. (2002a). *Case Studies in Bayesian Statistics, Volume V*. New York: Springer-Verlag.

Gatsonis, C., Kass, R.E., Carriquiry, A.L., Gelman, A., Higdon, D., Pauler, D.K., and Verdinelli, I., eds. (2002b). *Case Studies in Bayesian Statistics, Volume VI*. New York: Springer-Verlag.

Gaver, D.P. and O'Muircheartaigh, I.G. (1987). Robust empirical Bayes analyses of event rates. *Technometrics*, **29**, 1–15.

Geisser, S. (1993). *Predictive Inference*. London: Chapman and Hall.

Geisser, S. and Eddy, W.F. (1979). A predictive approach to model selection. *Journal of the American Statistical Association*, **74**, 153–160.

Gelfand, A.E. (1998). Approaches to Semiparametric Bayesian Regression. In *Asymptotics, Nonparametrics, and Time Series*, ed. S. Ghosh, New York: Marcel Dekker, pp. 615–638.

Gelfand, A.E. and Carlin, B.P. (1993). Maximum likelihood estimation for constrained or missing data models. *Canad. J. Statist.*, **21**, 303–311.

Gelfand, A.E., Carlin, B.P., and Trevisani, M. (2000). On computation using Gibbs sampling for multilevel models. Technical report, Department of Statistics, University of Connecticut.

Gelfand, A.E. and Dalal, S.R. (1990). A note on overdispersed exponential families. *Biometrika*, **77**, 55–64.

Gelfand, A.E. and Dey, D.K. (1994). Bayesian model choice: Asymptotics and exact calculations. *J. Roy. Statist. Soc., Ser. B*, **56**, 501–514.

Gelfand, A.E., Dey, D.K., and Chang, H. (1992). Model determination using predictive distributions with implementation via sampling-based methods (with discussion). In *Bayesian Statistics 4*, J.M. Bernardo, J.O. Berger, A.P. Dawid, and A.F.M. Smith, eds., Oxford: Oxford University Press, pp. 147–167.

Gelfand, A.E. and Ghosh, S.K. (1998). Model choice: A minimum posterior predictive loss approach. *Biometrika*, **85**, 1–11.

Gelfand, A.E., Hills, S.E., Racine-Poon, A., and Smith, A.F.M. (1990). Illustration of Bayesian inference in normal data models using Gibbs sampling. *J. Amer. Statist. Assoc.*, **85**, 972–985.

Gelfand, A.E. and Kottas, A. (2002). A computational approach for full Bayesian inference in single and multiple sample problems. *J. Comp. Graph. Statist.*, **11**, 289–305.

Gelfand, A.E., and Lee, T.-M. (1993). Discussion of the meeting on the Gibbs sampler and other Markov chain Monte Carlo methods. *J. Roy. Statist. Soc., Ser. B*, **55**, 72–73.

Gelfand A.E. and Mallick, B.K. (1995). Bayesian analysis of proportional hazards models built from monotone functions. *Biometrics*, **51**, 843–852.

Gelfand, A.E. and Sahu, S.K. (1994). On Markov chain Monte Carlo acceleration.

J. Comp. Graph. Statist., **3**, 261–276.

Gelfand, A.E. and Sahu, S.K. (1999). Identifiability, improper priors and Gibbs sampling for generalized linear models. *J. Amer. Statist. Assoc.*, **94**, 247–253.

Gelfand, A.E., Sahu, S.K., and Carlin, B.P. (1995). Efficient parametrizations for normal linear mixed models. *Biometrika*, **82**, 479–488.

Gelfand, A.E., Sahu, S.K., and Carlin, B.P. (1996). Efficient parametrizations for generalized linear mixed models (with discussion). In *Bayesian Statistics 5*, eds. J.M. Bernardo, J.O. Berger, A.P. Dawid, and A.F.M. Smith. Oxford: Oxford University Press, pp. 165–180.

Gelfand, A.E. and Smith, A.F.M. (1990). Sampling-based approaches to calculating marginal densities. *J. Amer. Statist. Assoc.*, **85**, 398–409.

Gelfand, A.E., Smith, A.F.M., and Lee, T-M. (1992). Bayesian analysis of constrained parameter and truncated data problems using Gibbs sampling. *J. Amer. Statist. Assoc.*, **87**, 523–532.

Gelman, A. (1996). Bayesian model-building by pure thought: Some principles and examples. *Statistica Sinica*, **6**, 215–232.

Gelman, A. (2004). Parameterization and Bayesian modeling. *J. Amer. Statist. Assoc.*, **99**, 537–545.

Gelman, A. (2006). Prior distributions for variance parameters in hierarchical models. *Bayesian Analysis*, **1**, 515–534.

Gelman, A., Carlin, J., Stern, H., and Rubin, D.B. (2004). *Bayesian Data Analysis*, 2nd ed. Boca Raton, FL: Chapman and Hall/CRC Press.

Gelman, A. and Hill, J. (2007). *Data Analysis using Regression and Multilevel/Hierarchical Models*. Cambridge: Cambridge University Press.

Gelman, A., van Mechelen, I., Verbeke, G., Heitjan, D.F., and Meulders, M. (2005). Multiple imputation for model checking: Completed-data plots with missing and latent data. *Biometrics*, **61**, 74–85.

Gelman, A., Meng, X.-L., and Stern, H.S. (1996). Posterior predictive assessment of model fitness via realized discrepancies (with discussion). *Statistica Sinica*, **6**, 733–807.

Gelman, A. and Price, P.N. (1999). All maps of parameter estimates are misleading. *Statistics in Medicine*, **18**, 3221–3234.

Gelman, A., Roberts, G.O., and Gilks, W.R. (1996). Efficient Metropolis jumping rules. In *Bayesian Statistics 5*, J.M. Bernardo, J.O. Berger, A.P. Dawid, and A.F.M. Smith, eds., Oxford: Oxford University Press, pp. 599–607.

Gelman, A. and Rubin, D.B. (1992). Inference from iterative simulation using multiple sequences (with discussion). *Statistical Science*, **7**, 457–511.

Geman, S. and Geman, D. (1984). Stochastic relaxation, Gibbs distributions and the Bayesian restoration of images. *IEEE Trans. on Pattern Analysis and Machine Intelligence*, **6**, 721-741.

Gentle, J.E. (1998). *Random Number Generation and Monte Carlo Methods*. New York: Springer-Verlag.

Genz, A. and Kass, R. (1997). Subregion adaptive integration of functions having a dominant peak. *J. Comp. Graph. Statist.*, **6**, 92–111.

George, E.I. and McCulloch, R.E. (1993). Variable selection via Gibbs sampling. *J. Amer. Statist. Assoc.*, **88**, 881–889.

Geweke, J. (1989). Bayesian inference in econometric models using Monte Carlo integration. *Econometrica*, **57**, 1317-1339.

Geweke, J. (1992). Evaluating the accuracy of sampling-based approaches to the calculation of posterior moments (with discussion). In *Bayesian Statistics 4*, J.M. Bernardo, J.O. Berger, A.P. Dawid, and A.F.M. Smith, eds., Oxford: Oxford University Press, pp. 169–193.

Geweke, J. (2007). Bayesian model comparison and validation. *AEA Papers and Procedings*, **97 (2)**, 60–64.

Geyer, C.J. (1992). Practical Markov Chain Monte Carlo (with discussion). *Statistical Science*, **7**, 473–511.

Geyer, C.J. (1993). Estimating normalizing constants and reweighting mixtures in Markov chain Monte Carlo. Technical report No. 568, School of Statistics, University of Minnesota.

Geyer, C.J. and Thompson, E.A. (1992). Constrained Monte Carlo maximum likelihood for dependent data (with discussion). *J. Roy. Statist. Soc., Ser. B*, **54**, 657–699.

Geyer, C.J. and Thompson, E.A. (1995). Annealing Markov chain Monte Carlo with applications to ancestral inference. *J. Amer. Statist. Assoc.*, **90**, 909–920.

Ghosh, M. (1992). Constrained Bayes estimation with applications. *J. Amer. Statist. Assoc.*, **87**, 533–540.

Ghosh, M. and Meeden, G. (1986). Empirical Bayes estimation in finite population sampling. *J. Amer. Statist. Assoc.*, **81**, 1058–1062.

Ghosh, M. and Rao, J.N.K. (1994). Small area estimation: An appraisal (with discussion). *Statistical Science*, **9**, 55–93.

Gilks, W.R. (1992). Derivative-free adaptive rejection sampling for Gibbs sampling. In *Bayesian Statistics 4*, J.M. Bernardo, J.O. Berger, A.P. Dawid, and A.F.M. Smith, eds., Oxford: Oxford University Press, pp. 641–649.

Gilks, W.R., Clayton, D.G., Spiegelhalter, D.J., Best, N.G., McNeil, A.J., Sharples, L.D., and Kirby, A.J. (1993). Modelling complexity: Applications of Gibbs sampling in medicine. *J. Roy. Statist. Soc., Ser. B*, **55**, 38–52.

Gilks, W.R., Richardson, S., and Spiegelhalter, D.J., eds. (1996). *Markov Chain Monte Carlo in Practice*. London: Chapman and Hall.

Gilks, W.R. and Roberts, G.O. (1996). Strategies for improving MCMC. In *Markov Chain Monte Carlo in Practice*, W.R. Gilks, S. Richardson, and D.J. Spiegelhalter, D.J., eds, London: Chapman and Hall, pp. 89–114.

Gilks, W.R., Roberts, G.O., and Sahu, S.K. (1998). Adaptive Markov chain Monte Carlo through regeneration. *J. Amer. Stat. Assoc.*, **93**, 1045–1054.

Gilks, W.R., Thomas, A., and Spiegelhalter, D.J. (1992) Software for the Gibbs sampler. *Computing Science and Statistics*, **24**, 439–448.

Gilks, W.R., Wang, C.C., Yvonnet, B., and Coursaget, P. (1993). Random-effects models for longitudinal data using Gibbs sampling. *Biometrics*, **49**, 441–453.

Gilks, W.R. and Wild, P. (1992). Adaptive rejection sampling for Gibbs sampling. *J. Roy. Statist. Soc., Ser. C (Applied Statistics)*, **41**, 337–348.

Givens, G.H. and Hoeting, J.A. (2005). *Computational Statistics*. New York: John Wiley & Sons.

Godsill, S.J. (2001). On the relationship between Markov chain Monte Carlo methods for model uncertainty. *J. Comp. Graph. Stats.*, **10**, 230–248.

Goel, P.K. (1988). Software for Bayesian analysis: Current status and additional needs (with discussion). In *Bayesian Statistics 3*, J.M. Bernardo, M.H. DeGroot, D.V. Lindley, and A.F.M. Smith, eds., Oxford: Oxford University Press, pp. 173–188.

Goldman, A.I., Carlin, B.P., Crane, L.R., Launer, C., Korvick, J.A., Deyton, L., and Abrams, D.I. (1996). Response of CD4$^+$ and clinical consequences to treatment using ddI or ddC in patients with advanced HIV infection. *Journal of Acquired Immune Deficiency Syndromes and Human Retrovirology*, **11**, 161–169.

Goldstein, H. (1995). *Kendall's Library of Statistics 3: Multilevel Statistical Models*, 2nd ed. London: Arnold.

Goldstein, H. and Spiegelhalter, D.J. (1996). League tables and their limitations: Statistical issues in comparisons of institutional performance (with discussion). *J. Roy. Statist. Soc., Ser. A* **159**, 385-443.

Goldstein, M. (1981). Revising previsions: A geometric interpretation. *J. Roy. Statist. Soc., Ser. B*, **43**, 105–130.

Good, I.J. (1965). *The Estimation of Probabilities.* Cambdrige, MA: MIT Press.

Grambsch, P. and Therneau, T.M. (1994). Proportional hazards tests and diagnostics based on weighted residuals. *Biometrika*, **81**, 515-526.

Green, P.J. (1995). Reversible jump Markov chain Monte Carlo computation and Bayesian model determination. *Biometrika*, **82**, 711–732.

Green, P.J. and Murdoch, D.J. (1999). Exact sampling for Bayesian inference: Towards general purpose algorithms (with discussion). In *Bayesian Statistics 6*, J.M. Bernardo, J.O. Berger, A.P. Dawid, and A.F.M. Smith, eds., Oxford: Oxford University Press, pp. 301–321.

Green, P.J. and O'Hagan, A. (1998). Model choice with MCMC on product spaces without using pseudo-priors. Research Report 98-1, Department of Statistics, University of Nottingham.

Greenhouse, J.B. and Wasserman, L.A. (1995). Robust Bayesian methods for monitoring clinical trials. *Statistics in Medicine*, **14**, 1379–1391.

Greenland, S. (2006). Bayesian perspectives for epidemiological research: I. Foundations and basic methods. *International Journal of Epidemiology*, **35**, 765–775.

Greenland, S. (2007a). Prior data for non-normal priors. *Statistics in Medicine*, **26**, 3578–3590.

Greenland, S. (2007b). Bayesian perspectives for epidemiological research: II. Regresssion analysis. *International Journal of Epidemiology*, **36**, 195–202.

Greenwood, M. (1931). On the statistical measure of infectiousness. *Journal of Hygiene, Cambridge*, **31**, 336–351.

Grenander, U. and Miller, M.I. (1994). Representation of knowledge in complex systems (with discussion). *J. Roy. Statist. Soc., Ser. B*, **56**, 549–603.

Gustafson, P. (1994). Hierarchical Bayesian analysis of clustered survival data. Technical Report No. 144, Department of Statistics, University of British Columbia.

Guttman, I. (1982). *Linear Models: An Introduction.* New York: John Wiley & Sons.

Haario, H., Saksman, E., and Tamminen, J. (2001). An adaptive Metropolis algorithm. *Bernoulli*, **7**, 223–242.

Hammersley, J.M. and Handscomb, D.C. (1964). *Monte Carlo Methods*. London: Methuen.

Han, C. and Carlin, B.P. (2001). Markov chain Monte Carlo methods for computing Bayes factors: A comparative review. *J. Amer. Statist. Assoc.*, **96**, 1122–1132.

Handcock, M.S. and Stein, M.L. (1993). A Bayesian analysis of kriging. *Technometrics*, **35**, 403–410.

Handcock, M.S. and Wallis, J.R. (1994). An approach to statistical spatial-temporal modeling of meteorological fields (with discussion). *J. Amer. Statist. Assoc.*, **89**, 368–390.

Hanson, T. (2006). Inference for mixtures of finite Polya tree models. *J. Amer. Statist. Assoc.*, **101**, 1548–1565.

Hanson, T. and Johnson, W.O. (2002). Modeling regression error with a mixture of Polya trees. *J. Amer. Statist. Assoc.*, **97**, 1020–1033.

Hartigan, J.A. (1983). *Bayes Theory*. New York: Springer-Verlag.

Hastings, W.K. (1970). Monte Carlo sampling methods using Markov chains and their applications. *Biometrika*, **57**, 97–109.

Hayden, F., Fritz, R., Lobo, M., Alvord, W., Strober, W. and Straus, S. (1998). Local and systemic cytokine responses during experimental human influenza A virus infection. *Journal of Clinical Investigation*, **101**, 643–649.

Heidelberger, P. and Welch, P.D. (1983). Simulation run length control in the presence of an initial transient. *Operations Research*, **31**, 1109–1144.

Higdon, D.M. (1998). Auxiliary variable methods for Markov chain Monte Carlo with applications. *J. Amer. Statist. Assoc.*, **93**, 585–595.

Hill, J.R. (1990). A general framework for model-based statistics. *Biometrika*, **77**, 115–126.

Hills, S.E. and Smith, A.F.M. (1992). Parametrization issues in Bayesian inference (with discussion). In *Bayesian Statistics 4*, eds. J.M. Bernardo, J.O. Berger, A.P. Dawid, and A.F.M. Smith, pp. 227–246, Oxford: Oxford University Press.

Hobert, J.P. and Casella, G. (1996). The effect of improper priors on Gibbs sampling in hierarchical linear mixed models. *J. Amer. Statist. Assoc.*, **91**, 1461–1473.

Hobert, J.P., Jones, G.L., Presnell, B. and Rosenthal, J.S. (2002). On the applicability of regenerative simulation in Markov chain Monte Carlo. *Biometrika*, **89**, 731–743.

Hodges, J.S. (1998). Some algebra and geometry for hierarchical models, applied to diagnostics (with discussion). *J. Roy. Statist. Soc., Series B*, **60**, 497–536.

Hodges, J.S. and Sargent, D.J. (2001). Counting degrees of freedom in hierarchical and other richly-parameterised models. *Biometrika*, **88**, 367–379.

Hogg, R.V. and Craig, A.T. (1978). *Introduction to Mathematical Statistics*, 4th ed. New York: Macmillan.

Hui, S.L. and Berger, J.O. (1983). Empirical Bayes estimation of rates in longitudinal studies. *J. Amer. Statist. Assoc.*, **78**, 753–759.

Hurn, M. (1997). Difficulties in the use of auxiliary variables in Markov chain Monte carlo methods. *Statistics and Computing*, **7**, 35–44.

Ibrahim, J.G., Chen, M.-H., and Sinha, D. (2001). *Bayesian Survival Analysis*.

New York: Springer.

Jacobson, M.A., Besch, C.L., Child, C., Hafner, R., Matts, J.P., Muth, K., Wentworth, D.N., Neaton, J.D., Abrams, D., Rimland, D., Perez, G., Grant, I.H., Saravolatz, L.D., Brown, L.S., Deyton, L., and the Terry Beirn Community Programs for Clinical Research on AIDS (1994). Primary prophylaxis with pyrimethamine for toxoplasmic encephalitis in patients with advanced human immunodeficiency virus disease: Results of a randomized trial. *J. Infectious Diseases*, **169**, 384–394.

Jain, S. and Neal, R.M. (2007). Splitting and merging components of a nonconjugate Dirichlet process mixture model (with discussion). *Bayesian Analysis*, **2**, 445–500.

James, W. and Stein, C. (1961). Estimation with quadratic loss. In *Proc. Fourth Berkeley Symp. on Math. Statist. and Prob.*, **1**, Berkeley, CA: Univ. of California Press, pp. 361–379.

Jeffreys, H. (1961). *Theory of Probability*, 3rd ed. Oxford: University Press.

Johnson, V.E. (1998). A coupling-regeneration scheme for diagnosing convergence in Markov chain Monte Carlo algorithms. *J. Amer. Statist. Assoc.*, **93**, 238–248.

Johnson, B., Carlin, B.P., and Hodges, J.S. (1999). Cross-study hierarchical modeling of stratified clinical trial data. *Journal of Biopharmaceutical Statistics*, **9**, 617–640.

Kackar, R.N. and Harville, D.A. (1984). Approximations for standard errors of estimators of fixed and random effects in mixed linear models. *J. Amer. Statist. Assoc.*, **79**, 853–862.

Kadane, J.B. (1986). Progress toward a more ethical method for clinical trials. *Journal of Medical Philosophy*, **11**, 385–404.

Kadane, J.B., ed. (1996). *Bayesian Methods and Ethics in a Clinical Trial Design*. New York: John Wiley & Sons.

Kadane, J.B., Chan, N.H., and Wolfson, L.J. (1996). Priors for unit root models. *Journal of Econometrics*, **75**, 99–111.

Kadane, J.B., Dickey, J.M., Winkler, R.L., Smith, W.S., and Peters, S.C. (1980). Interactive elicitation of opinion for a normal linear model. *J. Amer. Statist. Assoc.*, **75**, 845–854.

Kadane, J.B. and Wolfson, L.J. (1996). Priors for the design and analysis of clinical trials. In *Bayesian Biostatistics*, eds. D. Berry and D. Stangl, New York: Marcel Dekker, pp. 157–184.

Kadane, J.B. and Wolfson, L.J. (1998). Experiences in elicitation (with discussion). *The Statistician*, **47**, 1–20 (discussion, 55–68).

Kalbfleisch, J.D. (1978). Nonparametric Bayesian analysis of survival time data. *J. Roy. Statist. Soc., Ser B*, **40**, 214–221.

Kaluzny, S.P., Vega, S.C., Cardoso, T.P., and Shelly, A.A. (1998). *S+SpatialStats: User's Manual for Windows and UNIX*. New York: Springer-Verlag.

Kass, R.E., Carlin, B.P., Gelman, A., and Neal, R. (1998). Markov chain Monte Carlo in practice: A roundtable discussion. *The American Statistician*, **52**, 93–100.

Kass, R.E. and Raftery, A.E. (1995). Bayes factors. *J. Amer. Statist. Assoc.*, **90**, 773–795.

Kass, R.E. and Steffey, D. (1989). Approximate Bayesian inference in conditionally independent hierarchical models (parametric empirical Bayes models). *J. Amer. Statist. Assoc.*, **84**, 717–726.

Kass, R.E., Tierney, L., and Kadane, J. (1988). Asymptotics in Bayesian computation (with discussion). In *Bayesian Statistics 3*, J.M. Bernardo, M.H. DeGroot, A.P. Dawid, and A.F.M. Smith, eds., Oxford: Oxford University Press, pp. 261–278.

Kass, R.E., Tierney, L., and Kadane, J. (1989). Approximate methods for assessing influence and sensitivity in Bayesian analysis. *Biometrika*, 76, 663–674.

Kass, R.E. and Vaidyanathan, S. (1992). Approximate Bayes factors and orthogonal parameters, with application to testing equality of two binomial proportions. *J. Roy. Statist. Soc. B*, **54**, 129–144.

Kass, R.E. and Wasserman, L. (1996). The selection of prior distributions by formal rules. *J. Amer. Statist. Assoc.*, **91**, 1343–1370.

Killough, G.G., Case, M.J., Meyer, K.R., Moore, R.E., Rope, S.K., Schmidt, D.W., Schleien, B., Sinclair, W.K., Voillequé, P.G., and Till, J.E. (1996). Task 6: Radiation doses and risk to residents from FMPC operations from 1951–1988. Draft report, Radiological Assessments Corporation, Neeses, SC.

Kolassa, J.E. and Tanner, M.A. (1994). Approximate conditional inference in exponential families via the Gibbs sampler. *J. Amer. Statist. Assoc.*, **89**, 697–702.

Knorr-Held, L. and Rasser, G. (2000). Bayesian detection of clusters and discontinuities in disease maps. *Biometrics*, **56**, 13–21.

Kreft, I.G.G., de Leeuw, J., and van der Leeden, R. (1994). Review of five multilevel analysis programs: BMDP-5V, GENMOD, HLM, ML3, and VARCL. *The American Statistician*, **48**, 324–335.

Krewski, D., Leroux, B.G., Bleuer, S.R., and Broekhoven, L.H. (1993). Modeling the Ames *salmonella*/microsome Assay. *Biometrics*, **49**, 499–510.

Krige, D.G. (1951). A statistical approach to some basic mine valuation problems on the Witwatersrand. *J. Chemical, Metallurgical and Mining Society of South Africa*, **52**, 119–139.

Lacson, E., Teng, M., Lazarus, J.M., Lew, N., Lowrie, E.G., and Owen, W.F. (2001). Limitations of the facility-specific standardized mortality ratio for profiling health care quality in dialysis. *American Journal of Kidney Diseases*, **37**, 267–275.

Laird, N.M. (1978). Nonparametric maximum likelihood estimation of a mixing distribution. *J. Amer. Statist. Assoc.*, **73**, 805–811.

Laird, N.M., Lange, N., and Stram, D. (1987). Maximum likelihood computations with repeated measures: Application of the EM algorithm. *J. Amer. Statist. Assoc.*, **82**, 97–105.

Laird, N.M. and Louis, T.A. (1982). Approximate posterior distributions for incomplete data problems. *J. Roy. Statist. Soc., Ser. B*, **44**, 190–200.

Laird, N.M. and Louis, T.A. (1987). Empirical Bayes confidence intervals based on bootstrap samples (with discussion). *J. Amer. Statist. Assoc.*, **82**, 739–757.

Laird, N.M. and Louis, T.A. (1989). Empirical Bayes confidence intervals for a series of related experiments. *Biometrics*, **45**, 481–495.

Laird, N.M. and Louis, T.A. (1989). Empirical Bayes ranking methods. *J. Edu-*

cational Statistics, **14**, 29–46.

Laird, N.M. and Louis, T.A. (1991). Smoothing the non-parametric estimate of a prior distribution by roughening: A computational study. *Computational Statistics and Data Analysis*, **12**, 27–37.

Laird, N.M. and Ware, J.H. (1982). Random-effects models for longitudinal data. *Biometrics*, **38**, 963–974.

Lan, K.K.G. and DeMets, D.L. (1983). Discrete sequential boundaries for clinical trials. *Biometrika*, **70**, 659–663.

Landrum, M.B., Normand, S.T., and Rosenheck, R.A. (2003). Selection of related multivariate means: Monitoring psychiatric care in the Department of Veterans Affairs. *J. Amer. Statist. Assoc.*, **98**, 7–16.

Lange, N., Carlin, B.P., and Gelfand, A.E. (1992). Hierarchical Bayes models for the progression of HIV infection using longitudinal CD4 T-cell numbers (with discussion). *J. Amer. Statist. Assoc.*, **87**, 615–632.

Lange, N. and Ryan, L. (1989). Assessing normality in random effects models. *Annals of Statistics*, **17**, 624–642.

Laplace, P.S. (1986). Memoir on the probability of the causes of events (English translation of the 1774 French original by S.M. Stigler). *Statistical Science*, **1**, 364–378.

Laud, P. and Ibrahim, J. (1995). Predictive model selection. *J. Roy. Statist. Soc., Ser. B*, **57**, 247–262.

Lauritzen, S.L. and Spiegelhalter, D.J. (1988). Local computations with probabilities on graphical structures and their application to expert systems (with discussion). *J. Roy. Statist. Soc., Ser. B*, **50**, 157–224.

Lavine, J. and Schervish, M.J. (1999). Bayes factors: What they are and what they are not. *The American Statistician*, **53**, 119–123.

Le, N.D., Sun, W., and Zidek, J.V. (1997). Bayesian multivariate spatial interpolation with data missing by design. *J. Roy. Statist. Soc., Ser. B*, **59**, 501–510.

Lee, P.M. (1997). *Bayesian Statistics: An Introduction*, 2nd ed. London: Arnold.

Lekone, P.E. and Finkenstädt, B. (2006). Statistical inference in a stochastic epidemic seir model with control intervention: Ebola as a case study. *Biometrics*, **62**, 1170–1177.

Leonard, T. and Hsu, J.S.J. (1999). *Bayesian Methods*. Cambridge: Cambridge University Press.

Li, Y. and Lin, X. (2006). Semiparametric normal transformation models for spatially correlated survival data. *J. Amer. Statist. Assoc.*, **101**, 591–603.

Liang, K.-Y. and Waclawiw, M.A. (1990). Extension of the Stein estimating procedure through the use of estimating functions. *J. Amer. Statist. Assoc.*, **85**, 435–440.

Lin, D.Y., Fischl, M.A., and Schoenfeld, D.A. (1993). Evaluating the role of CD4-lymphocyte counts as surrogate endpoints in human immunodeficiency virus clinical trials. *Statist. in Med.*, **12**, 835–842.

Lin, D.Y. and Wei, L.J. (1989). The robust inference for the Cox proportional hazards model. *J. Amer. Statist. Assoc.*, **84**, 1074–1078.

Lin, R., Louis, T.A., Paddock, S.M., and Ridgeway, G. (2006). Loss function based ranking in two-stage, hierarchical models. *Bayesian Analysis*, **1**, 915–946.

Lindley, D.V. (1972). *Bayesian statistics: A review.* Philadelphia: SIAM.

Lindley, D.V. (1983). Discussion of "Parametric empirical Bayes inference: theory and applications," by C.N. Morris. *J. Amer. Statist. Assoc.*, **78**, 61–62.

Lindley, D.V. (1990). The 1988 Wald memorial lectures: The present position in Bayesian statistics (with discussion). *Statistical Science*, **5**, 44–89.

Lindley, D.V. and Phillips, L.D. (1976). Inference for a Bernoulli process (a Bayesian view). *Amer. Statist.*, **30**, 112–119.

Lindley, D.V. and Smith, A.F.M. (1972). Bayes estimates for the linear model (with discussion). *J. Roy. Statist. Soc., Ser. B*, **34**, 1-41.

Lindsay, B.G. (1983). The geometry of mixture likelihoods: A general theory. *Ann. Statist.*, **11**, 86–94.

Lindstrom, M.J. and Bates, D.M. (1990). Nonlinear mixed effects models for repeated measures data. *Biometrics*, **46**, 673–687.

Liseo, B. (1993). Elimination of nuisance parameters with reference priors. *Biometrika*, **80**, 295–304.

Little, R.J.A. (2006). Calibrated Bayes: A Bayes/frequentist roadmap. *The American Statistician*, **60**, 213–223.

Little, R.J.A. and Rubin, D.B. (1987). *Statistical Analysis with Missing Data.* New York: John Wiley & Sons.

Liu, C.H., Rubin, D.B., and Wu, Y.N. (1998). Parameter expansion to accelerate EM: the PX-EM algorithm. *Biometrika*, **85**, 755–770.

Liu, J., Louis, T., Pan, W., Ma, J., and Collins, A. (2004). Methods for estimating and interpreting provider-specific, standardized mortality ratios. *Health Services and Outcomes Research Methodology*, **4**, 135–149.

Liu, J.S. (1994). The collapsed Gibbs sampler in Bayesian computations with applications to a gene regulation problem. *J. Amer. Statist. Assoc.*, **89**, 958–966.

Liu, J.S. (2001). *Monte Carlo Strategies in Scientific Computing.* New York: Springer.

Liu, J.S., Wong, W.H., and Kong, A. (1994). Covariance structure of the Gibbs sampler with applications to the comparisons of estimators and augmentation schemes. *Biometrika*, **81**, 27–40.

Liu, J.S. and Wu, Y.N. (1999). Parameter expansion for data augmentation. *J. Amer. Statist. Assoc.*, **94**, 1264–1274.

Lockwood, J., Louis, T., and McCaffrey, D. (2002). Uncertainty in rank estimation: Implications for value-added modeling accountability systems. *Journal of Educational and Behavioral Statistics*, **27**, 255–270.

Longini, I., Halloran, M., Nizam, A., and Yang, Y. (2004). Containing pandemic infuenza with antiviral agents. *American Journal of Epidemiology*, **159**, 623–633.

Longmate, J.A. (1990). Stochastic approximation in bootstrap and Bayesian approaches to interval estimation in hierarchical models. In *Computing Science and Statistics: Proc. 22nd Symposium on the Interface*, New York: Springer-Verlag, pp. 446–450.

Louis, T.A. (1981a). Confidence intervals for a binomial parameter after observing no successes. *The American Statistician*, **35**, 151–154.

Louis, T.A. (1981b). Nonparametric analysis of an accelerated failure time model.

Biometrika, **68**, 381–390.

Louis, T.A. (1982). Finding the observed information matrix when using the EM algorithm. *J. Roy. Statist. Soc., Ser. B*, **44**, 226–233.

Louis, T.A. (1984). Estimating a population of parameter values using Bayes and empirical Bayes methods. *J. Amer. Statist. Assoc.*, **79**, 393–398.

Louis, T.A. (1991). Using empirical Bayes methods in biopharmaceutical research (with discussion). *Statistics in Medicine*, **10**, 811–829.

Louis, T.A. and Bailey, J.K. (1990). Controlling error rates using prior information and marginal totals to select tumor sites. *J. Statistical Planning and Inference*, **24**, 297–316.

Louis, T.A. and DerSimonian, R. (1982). Health statistics based on discrete population groups. In *Regional Variations in Hospital Use*, D. Rothberg, ed., Boston: D.C. Heath & Co.

Louis, T.A. and Shen, W. (1999). Innovations in Bayes and empirical Bayes methods: Estimating parameters, populations and ranks. *Statistics in Medicine*, **18**, 2493–2505.

Lu, H., Hodges, J.S., and Carlin, B.P. (2007). Measuring the complexity of generalized linear hierarchical models. *Canadian Journal of Statistics*, **35**, 69–87.

MacArthur, R.D., Chen, L., Mayers, D.L., Besch, C.L., Novak, R., Berg-Wolf, M., Yurik, T., Peng, G., Schmetter, B., Brizz, B., and Abrams, D. for the CPCRA 058 FIRST Trial Study Team (2001). The Rationale and Design of the CPCRA (Terry Beirn Community Programs for Clinical Research on AIDS) 058 FIRST (Flexible Initial Retrovirus Suppressive Therapies) Trial. *Controlled Clinical Trials*, **22**, 176–190.

MacEachern, S.N. and Berliner, L.M. (1994). Subsampling the Gibbs sampler. *The American Statistician*, **48**, 188–190.

MacEachern, S.N. and Müller, P. (1994). Estimating mixture of Dirichlet process models. Discussion Paper 94-11, Institute for Statistics and Decision Sciences, Duke University.

MacEachern, S.N. and Peruggia, M. (2000a). Subsampling the Gibbs sampler: Variance reduction. *Statisitcs and Probability Letters*, **47**, 91–98.

MacEachern, S.N. and Peruggia, M. (2000b). Importance link function estimation for Markov chain Monte Carlo methods. *Journal of Computational and Graphical Statistics*, **9**, 99–121.

MacGibbon, B. and Tomberlin, T.J. (1989). Small area estimates of proportions via empirical Bayes techniques. *Survey Methodology*, **15**, 237–252.

Madigan, D. and Raftery, A.E. (1994). Model selection and accounting for model uncertainty in graphical models using Occam's window. *J. Amer. Statist. Assoc.*, **89**, 1535–1546.

Madigan, D. and York, J. (1995). Bayesian graphical models for discrete data. *International Statistical Review*, **63**, 215–232.

Manton, K.E., Woodbury, M.A., Stallard, E., Riggan, W.B., Creason, J.P., and Pellom, A.C. (1989). Empirical Bayes procedures for stabilizing maps of U.S. cancer mortality rates. *J. Amer. Statist. Assoc.*, **84**, 637–650.

Manu, P., Louis, T.A., Lane, T.J., Gottlieb, L., Engel, P., and Rippey, R.M. (1988). Unfavorable outcomes of drug therapy: Subjective probability versus confidence intervals. *J. Clin. Pharm. and Therapeutics*, **13**, 213–217.

Marin, J.-M. and Robert, C.P. (2007) *Bayesian Core: A Practical Approach to Computational Bayesian Statistics*. New York: Springer.

Maritz, J.S. and Lwin, T. (1989). *Empirical Bayes Methods*, 2nd ed. London: Chapman and Hall.

Marriott, J., Ravishanker, N., Gelfand, A., and Pai, J. (1996). Bayesian analysis of ARMA processes: Complete sampling-based inference under exact likelihoods. In *Bayesian Analysis in Statistics and Econometrics: Essays in Honor of Arnold Zellner*, eds. D. Berry, K. Chaloner, and J. Geweke, New York: John Wiley & Sons, pp. 243–256.

Massad, E., Burattini, M.N., Coutinho, F.A.B., and Lopez, L.F. (2006). The 1918 influenza A epidemic in the city of São Paulo, Brazil. *Medical Hypotheses*, **68**, 442–445.

McCullagh, P. and Nelder, J.A. (1989). *Generalized Linear Models*, 2nd ed. London: Chapman and Hall.

McNeil, A.J. and Gore, S.M. (1996). Statistical analysis of zidovudine (AZT) effect on CD4 cell counts in HIV disease. *Statistics in Medicine*, **15**, 75–92.

Meeden, G. (1972). Some admissible empirical Bayes procedures. *Ann. Math. Statist.*, **43**, 96–101.

Meilijson, I. (1989). A fast improvement of the EM algorithm on its own terms. *J. Roy. Statist. Soc., Ser B*, **51**, 127–138.

Meng, X.-L. (1994). Posterior predictive *p*-values. *Ann. Statist.*, **22**, 1142–1160.

Meng, X.-L. and Rubin, D.B. (1991). Using EM to obtain asymptotic variance-covariance matrices: The SEM algorithm. *J. Amer. Statist. Assoc.*, **86**, 899–909.

Meng, X.-L. and Rubin, D.B. (1992). Recent extensions to the EM algorithm (with discussion). In *Bayesian Statistics 4*, J.M. Bernardo, J.O. Berger, A.P. Dawid, and A.F.M. Smith, eds., Oxford: Oxford University Press, pp. 307–315.

Meng, X.-L. and Rubin, D.B. (1993). Maximum likelihood estimation via the ECM algorithm: A general framework. *Biometrika*, **80**, 267–278.

Meng, X.-L. and Van Dyk, D. (1997). The EM algorithm – an old folk-song sung to a fast new tune (with discussion). *J. Roy. Statist. Soc., Ser. B*, **59**, 511–567.

Meng, X.-L. and Van Dyk, D. (1999). Seeking efficient data augmentation schemes via conditional and marginal augmentation. *Biometrika*, **86**, 301–320.

Mengersen, K.L., Robert, C.P., and Guihenneuc-Jouyaux, C. (1999). MCMC convergence diagnostics: A reviewww (with discussion). In *Bayesian Statistics 6*, J.M. Bernardo, J.O. Berger, A.P. Dawid, and A.F.M. Smith, eds., Oxford: Oxford University Press, pp. 415–440.

Metropolis, N., Rosenbluth, A.W., Rosenbluth, M.N., Teller, A.H., and Teller, E. (1953). Equations of state calculations by fast computing machines. *J. Chemical Physics*, **21**, 1087–1091.

Meyn, S.P. and Tweedie, R.L. (1993). *Markov Chains and Stochastic Stability*. London: Springer-Verlag.

Mills, C.E., Robins, J.M., and Lipsitch, M. (2004). Transmissibility of 1918 pandemic influenza. *Nature*, **432**, 904.

Mira, A. (1998). *Ordering, Slicing and Splitting Monte Carlo Markov Chains*. Unpublished Ph.D. thesis, Department of Statistics, University of Minnesota.

Mira, A., Møller, J., and Roberts, G.O. (1999). Perfect slice samplers. *J. Roy.*

Statist. Soc., Ser. B, **63**, 593–606.

Mira, A. and Sargent, D. (2003). A new strategy for speeding Markov chain Monte Carlo algorithms. *Statistical Methods and Applications,* **1**, 49–60.

Mira, A. and Tierney, L. (2001). Efficiency and convergence properties of slice samplers. *Scandinavian Journal of Statistics,* **29**, 1–12.

Mood, A.M., Graybill, F.A., and Boes, D.C. (1974). *Introduction to the Theory of Statistics,* 3rd ed., New York: McGraw-Hill.

Morris, C.N. (1983a). Parametric empirical Bayes inference: Theory and applications. *J. Amer. Statist. Assoc.,* **78**, 47–65.

Morris, C.N. (1983b). Natural exponential families with quadratic variance functions: Statistical theory. *Ann. Statist.,* **11**, 515–529.

Morris, C.N. (1988). Determining the accuracy of Bayesian empirical Bayes estimators in the familiar exponential families. In *Statistical Decision Theory and Related Topics,* eds. S.S. Gupta and J.O. Berger, New York: Springer-Verlag, pp. 327–344.

Morris, C.N. and Normand, S.-L. (1992). Hierarchical models for combining information and for meta-analyses (with discussion). In *Bayesian Statistics 4,* J.M. Bernardo, J.O. Berger, A.P. Dawid, and A.F.M. Smith, eds., Oxford: Oxford University Press, pp. 321–344.

Mosteller, F. and Wallace, D.L. (1964). *Inference and Disputed Authorship: The Federalist.* Reading, MA: Addison-Wesley.

Mukhopadhyay, S. and Gelfand, A.E. (1997). Dirichlet process mixed generalized linear models. *J. Amer. Statist. Assoc.,* **92**, 633–639.

Mugglin, A.S., Carlin, B.P., and Gelfand, A.E. (2000). Fully model based approaches for spatially misaligned data. *J. Amer. Statist. Assoc.,* **95**, 877–887.

Müller, P. (1991). A generic approach to posterior integration and Gibbs sampling. Technical Report 91-09, Department of Statistics, Purdue University.

Müller, P., Berry, D., Grieve, A., Smith, M., and Krams, M. (2007). Simulation-based sequential Bayesian design. *Journal of Statistical Planning and Inference,* **137**, 3140–3150.

Müller, P. and Parmigiani, G. (1995). Optimal design via curve fitting of Monte Carlo experiments. *J. Amer. Statist. Assoc.,* **90**, 1322–1330.

Myers, L.E., Adams, N.H., Hughes, T.J., Williams, L.R., and Claxton, L.D. (1987). An interlaboratory study of an EPA/Ames/salmonella test protocol. *Mutation Research,* **182**, 121–133.

Mykland, P., Tierney, L., and Yu, B. (1995). Regeneration in Markov chain samplers. *J. Amer. Statist. Assoc.,* **90**, 233–241.

Natarajan, R. and McCulloch, C.E. (1995). A note on the existence of the posterior distribution for a class of mixed models for binomial responses. *Biometrika,* **82**, 639–643.

Natarajan, R. and Kass, R.E. (2000). Reference Bayesian methods for generalized linear mixed models. *J. Amer. Statist. Assoc.,* **95**, 227–237.

National Research Council (2000). *Small-Area Income and Poverty Estimates: Priorities for 2000 and Beyond.* Panel on Estimates of Poverty for Small Geographic Areas, Committee on National Statistics, C.F. Citro and G. Kalton, eds. Washington, D.C.: National Academy Press.

Neal, R.M. (1996a). Sampling from multimodal distributions using tempered transitions. *Statistics and Computing*, **6**, 353–366.

Neal, R.M. (1996b). *Bayesian Learning for Neural Networks*, New York: Springer-Verlag.

Neal, R.M. (1997). Markov chain Monte Carlo methods based on 'slicing' the density function. Technical Report No. 9722, Dept. of Statistics, University of Toronto.

Neal, R.M. (1998). Suppressing random walks in Markov Chain Monte Carlo using ordered overrelaxation. In *Learning in Graphical Models*, M.I. Jordan, ed., Dordrecht: Kluwer Academic Publishers, pp. 205–228.

Neaton, J.D., Normand, S.-L., Gelijns, A., Starling, R.C., Mann, D.L., and Konstam, M.A. for the HFSA Working Group (2007). Designs for mechanical circulatory support device studies. *J. Cardiac Failure*, **13**, 63–74.

Newton, M.A., Guttorp, P., and Abkowitz, J.L. (1992). Bayesian inference by simulation in a stochastic model from hematology. *Computing Science and Statistics*, **24**, 449–455.

Newton, M.A. and Raftery, A.E. (1994). Approximate Bayesian inference by the weighted likelihood bootstrap (with discussion). *J. Roy. Statist. Soc., Ser. B*, **56**, 1–48.

Nummelin, E. (1984). *General Irreducible Markov Chains and Non-Negative Operators*. Cambridge: Cambridge University Press.

Núñez, M., Soriano, V., Martín-Carbonero, L., Barrios, A., Barreiro, P., Blanco, F., García-Benayas, T., and González-Lahoz, J. (2002). SENC (Spanish Efavirenz vs. Nevirapine Comparison) Trial: A randomized, open-label study in HIV-infected naive individuals. *HIV Clinical Trials*, **3**, 186–194.

Oakley, J.E. and O'Hagan, A. (2007). Uncertainty in prior elicitations: a nonparametric approach. *Biometrika*, **94**, 427–441.

O'Brien, P.C. and Fleming, T.R., (1979). A multiple testing procedure for clinical trials. *Biometrics*, **35**, 549–556.

Odell, P.L. and Feiveson, A.H. (1966). A numerical procedure to generate a sample covariance matrix. *J. Amer. Statist. Assoc.*, **61**, 198–203.

O'Hagan, A. (1995). Fractional Bayes factors for model comparison (with discussion). *J. Roy. Statist. Soc., Ser. B*, **57**, 99–138.

O'Hagan, A. (1997). Properties of intrinsic and fractional Bayes factors. *Test*, **6**, 101–118.

O'Hagan, A. (2003). HSSS model criticism (with discussion). In *Highly Structured Stochastic Systems*, eds. P.J. Green, N.L. Hjort, and S.T. Richardson, Oxford: Oxford University Press, pp. 423–445.

O'Hagan, A. and Berger, J.O. (1988). Ranges of posterior probabilities for quasi-unimodal priors with specified quantiles. *J. Am. Statist. Assoc.*, **83**, 503–508.

O'Hagan, A., Buck, C.E., Daneshkhah, A., Eiser, J.R., Garthwaite, P.H., Jenkinson, D.J., Oakley, J.E. and Rakow, T. (2006). *Uncertain Judgements: Eliciting Experts' Probabilities*. Chichester: John Wiley & Sons.

O'Hagan, A. and Forster, J. (2004). *Bayesian Inference: Kendall's Advanced Theory of Statistics Volume 2B*, 2nd ed. London: Edward Arnold.

O'Hagan, A. and Stevens, J.W. (2001). Bayesian assessment of sample size for clinical trials of cost-effectiveness. *Medical Decision Making*, **21**, 219–230.

Olschewski, M., Schumacher, M., and Davis, K.B. (1992). Analysis of randomized and non-randomized patients in clinical trials using the comprehensive cohort-follow-up study design. *Controlled Clinical Trials*, **13**, 226–239.

O'Neill, P. (2002). A tutorial introduction to Bayesian inference for stochastic epidemic models using Markov chain Monte Carlo methods. *Mathematical Biosciences*, **180**, 103–114.

Paddock, S.M., Ridgeway, G., Lin, R., and Louis, T.A. (2006). Flexible distributions for triple-goal estimates in two-stage hierarchical models. *Computational Statistics and Data Analysis*, **50**, 3243–3262.

Pauler, D.K. (1998). The Schwarz criterion and related methods for normal linear models. *Biometrika*, **85**, 13–27.

Phillips, D.B. and Smith, A.F.M. (1996). Bayesian model comparison via jump diffusions. In *Markov Chain Monte Carlo in Practice*, eds. W.R. Gilks, S. Richardson, and D.J. Spiegelhalter, London: Chapman and Hall, pp. 215–239.

Pinheiro, J.C. and Bates, D.M. (2000). *Mixed-Effects Models in S and S-PLUS*. New York: Springer-Verlag.

Pocock, S.J. (1976). The combination of randomized and historical controls in clinical trials. *J. Chronic Disease*, **29**, 175–188.

Pocock, S.J. (1977). Group sequential methods in the design and analysis of clinical trials. *Biometrika*, **64**, 191–199.

Pole, A., West, M., and Harrison, P.J. (1994). *Applied Bayesian Forecasting and Time Series Analysis*. London: Chapman and Hall.

Polson, N.G. (1996). Convergence of Markov chain Monte Carlo algorithms (with discussion). In *Bayesian Statistics 5*, eds. J.M. Bernardo, J.O. Berger, A.P. Dawid, and A.F.M. Smith, Oxford: Oxford University Press, pp. 297–322.

Prentice, R.L. (1976). A generalization of the probit and logit model for dose response curves. *Biometrics*, **32**, 761–768.

Prentice, R.L., Langer, R.D., Stefanick, M.L., Howard, B.V., Pettinger, M., Anderson, G.L., Barad, D., Curb, J.D., Kotchen, J., Kuller, L., Limacher, M., and Wactawski-Wende, J. for the Women's Health Initiative Investigators (2006). Combined analysis of Women's Health Initiative observational and clinical trial data on postmenopausal hormone treatment and cardiovascular disease. *Amer. J. Epid.*, **163**, 589–599.

Press, S.J. (1982). *Applied Multivariate Analysis: Using Bayesian and Frequentist Methods of Inference*, 2nd ed. New York: Krieger.

Propp, J.G. and Wilson, D.B. (1996). Exact sampling with coupled Markov chains and applications to statistical mechanics. *Random Structures and Algorithms*, **9**, 223–252.

R Development Core Team (2006). *R: A Language and Environment for Statistical Computing*. Vienna, Austria: R Foundation for Statistical Computing. ISBN 3-900051-07-0.

Racine, A., Grieve, A.P., Fluher, H., and Smith, A.F.M. (1986). Bayesian methods in practice: Experiences in the pharmaceutical industry. *Applied Statistics*, **35**, 93–120.

Raftery, A.E. (1996). Approximate Bayes factors and accounting for model uncertainty in generalised linear models. *Biometrika*, **83**, 251–266.

Raftery, A.E. and Lewis, S. (1992). How many iterations in the Gibbs sampler? In *Bayesian Statistics 4*, J.M. Bernardo, J.O. Berger, A.P. Dawid, and A.F.M.

Smith, eds., Oxford: Oxford University Press, pp. 763–773.

Raftery, A.E., Madigan, D., and Hoeting, J.A. (1997). Bayesian model averaging for linear regression models. *J. Amer. Statist. Assoc.*, **92**, 179–191.

Ratkowski, D.A. (1983). *Non-linear Regression Modeling.* New York: Marcel Dekker.

Reich, B.J., Hodges, J.S., and Carlin, B.P. (2007). Spatial analysis of periodontal data using conditionally autoregressive priors having two classes of neighbor relations. *J. Amer. Statist. Assoc.*, **102**, 44–55.

Ribeiro, P. and Diggle, P. (2001). geoR: A package for geostatistical analysis. *R News*, **1**, 14–18.

Richardson, S. and Green, P.J. (1997). Bayesian analysis of mixtures with an unknown number of components. *J. Roy. Statist. Soc., Ser. B* **59**, 731–758.

Ripley, B.D. (1987). *Stochastic Simulation.* New York: John Wiley & Sons.

Ritter, C. and Tanner, M.A. (1992). Facilitating the Gibbs sampler: The Gibbs stopper and the griddy Gibbs sampler. *J. Amer. Statist. Assoc.*, **87**, 861–868.

Robbins, H. (1955). An empirical Bayes approach to statistics. In *Proc. 3rd Berkeley Symp. on Math. Statist. and Prob.*, **1**, Berkeley, CA: Univ. of California Press, pp. 157–164.

Robbins, H. (1983). Some thoughts on empirical Bayes estimation. *Ann. Statist.*, **1**, 713–723.

Robert, C.P. (2001). *The Bayesian Choice*, 2nd ed. New York: Springer-Verlag.

Robert, C.P. (1995). Convergence control methods for Markov chain Monte Carlo algorithms. *Statistical Science*, **10**, 231–253.

Robert, C.P. and Casella, G. (1999). *Monte Carlo Statistical Methods.* New York: Springer-Verlag.

Robert, C.P. and Marin, J.-M. (2008). On some difficulties with a posterior probability approximation technique. To appear in *Bayesian Analysis*.

Roberts, G.O. (1996). Markov chain concepts related to sampling algorithms. In *Markov Chain Monte Carlo in Practice*, eds. W.R. Gilks, S. Richardson, and D.J. Spiegelhalter, London: Chapman and Hall, pp. 45–57.

Roberts, G.O. and Rosenthal, J.S. (1998). Optimal scaling of discrete approximations to Langevin diffusions. *J. Roy. Statist. Soc., Ser. B*, **60**, 255–268.

Roberts, G.O. and Rosenthal, J.S. (1999). Convergence of slice samper Markov chains. *J. Roy. Statist. Soc., Ser. B*, **61**, 643–660.

Roberts, G.O. and Rosenthal, J.S. (2001). Optimal scaling for various Metropolis-Hastings algorithms. *Statistical Science*, **16**, 351–367.

Roberts, G.O. and Rosenthal, J.S. (2002). The polar slice sampler. *Stochastic Models* **18**, 257–280.

Roberts, G.O. and Rosenthal, J.S. (2007). Coupling and ergodicity of adaptive Markov chain Monte Carlo algorithms. *J. Applied Probability*, **44**, 458–475.

Roberts, G.O. and Sahu, S.K. (1997). Updating schemes, correlation structure, blocking and parameterization for the Gibbs sampler. *J. Roy. Statist. Soc., Ser. B*, **59**, 291–317.

Roberts, G.O. and Smith, A.F.M. (1993). Simple conditions for the convergence of the Gibbs sampler and Metropolis-Hastings algorithms. *Stochastic Processes and their Applications*, **49**, 207–216.

Roberts, G.O. and Stramer, O. (2003) Langevin diffusions and Metropolis-

Hastings algorithm. *Methodology and Computing in Applied Probability*, **4**, 337–358.

Roberts, G.O. and Tweedie, R.L. (1996a). Exponential convergence of Langevin diffusions and their discrete approximations. *Bernoulli*, **2**, 341–363.

Roberts, G.O. and Tweedie, R.L. (1996b). Geometric convergence and central limit theorems for multidimensional Hastings and Metropolis algorithms. *Biometrika*, **83**, 95–110.

Robertson, T., Wright, F.T., and Dykstra, R.L. (1988). *Order Restricted Statistical Inference*. Chichester: John Wiley & Sons.

Rose, E.A., Gelijns, A.C., Moskowitz, A.J., Heitjan, D.F., Stevenson, L.W., Dembitsky, W., Long, J.W., Ascheim, D.D., Tierney, A.R., Levitan, R.G., Watson, J.T., and Meier, P. for the Randomized Evaluation of Mechanical Assistance for the Treatment of Congestive Heart Failure study group (2001). Long-term use of a left ventricular assist device for end-stage heart failure. *New England J. Med.*, **345**, 1435–43.

Rosenthal, J.S. (1993). Rates of convergence for data augmentation on finite sample spaces. *Ann. App. Prob.*, **3**, 819–839.

Rosenthal, J.S. (1995a). Rates of convergence for Gibbs sampling for variance component models. *Ann. Statist.*, **23**, 740–761.

Rosenthal, J.S. (1995b). Minorization conditions and convergence rates for Markov chain Monte Carlo. *J. Amer. Statist. Assoc.*, **90**, 558–566.

Rosenthal, J.S. (1996). Analysis of the Gibbs sampler for a model related to James-Stein estimators. *Statistics and Computing*, **6**, 269–275.

Rossell, D., Müller, P. and Rosner, G. (2007). Screening designs for drug development. *Biostatistics*, **8**, 595–608.

Rubin, D.B. (1980). Using empirical Bayes techniques in the law school validity studies (with discussion). *J. Amer. Statist. Assoc.*, **75**, 801–827.

Rubin, D.B. (1984). Bayesianly justifiable and relevant frequency calculations for the applied statistician. *Ann. Statist.*, **12**, 1151–1172.

Rubin, D.B. (1988). Using the SIR algorithm to simulate posterior distributions (with discussion). In *Bayesian Statistics 3*, J.M. Bernardo, M.H. DeGroot, D.V. Lindley, and A.F.M. Smith, eds., Oxford: Oxford University Press, pp. 395–402.

Samaniego, F.J. and Reneau, D.M. (1994). Toward a reconciliation of the Bayesian and frequentist approaches to point estimation. *J. Amer. Statist. Assoc.*, **89**, 947–957.

Sargent, D.J. (1995). A general framework for random effects survival analysis in the Cox proportional hazards setting. Research Report 95–004, Division of Biostatistics, University of Minnesota.

Sargent, D.J. and Carlin, B.P. (1996). Robust Bayesian design and analysis of clinical trials via prior partitioning (with discussion). In *Bayesian Robustness*, IMS Lecture Notes – Monograph Series, **29**, eds. J.O. Berger et al., Hayward, CA: Institute of Mathematical Statistics, pp. 175–193.

Sargent, D.J., Hodges, J.S., and Carlin, B.P. (2000). Structured Markov chain Monte Carlo. *Journal of Computational and Graphical Statistics*, **9**, 217–234.

Schervish, M.J. and Carlin, B.P. (1992). On the convergence of successive substitution sampling. *J. Computational and Graphical Statistics*, **1**, 111–127.

Schluchter, M.D. (1988). "5V: Unbalanced repeated measures models with structured covariance matrices," in *BMDP Statistical Software Manual, Vol. 2*, W.J. Dixon, ed., University of California Press: pp. 1081–1114.

Schmeiser, B. and Chen, M.H. (1991). On random-direction Monte Carlo sampling for evaluating multidimensional integrals. Technical Report #91-39, Department of Statistics, Purdue University.

Schmittlein, D.C. (1989). Surprising inferences from unsurprising observations: Do conditional expectations really regress to the mean? *The American Statistician*, **43**, 176–183.

Schumacher, M., Olschewski, M., and Schmoor, C. (1987). The impact of heterogeneity on the comparison of survival times. *Statist. in Med.*, **6**, 773–784.

Schwarz, G. (1978). Estimating the dimension of a model. *Ann. Statist.*, **6**, 461–464.

Scott, S.L. (2002). Bayesian methods for hidden Markov models: Recursive computing in the 21st century. *J. Amer. Statist. Assoc.*, **97**, 337–351.

Searle, S.R., Casella, G., and McCulloch, C.E. (1992). *Variance Components*. New York: John Wiley & Sons.

Sertsou, G., Wilson, N., Baker, M., Nelson, P., and Roberts, M. (2006). Key transmission parameters of an institutional outbreak during the 1918 influenza pandemic estimated by mathematical modelling. *Theoretical Biology and Medical Modelling*, **3(1)**, 38.

Shen, W. and Louis, T.A. (1998). Triple-goal estimates in two-stage, hierarchical models. *J. Roy. Statist. Soc., Ser. B*, **60**, 455–471.

Shen, W. and Louis, T.A. (1999). Empirical Bayes estimation via the smoothing by roughening approach. *J. Comp. Graph. Statist.*, **8**, 800–823.

Shen, W. and Louis, T.A. (2000). Triple-goal estimates for disease mapping. *Statistics in Medicine*, **19**, 2295–2308.

Shih, J. (1995). Sample size calculation for complex clinical trials with survival endpoints. *Controlled Clinical Trials*, **16**, 395–407.

Silverman, B.W. (1986). *Density Estimation for Statistics and Data Analysis*. London: Chapman and Hall.

Simar, L. (1976). Maximum likelihood estimation of a compound Poisson process. *Ann. Statist.*, **4**, 1200–1209.

Sinha, D. (1993). Semiparametric Bayesian analysis of multiple event time data. *J. Amer. Statist. Assoc.*, **88**, 979–983.

Sinha, D. and Dey, D.K. (1997). Semiparametric Bayesian analysis of survival data. *J. Amer. Statist. Assoc.*, **92**, 1195-1212.

Sisson, S.A. (2005). Transdimensional Markov chains: A decade of progress and future perspectives. *J. Amer. Statist. Assoc.*, **100**, 1077–1089.

Smith, A.F.M. and Gelfand, A.E. (1992). Bayesian statistics without tears: A sampling-resampling perspective. *The American Statistician*, **46**, 84–88.

Smith, A.F.M. and Roberts, G.O. (1993). Bayesian computation via the Gibbs sampler and related Markov chain Monte Carlo methods. *J. Roy. Statist. Soc., Ser. B*, **55**, 1–24.

Snedecor, G.W. and Cochran, W.G. (1980). *Statistical Methods*, 7th ed. Ames, Iowa: The Iowa State University Press.

Spiegelhalter, D.J. (2000). Bayesian methods for cluster randomised trials with continuous responses. Technical report, Medical Research Council Biostatistics Unit, Institute of Public Health, Cambridge University.

Spiegelhalter, D.J., Abrams, K.R., and Myles, J.P. (2004). *Bayesian Approaches to Clinical Trials and Health-Care Evaluation*. Chichester: John Wiley & Sons.

Spiegelhalter, D.J., Best, N., Carlin, B.P., and van der Linde, A. (2002). Bayesian measures of model complexity and fit (with discussion). *J. Roy. Statist. Soc., Ser. B*, **64**, 583–639.

Spiegelhalter, D.J., Best, N.G., Thomas, A., and Lunn, D.J. (2008). *Bayesian Analysis using BUGS: A Practical Introduction*. London: Chapman and Hall/CRC Press.

Spiegelhalter, D.J., Dawid, A.P., Lauritzen, S.L., and Cowell, R.G. (1993). Bayesian analysis in expert systems (with discussion). *Statistical Science*, **8**, 219–283.

Spiegelhalter, D.J., Freedman, L.S., and Parmar, M.K.B. (1994). Bayesian approaches to randomised trials (with discussion). *J. Roy. Statist. Soc., Ser. A*, **157**, 357–416.

Spiegelhalter, D.J., Myles, J.P., Jones, D.R., and Abrams, K.R. (2000). Bayesian methods in health technology assessment: A review. *Health Technology Assessment*, **4**, No. 38.

Spiegelhalter, D.J., Thomas, A., Best, N., and Gilks, W.R. (1995a). BUGS: Bayesian inference using Gibbs sampling, Version 0.50. Technical report, Medical Research Council Biostatistics Unit, Institute of Public Health, Cambridge University.

Spiegelhalter, D.J., Thomas, A., Best, N., and Gilks, W.R. (1995b). BUGS examples, Version 0.50. Technical report, Medical Research Council Biostatistics Unit, Institute of Public Health, Cambridge University.

Spiegelhalter, D.J., Thomas, A., Best, N., and Lunn, D. (2004). *WinBUGS User Manual, Version 1.4.1*. Cambridge, UK: Medical Research Council Biostatistics Unit, Institute of Public Health, Cambridge University.

Steensma, D.P, Molina, R., Sloan, J.A., Nikcevich, D.A., Schaefer, P.L., Rowland, K.M., Dentchev, T., Tschetter, L.K., Novotny, P.J., and Loprinzi, C.L. for the North Central Cancer Treatment Group. (2005). A Phase III Randomized Trial of Two Different Dosing Schedulesof Erythropoietin (EPO) in Patients With Cancer-Associated Anemia: North Central Cancer Treatment. *Journal of Clinical Oncology, 2005 ASCO Annual Meeting Proceedings*, **23**, No. 16S, Part I of II (June 1 Supplement), 8031.

Stein, C. (1955). Inadmissibility of the usual estimator for the mean of a multivariate normal distribution. In *Proc. Third Berkeley Symp. on Math. Statist. and Prob.*, **1**, Berkeley, CA: Univ. of California Press, pp. 197–206.

Stein, C. (1981). Estimation of the parameters of a multivariate normal distribution: I. Estimation of the means. *Annals of Statistics*, **9**, 1135–1151.

Stephens, D.A. and Smith, A.F.M. (1992). Sampling-resampling techniques for the computation of posterior densities in normal means problems. *Test*, **1**, 1–18.

Stern, H.S. and Cressie, N. (1999). Inference for extremes in disease mapping. In *Disease Mapping and Risk Assessment for Public Health*, A. Lawson, A. Biggeri, D. Böhning, E. Lesaffre, J.-F. Viel, and R. Bertollini, eds., New York:

John Wiley & Sons, pp. 63–84.

Stigler, S.M. (2005). Fisher in 1921. *Statistical Science*, **20**, 32–49.

Stoffer, D.S., Scher, M.S., Richardson, G.A., Day, N.L., and Coble, P.A. (1988). A Walsh-Fourier analysis of the effects of moderate maternal alcohol consumption on neonatal sleep-state cycling. *J. Amer. Statist. Assoc.*, **83**, 954–963.

Swendsen, R.H. and Wang, J.-S. (1987). Nonuniversal critical dynamics in Monte Carlo simulations. *Phys. Rev. Letters*, **58**, 86–88.

Tamura, R.N. and Young, S.S. (1986). The incorporation of historical control information in tests of proportions: Simulation study of Tarone's procedure. *Biometrics*, **42**, 343–349.

Tan, M., Tian, G.-L., and Ng, K. (2008). *Bayesian Missing Data Problems: EM, Data Augmentation, and Noniterative Computation*. Boca Raton, FL: Chapman and Hall/CRC Press.

Tanner, M.A. (1993). *Tools for Statistical Inference: Methods for the Exploration of Posterior Distributions and Likelihood Functions*, 2nd ed. New York: Springer-Verlag.

Tanner, M.A. and Wong, W.H. (1987). The calculation of posterior distributions by data augmentation (with discussion). *J. Amer. Statist. Assoc.*, **82**, 528–550.

Tarone, R.E. (1982). The use of historical control information in testing for a trend in proportions. *Biometrics*, **38**, 215–220.

Ten Have, T.R. and Localio, A.R. (1999). Empirical Bayes estimation of random effects parameters in mixed effects logistic regression models. *Biometrics*, **55**, 1022–1029.

Thall, P.F., Simon, R.M., and Estey, E.H. (1995). Bayesian sequential monitoring designs for single-arm clinical trials with multiple outcomes. *Statistics in Medicine*, **14**, 357–379.

Thisted, R.A. (1988). *Elements of Statistical Computing*. London: Chapman and Hall.

Tierney, L. (1990). *LISP-STAT – An Object-Oriented Environment for Statistical Computing and Dynamic Graphics*. New York: John Wiley & Sons.

Tierney, L. (1994). Markov chains for exploring posterior distributions (with discussion). *Ann. Statist.*, **22**, 1701–1762.

Tierney, L. and Kadane, J.B. (1986). Accurate approximations for posterior moments and marginal densities. *J. Amer. Statist. Assoc.*, **81**, 82-86.

Tierney, L., Kass, R.E., and Kadane, J.B. (1989). Fully exponential Laplace approximations to expectations and variances of nonpositive functions. *J. Amer. Statist. Assoc.*, **84**, 710–716.

Titterington, D.M., Makov, U.E. and Smith, A.F.M. (1985). *Statistical Analysis of Finite Mixture Distributions*. Chichester: John Wiley & Sons.

Tolbert, P., Mulholland, J., MacIntosh, D., Xu, F., Daniels, D., Devine, O., Carlin, B.P., Butler, A., Nordenberg, D., and White, M. (2000). Air quality and pediatric emergency room visits for asthma in Atlanta. *Amer. J. Epidemiology*, **151:8**, 798–810.

Tomberlin, T.J. (1988). Predicting accident frequencies for drivers classified by two factors. *J. Amer. Statist. Assoc.*, **83**, 309–321.

Treloar, M.A. (1974). Effects of Puromycin on galactosyltransferase of golgi membranes. Unpublished Master's thesis, University of Toronto.

Troughton, P.T. and Godsill, S.J. (1998). A reversible jump sampler for autoregressive time series, employing full conditionals to achieve efficient model space moves. In *Proceedings of the IEEE International Conference on Acoustic, Speech and Signal Processing*, **4**, 2257–2260.

Tsiatis, A.A. (1990). Estimating regression parameters using linear rank tests for censored data. *Annals of Statistics*, **18**, 354–372.

Tsutakawa, R.K., Shoop, G.L., and Marienfeld, C.J. (1985). Empirical Bayes estimation of cancer mortality rates. *Statist. in Med.*, **4**, 201–212.

Tukey, J.W. (1974). Named and faceless values: An initial exploration in memory of Prasanta C. Mahalanobis. *Sankhya, Series A*, **36**, 125-176.

United States Renal Data System (2005). Annual Data Report: Atlas of end-stage renal disease in the United States. Technical report, Health Care Financing Administration.

van der Linde, A. (2004). On the association between a random parameter and an observable. *Test*, **13**, 85–111.

van der Linde, A. (2005). DIC in variable selection. *Statistica Neerlandica*, **59**, 45–56.

van Houwelingen, J.C. (1977). Monotonizing empirical Bayes estimators for a class of discrete distributions with monotone likelihood ratio. *Statist. Neerl.*, **31**, 95–104.

van Houwelingen, J.C. and Thorogood, J. (1995). Construction, validation and updating of a prognostic model for kidney graft survival. *Statistics in Medicine*, **14**, 1999–2008.

van Leth, F., Phanuphak, P., Ruxrungtham, K., Baraldi, E., Miller, S., Gazzard, B., Cahn, P., Lalloo, U.G., van der Westhuizen, I.P., Malan, D.R., Johnson, M.A., Santos, B.R., Mulcahy, F., Wood, R., Levi, G.C., Reboredo, G., Squires, K., Cassetti, I., Petit, D., Raffi, F., Katlama, C., Murphy, R.L., Horban, A., Dam, J.P., Hassink, E., van Leeuwen, R., Robinson, P., Wit, F.W., and Lange, J.M.A. for the 2NN Study team (2004). Comparison of first-line antiretroviral therapy with regimens including nevirapine, efavirenz, or both drugs, plus stavudine and lamivudine: a randomised open-label trial, the 2NN Study. *Lancet*, **363**, 1253–1263.

van Ryzin, J. and Susarla, J.V. (1977). On the empirical Bayes approach to multiple decision problems. *Ann. Statist.*, **5**, 172–181.

Vaupel, J.W., Manton, K.G., and Stallard, E. (1979). The impact of heterogeneity in individual frailty on the dynamics of mortality. *Demography*, **16**, 439–454.

Vaupel, J.W. and Yashin, A.I. (1985). Heterogeneity's ruses: Some surprising effects of selection on population dynamics. *The American Statistician*, **39**, 176–185.

Venables, W.N. and Ripley, B.D. (2002). *Modern Applied Statistics with S*, 4th ed. New York: Springer.

Verdinelli, I. (1992). Advances in Bayesian experimental design (with discussion). In *Bayesian Statistics 4*, J.M. Bernardo, J.O. Berger, A.P. Dawid, and A.F.M. Smith, eds., Oxford: Oxford University Press, pp. 467–481.

Volinsky, C.T. and Raftery, A.E. (2000). Bayesian information criterion for censored survival models. *Biometrics*, **56**, 256–262.

Von Mises, R. (1942). On the correct use of Bayes' formula. *Ann. Math. Statist.*,

13, 156–165.

Wagener, D.K. and Williams, D.R. (1993). Equity in environmental health: Data collection and interpretation issues. *Toxicology and Industrial Health*, **9**, 775–795.

Wakefield, J. (1996). The Bayesian analysis of population pharmacokinetic models. *J. Amer. Statist. Assoc.*, **91**, 62–75.

Wakefield, J.C. (1998). Discussion of "Some algebra and geometry for hierarchical models, applied to diagnostics," by J.S. Hodges. *J. Roy. Stat. Soc., Ser. B*, **60**, 523–525.

Wakefield, J.C., Gelfand, A.E., and Smith, A.F.M. (1991). Efficient generation of random variates via the ratio-of-uniforms method. *Statistics and Computing*, **1**, 129–133.

Wakefield, J.C., Smith, A.F.M., Racine-Poon, A., and Gelfand, A.E. (1994). Bayesian analysis of linear and nonlinear population models using the Gibbs sampler. *J. Roy. Statist. Soc., Ser. C (Applied Statistics)*, **43**, 201–221.

Walker, S. (1995). Generating random variates from *D*-distributions via substitution sampling. *Statistics in Computing*, **5**, 311–315.

Waller, L.A., Carlin, B.P., Xia, H., and Gelfand, A.E. (1997). Hierarchical spatio-temporal mapping of disease rates. *J. Amer. Statist. Assoc.*, **92**, 607–617.

Waller, L.A., Turnbull, B.W., Clark, L.C., and Nasca, P. (1994). Spatial pattern analyses to detect rare disease clusters. In *Case Studies in Biometry*, N. Lange, L. Ryan, L. Billard, D. Brillinger, L. Conquest, and J. Greenhouse, eds., New York: John Wiley & Sons, pp. 3–23.

Walter, S.D. and Birnie, S.E. (1991). Mapping mortality and morbidity patterns: An international comparison. *International J. Epidemiology*, **20**, 678–689.

Ware, J.H. (1989). Investigating therapies of potentially great benefit: ECMO (with discussion). *Statistical Science*, **4**, 298–340.

Wasserman, L. and Kadane, J.B. (1992). Computing bounds on expectations. *J. Amer. Statist. Assoc.*, **87**, 516–522.

Waternaux, C., Laird, N.M., and Ware, J.A. (1989). Methods for analysis of longitudinal data: Blood lead concentrations and cognitive development. *J. Amer. Statist. Assoc.*, **84**, 33–41.

Wei, G.C.G. and Tanner, M.A. (1990). A Monte Carlo implementation of the EM algorithm and the poor man's data augmentation algorithms. *J. Amer. Statist. Assoc.*, **85**, 699–704.

West, M. (1992). Modelling with mixtures (with discussion). In *Bayesian Statistics 4*, J.M. Bernardo, J.O. Berger, A.P. Dawid, and A.F.M. Smith, eds., Oxford: Oxford University Press, pp. 503–524.

West, M., Müller, P., and Escobar, M.D. (1994). Hierarchical priors and mixture models, with application in regression and density estimation. In *Aspects of Uncertainty: A Tribute to D.V. Lindley*, eds. A.F.M. Smith and P.R. Freeman, London: John Wiley & Sons.

West, M. and Harrison, P.J. (1989). *Bayesian Forecasting and Dynamic Models*. New York: Springer-Verlag.

West, M., Harrison, P.J., and Pole, A. (1987). BATS: Bayesian Analysis of Time Series. *The Professional Statistician*, **6**, 43–46.

Williams, E. (1959). *Regression Analysis*. New York: John Wiley & Sons.

Winkler, R.L. (1967). The assessment of prior distributions in Bayesian analysis. *J. Amer. Statist. Assoc.*, **62**, 776–800.

Whittemore, A.S. (1989). Errors-in-variables regression using Stein estimates. *The American Statistician*, **43**, 226–228.

Wolfinger, R. (1993). Laplace's approximation for nonlinear mixed models. *Biometrika*, **80**, 791–795.

Wolfinger, R. and Kass, R.E. (2000). Nonconjugate Bayesian analysis of variance component models. *Biometrics*, **56**, 768–774.

Wolfinger, R. and O'Connell, M. (1993). Generalized linear mixed models: A pseudo-likelihood approach. *Journal of Statistical Computation and Simulation*, **48**, 233-243.

Wolfinger, R.D. and Rosner, G.L. (1996). Bayesian and frequentist analyses of an *in vivo* experiment in tumor hemodynamics. In *Bayesian Biostatistics*, eds. D. Berry and D. Stangl, New York: Marcel Dekker, pp. 389–410.

Wolpert, R.L. and Ickstadt, K. (1998). Poisson/gamma random field models for spatial statistics. *Biometrika*, **85**, 251–267.

Wooff, D. (1992). [B/D] works. In *Bayesian Statistics 4*, J.M. Bernardo, J.O. Berger, A.P. Dawid, and A.F.M. Smith, eds., Oxford: Oxford University Press, pp. 851–859.

Wright, D.L., Stern, H.S., and Cressie, N. (2003). Loss functions for estimation of extrema with an application to disease mapping. *Canad. J. Statist.*, **31**, 251–266.

Xia, H., Carlin, B.P., and Waller, L.A. (1997). Hierarchical models for mapping Ohio lung cancer rates. *Environmetrics*, **8**, 107–120.

Ye, J. (1998). On measuring and correcting the effects of data mining and model selection. *J. Amer. Statist. Assoc.*, **93**, 120–131.

Yin, Y., Cummings, D.A., Burke, D.S., and Louis, T.A. (2007). Bayesian analysis of infectious disease time series data. Technical report, Department of Biostatistics, Johns Hopkins University.

Zaslavsky, A.M. (1993). Combining census, dual-system, and evaluation study data to estimate population shares. *J. Amer. Statist. Assoc.*, **88**, 1092–1105.

Zeger, S.L. and Karim, M.R. (1991). Generalized linear models with random effects; a Gibbs sampling approach. *J. Amer. Statist. Assoc.*, **86**, 79–86.

Zhao, L., Hanson, T., and Carlin, B.P. (2006). Mixtures of Polya trees for flexible spatial frailty modeling. Research Report 2006–018, Division of Biostatistics, University of Minnesota.

Zhou, X.-H., Perkins, A.J., and Hui, S.L. (1999). Comparisons of software packages for generalized linear multilevel models. *The American Statistician*, **53**, 282–290.

Zhu, L. and Carlin, B.P. (2000). Comparing hierarchical models for spatio-temporally misaligned data using the Deviance Information Criterion. *Statistics in Medicine*, **19**, 2265–2278.

Author index

Subject index

Printed in the United States
by Baker & Taylor Publisher Services